College Algebra

ALTERNATE EDITION

PRENTICE-HALL MATHEMATICS SERIES

Albert A. Bennett, Editor

College Algebra

ALTERNATE EDITION

M. Richardson, Ph.D.

PROFESSOR OF MATHEMATICS
BROOKLYN COLLEGE

Prentice-Hall, Inc.

ENGLEWOOD CLIFFS, N.J.

Library of Congress Catalog Card No.: 58-13417

First printing..........February, 1958
Second printing.......September, 1958
Third printing..........January, 1959

PRINTED IN THE UNITED STATES OF AMERICA

14174

To E., N., and L

Preface to the alternate edition

THE MAJOR CHANGES in this edition are the following:

1. The exercise lists have been reworked and improved.

2. The chapter on probability has been completely rewritten to avoid giving the impression that the assumption of equally likely cases is a necessity.

3. A section on non-decimal scales of notation has been included.

4. A section on elementary problems in linear programming has been included in the chapter on inequalities. This modern application of mathematics to economics is well within reach of college freshmen if the examples are kept small.

5. Expansion of determinants by minors, even for third order, has been stressed, rather than special schemes.

The order of the chapters has been rearranged, and minor changes, many of which have been suggested by friendly and helpful users of the first edition, have been made. To all these contributors, I address my sincere thanks. I am indebted to Mr. Roy Frieden for typing the manuscript, and to Miss Frieda Zames for checking the answers.

M.R.

Preface to the first edition

IN THIS TEXTBOOK the author has attempted to combine lucid explanation of procedures with reasonable motivation for and justification of the processes, at the student's level.

The early chapters constitute a thorough review of elementary algebra arranged as a self-contained development from first principles, with careful explanations of the most elementary matters. No accurate recollection of high school algebra is required if the book is taken from the beginning. However, the review is presented in a more mature way than would be possible with first-year high school students; this plan enables the instructor to take up the material with a group of students having widely varying degrees of preparation without boring the better-prepared students.

The subject matter is for the most part conventional. Chapter I on the number system seems to the author to be a much neglected necessity. The usual theorems in the theory of equations, or even the elementary problems on the character of the roots of a quadratic, can have little meaning to a student who has not had some discussion of rational, irrational, real, and imaginary numbers such as that in Chapter I. *A suitable starting point in the book can be found for almost any class regardless of the degree of preparation. However, the inclusion of much, if not all, of Chapter I is recommended, no matter at what point in the book the course is begun.*

Technical terms are carefully defined, and ample lists of exercises are provided. *Starred exercises are more difficult than others, and starred sections may be omitted in a short course without disturbing the continuity of the book.*

Stress is laid upon explanation of fundamental concepts and reasoning which the mathematics and science major is often over-optimistically expected to absorb by osmosis. The author is committed to the belief that a process understood will be remembered bet-

ter and longer than one which is merely memorized. The mathematics itself is, it is hoped, accurate. This does not mean that rigorous proofs are presented even when above the freshman level, but rather that bad proofs are not presented as good ones. When a correct proof is thought to be too difficult for the student, an informal, heuristic discussion, designed to make the result plausible, is presented but not falsely called a proof.* For a class which is going to omit a certain proof, it will be no more difficult to omit a correct proof than an inadequate one; but to teach incorrect reasoning as though it were correct can only undermine the student's future progress in and understanding of mathematics. Despite the inevitable compromise with rigor in an elementary book, the author has tried to avoid statements which would have to be unlearned by the student at a later stage.

Some topics are treated more thoroughly than is usual, as the table of contents indicates. It is hoped that this thorough treatment will increase the value of the book as a reference work for future study for the student who goes on with mathematics. For example, the student of integral calculus may be pleased to find the simple proof of the theorem on partial fractions, which is seldom found in textbooks on either college algebra or calculus.

Many practical applications are included in the exercises. But motivation does not necessarily mean introducing the subject by means of so-called "practical" problems that are either artificially constructed for the purpose or beyond the student's background; it may be purely mathematical. See, for example, the discussion preceding the standard theorem on rational roots of equations with integral coefficients. Due attention is paid to historical matters and to the relevance of the subject matter to other sciences and to civilization.

The later chapters are arranged as far as possible to be independent of each other, to provide for maximum elasticity in the choice of material. Where earlier material is prerequisite, cross references are provided. *The various chapters need not always be taken*

* In avoiding certain errors which appear to have become standardized, it is possible that the author has committed others; he will be glad to have his attention called to these.

in the order of their appearance. For example, the instructor may prefer to omit Chapter I at the start and take parts of it as needed. Similarly, the chapter on logarithms may be taken immediately after the chapter on exponents. The chapter on progressions may be inserted much earlier, provided the starred sections are omitted.

The author is indebted to Professor Harris F. MacNeish for permission to include selected Brooklyn College final examination questions among the exercises; to The Macmillan Company, for permission to include approximately eight pages of scattered, miscellaneous material taken verbatim from his book *Fundamentals of Mathematics,* of which they are the publishers, and for permission to adapt Tables I, IV, V, VI, and VII from the *Macmillan Logarithmic and Trigonometric Tables,* Revised Edition, edited by E. R. Hedrick; to Miss Helen Halpern, for her able and conscientious assistance in checking the answers to exercises and typing the manuscript; and to Prentice-Hall, Inc., for their co-operation and efficiency in publishing this book expeditiously during a trying period.

M.R.

Contents

★ Sections starred may be omitted without disturbing the continuity of the chapter.

Preface for the student

1. A Brief Lesson in Logic

Mathematics is logical in nature, and mathematical writing is precise, compact, and logical in style. Therefore a brief review of certain elementary but fundamental logical concepts and terms is given here.

When two statements are so related that the second *must* be true *if* the first is true, we say that the second **follows from** the first, or the second is a **logical consequence** of the first, or the first **implies** the second. The first statement is called the **hypothesis** and the second is called the **conclusion.**

Example 1. Hypothesis: $x = 2$ and $y = 5$.
Conclusion: $x + y = 7$.

If the conclusion of an argument really follows inescapably from the hypothesis, the argument is called **valid.** The process of drawing inescapable conclusions from given hypotheses is called **deduction** or **deductive reasoning.**

The proposition of example 1 may be written as: (*a*) *If* $x = 2$ *and* $y = 5$, *then* $x + y = 7$; or (*b*) "$x = 2$ *and* $y = 5$" *implies* "$x + y = 7$."

In a statement of the form "If A is true, then B is true," or "A implies B," A is the *hypothesis* and B is the *conclusion.* This statement does not assert that A is true or that B is true. It asserts merely that *if A were true, then B would have to be true.* That is, *in a valid argument, the truth of the hypothesis guarantees the truth of the conclusion.*

Example 2. "If today were Election Day,* then tomorrow would be Wednesday" is a valid argument even if today is neither Election Day nor any other Tuesday.

* We assume that Election Day in the U.S.A. is, by definition, the Tuesday after the first Monday of November.

Note that the conclusion may be true even if the hypothesis is false. Thus, in example 2, the conclusion would be true if today were any Tuesday in the year.

Therefore, *in a valid argument, the truth of the conclusion does not guarantee the truth of the hypothesis. But the falsity of the conclusion does guarantee the falsity of the hypothesis.*

Example 3. Granting the axiom that "if equals are divided by equals the results are equal," we may reason as follows:

Hypothesis:	$3 = 7.$	
Conclusion:	$1 = 1.$	
Proof:	$3 = 7$	(by hypothesis)
	$3 = 7$	(by hypothesis).
Therefore,	$\frac{3}{3} = \frac{7}{7}$	(foregoing axiom)
or,	$1 = 1.$	

The truth of this conclusion does not imply the truth of the hypothesis. Also, *the falsity of the hypothesis does not guarantee the falsity of the conclusion.*

The **converse** of the proposition "If A then B" or "A implies B" is the proposition "If B then A" or "B implies A." That is, *the converse of a proposition is formed by interchanging hypothesis and conclusion.*

The converse of a correct proposition need not be correct.

Example 4. The converse of the proposition of example 1 is "If $x + y = 7$, then $x = 2$ and $y = 5$." This is not valid, for the conclusion is not inescapable; we might have $x = 3$ and $y = 4$, or some other combination.

Example 5. The converse of the proposition of example 2 is "If tomorrow were Wednesday, then today would be Election Day," which is clearly not valid.

If both "A implies B" and its converse "B implies A" are valid, then the statements A and B are called **equivalent** to each other.

If A implies B, we sometimes say that A is a **sufficient condition** for B, and B is a **necessary condition** for A; or that B is true **if** A is true, and A is true **only if** B is true. If A and B are equivalent, we sometimes say that A is a **necessary and sufficient condition** for B, or that A is true **if and only if** B is true.

Example 6. "$x = 2$ and $y = 5$" is a sufficient condition for "$x + y = 7$" but not a necessary condition.

Example 7. "Today is Tuesday" is a necessary and sufficient condition for "Tomorrow is Wednesday." These two statements and equivalent; or "Today is Tuesday if and only if tomorrow is Wednesday."

The statement "Some A's are B's" means "At least one A is a B," and is better stated in the latter form.

The distinction between a proposition and its converse is of paramount importance. Another distinction which should be carefully noted is that between the proposition (1) "Not all A's are B's" and (2) "All A's are not B's." These are often carelessly confused. The first is the denial of the proposition "All A's are B's" and really means "Some A's are not B's" or "At least one A is not a B" and is better written in the latter form. The second means "No A is a B" and is better written in the latter form.

Every proof of a mathematical theorem must be an example of deductive reasoning; that is, the conclusion must be a necessary (inescapable) consequence of the hypothesis.

Note 1. While only deductive reasoning is employed in a finished chapter of mathematics, other forms of reasoning are of great value in experimental science and in everyday life. Thus analogies and good guesses may be useful in conjecturing what may be the case, although such a conjecture should not be relied on without further proof. Of particular importance in experimental science and everyday life is **induction** or **inductive reasoning**; this is the name given to the process of coming to a probable conclusion on the basis of (many) particular instances. For example, a one-year-old scientist, after dropping objects over the side of his high chair many times, may come to the conclusion that an object released without support in the air will fall to the ground. But this conclusion is merely probable, not inescapable, as he might find out when, to his surprise, a gas-filled balloon "falls" to the ceiling. In fact, however, the young scientist's conclusion has a high probability, in the sense that, in his experience, it will prove correct most of the time. Inductive reasoning is thus the backbone of experimental science. It is used by mathematicians as well, but only as a means of suggesting what results may be proved by deduction.

No mathematical theorem is considered proved on the evidence of induction alone; deductive proof is required.

Note 2. People whose intention is to persuade rather than to arrive at the truth, such as orators, debaters, demagogues, editors, et al., often use bad reasoning dressed up to seem good. Such reasoning may, for lack of a better term, be called *seductive reasoning*. Our only interest in it should be to detect and expose it.

EXERCISES

1. Assuming that the conclusion follows from the hypothesis by valid reasoning, complete each of the following sentences with one of the phrases "must be true," "must be false," "may be true or false":

(*a*) If the hypothesis is true, then the conclusion ...
(*b*) If the hypothesis is false, then the conclusion ...
(*c*) If the conclusion is true, then the hypothesis ...
(*d*) If the conclusion is false, then the hypothesis ...

2. Write five correct statements of the "if ... then ... " type which have incorrect converses.

3. Assuming that statement (*a*) below is correct, decide which of the other statements necessarily follows from it:

(*a*) If Jones is intelligent, then he can pass mathematics.
(*b*) If Jones is not intelligent, then he cannot pass mathematics.
(*c*) If Jones can pass mathematics, then he is intelligent.
(*d*) If Jones cannot pass mathematics, then he is not intelligent.
(*e*) Jones can pass mathematics if he is intelligent.

4. Which of the statements in exercise 3 is the converse of (*a*)?

5. According to the statement (*a*) of exercise 3, is "Jones is intelligent" a sufficient or a necessary condition for "he can pass mathematics"?

6. The following statements are correct. Write the converse of each statement and decide whether or not the converse is necessarily correct:

(*a*) If $x = 2$, then $(x - 2)(x - 3) = 0$.
(*b*) If $x = -3$, then $x^2 = 9$.
(*c*) If $x = y$, then $x^2 = y^2$.
(*d*) If $2x = 4$, then $x = 2$.

Decide whether or not each of the proposed conclusions really follows from the hypothesis by valid (deductive) reasoning:

7. *Hypothesis:* All polynomials are rational expressions; all rational expressions are algebraic expressions.
Conclusion: All polynomials are algebraic expressions.

8. *Hypothesis:* All men are vegetables; all vegetables are mortal.
Conclusion: All men are mortal.

9. *Hypothesis:* No men are vegetables; no vegetables are beasts.
Conclusions: (*a*) All men are beasts.
(*b*) Some men are beasts.

10. *Hypothesis:* All triangles are polygons; no quadrilaterals are triangles.
Conclusion: No quadrilaterals are polygons.

11. *Hypothesis:* All freshmen are human; all students are human.
Conclusions: (*a*) All freshmen are students.
(*b*) Some freshmen are students.
(*c*) Some students are not freshmen.
(*d*) No freshmen are students.

12. All honest congressmen voted for this bill. I voted for this bill. Therefore I am an honest congressman.

13. If the problem was worked correctly, the answer obtained was 7. I obtained the answer 7. Therefore I worked the problem correctly.

14. All good cars are expensive. This car is expensive. Therefore this car is good.

15. The sum of two even numbers is always even. Neither 7 nor 11 is even. Therefore the sum of 7 and 11 is not even.

16. If x is even, then $2x$ is even. We know that $2x$ is even. Therefore x is even.

17. Every rational number can be expressed as a quotient of two integers. $\sqrt{2}$ cannot be expressed as a quotient of two integers. Therefore no rational number is $\sqrt{2}$.

2. Glossary of Technical Symbols

For easy reference we list the symbols used in this book.

Symbol	*Definition*
$+$	*plus.*
$-$	*minus.*
$\times$ or $\cdot$	signs of multiplication, read *times.*
$\div$ or $/$	*divided by.*
$=$	*is equal to.*
$\neq$	*is not equal to,* as $5 \neq 3$.
$>$	*is greater than,* as $5 > 3$.
$<$	*is less than,* as $3 < 5$.
$\geqq$	*is greater than or equal to.*

$\leqq$	*is less than or equal to.*
$\lvert x \rvert$	*the absolute value of x.*
( )	*parentheses;* [ ], *brackets;* { }, *braces.*
$\sqrt[n]{a}$	the *radical sign,* meaning *the (principal)* nth *root of a;* in particular $\sqrt{a}$ means *the (principal) square root of a.*
$n!$	*factorial n.*
$C(n,r)$	*the number of combinations of n distinct things taken r at a time.*
$P(n,r)$	*the number of permutations of n distinct things taken r at a time.*
$f(x)$	*a function of x,* or *the f-function of x;* similarly $d(x)$ is *the d-function of x,* etc.
(x,y)	denotes a *point whose rectangular coordinates are x and y.*
$\rightarrow$	read *approaches.*
$\lim_{x \to a} f(x)$	read *limit of f(x) as x approaches a.*
$\lim_{n \to \infty} a_n$	read *limit of a_n as n increases indefinitely,* or *as n approaches infinity.*
Δ	*difference of first order,* or *first difference.*
Δ^r	*difference of* rth *order,* or rth *difference.*
$s_{\overline{n}\rceil}$	*amount of an annuity of 1 for n periods.*
$a_{\overline{n}\rceil}$	*present value of an annuity of 1 for n periods.*

Greek Alphabet

Letters		*Names*	*Letters*		*Names*	*Letters*		*Names*
Α	α	Alpha	Ι	ι	Iota	Ρ	ρ	Rho
Β	β	Beta	Κ	κ	Kappa	Σ	σ	Sigma
Γ	γ	Gamma	Λ	λ	Lambda	Τ	τ	Tau
Δ	δ	Delta	Μ	μ	Mu	Υ	υ	Upsilon
Ε	ϵ	Epsilon	Ν	ν	Nu	Φ	ϕ	Phi
Ζ	ζ	Zeta	Ξ	ξ	Xi	Χ	χ	Chi
Η	η	Eta	Ο	ο	Omicron	Ψ	ψ	Psi
Θ	θ	Theta	Π	π	Pi	Ω	ω	Omega

Chapter I

The number system of algebra*

3. Introduction

The word *algebra* comes from the Arabic *al-jebr* signifying transposition of negative terms from one side of an equation to the other. Thus $2x - 3 = 7$ becomes $2x = 7 + 3$ by *al-jebr*. The ancients, such as the Egyptians† (about 1700 B.C.), Greeks (6th century B.C.–3rd century A.D.), Chinese (2nd–13th centuries A.D.), Hindus (5th–12th centuries A.D.), Arabs (9th–15th centuries A.D.), and others, had some rudimentary ideas of algebra. But the subject as we know it now really began to develop in the Middle Ages in Europe, beginning about the 13th century. Much of the algebra in this book dates from the 16th and 17th centuries although some of it is as recent as the 19th. During this long period of growth, the word *algebra* has come to represent an extensive body of knowledge concerning the processes of reckoning with numbers. The ancients were usually content to solve isolated problems and were usually satisfied with a single solution of a given problem. But the essence of the spirit of modern algebra is the desire to find all solutions of a given problem and especially to find general methods for coping with all problems of a given kind. This desire for generality will be seen to be the source of much of our modern progress in mathematics. As will be seen as we proceed, the superiority

* This chapter may be omitted on first reading, if desired, and referred to as needed.

† Our knowledge of Egyptian mathematics is obtained largely from the Rhind papyrus by Ahmes (about 1700 B.C.), somewhat ambitiously entitled *Directions for Obtaining the Knowledge of All Dark Things.*

of algebraic methods over elementary arithmetic in solving problems springs, in large measure, from the use of letters to represent unspecified or unknown numbers. But since the letters used in elementary algebra do stand for numbers, it is necessary to have some understanding of the various kinds of numbers that they represent. In this chapter, we shall survey rapidly the long, gradual development of the concept of number, from prehistoric beginnings to the complex number system now in use.

4. The Natural Numbers. Addition and Multiplication

The earliest arithmetical discovery made by a primitive man or a young child would doubtless be that two apples, two fingers, two people, etc., have something in common, and can be distinguished readily from one apple, one finger, one person, etc. Thus the first numbers to be invented were probably 1 and 2. Some primitive races are said to find it too much of a strain to engage in further numerical subtleties, and have in their language only three words of number—*one, two,* and *many.* This would provide a charmingly simple system of arithmetic of which lazy students might approve, but which would be inadequate for the purposes of modern science. Our prehistoric ancestors surely invented the process of counting objects and thus developed the endless succession of the so-called **natural numbers** 1, 2, 3, 4, . . . ,* which are used in counting. The natural numbers are also often called **positive integers** for a reason which will be explained later. Experience in counting objects yields the tables of addition which we memorize at an early age and use in calculating sums. It is intuitively clear that the system of all the natural numbers obeys the following basic law.

I. Law of closure for addition. *Given any pair of numbers a and b in the system, there exists a unique*† *number in the system called the* **sum** *of a and b. The sum of a and b is denoted by $a + b$.*

The numbers a and b, separated by a plus sign, are themselves called **terms** of the sum.

The significance of the word *closure* is that the system of all

* Three dots occurring in succession mean "and so on."

† That is, there is one and only one such number.

natural numbers is *closed under addition* in the sense that the sum of two numbers in the system is itself in the system. It is as though the natural numbers are contained in an enclosure, and performing the operation of addition never takes us outside the enclosure. On the other hand, the system consisting of the natural numbers from 1 to 10 inclusive, and no others, is not closed under addition, for there are numbers in this system, 5 and 7 for example, whose sum is not in this system.

Experience also suggests that the system of natural numbers obeys the following basic law.

II. Commutative law for addition. *If a and b are any numbers in the system, $a + b = b + a$.*

This means that we may commute or interchange the order of the terms in the sum of two numbers.

Recall that parentheses, brackets, etc., are punctuation marks in the written language of algebra which are used to group together whatever is in them. Thus the symbol $3 + (4 + 5)$ indicates that we are to add 3 to the sum of 4 and 5, obtaining $3 + 9$ or 12; while $(3 + 4) + 5$ indicates that we are to add the sum of 3 and 4 to 5, obtaining $7 + 5$ or 12. The fact that we get the same result both times, although the operations performed are different, suggests that the system of natural numbers obeys the following basic law.

III. Associative law for addition. *If a, b, and c are any numbers in the system, $a + (b + c) = (a + b) + c$.*

Extended statements of the three laws. It is intuitively clear that the three laws just stated apply no matter how many terms are involved in a sum. Thus *the sum of any number of natural numbers is a natural number and the order in which they are added is immaterial, as is the manner in which they are grouped by parentheses.** That all this is familiar is evident from the fact that shopkeepers will add a column of figures from the top down and then check by adding from the bottom up, expecting to get the

* This can be proved by means of mathematical induction, which will be taken up in a later chapter.

same result. For adding the column

$$\begin{array}{r} 3 \\ 4 \\ \underline{5} \\ 12 \end{array}$$

from the top down means calculating the value of the expression $(3 + 4) + 5$, while adding from the bottom up amounts to calculating the value of the expression $(5 + 4) + 3$. Hence *in a sum like $a + b + c + d$ we may freely rearrange the order of the terms and insert or remove parentheses in any way that makes sense;* thus we may write

$$a + b + c + d = (a + b) + (c + d) = [a + (b + c)] + d$$
$$= a + [(b + c) + d]$$

and so on.

Early in history, someone may have noticed that it was more efficient to "count by threes" than to "count by ones." This leads to the definition of multiplication as repeated addition.

DEFINITION 1. *If a and b are any natural numbers, we define the* **product** *of a and b to mean $b + b + \ldots + b$ where there are* ***a*** *terms in the sum. The product of a and b is denoted by ab, $a \cdot b$, or more rarely by $a \times b$. The numbers a and b are themselves called* **factors** *of the product. To* **multiply** *a and b means to find the product ab. In particular, $1 \cdot b = b$ for any number b.*

Examples. $4 \cdot 3 = 3 + 3 + 3 + 3 = 12$. Also $3b = b + b + b$; in particular, $3 \cdot 4 = 4 + 4 + 4 = 12$.

It is clear that the system of natural numbers is closed under multiplication. That is, it obeys the following basic law.

IV. Law of closure for multiplication. *If a and b are any numbers of the system, there is a unique number of the system called their product and denoted by ab.*

The examples above show that although $4 \cdot 3$ and $3 \cdot 4$ indicate different operations, they yield the same result. This suggests that the system of natural numbers obeys the following basic law.

V. Commutative law for multiplication. *If a and b are any numbers of the system, $ab = ba$.*

Similarly $(2 \cdot 3) \cdot 4 = 6 \cdot 4 = 24$ and $2 \cdot (3 \cdot 4) = 2 \cdot 12 = 24$ indicate different operations but yield the same result. This suggests that the system of natural numbers obeys the following basic law.

VI. Associative law for multiplication. *If a, b, and c are any numbers of the system, then $a(bc) = (ab)c$.*

We agree that *in a succession of additions and multiplications, the multiplications are to be done first except where otherwise indicated by parentheses.*

Example. $2 + 3 \cdot 4 = 2 + 12 = 14$, while $(2 + 3) \cdot 4 = 5 \cdot 4 = 20$.

Thus $2 \cdot (3 + 4) = 2 \cdot 7 = 14$, while $2 \cdot 3 + 2 \cdot 4 = 6 + 8 = 14$. The fact that these operations are different but yield the same result suggests that the system of natural numbers obeys the following basic law.

VII. Distributive law. *If a, b, and c are any numbers of the system, $a(b + c) = ab + ac$.*

That is, the effect of the multiplier a is distributed between the terms b and c. This law is often referred to as "multiplying out the parentheses" or "removing the parentheses" if read from left to right, and as "taking out the common factor" if read from right to left.

A statement of equality, like $a = b$, is called an **equation.** The expressions on either side of the equals sign are called the **left member** and **right member** of the equation, respectively. If $a = b$, we may substitute a for b or b for a in any equation; this is known as the **principle of substitution.** It is related to the following obvious basic laws of equality.

VIII. $a = a$.

IX. *If $a = b$, then $b = a$.*

X. *If $a = b$ and $b = c$, then $a = c$. That is, things equal to the same thing are equal to each other.*

XI. *If $a = b$ and $c = d$, then $a + c = b + d$. That is, if equals are added to equals, the results are equal.*

XII. *If $a = b$ and $c = d$, then $ac = bd$. That is, if equals are multiplied by equals, the results are equal.*

Extended statements of the laws. Just as in the case of laws I–III, the laws for multiplication and equality can be applied to any number of terms. For example, we may extend the distributive law to read

$$a(b + c + d + \ldots + k) = ab + ac + ad + \ldots + ak.$$

We shall not write out these extended statements here in full, since they are obvious.

EXERCISES

Name the principal law or laws which justify each of the following statements. All letters represent natural numbers.

1. $x + y$ is a natural number.
2. xy is a natural number.
3. $x + y = y + x$.
4. $xy = yx$.
5. $x + (y + z) = (x + y) + z$.
6. $x(yz) = (xy)z$.
7. $x(y + z) = xy + xz$.
8. $(y + z)x = xy + xz$.
9. $2(x + 3) = 2x + 6$.
10. $2x + 3x = (2 + 3)x = 5x$.

Evaluate each of the following.

11. $2 + 3 \cdot 4$.
12. $(2 + 3) \cdot 4$.
13. $2 + 3 \cdot 4 + 5$.
14. $(2 + 3) \cdot 4 + 5$.
15. $(2 + 3) \cdot (4 + 5)$.
16. $2 + 3 \cdot (4 + 5)$.
17. $2[3 + 4(5 + 6)]$.
18. $2(3 + 4 \cdot 5 + 6)$.
19. $2[(3 + 4) \cdot 5 + 6]$
20. $2(3 + 4)(5 + 6)$.

Use the basic laws I–XII to prove the following theorems, in which all letters represent natural numbers.

★21. $(a + b) + c = a + (c + b)$. (*Hint:* use II and III.)
★22. $(a + b) + c = c + (a + b)$.
★23. $(x + y) + z = z + (x + y)$. (*Hint:* use I and II.)
★24. $(ab)c = a(cb)$. (*Hint:* use V and VI.)
★25. $(ab)c = b(ac)$.
★26. $(ab)(cd) = (cd)(ab)$. (*Hint:* use IV and V.)
★27. $a(b + c) = ca + ba$.
★28. $a(b + c) = ba + ca$.
★29. $(a + b)c = ca + bc$.
★30. $(a + 3)(b + 2) = ab + 3b + 2a + 6$.

5. Subtraction and Division of Natural Numbers

Children are usually taught to subtract 2 from 5 by asking themselves, "What must be added to 2 in order to get 5?"

★ May be omitted without disturbing continuity.

Formalizing this question, we define subtraction in terms of addition, as follows.

DEFINITION 1. *If a and b are any two numbers of the system, the* ***difference*** $a - b$ *stands for the number x such that $b + x = a$, provided such a number exists in the system. To* **subtract** b *from a means to find $a - b$.*

Thus $5 - 2$ is found by asking, "$2 +$ what $= 5$?" and looking in the "two plus" table for the answer 3. The proviso in the definition is necessary because, in the system of natural numbers, the difference does not exist in some cases. For example, $2 - 5$ does not exist within the system of natural numbers, since there is no natural number x such that $5 + x = 2$.

A system of numbers is called **closed under a given operation** if performing that operation upon numbers of the system always yields a number of the system as a result. Thus, *the system of all natural numbers is not closed under subtraction, although it is closed under addition and multiplication.*

To prove that a first quantity, a, minus a second quantity, b, is equal to a third quantity, x, we have only to use Definition 1 and verify that the second quantity, b, plus the third, x, is equal to the first, a. This process is often called "checking" the subtraction.

> *Example 1.* To verify that $(q + p) - p = q$ it is sufficient to show that the second quantity, p, plus the third quantity, q, is equal to the first quantity, $q + p$. But $p + q = q + p$ by the commutative law for addition. This completes the proof.

Example 1 shows that subtraction "undoes" addition, in an obvious sense. Therefore subtraction is called the *inverse* of addition.

DEFINITION 2. *If a and b are natural numbers, and if there exists a natural number x such that $b + x = a$, then a is said to be* ***greater than*** b *or b* ***less than*** a. *In symbols, $a > b$ or $b < a$.*

For example, $5 > 2$ because $2 + 3 = 5$.

Similarly, children are taught to divide 6 by 2 by asking, "By what must we multiply 2 in order to get 6?" Formalizing this question, we define division in terms of multiplication, as follows.

DEFINITION 3. *If a and b are any two numbers of the system, the* ***quotient*** $a \div b$ *stands for the number x such that $bx = a$, pro-*

vided such a number x exists in the system. To ***divide*** *a by b means to find the quotient $a \div b$.*

Thus $6 \div 2$ is found by asking, "2 times what equals 6?" and looking in the "2 times" table for the quotient 3. The proviso in the definition is necessary because, in the system of natural numbers, the quotient does not exist in some cases. For example, $5 \div 2$ does not exist within the system, since there is no natural number x such that $2x = 5$. Therefore *the system of natural numbers is not closed under division.*

To prove that a first quantity, a, divided by a second quantity, b, is equal to a third quantity, x, we have only to use Definition 3 and verify that the second, b, times the third, x, equals the first, a. This process is often called "checking" the division.

Example 2. To verify that $(qp) \div p = q$ we have only to show that the second quantity, p, times the third, q, is equal to the first, qp. But $qp = pq$ by the commutative law for multiplication. This completes the proof.

Example 2 shows that division "undoes" multiplication in an obvious sense. Therefore division is called the *inverse* of multiplication.

DEFINITION 4. *If a and b are any natural numbers, and if there exists a natural number x such that $bx = a$, then b is called a* ***factor*** *of a, a is said to be a* ***multiple*** *of b, or a is said to be* ***divisible*** *by b.*

For example, 6 is divisible by 2, 6 is a multiple of 2, and 2 is a factor of 6 because $2 \cdot 3 = 6$.

In a sequence of additions, subtractions, multiplications, and divisions it is understood that multiplications and divisions are to be done before additions and subtractions, except where otherwise indicated by parentheses. For example, $2 \cdot 5 - 2 \cdot 2 = 10 - 4$, while $2 \cdot (5 - 2) = 2 \cdot 3$. The fact that both answers are equal suggests that the distributive law can be extended to include subtraction. Thus, it is possible to prove that

$$a(b + c - d - e + \ldots + k) = ab + ac - ad - ae + \ldots + ak$$

whenever these expressions have a meaning. We shall not prove this here.

EXERCISES

1. Explain what is meant by the statement $8 \div 2 = 4$ in terms of definition 3.

2. Explain what is meant by the statement $8 - 2 = 6$ in terms of definition 1.

3. Which of the following expressions are meaningless within the system of natural numbers? Explain. (*a*) $11 \div 2$. (*b*) $6 \div 2$. (*c*) $3 - 5$. (*d*) $3 \div 5$. (*e*) $4 \div 8$. (*f*) $8 - 2$.

4. Assuming that a and b are natural numbers, tell whether the following expressions always represent a natural number: (*a*) $a + b$. (*b*) ab. (*c*) $a - b$. (*d*) $a \div b$.

5. If your answer to parts of exercise 4 is "no," what restriction must be placed on a and b in order for these expressions to represent natural numbers?

Prove each of the following, justifying each step by means of a basic law, definition, or previously proved theorem, all letters representing natural numbers.

★6. $(n + m) - n = m$.

★7. $(nm) \div n = m$.

★8. If $a > b$ and $b > c$, then $a > c$.

★9. If c is a factor of b and b is a factor of a, then c is a factor of a.

★10. If $a < b$, then $a + c < b + c$.

★11. If $a < b$, then $ac < bc$.

6. Even, Odd, and Prime Numbers. H.C.F. and L.C.M.

A natural number is called **even** if it is divisible by 2, or has 2 as a factor. That is, *a natural number, a, is even if it can be expressed as $2x$ where x is some natural number.* For example, 6 is even since it is $2 \cdot 3$. But 5 is not even, for there is no natural number x such that $5 = 2x$. A natural number which is not even is called **odd.** It can be proved that *every odd number is one less than some even number.*

A natural number is called **prime** if it is greater than 1 and has no factors except itself and 1.

By trial we find that 2, 3, 5, 7, 11, 13, 17, 19, 23, 29 are the first ten prime numbers. Clearly 2 is the only even prime, because any other even number has the factor 2.

★ May be omitted without disturbing continuity.

All primes less than a given number may be found by a simple process known as the **sieve of Eratosthenes** (Greek, 275–194 B.C.). We illustrate the process by finding all primes less than 50. First write in succession all natural numbers from 2 up to 50:

2 3 ~~4~~ 5 ~~6~~ 7 ~~8~~ ~~9~~ ~~10~~ 11 ~~12~~ 13 ~~14~~
~~15~~ ~~16~~ 17 ~~18~~ 19 ~~20~~ ~~21~~ ~~22~~ 23 ~~24~~ ~~25~~ ~~26~~ ~~27~~
~~28~~ 29 ~~30~~ 31 ~~32~~ ~~33~~ ~~34~~ ~~35~~ ~~36~~ 37 ~~38~~ ~~39~~ ~~40~~
41 ~~42~~ 43 ~~44~~ ~~45~~ ~~46~~ 47 ~~48~~ ~~49~~ ~~50~~.

Then leave 2, but cross out every second number thereafter. Then leave the next uncrossed number, namely 3, but cross out every third number thereafter. Then leave the next uncrossed number, namely 5, but cross out every fifth number thereafter. And so on. The numbers left uncrossed at the end are primes.

It can be proved that *every natural number greater than 1 can be expressed as a product of prime factors in one and only one way, apart from the order in which the factors are written*. This is known as the **unique factorization theorem.** We assume it here. For example, $60 = 2 \cdot 2 \cdot 3 \cdot 5$, $72 = 2 \cdot 2 \cdot 2 \cdot 3 \cdot 3$, $12 = 2 \cdot 2 \cdot 3$, and $550 = 2 \cdot 5 \cdot 5 \cdot 11$. *To express any natural number as a product of its prime factors, first divide it by 2 if possible. Then divide the quotient by 2 again, if possible. Do this until the resulting quotient is not divisible by 2. Then divide by the next prime, 3. And so on.*

The **highest common factor (H.C.F.)** of several natural numbers is the largest natural number which is a factor of all of the given numbers. *It may be found by taking the product of all different prime factors common to the given numbers, each taken the smallest number of times that it occurs in any of these numbers. If the given numbers have no prime factors in common, the H.C.F. is defined to be 1; in this case, the given numbers are called* ***relatively prime.***

Example 1. The H.C.F. of $60 = 2 \cdot 2 \cdot 3 \cdot 5$ and $72 = 2 \cdot 2 \cdot 2 \cdot 3 \cdot 3$ is $2 \cdot 2 \cdot 3 = 12$.

The **lowest common multiple (L.C.M.)** of several natural numbers is the smallest natural number of which each of the given numbers is a factor. *It may be found by taking the product* *of all the different prime factors in these numbers, each taken the greatest number of times that it occurs in any of these numbers.*

Example 2. The L.C.M. of $60 = 2 \cdot 2 \cdot 3 \cdot 5$ and $72 = 2 \cdot 2 \cdot 2 \cdot 3 \cdot 3$ is $2 \cdot 2 \cdot 2 \cdot 3 \cdot 3 \cdot 5 = 360$.

★THEOREM 1. *If H is the H.C.F. of the natural numbers a and b, and if L is their L.C.M., then $LH = ab$.*

Proof. We have $a = Hx$ and $b = Hy$ where the natural numbers x and y are relatively prime. Then, clearly, $L = Hxy$. Multiplying both sides of the latter equation by H, we have $LH = (Hx)(Hy)$, or $LH = ab$. This completes the proof.

Therefore, if the H.C.F. of a and b is known to be H, the L.C.M. can be found from the relation $L = (ab) \div H$.

*THEOREM 2. *If a, b, and c are natural numbers and if a and b are relatively prime and a is a factor of bc, then a is a factor of c.*

Proof. Express each of the numbers a, b, and c as the (unique) product of its prime factors. Since a and b have no prime factor in common and a is a factor of bc, it follows that every prime factor of a is a prime factor of c and, hence, that a is a factor of c.†

Note the necessity for the hypothesis that a and b are relatively prime. For example, if $a = 6$, $b = 10$, $c = 21$, then a is a factor of $bc = 210$, but a is not a factor of either b or c; note that neither a and b nor a and c are relatively prime.

Note. It may astonish the student to learn that at this seemingly elementary stage of the development of mathematics, there are already unsolved problems at hand which have resisted solution stubbornly. But this is the case concerning natural numbers and prime numbers. Among the questions which are easy to ask but remain at present unanswered are the following:‡

(*a*) Is it true that every even number greater than 2 is the sum of two primes? It seems plausible on the basis of experi-

★ Theorem 1 may be omitted without disturbing the continuity of the section.

* Theorem 2 will be needed in section 122.

† If the assumption of the unique factorization theorem is too strong for the instructor's taste, another proof of this important theorem, free of this assumption, is given in Chapter XXV, section 171, theorem 4.

‡ Other such unsolved problems are discussed in M. Richardson, *Fundamentals of Mathematics,* Macmillan, 1941, Chapter VIII.

ment, for $4 = 2 + 2$, $6 = 3 + 3$, $8 = 3 + 5$, $10 = 5 + 5$, $12 = 7 + 5$, and so on. This conjecture was made by Goldbach in 1742 but no proof has yet been given.

(*b*) Is the number of pairs of primes which differ by 2 limited or unlimited? For example, 3 and 5, 11 and 13, 17 and 19, and so on.

(*c*) Are there any natural numbers a, b, m, and n, where m and n are greater than 1, such that a^m and b^n differ by 1, except for 3^2 and 2^3?

EXERCISES

1. Use the sieve of Eratosthenes to find all prime numbers less than (*a*) 100; (*b*) 200.

Express each of the following numbers as a product of primes or else show it is prime:

2. 28. **3.** 24. **4.** 86. **5.** 153. **6.** 361.
7. 401. **8.** 2844. **9.** 3135. **10.** 353.

Find (a) the H.C.F., (b) the L.C.M. of each of the following sets of numbers:

11. 60 and 80. **12.** 80 and 75. **13.** 24 and 36.
14. 28 and 75. **15.** 86 and 75. **16.** 153 and 189.
17. 24, 36, and 60. **18.** 8, 20, and 24. **19.** 60, 80, and 40.

20. If $ab = 192$ and the H.C.F. of a and b is 4, find their L.C.M.

21. if $ab = 378$ and the L.C.M. of a and b is 126, find their H.C.F.

22. Show that if a and b are both even, then $a + b$ is even.

23. Show that if a and b are both odd, then $a + b$ is even.

24. Show that if a is even and b is any natural number, then ab is even.

25. Are the following systems closed under addition? Explain. (*a*) The system of all even numbers. (*b*) The system of all odd numbers. (*c*) The system of all multiples of 3.

26. Are the systems of exercise 25 closed under multiplication? Explain.

27. If a is any natural number, show that the system of all multiples of a is closed under addition and multiplication.

28. A natural number is called **perfect** if it is equal to the sum of all its factors except itself. For example, 6 is perfect, since $6 = 1 + 2 + 3$. Show that 28 is perfect.*

* Whether or not there exists an odd perfect number is at present unknown.

29. The numbers a and b are called **amicable** if each is the sum of all the factors of the other except the other itself. Show that 220 and 284 are amicable.

7. Fractions. Multiplication and Division

Early in history, fractions were invented to facilitate calculations involving things divided into equal parts, such as real estate or harvests of grain. From a more sophisticated point of view we may credit motivation of the invention of fractions to our natural desire to have our system of numbers closed under all our operations. The system of natural numbers is not closed under division. That is, we cannot always divide one natural number by another and get a natural number. For example, 6 cannot be divided by 4 within the system of natural numbers. This unpleasant defect will be remedied by the introduction of fractions. Looking at it naïvely, a fraction, like 6/4, is merely a symbol consisting of two natural numbers. Thus we make the following definition:

DEFINITION 1. *A **fraction** is a symbol* $\frac{a}{b}$ *or* a/b *where* a *and* b *are natural numbers. We call* a *the **numerator** and* b *the **denominator** of the fraction* a/b.

The symbol a/b, read "a over b," is used to represent a of b equal parts of something. Thus 2/3 represents 2 of 3 equal parts of something. Because of this application we wish to regard 2/3, 4/6, 6/9, etc., as equal (Fig. 1), although they are recognizably

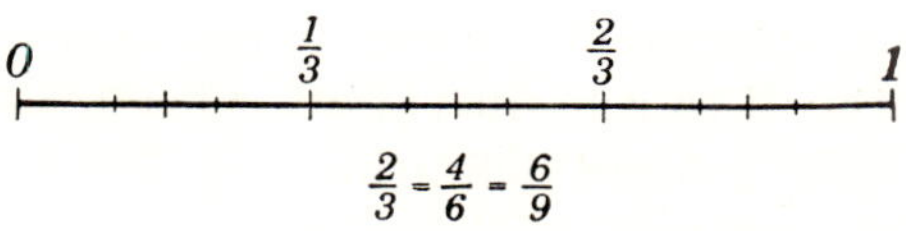

Fig. 1.

different symbols. Fortunately, fractions are our own creations and we are free to decide what we wish to mean by equal fractions, and by the sum, product, etc., of two fractions, so long as our decision does not involve logical inconsistency. If, instead of dividing something into three equal parts we double the number of parts, we must take twice as many of the smaller parts, or 4/6, to get the same quantity as 2/3. That is,

$$\frac{4}{6} = \frac{2 \cdot 2}{3 \cdot 2} = \frac{2}{3}.$$

This suggests that, in general, we adopt the following preliminary definition.

DEFINITION 2. $\frac{a}{b} = \frac{ax}{bx}$ *where a, b, and x are any natural numbers. That is, a common factor may be introduced into, or removed from, the numerator and denominator of a fraction without altering its value.*

Removing a common factor from numerator and denominator of a fraction is often called **cancellation.** Note that only *factors* which are common to numerator and denominator may be cancelled. When all possible common factors have been cancelled, the fraction is said to have been **reduced to simplest form** or **lowest terms.** For example,

$$\frac{12}{18} = \frac{2 \cdot 6}{3 \cdot 6} = \frac{2}{3}.$$

Now 2/4 and 5/10 are not related directly by definition 2 since there is no natural number x such that $2x = 5$ or $4x = 10$. But both 2/4 and 5/10 can be expressed as equal fractions having a common denominator, by means of definition 2. Thus

$$\frac{2}{4} = \frac{2 \cdot 10}{4 \cdot 10} = \frac{20}{40} \quad \text{and} \quad \frac{5}{10} = \frac{5 \cdot 4}{10 \cdot 4} = \frac{20}{40}.$$

Since we wish the statement "things equal to the same thing are equal to each other" to remain valid for fractions, we are led to make the following general definition.

DEFINITION 3. *Two fractions are* ***equal*** *if and only if when both are expressed as equivalent fractions having a common denominator (by definition 2 if necessary) they then have the same numerator as well.*

For example, $\frac{2}{4} = \frac{5}{10}$ since both are equal to $\frac{20}{40}$.

THEOREM 1. $\frac{a}{b} = \frac{c}{d}$ *if and only if $ad = bc$.*

Proof. By definition 2, $\frac{a}{b} = \frac{ad}{bd}$ and $\frac{c}{d} = \frac{bc}{bd}$. Hence they are equal to each other if and only if the numerators ad and bc are equal.

The operation of passing from $\frac{a}{b} = \frac{c}{d}$ to $ad = bc$ is sometimes referred to as "cross-multiplying."

DEFINITION 4. *The* ***product*** *of two fractions is the fraction whose numerator is the product of the two given numerators and whose denominator is the product of the two given denominators. In symbols*

$$\frac{a}{b} \cdot \frac{c}{d} = \frac{ac}{bd}.$$

The choice of this definition may be motivated by considering rectangular areas.* A rectangle with length 3 inches and width 2 inches clearly has an area of 6 square inches (Fig. 2).

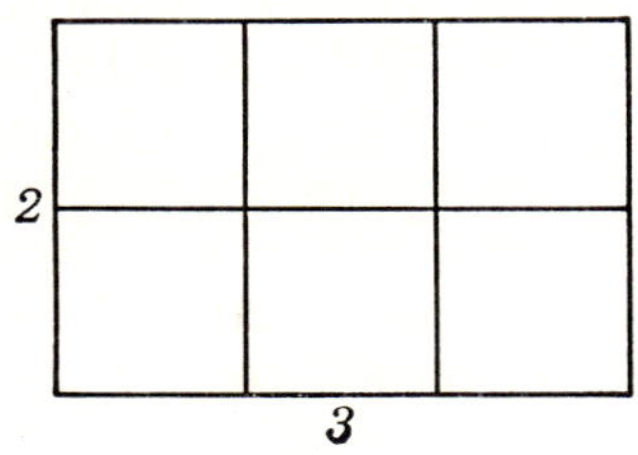

Fig. 2.

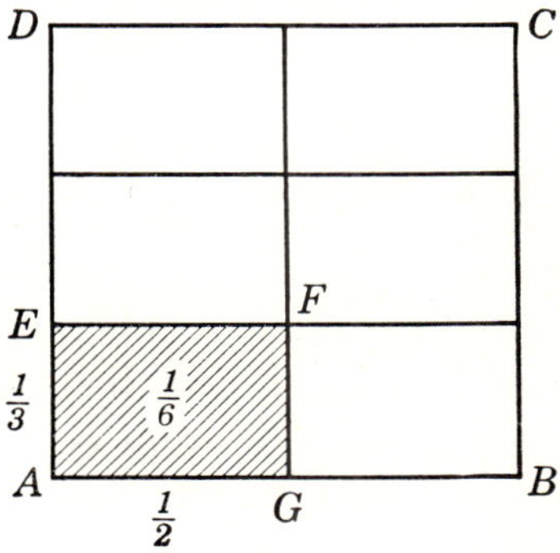

Fig. 3.

The square $ABCD$ with side 1 inch long has an area of 1 square inch (Fig. 3). This suggests the formula $A = lw$ for the area of a rectangle. If we wish this formula to remain valid for fractional lengths and widths, we must define the product of two fractions as above, so that the area $AGFE$, for example, in Fig. 3, will be $\frac{1}{2} \cdot \frac{1}{3} = \frac{1}{6}$.

Whenever we introduce or invent new kinds of numbers, we want them, as far as possible, to obey the same basic laws and follow the

* Other considerations leading to the same definition may be found in M. Richardson, *Fundamentals of Mathematics*, Macmillan, 1941.

same definitions as the preceding numbers did. Hence, for division of fractions we use definition 3 of section 5. That is, the **quotient** of two fractions $\frac{a}{b} \div \frac{c}{d}$ shall be the fraction $\frac{x}{y}$ such that $\frac{c}{d} \cdot \frac{x}{y} = \frac{a}{b}$, provided such a fraction exists. But we shall now prove that *in the system of fractions, the quotient* $\frac{a}{b} \div \frac{c}{d}$ *always exists and is the fraction obtained by inverting* $\frac{c}{d}$ *and multiplying.*

THEOREM 2. $\frac{a}{b} \div \frac{c}{d} = \frac{a}{b} \cdot \frac{d}{c}$.

Proof. The right member is equal to $\frac{ad}{bc}$ by definition 4. We have only to show that $\frac{c}{d} \cdot \frac{ad}{bc} = \frac{a}{b}$. But $\frac{c}{d} \cdot \frac{ad}{bc} = \frac{cad}{dbc}$ by definition 4, and $\frac{cad}{dbc} = \frac{a}{b}$ by cancellation. This completes the proof.

Example. $\frac{2}{3} \div \frac{4}{9} = \frac{2}{3} \cdot \frac{9}{4} = \frac{18}{12} = \frac{3}{2}$.

We would like now to identify the fraction a/b with the quotient $a \div b$ so that division of one natural number by another would always be possible, according to our original aim in introducing fractions. But then we would have $3 \div 1 = 3/1$ and also, by definition 3, section 5, $3 \div 1 = 3$. Thus we are led to the following agreement: *we shall identify fractions a/1 having denominator 1 with the natural number a in the numerator.* Thus we write $3/1 = 3$. We can now prove the following theorem.

THEOREM 3. *If a and b are natural numbers, then* $a \div b = a/b$.

Proof. By agreement $a = a/1$ and $b = b/1$. Therefore $a \div b = \frac{a}{1} \div \frac{b}{1} = \frac{a}{1} \cdot \frac{1}{b}$ by theorem 2. Hence $a \div b = \frac{a}{b}$ by definition 4.

We shall henceforth use the division sign $\div$ *and the fraction line interchangeably.* Clearly, *the system of fractions is closed under both multiplication and division.*

We could now prove that the basic laws of equality, such as

"if equals are multiplied or divided by equals, the results are equal," are valid for fractions.

Note that the equality $\frac{a}{b} = \frac{ax}{bx}$ in definition 2 may now be understood in terms of multiplication. For

$$\frac{ax}{bx} = \frac{a}{b} \cdot \frac{x}{x} = \frac{a}{b} \cdot \frac{1}{1} = \frac{a}{b} \cdot 1 = \frac{a}{b},$$

which is in harmony with the intuitive idea that multiplication of a quantity by 1 should not change its value.

EXERCISES

All letters represent natural numbers.
Reduce each of the following fractions to lowest terms:

1. $\frac{28}{21}$. **2.** $\frac{36}{30}$. **3.** $\frac{17}{51}$. **4.** $\frac{57}{95}$. **5.** $\frac{45abc}{63bcd}$.

6. $\frac{32xyz}{48xzv}$. **7.** $\frac{3m + 9n}{3mn}$. **8.** $\frac{3m + 9n}{5m + 15n}$.

9. Write a fraction equal to $\frac{4}{7}$ having the denominator 35.
10. Write a fraction equal to $\frac{2}{3}$ having the denominator 18.
11. Write a fraction equal to $\frac{a}{b}$ having the denominator bd.
12. Write a fraction equal to $\frac{a}{b}$ having the denominator $bx + by$.

Calculate each of the following and reduce the answer to simplest form:

13. $\frac{5}{4} \cdot \frac{2}{15}$. **14.** $\frac{4}{3} \cdot \frac{6}{4}$. **15.** $\frac{5}{16} \div \frac{35}{4}$.

16. $\frac{12}{35} \div \frac{3}{10}$. **17.** $\frac{xy}{zw} \div \frac{y}{z}$. **18.** $\frac{3a}{4b} \div \frac{6}{8bc}$.

19. $\frac{ab + ac}{b} \div \frac{b + c}{3b}$. **20.** $\frac{ax + ay}{bc + bd} \div \frac{mx + my}{3c + 3d}$.

8. Fractions. Addition and Subtraction

Since fractions are symbols invented by us, we may decide what we want to mean by the sum of two fractions. To add coins of the same denomination we add the number of coins; for example, 2 quarters plus 1 quarter yields 3 quarters. Analogously, we adopt the following definition.

DEFINITION 1. *The* ***sum*** *of two fractions having the same denominator shall be the fraction whose numerator is the sum of the* *given numerators and whose denominator is the given common denominator. In symbols,*

$$\frac{a}{c} + \frac{b}{c} = \frac{a+b}{c}.$$

For example, $$\frac{2}{4} + \frac{1}{4} = \frac{3}{4}.$$

To add coins of different denominations, we first express them as equivalent amounts with a common denomination, and then add as above; for example, 2 dimes and 1 quarter would be expressed as 20 cents and 25 cents, respectively, and then added, obtaining 45 cents. Analogously, we adopt the following definition.

DEFINITION 2. *The* **sum** *of two fractions having different denominators shall be the fraction obtained by expressing them as equal fractions having a common denominator (by means of definition 2, section 7), and then adding by means of definition 1.*

Example 1. $\frac{1}{2} + \frac{1}{3} = \frac{1 \cdot 3}{2 \cdot 3} + \frac{1 \cdot 2}{3 \cdot 2} = \frac{3}{6} + \frac{2}{6} = \frac{5}{6}.$ (See Fig. 4.)

Example 2. $\frac{a}{b} + \frac{c}{d} = \frac{ad}{bd} + \frac{bc}{bd} = \frac{ad+bc}{bd}.$

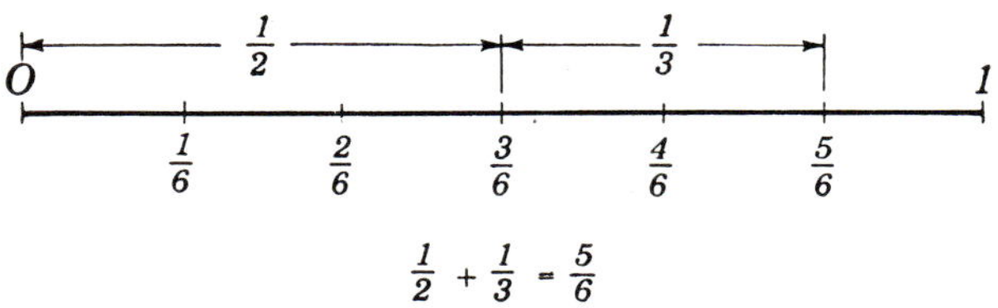

Fig. 4.

It is often desirable, although not essential, to use the least common denominator. By the **least common denominator** of several given fractions is meant the least common multiple of the given denominators.

Subtraction is defined as the inverse of addition, by means of definition 1, section 5, just as division was defined as the inverse of multiplication. That is, the **difference** of two fractions $\frac{a}{b} - \frac{c}{d}$ shall be the fraction $\frac{x}{y}$ such that $\frac{c}{d} + \frac{x}{y} = \frac{a}{b}$, provided such a fraction exists. If such a fraction does exist, we shall say that $\frac{a}{b}$ **is greater than** $\frac{c}{d}$; in symbols $\frac{a}{b} > \frac{c}{d}$ or $\frac{c}{d} < \frac{a}{b}$.

We now prove the following theorem.

THEOREM 1. *If* $ad > bc$, *then* $\frac{a}{b} - \frac{c}{d} = \frac{ad - bc}{bd}$.

Proof. According to the definition of subtraction, we have only to show that

$$\frac{c}{d} + \frac{ad - bc}{bd} = \frac{a}{b}.$$

Adding the fractions in the left member, we obtain

$$\frac{bc + (ad - bc)}{bd}.$$

By definition $(ad - bc)$ stands for a number which, when added to bc, will yield ad. Hence, $bc + (ad - bc) = ad$. Therefore the left member reduces to $\frac{ad}{bd}$, which becomes $\frac{a}{b}$ by cancellation of the common factor d. This completes the proof.

In practice, this theorem asserts that *to subtract one fraction from another we may express them as equal fractions having a common denominator and then write the difference of the numerators over the common denominator.* For example,

$$\frac{1}{2} - \frac{1}{3} = \frac{1 \cdot 3}{2 \cdot 3} - \frac{2 \cdot 1}{2 \cdot 3} = \frac{3}{6} - \frac{2}{6} = \frac{3 - 2}{6} = \frac{1}{6}.$$

It is clear that this difference $\frac{ad - bc}{bd}$ exists if and only if $ad > bc$, for otherwise the numerator $ad - bc$ has no meaning within the system of natural numbers. Hence, $\frac{a}{b} > \frac{c}{d}$ if and only if $ad > bc$. Thus, *to decide whether one fraction is greater than, less than, or equal to another, we may express them as equal fractions having a common denominator, and compare the numerators.*

It can be shown* that the system of fractions obeys the asso-

* See M. Richardson, *Fundamentals of Mathematics,* Macmillan, 1941, for further details.

ciative, commutative, and distributive laws, and the usual laws of equality and inequality. The system of fractions is closed under addition, multiplication, and division, but not under subtraction, since it is still impossible to subtract a larger number from a smaller one.

EXERCISES

Calculate each of the following expressions, and reduce the answer to simplest form:

1. $\frac{1}{4} + \frac{1}{6}$. **2.** $\frac{1}{4} + \frac{2}{3}$. **3.** $\frac{3}{4} + \frac{1}{8}$. **4.** $\frac{1}{6} + \frac{5}{12}$.

5. $\frac{1}{3} - \frac{1}{4}$. **6.** $\frac{3}{16} - \frac{1}{8}$. **7.** $\frac{7}{8} - \frac{3}{4}$. **8.** $\frac{3}{4} - \frac{1}{8}$.

9. $\frac{3}{4} + \left(\frac{1}{4} + \frac{5}{12}\right)$. **10.** $\left(\frac{3}{4} + \frac{1}{4}\right) - \frac{5}{12}$.

11. $\left(\frac{2}{3} + \frac{1}{2}\right) - \left(\frac{1}{4} + \frac{5}{12}\right)$. **12.** $\left(\frac{3}{4} + \frac{5}{12}\right) \div \left(\frac{5}{12} - \frac{1}{4}\right)$.

13. $\frac{1}{x} + \frac{1}{3}$. **14.** $\left(\frac{3}{x} + \frac{3}{y}\right) \div \left(\frac{4}{x} + \frac{4}{y}\right)$.

15. $\frac{a}{3b} + \frac{c}{3}$. **16.** $\left(\frac{1}{x} - \frac{1}{y}\right) \div \left(\frac{1}{x} + \frac{1}{y}\right)$ where $x < y$.

In each of the following decide whether the first fraction is equal to, less than, or greater than the second:

17. $\frac{17}{19}, \frac{21}{23}$. **18.** $\frac{38}{57}, \frac{34}{51}$. **19.** $\frac{31}{33}, \frac{35}{39}$. **20.** $\frac{31}{33}, \frac{35}{38}$.

21. If x and y represent (positive) fractions, which of the expressions (*a*) $x + y$, (*b*) $x - y$, (*c*) xy, (*d*) $x \div y$, does not always represent a (positive) fraction? What restriction must be placed on x and y to insure the existence of this number within the system of fractions?

9. Directed Numbers. Addition and Multiplication

Within the system of (positive) fractions it is not possible to deal with the expression $a - b$ without first ascertaining that $a > b$. This inconvenience will be removed by the invention of directed or signed numbers. Corresponding to each number already in existence, such as 3, we invent two new symbols, such as $+3$ and -3. The symbols preceded by a $+$ sign are called **positive** numbers, those preceded by a $-$ sign are called **negative** numbers. We also invent a new symbol 0, called **zero,** which is considered to be neither positive nor negative. The numbers 0, +1, +2, +3, . . . and −1, −2, −3, . . . are called **integers*** or **whole numbers.** We speak of positive and negative integers and of positive and negative fractions. All these new numbers are called **signed numbers** or **directed numbers** because they may be conveniently interpreted in terms of direction, as follows. We choose a straight line, a unit of length, and a point on the line which we shall call the **origin.** To the origin we attach the number zero. Marking off the unit of length an indefinite number of times in both directions from the origin,

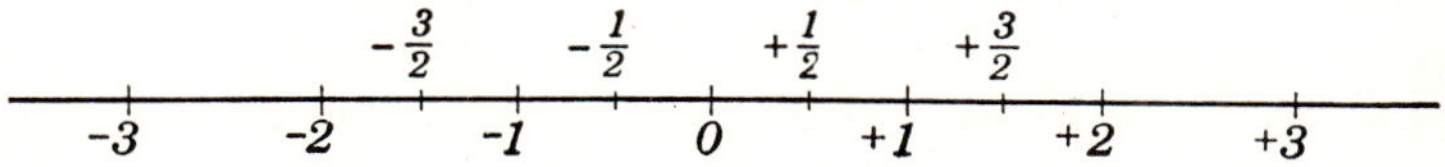

Fig. 5.

we attach the positive integers to marked points in one direction, and negative integers in the other. The other positive and negative fractions are attached to points on the line in the obvious way (Fig. 5). It is customary to use the right side of the line for positive numbers and the left side for negative numbers.† The numbers studied in preceding sections are called **unsigned numbers** to distinguish them from the new signed numbers.

The idea of negative numbers struggled for recognition for centuries, and was accepted with considerable reluctance as late as the beginning of the 17th century, even by mathematicians. Negative numbers were often called "false" or "fictitious" numbers. Now it is true that no one ever saw −3 books on a table, except possibly at a séance. But to try to interpret −3 as referring to the ghosts of three departed books, or to the operation of taking three books away from an empty table, is to miss the point completely. The point is that directed numbers are not to be interpreted as quantity or magnitude alone, but rather as quantity or magnitude together with direction. Hence they are convenient for such applications as east and west, up and down, profit and loss, future and past, temperature above and below zero, etc. They are called numbers largely because they obey

* The Latin word *integer* means *whole.*

† This is, of course, merely a convention. We could equally well put the negative numbers on the right. But there is no particular advantage in flouting a harmless convention.

the basic laws of calculation which all other numbers obey, such as the laws of equality, the associative, commutative, and distributive laws, etc.

Note also that the + and − signs in the symbols +3 and −3 are not intended to indicate the operations of addition and subtraction but are merely marks or signs to distinguish one direction from another. It would be better, from a logical standpoint, to reserve the + and − signs for the operations of addition and subtraction, and to use black and red ink, or other signs like ${}_R3$ and ${}_L3$ to indicate the right-hand 3 and the left-hand 3, respectively. However, we shall adhere to the traditional notation to avoid confusion.

The directed numbers, like the fractions, are our own invention, and it is now up to us to decide what we want to mean by the sum, product, etc., of two directed numbers.

Our definition of the sum of two directed numbers can be best understood by referring to Fig. 5. Let us imagine ourselves standing at the origin, facing in the positive direction, and let us interpret each signed number as a marching order. The directed number +3 shall mean "march forward, that is, to the right, 3 units," while the directed number −3 shall mean "march backward, that is, to the left, 3 units." Zero shall mean "do not march at all." The plus sign for addition shall be translated as "and then." *The number attached to the point at which we arrive after carrying out all our marching orders is called the* ***sum*** *of the given directed numbers.*

Example 1. (+3) + (+2) means walk right 3 units and then right 2 units, arriving at +5. Hence (+3) + (+2) = +5.

Notice that the plus signs within each pair of parentheses indicate positive numbers while the plus sign between the two pairs of parentheses indicates the operation of addition. As remarked above, it would be preferable, from a logical standpoint, to avoid this ambiguity by writing ${}_R3 + {}_R2 = {}_R5$, for example.

Example 2. (+3) + (−2) means walk right 3 units and then left 2 units, arriving at +1. Hence (+3) + (−2) = +1.

Example 3. (+2) + (−3) means walk right 2 units and then left 3 units, arriving at −1. Hence (+2) + (−3) = −1.

Example 4. $(+3) + (-3)$ means walk right 3 units and then left 3 units, arriving at 0. Hence $(+3) + (-3) = 0$.

Example 5. $(-3) + (-2)$ means walk left 3 units and then left 2 units, arriving at -5. Hence $(-3) + (-2) = -5$.

Example 6. Any number plus zero will be the given number again, since zero means "do not march at all." Thus $(+3) + 0 = +3$.

This intuitive idea of walking back and forth on the line of Fig. 5 is all that the student needs to understand in order to calculate sums of signed numbers successfully. For the sake of completeness, however, we append a formal definition, which may well be omitted on first reading, and which is cumbersome because it is necessary to use several cases in order to define the sum of signed numbers in terms of operations on unsigned numbers, which have been previously defined.

DEFINITION 1. *If a and b are any unsigned numbers, then:*

Case 1. $(+a) + (+b) = +(a + b)$.
Case 2.

$$(+a) + (-b) = (-b) + (+a) = \begin{cases} +(a - b) & \text{if } a > b. \\ -(b - a) & \text{if } a < b. \\ 0 & \text{if } a = b. \end{cases}$$

Case 3. $(-a) + (-b) = -(a + b)$.
Case 4. $(+a) + 0 = 0 + (+a) = +a$.
Case 5. $(-a) + 0 = 0 + (-a) = -a$.
Case 6. $0 + 0 = 0$.

It is convenient to identify the positive (signed) numbers with the corresponding unsigned numbers. We do this henceforth. Thus we shall write 3 and +3 interchangeably.

By the **absolute value** of a non-zero directed number is meant the corresponding positive number; thus the absolute value of 3 is 3 and the absolute value of -3 is 3. The **absolute value** of zero is zero. The first three cases of definition 1 may be put as follows:

To add two directed numbers of the same sign, add their absolute values and prefix their common sign. To add two directed numbers of opposite sign, subtract the smaller absolute value from the larger and prefix the sign attached to the number of larger absolute value.

Our definition of multiplication of signed numbers will be motivated by our desire to have directed numbers behave as much like the previous unsigned numbers as possible. For example, we would like them to obey the basic commutative and distributive laws. By definition 1, section 4, $3 \cdot 2$ stands for $2 + 2 + 2$ or the sum of three 2's. Therefore we would like $3 \cdot (-2)$ to mean the sum of three (-2)'s, or $(-2) + (-2) + (-2)$, or -6. Since we want to preserve the commutative law for multiplication, we want $(-2) \cdot 3$ to mean the same thing as $3 \cdot (-2)$, or -6, even though $(-2) \cdot 3$ cannot be interpreted as the sum of -2 threes. Similarly, we want $3 \cdot 0$ to mean $0 + 0 + 0$ or 0, and, because we wish to preserve the commutative law, we want $0 \cdot 3$ to mean 0 also. In general, we define $0 \cdot a = a \cdot 0 = 0$ where a is any directed number.

Our desire to preserve the distributive law can be the motivation of the mysterious case $(-3)(-2) = +6$.* Consider the expression $(-3)[(-2) + 2]$. We want the distributive law to hold so that

$$(-3)[(-2) + 2] = (-3)(-2) + (-3) \cdot 2.$$

The left member is $(-3) \cdot [0]$ which we have already decided should be 0. The last term of the right member, as we have already decided, should be -6. Hence we have

$$0 = (-3)(-2) + (-6).$$

If the distributive law is to hold, we must define $(-3)(-2) = +6$. We are thus led to the following definition of the product of signed numbers.

DEFINITION 2. *If a and b are any unsigned numbers, then:*

Case 1. $(+a)(+b) = (-a)(-b) = +(ab)$. (*The product of two numbers with like signs is positive.*)

Case 2. $(+a)(-b) = (-b)(+a) = -(ab)$. (*The product of two numbers with unlike signs is negative.*)

Case 3. $(+a) \cdot 0 = 0 \cdot (+a) = (-a) \cdot 0 = 0 \cdot (-a) = 0 \cdot 0 = 0$. (*The product of any number with zero is zero.*)

* Other considerations which also serve as motivations in this case will be found in M. Richardson, *Fundamentals of Mathematics*, Macmillan, 1941.

This definition is sometimes known as the **rule of signs.**

EXERCISES

Calculate each of the following:

1. $7 + (-1)$. **2.** $4 + (-7)$. **3.** $(-2) + (-7)$.
4. $(-5) + 0$. **5.** $(+12) + (-5) + (-17)$.
6. $(+11) + (-7) + (-20) + (+5)$.
7. $(-13) + (+10) + (-14) + (+8)$.
8. $(-13) + (-11) + (+38) + (-1)$.
9. $(+5)(-2)$. **10.** $(-6)(-2)$. **11.** $(-5) \cdot 0$.
12. $(-3)(-4)(+5)$. **13.** $(-2)(+5)(-3)$.
14. $(-2)(+2)(-2)$. **15.** $(-1)(+2)(-3)(-1)$.
16. $(+22)(-3)(-17) \cdot 0$. **17.** $(-2)(-3) + (-4)(+3)$.
18. $(-4)(+7) + (-6)(-4)$. **19.** $(-3)(-5) + (+6)(-3)$.
20. $(-1)(+2)(-3) + (-2)(0)(-3) + (-2)(-3)(-1) + (-2)(-7)$.

10. Directed Numbers. Subtraction and Division

For subtraction we use definition 1, section 5, as before.

Definition 1. *If a and b are any directed numbers, $a - b$ shall stand for the directed number x such that $b + x = a$, provided such a number x exists.*

For example, $(-2) - (-3)$ stands for the number x such that $(-3) + x = (-2)$; hence $x = 1$ since $(-3) + 1 = -2$. In a like manner $2 - (-3)$ stands for the number x such that $(-3) + x = 2$; hence $x = +5$ since $(-3) + (+5) = (+2)$.

It can be proved that in the system of all directed numbers, $a - b$ always exists. That is, *the system is closed under subtraction.* Hence the definition of $>$ must be altered as follows.

Definition 2. *If a and b are directed numbers, and if there exists a* ***positive*** *number x such that $b + x = a$, then we write $a > b$ or $b < a$. In other words, $a > b$ if and only if $a - b$ is positive.*

Thus $-2 > -5$ and $2 > -3$. Intuitively $a > b$ means that a is to the right of b on the line of Fig. 5.

Division is defined as in definition 3, section 5.

DEFINITION 3. *If a and b are any directed numbers, $a \div b$ stands for the unique directed number x such that $bx = a$, provided such a number x exists.*

It can be shown that, except for the case $b = 0$, to be discussed in the next section, $a \div b$ always exists.

For example, $(-6) \div 2$ stands for the number x such that $2x = -6$; hence $x = -3$ because $2(-3) = -6$. Similarly $6 \div (-2)$ stands for the number x such that $(-2) \cdot x = 6$; hence $x = -3$ because $(-2)(-3) = +6$. Also $(-6) \div (-2)$ stands for the number x such that $(-2) \cdot x = -6$; hence $x = +3$ because $(-2)(+3) = -6$.

These examples suggest that the rule of signs for division is like that for multiplication: *the quotient of two numbers with like signs is positive, while the quotient of two numbers with unlike signs is negative.*

DEFINITION 4. *The* **negative** *of a directed number x shall be the directed number which must be added to x in order to obtain the sum 0. The negative of x is denoted by $-x$.*

For example, the negative of $+3$ is -3 since $(+3) + (-3) = 0$. Hence $-(+3) = -3$. Similarly, the negative of -3 is $+3$ because $(-3) + (+3) = 0$. Hence $-(-3) = +3$. That is, $+3$ and -3 are negatives of each other. This is a third use of the minus sign. It can be shown that no trouble arises from this triple ambiguity.

We could now prove that directed numbers obey all the familiar rules of algebra, such as the basic associative, commutative, and distributive laws. Also all rules for manipulating fractions carry over in an obvious way. For example,

$$\frac{a}{b} = \frac{ax}{bx} \tag{1}$$

where a, b, and x are any directed numbers, provided b and x are not zero. As illustrations, we shall indicate a few theorems each of which justifies a familiar but important manipulation.

THEOREM 1. *If a and b are any directed numbers, then $a - b = a + (-b)$.*

Proof. By definition 1 we have only to show that b plus the right member equals a. But

$$
\begin{aligned}
b + [a + (-b)] &= b + [(-b) + a] && \text{(commutative law for addition)} \\
&= [b + (-b)] + a && \text{(associative law for addition)} \\
&= 0 + a && \text{(definition 4)} \\
&= a && \text{(definition 1, section 9).}
\end{aligned}
$$

This completes the proof.

THEOREM 2. *If a and b are any directed numbers, then* $a - (-b) = a + b$.

Proof. Similar to theorem 1. We leave this as an exercise for the reader.

Theorems 1 and 2 tell us that *subtracting a directed number is equivalent to adding its negative.*

A series of numbers separated by $+$ and $-$ signs may be called an **algebraic sum.** One of the individual numbers separated by these $+$ and $-$ signs, taken together with the preceding sign, constitutes a **term** of this algebraic sum. Thus in the expression

$$(1) \qquad 2 + 3 - 4 + a - b,$$

the terms are $+2$, $+3$, -4, $+a$, $-b$. By virtue of theorem 1, the expression (1) may be interpreted as $2 + 3 + (-4) + a + (-b)$. It follows from theorems 1 and 2 that the extended form of the commutative law for addition may be applied to algebraic sums in the following sense. *The order of the terms in an algebraic sum may be rearranged without affecting the value of the sum.*

For example, (1) may be rewritten as $-4 + 2 - b + a + 3$, or $+3 - 4 - b + a + 2$, and so on.

We could also prove* that if a and b are any directed numbers, then

$$(2) \qquad a(-b) = -(ab) \quad \text{and} \quad (-a)(-b) = +(ab).$$

This may also be called the **rule of signs for multiplication.** In particular,

$$(3) \qquad -a = (-1)a \quad \text{and} \quad (-1)(-a) = a.$$

* Proofs may be found in M. Richardson, *Fundamentals of Mathematics*, Macmillan, 1941. The appropriateness of these definitions in certain geometric and practical situations is also discussed there.

That is, *multiplication by −1 reverses the sign of a directed number.*

Also,

$$(4) \qquad \frac{-a}{b} = \frac{a}{-b} = -\frac{a}{b}, \quad \text{while} \quad \frac{-a}{-b} = \frac{a}{b}.$$

This may be called the **rule of signs for division.**

THEOREM 3. $-(a - b) = -a + b$.

First Proof. By (3),

$$\begin{aligned} -(a-b) &= (-1)(a-b) \\ &= (-1)[a + (-b)] \\ &= (-1)a + (-1)(-b) \quad \text{(distributive law)} \\ &= -a + b \quad \text{(by (3)).} \end{aligned}$$

Second Proof. By definition 4, we have only to prove that $(a - b) + (-a + b) = 0$. But the left member is $a - a + (-b) + b = 0 + 0 = 0$. This completes the proof.

THEOREM 4. $-(a + b) = -a - b$.

Proof. Left to the reader as an exercise. Compare theorem 3.

EXERCISES

Calculate each of the following:

1. $(-7) - (+2)$. **2.** $(+7) - (-2)$.

3. $(-2) - (-7)$. **4.** $(+2) - (+7)$.

5. $(-5) - (-3) + (-1) - (+2)$.

6. $(-7) + (+3) - (+2) - (-1)$.

7. $(-9) \div 3$. **8.** $(-9) \div (-3)$. **9.** $9 \div (-3)$.

10. $-\frac{(-2)(-7)}{-4}$. **11.** $-\frac{(-2)(+4)}{(-8)(-1)}$.

12. $-\left(\frac{-6}{+2} + \frac{16}{-4}\right)$. **13.** $\frac{5 - (-5)}{3 - (-1)}$.

14. $\frac{12}{-2} - \frac{18}{-3}$. **15.** $\left(\frac{15}{-4} \div \frac{-3}{20}\right) - \left(\frac{-9}{-2} \div \frac{+3}{-10}\right)$.

16. $\frac{1}{2} - \frac{-3}{5}$. **17.** $\frac{-2}{3} + \frac{2}{5}$. **18.** $-\frac{1}{2} + \frac{2}{5}$.

19. $\frac{(-4) + (-3)}{(+2) + (-3)} - \frac{(-1) - 1}{(-2) - (+3)}$.

20. $\frac{(+5) - (-3)}{-2 - 3} \div \frac{8 - (-3) + (-6)}{-3 - (-13)}$.

21. Subtract the quotient of (-21) divided by $(+14)$ from the product of (-2) and (-6).

22. Divide the quotient of $(+16)$ divided by (-2) by the sum of (-3) and $(+1)$.

11. The System of Rational Numbers. Properties of Zero

All the numbers introduced so far, namely zero, the positive and negative integers, and the positive and negative fractions, and no others, are called **rational numbers.** The word "rational" does not mean "reasonable"; it comes from the word "ratio." *A number is **rational** if and only if it can be expressed as the quotient or ratio of two integers.* Thus $\frac{5}{2}$, $\frac{3}{-7}$, and $\frac{2}{1}$ are rational numbers. *A rational number which can be expressed with denominator 1 is an **integer**.* For example, $-2 = \frac{-2}{1}$ and $0 = \frac{0}{1}$ are integers. Thus the system of rational numbers may be classified as follows:

Rational numbers (positive and negative fractions and zero)
- *Integers* (rational numbers which can be expressed with denominator = 1)
  - *Negative integers*
  - *Zero*
  - *Positive integers or natural numbers.*

The system of rational numbers is closed under addition, subtraction, multiplication, and division, with the single exception that division by zero must be excluded. Why this exception must be made will be seen from the definition of division: the quotient $a \div b$ stands for the unique number x such that $bx = a$, if such a number exists. Let us try to apply this definition if $b = 0$. Either $a \neq 0$ or $a = 0$.

Case 1. Suppose $a \neq 0$. Then $a \div 0$ stands for the number x such that $0 \cdot x = a$. But $0 \cdot x = 0$ no matter what number x is, and $a \neq 0$. Hence no such number x exists. For example, $5/0$ stands for the number x such that $0 \cdot x = 5$, but no number x can satisfy this requirement. Thus $5/0$ is a meaningless symbol.

Case 2. Suppose $a = 0$. Then $a \div 0$ or $0 \div 0$ stands for the number x such that $0 \cdot x = 0$. But this equation is satisfied no matter what number we choose for x, and is therefore quite useless.

Therefore, division by zero is excluded in all cases.

However, $0 \div b$ does have a definite value if $b \neq 0$. It stands for the number x such that $bx = 0$. There is one and only one such value of x, namely zero. *Hence, $0/b = 0$ if $b \neq 0$.* For example, $0/3 = 0$.

Notice that $0/1 = 0$ while $1/0$ is a meaningless symbol. The statement that a symbol has no meaning is very different from the statement that it has the value zero. A student who is not registered for this course receives no grade for it; a student must register for the course before he can aspire to the grade of zero.

The system of integers is closed under addition, subtraction, and multiplication, but not division. All definitions concerning factors, even and odd numbers, etc., may be extended from natural numbers to integers, if desired. For example, an integer b is said to be a **factor** of an integer a if there exists an integer x such that $a = bx$. An **even** integer is one that has 2 as a factor. For example $+6$ and -6 are even. Zero is even since $0 = 0 \cdot 2$; zero has every integer as a factor.

Since the system of rational numbers is closed under division, except for division by zero, it follows that $1/a$ is a rational number, provided a is a rational number different from zero.

DEFINITION. *If $a \neq 0$, the number $1/a$ is called the* ***reciprocal*** *of a.*

Notice that *division by a is equivalent to multiplication by the reciprocal of a.*

The number zero has the following important property, which will be needed later:

THEOREM 1. *If $ab = 0$ and $a \neq 0$, then $b = 0$.*

Proof. Since $a \neq 0$, $1/a$ exists. By hypothesis,

$$ab = 0.$$

Multiplying both sides by $1/a$, we have

$$\frac{1}{a} \cdot ab = \frac{1}{a} \cdot 0$$

because if equals are multiplied by equals, the results are equal. But the right member is zero and the left member becomes $\dfrac{ab}{a}$ or $\dfrac{b}{1}$ or b. Hence $b = 0$. This completes the proof.

It follows from theorem 1 that *the product of any number of non-zero numbers cannot be zero*. Or, *the product of several numbers is zero if and only if at least one of the factors is itself zero.*

The system of rational numbers, extensive as it is, will be seen to be inadequate for the purposes of both algebra and geometry. First we need some preliminary considerations, to be taken up in the next sections.

Note. Amusing results can be obtained if division by zero is overlooked. (We assume for the moment that the reader recalls some elementary algebra.) For example, let Alice be a years old and let Betty be b years old, and suppose they are of the same age. Then $a = b$. Multiplying both sides of this equation by b, we obtain $ab = b^2$. Subtracting a^2 from both sides, we have $ab - a^2 = b^2 - a^2$. Factoring, we get $a(b - a) = (b + a)(b - a)$. Dividing both sides by $b - a$, we have $a = b + a$. Since $a = b$, this implies $a = b + b$, by substitution. Therefore $a = 2b$, and Alice discovers, to her chagrin, that she is twice as old as Betty. The question is, what made Alice age so rapidly? It was not caused by overstudy of mathematics, but rather by division by $b - a$ which is zero, since $a = b$ by hypothesis. This absurdity can be pushed further. From the statement $a = 2b$ above, we get $a = 2a$, since $b = a$. Dividing both sides by a, we have $1 = 2$. But $3 = 2 + 1 = 1 + 1 = 2 = 1$, $4 = 3 + 1 = 1 + 1 = 2 = 1$, and so on. We thus arrive at an oversimplified system of arithmetic in which all numbers are equal to each other, and there can be no such thing as a wrong answer.

EXERCISES

If a and b are any two rational numbers, does each of the following expressions always represent a rational number?

1. $a + b$. **2.** $a - b$. **3.** $a \cdot b$. **4.** a/b.

If a and b are any two integers, does each of the following expressions always represent an integer?

5. $a + b$. **6.** $a - b$. **7.** $a \cdot b$. **8.** a/b.

12. Positive Integral Powers and Roots

The following definition is nothing more than an abbreviation.

DEFINITION 1. *If n is a positive integer, x^n shall stand for $x \cdot x \ldots x$ with n factors.*

For example, $x^3 = xxx$. The **exponent** n is merely the number of factors. We call x^n the **nth power of x.** x^2 is called the **square** of x, and x^3 is called the **cube** of x. We need the following theorems.

THEOREM 1. *If n is any positive integer, then $(xy)^n = x^n y^n$.*

Proof. By definition 1, $(xy)^n = (xy)(xy) \ldots (xy)$ where there are n parentheses. By the associative law for multiplication, we may remove parentheses, so that

$$(xy)^n = xyxy \ldots xy.$$

By the commutative law for multiplication, we may rearrange the factors, writing

$$(xy)^n = xx \ldots xyy \ldots y.$$

By the associative law, we may regroup them, writing

$$(xy)^n = (xx \ldots x)(yy \ldots y).$$

By definition 1, $(xx \ldots x) = x^n$ and $(yy \ldots y) = y^n$. Hence

$$(xy)^n = x^n y^n.$$

This proves the theorem.

It is clear that this theorem can be extended to more than 2 factors. Thus,

$$(1) \qquad (xyzw \ldots)^n = x^n y^n z^n w^n \ldots$$

THEOREM 2. $(a + b)^2 = a^2 + 2ab + b^2$.

Proof. By definition 1, $(a + b)^2 = (a + b)(a + b)$. By the law of closure for addition, we may regard $(a + b)$ as a single number. Hence, by the distributive law,

$$(a + b)(a + b) = (a + b)a + (a + b)b.$$

Using the distributive law again, and other laws (which ones?), the right member becomes $a^2 + ab + ab + b^2$, and finally $a^2 + 2ab + b^2$. This proves the theorem.

It is clear that this theorem can be extended to more than 2 terms. Thus,

$$(2) \qquad (a + b + c)^2 = a^2 + b^2 + c^2 + 2ab + 2ac + 2bc.$$

DEFINITION 2. *If there exists a number x such that $x^n = a$, where n is a positive integer greater than one, then x is called an*

nth root of a. *In particular if* $x^2 = a$, *x is called a* **square root of a**; *and if* $x^3 = a$, *x is called a* **cube root of a.**

For example, 3 is a square root of 9 since $3^2 = 9$. Also, -3 is a square root of 9 since $(-3)^2 = 9$. Both 2 and -2 are fourth roots of 16 since $2^4 = 16$ and $(-2)^4 = 16$. A cube root of 8 is 2 because $2^3 = 8$; a cube root of -8 is -2 because $(-2)^3 = -8$.

It is clear that a number may have more than one nth root; in fact, in a later chapter it will be shown that, in the system of complex numbers, every number has n distinct nth roots; that is, two square roots, three cube roots, and so on. To avoid the unpleasant ambiguity attached to the idea of nth root, we agree to reserve the **radical sign** $\sqrt[n]{a}$ to denote a single one of these nth roots, to be called the **principal *n*th root.** *If a is positive, the principal nth root will be the positive one. If a is negative and n is odd, the principal nth root will be the negative one.* No other cases are needed here. For example, $\sqrt{9}$ stands for 3 but not for -3; if we wish to indicate -3, we write $-\sqrt{9}$. Also $\sqrt[3]{8} = 2$ while $\sqrt[3]{-8} = -2$. This agreement is made to avoid the confusion that would arise if we permitted the symbol $\sqrt{9}$ to stand for either 3 or -3 ambiguously. The principal nth root of a number is uniquely determined. The number n written above the radical sign is called the **index** of the root.

We make the linguistic agreement that *powers shall take precedence over multiplication and division, except where otherwise indicated by parentheses.* For example, $5 \cdot 3^2 = 5 \cdot 9 = 45$ while $(5 \cdot 3)^2 = 15^2 = 225$. *The radical sign, however, acts as a parenthesis; all indicated operations under the radical sign are to be done before extracting the root.* For example, $\sqrt{9 + 16} = \sqrt{25} = 5$, while $\sqrt{9} + \sqrt{16} = 3 + 4 = 7$.

By definition, $\sqrt[n]{a}$ stands for a number which yields a when raised to the nth power; hence

$$(3) \qquad (\sqrt[n]{a})^n = a.$$

THEOREM 3. *If a and b are positive, then* $\sqrt[n]{a}\,\sqrt[n]{b} = \sqrt[n]{ab}$.

Proof. Let $x = \sqrt[n]{a}$ and $y = \sqrt[n]{b}$. By (3), $x^n = a$ and $y^n = b$. By theorem 1, $(xy)^n = x^n y^n$. Hence $(xy)^n = ab$. By definition, $xy = \sqrt[n]{ab}$, since xy is also positive. By substitution, $\sqrt[n]{a}\,\sqrt[n]{b} = \sqrt[n]{ab}$. This completes the proof.

For example, $\sqrt{4}\,\sqrt{9} = \sqrt{4 \cdot 9} = \sqrt{36}$, or $2 \cdot 3 = 6$.

Note that this theorem is not true for negative a and b. For example, $\sqrt{-2}\sqrt{-2}$ is not $\sqrt{4}$, because $\sqrt{-2}\sqrt{-2} = (\sqrt{-2})^2 = -2$ by (3), while $\sqrt{4} = +2$.

It is obvious that *if* $a = b$, *then* $a^2 = b^2$. *But the converse is not true. From* $a^2 = b^2$ *we cannot infer that* $a = b$; for example, from $(-3)^2 = (+3)^2$ we cannot infer that $-3 = +3$. It is correct to say, however, that *if* $a^2 = b^2$, *then a is equal to either* $+b$ *or* $-b$; *that is, from* $a^2 = b^2$ *we may conclude that* $a = \pm b$.

EXERCISES

By inspection, find the value of each of the following:

1. $\sqrt[3]{-27}$. **2.** $\sqrt[3]{64}$. **3.** $\sqrt[3]{125}$. **4.** $\sqrt{625}$.
5. $\sqrt[4]{16}$. **6.** $\sqrt{36}$. **7.** $\sqrt[5]{-32}$. **8.** $\sqrt[6]{64}$.
9. $\sqrt[4]{81}$. **10.** $\sqrt[3]{-1000}$.

11. (*a*) Is it true that $\sqrt{a^2 + b^2} = a + b$? Explain. (*b*) Of what is $a + b$ a square root?

12. (*a*) Is it true that $\sqrt{a^2 - b^2} = a - b$? Explain. (*b*) Of what is $a - b$ a square root?

13. Is it true that $\sqrt{x + y} = \sqrt{x} + \sqrt{y}$? Explain.

14. Is it true that $\sqrt{x - y} = \sqrt{x} - \sqrt{y}$? Explain.

Simplify each of the following, assuming that all letters represent positive numbers:

15. $(\sqrt{5})^2$. **16.** $(\sqrt[3]{-5})^3$. **17.** $(\sqrt[4]{a})^4$.
18. $(\sqrt[5]{a^3})^5$. **19.** $(\sqrt[n]{x})^n$. **20.** $\sqrt{64 + 36}$.
21. $\sqrt{1 - \left(\frac{\sqrt{3}}{2}\right)^2}$. **22.** $3 \cdot 2^2$. **23.** $(3 \cdot 2)^2$.
24. $4 + 3 \cdot 2^2$. **25.** $(4 + 3) \cdot 2^2$. **26.** $4 + (3 \cdot 2)^2$.
27. $(3 + 4 \cdot 2)^2$. **28.** $[(4 + 3) \cdot 2]^2$. **29.** $3(4 + 2)^2$.

★**30.** Find the fallacy in the following "proof" that all numbers are equal to each other:

Let a and b be any two unequal numbers. Let c be their average, that is, $c = \frac{a + b}{2}$. Then $2c = a + b$. Multiplying both sides by $(a - b)$ we have $2c(a - b) = (a + b)(a - b)$, or $2ac - 2bc = a^2 - b^2$. Transposing, we get $b^2 - 2bc = a^2 - 2ac$. Adding c^2 to both sides, we get $b^2 - 2bc + c^2 = a^2 - 2ac + c^2$, or $(b - c)^2 = (a - c)^2$. Therefore $b - c = a - c$. Hence $b = a$.

★ May be omitted without disturbing continuity.

13. The Square Root of 2

We shall prove the following surprising theorem.

THEOREM 1. *No rational number is a square root of 2.*

Before proving the theorem, let us see why it is surprising. It is clear that zero is not a square root of 2. Since $(-a)^2 = (+a)^2$ it is also clear that we may confine our attention to positive rational numbers. Recall that a rational number is one that can be expressed as a quotient of two integers. Hence we can arrange all the positive rational numbers in the following array, placing all those with denominator 1 in the first horizontal row, all those with denominator 2 in the second row, and so on:

$$\begin{array}{ccccc}
\frac{1}{1} & \frac{2}{1} & \frac{3}{1} & \frac{4}{1} & \cdots \\
\frac{1}{2} & \frac{2}{2} & \frac{3}{2} & \frac{4}{2} & \cdots \\
\frac{1}{3} & \frac{2}{3} & \frac{3}{3} & \frac{4}{3} & \cdots \\
\frac{1}{4} & \frac{2}{4} & \frac{3}{4} & \frac{4}{4} & \cdots \\
\cdot & \cdot & \cdot & \cdot & \\
\cdot & \cdot & \cdot & \cdot & \\
\cdot & \cdot & \cdot & \cdot &
\end{array}$$

This scheme, extending endlessly to the right and down, surely includes all positive rational numbers; for if you name any positive rational number, we can say exactly where it is in this array. For example, 159/357 is to be found in the 159th vertical column and the 357th horizontal row.

Let us imagine the points corresponding to these numbers marked off on a line, one horizontal row at a time, the number attached to each point representing its distance from the origin or zero point. As each succeeding row is used, the marked points clearly become more and more crowded together (*densely distributed* is a technical term for this). See Fig. 6. By the time the thousandth row is reached, we are marking off 1/1000, 2/1000, 3/1000, and so on. Now imagine all the rational numbers already marked off. Your intuition might tell you that surely all points on the line must have been marked off, *but this is not the case.*

There are still points which are not marked off by any rational number. One such point may be constructed as follows. Take an isosceles right triangle with legs 1 unit long (Fig. 7). Then, by the Pythagorean theorem, the length of the hypotenuse is $\sqrt{1^2 + 1^2} = \sqrt{2}$. This length may be laid off from the origin,

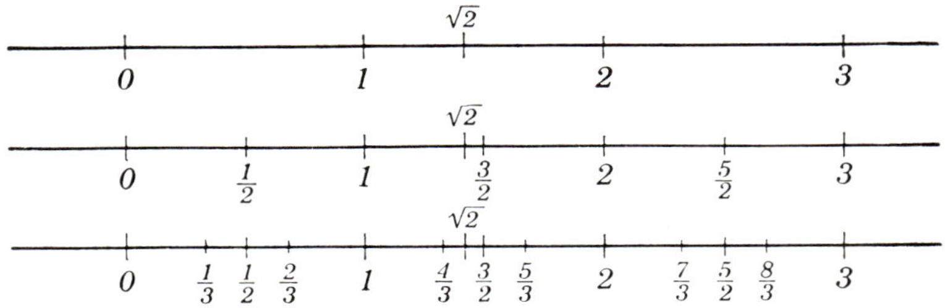

Fig. 6.

and the point thus marked off cannot have been marked by any rational number if our theorem is correct. If you recall that all rational numbers are supposed to be marked off, including all millionths, all trillionths, etc., it may seem remarkable that there are still unmarked points left (Fig. 6).

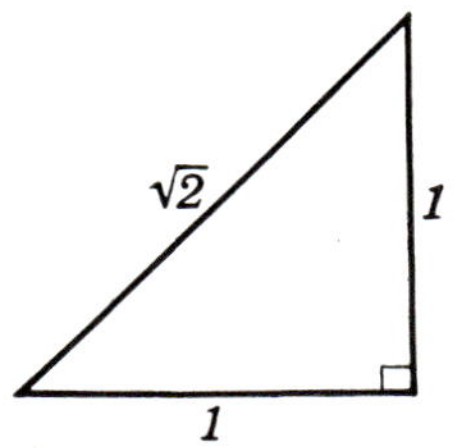

Fig. 7.

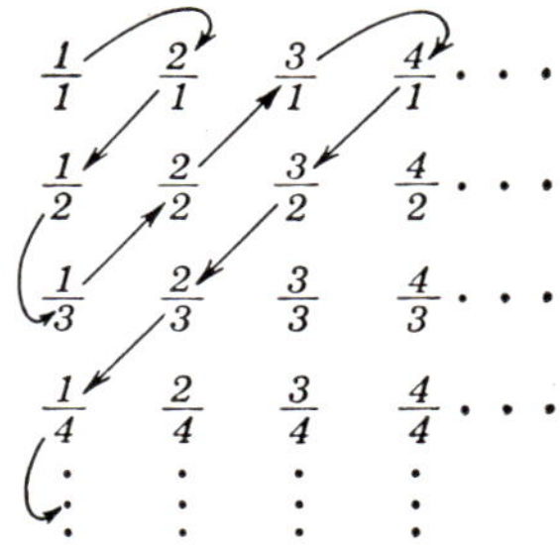

Fig. 8.

To prove our theorem we might at first try squaring all the rational numbers, one at a time, to see if any of them will yield 2 when squared. Of course, if we started out along the first horizontal row we would never get out of the first row, since it is endless. Therefore we might try to sweep up and down the diagonals, following the arrows in Fig. 8; that is, we would try squaring the rational numbers in the following order:

$$\frac{1}{1}, \frac{2}{1}, \frac{1}{2}, \frac{1}{3}, \frac{2}{2}, \frac{3}{1}, \frac{4}{1}, \frac{3}{2}, \frac{2}{3}, \frac{1}{4}, \cdots$$

If only the number of rational numbers were limited, we could indeed establish our theorem by this naïve method. However, since there is no end to the sequence of rational numbers, this method is absolutely hopeless. For if we squared the first 1000 rational numbers in this sequence, and failed to obtain 2 as the result, this would not prove that we might not get 2 as the result at some later trial. Therefore we must do something more clever to prove the theorem. We shall need the following preliminary theorem.

THEOREM A. *If a natural number n is the product of the prime factors $p, q, r, \ldots$, then n^2 is the product of the same prime factors, each occurring twice as often.*

Proof. Suppose $n = p \cdot q \cdot r \ldots$. Then, by (1), section 12, $n^2 = p^2q^2r^2 \ldots = p \cdot p \cdot q \cdot q \cdot r \cdot r \ldots$. This completes the proof.

For example, $12 = 2 \cdot 2 \cdot 3$ and $12^2 = 144 = 2 \cdot 2 \cdot 2 \cdot 2 \cdot 3 \cdot 3$.

Proof of Theorem 1. Suppose there were a (positive) rational number p/q (where p and q are natural numbers), whose square is 2. Then $p^2/q^2 = 2$, or, multiplying both sides by q^2,

$$p^2 = 2q^2. \tag{1}$$

By theorem A, both p^2 and q^2 have 2 as a factor an even number of times, if at all. But then the right member of (1) has 2 as a factor an odd number of times while the left member has it an even number of times. This is absurd because it contradicts the fact that a natural number can be expressed as a product of prime factors in only one way (section 6). Hence the supposition that there is a rational number whose square is 2 is false. This completes the proof.

The fact* that there is no rational number whose square is 2 was probably known to Pythagoras in the 6th century B.C. It is said to have disturbed him so much that he tried to suppress the information lest it discredit mathematicians in the eyes of the general public.

* For the relation between this fact and the so-called incommensurable magnitudes of geometry, see M. Richardson, *Fundamentals of Mathematics*, Macmillan, 1941. Another proof of the theorem above may be found there as well.

Note. In the proof of theorem 1 we used the method of **indirect proof** or ***reductio ad absurdum*** which you first met in your study of plane geometry. That is, we show that if the supposition that statement S is true leads logically to a false conclusion, then S must be false.

EXERCISES

Prove that no rational number is equal to:

1. $\sqrt{3}$. **2.** $\sqrt{5}$. **3.** $\sqrt{7}$.
4. $\sqrt{p}$, where p is a prime. **5.** $\sqrt[3]{2}$. **6.** $\sqrt[3]{3}$.
7. $\sqrt[n]{p}$, where p is a prime.
★8. $\sqrt{m}$ where m is a positive integer which is not the square of an integer.
★9. $\sqrt[n]{m}$ where m is a positive integer which is not the nth power of an integer.

14. The System of Real Numbers. Decimal Notation

The fact that no rational number is exactly a square root of 2 means, from the point of view of algebra, that the system of rational numbers is not closed under the operation of taking square roots. That is, we cannot apply the operation of taking square roots to any number of the system, 2 for example, and always get a number of the system. From the point of view of geometry, it means that there are lengths, such as the hypotenuse of the isosceles right triangle with unit legs, which are not exactly described by any rational number. From either point of view, the system of rational numbers, extensive though it is, is insufficient for our needs. We therefore invent **irrational numbers,** such as $\sqrt{2}$. The word "irrational" must not be taken to mean "unreasonable"; it means not expressible as a ratio of two integers. We shall give no accurate definition of irrational number here because of the technical difficulties involved. Indeed, while the fact that $\sqrt{2}$ is irrational was probably known to Pythagoras (6th century B.C.), a completely logical treatment of irrational numbers was not presented until about 1870, when G. Cantor and R. Dedekind, two German mathematicians, overcame the difficulties by different methods. However, we can discuss three of the essential characteristics of irrational numbers.

★ May be omitted without disturbing continuity.

The first point is that the irrational numbers fill up all the "gaps," that is, unmarked points, on the line (Fig. 6), so that every point on the line has a number attached to it, either rational or irrational. All these numbers, attached to points on the line, are called **real.** The system of real numbers is closed under addition, multiplication, subtraction, and division, except for division by zero, and obeys all the basic laws of algebra such as the associative, commutative, and distributive laws.

The second point is that irrational numbers are by no means rare, even though, in the preceding section, we proved only that one specimen, $\sqrt{2}$, was irrational. In fact, it can be shown that the square root of any positive integer which is not a perfect square (that is, the square of an integer) is irrational; for example, $\sqrt{3}$, $\sqrt{5}$, $\sqrt{6}$, etc. Furthermore, $\sqrt[n]{a}$ is irrational if a is a positive integer which is not the nth power of an integer. For example, $\sqrt[3]{2}$, $\sqrt[3]{7}$, $\sqrt[3]{9}$, etc. There are many other irrational numbers, such as $\pi = 3.14159265 \ldots$, which cannot be expressed in terms of nth roots of integers at all. To demonstrate further the plentifulness of irrational numbers, we prove the following two theorems.

THEOREM 1. *If a rational number R is added to an irrational one I, the sum S is again irrational.*

Proof. $S = R + I$ implies $I = S - R$. Now S is real and therefore either rational or irrational. If S were rational then $S - R$ would be rational since the system of rational numbers is closed under subtraction. But then I would be rational, contrary to hypothesis. Therefore S must be irrational.

Examples. It follows that $2 + \sqrt{2}$, $3 + \sqrt{2}$, $-\frac{1}{2} + \sqrt{2}$, etc., are irrational.

THEOREM 2. *If an irrational number I is multiplied by a rational number $R \neq 0$, then the product P is irrational.*

Proof. $P = RI$ implies $I = P/R$ since $R \neq 0$. Now P is real and therefore either rational or irrational. If P were rational, then P/R would be rational since the system of rational numbers is closed under division except for division by zero. Then I would be rational contrary to hypothesis. Therefore P is irrational.

Examples. $2\sqrt{2}$, $3\sqrt{2}$, $\frac{1}{2}\sqrt{2}$, $\dfrac{1 + \sqrt{2}}{2}$, etc., are irrational.

The third point concerns the decimal expression of real numbers. Recall that the usual decimal notation for numbers is based on powers of 10. For example, 3425.678 really stands for

$$3 \cdot 10^3 + 4 \cdot 10^2 + 2 \cdot 10 + 5 + \frac{6}{10} + \frac{7}{10^2} + \frac{8}{10^3}.$$

Let us try to express $\sqrt{2}$ as a decimal. If $x = \sqrt{2}$, then $x^2 = 2$. Hence we try $x = 1$ and find it too small since $x^2 = 1 < 2$; but $x = 2$ is too large since $x^2 = 4 > 2$. Hence $1 < \sqrt{2} < 2$, or $\sqrt{2} = 1.\ \ldots$. That is, we have located $\sqrt{2}$ between 1 and 2. We then get a better approximation by splitting the interval between 1 and 2 into tenths and trying them. We find $(1.1)^2 = 1.21$, $(1.2)^2 = 1.44$, $(1.3)^2 = 1.69$, $(1.4)^2 = 1.96$ all too small, while $(1.5)^2 = 2.25$ is too large. Hence $1.4 < \sqrt{2} < 1.5$, or $\sqrt{2} = 1.4\ \ldots$. (See Fig. 9.) Then we split the interval from 1.4 to 1.5 into hundredths and find by trial that $(1.41)^2 = 1.9881 < 2 < (1.42)^2 = 2.0164$. Hence $\sqrt{2} = 1.41\ \ldots$. Then trying successive thousandths, we find $(1.414)^2 = 1.999396 < 2 < (1.415)^2 = 2.002225$; hence $\sqrt{2} = 1.414\ \ldots$. At each stage of this successive approximation we fail to get an exact square root of 2. Thus the decimal expression for $\sqrt{2}$ does not seem to stop. The mere fact that we find that it doesn't stop for the first thousand decimal places, however, does not prove that it will not stop thereafter. But we can, in fact, prove that it never stops. For if it did stop, we would have a terminating decimal expression for $\sqrt{2}$. Now the following theorem is obvious.

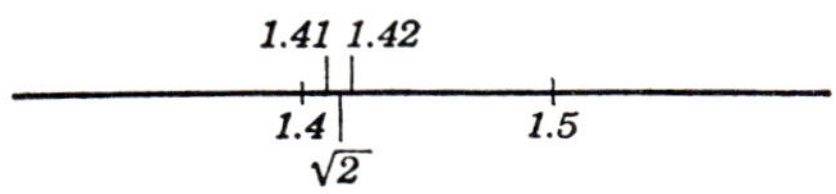

Fig. 9.

THEOREM 3. *If a decimal terminates, it represents a rational number.*

For example, 3.426 is a rational number since it is a quotient of two integers, namely 3426/1000.

But $\sqrt{2}$ is not rational. *Hence its decimal expression cannot terminate.* However, each of the stages of our process of successive approximations yields a rational number whose square becomes closer and closer to 2. That is, $\sqrt{2}$ *can be approximated as closely as we please (that is, to the nearest thousandth, millionth, etc.) by rational numbers,* by carrying the process far enough. This is one essential characteristic of irrational numbers.

The converse of theorem 3 is false, for a number may be rational without having a terminating decimal expression. For example we find by division, that

$$\frac{2}{3} = .666 \ldots \quad \text{and} \quad \frac{1}{7} = .142857142857142857 \ldots .$$

However, both of these decimal expressions "repeat in blocks"; in technical language they are **periodic** or **recurring.** In the case of 1/7, notice that the only remainders which can occur in the process of long division are 0, 1, 2, 3, 4, 5, 6. If 0 occurred, the process would terminate. If 0 does not occur, then, after enough steps (seven or fewer), we must get one of the remainders 1, 2, 3, 4, 5, 6 back again for the second time, whereupon the whole process repeats itself exactly, thus producing a periodic decimal. In the same way, the following theorem can be proved.

THEOREM 4. *The decimal expression of a rational number either terminates or, if it is non-terminating, is periodic.*

After the subject of geometric progressions has been taken up, it will be possible for us to prove the following.

THEOREM 5. *Every periodic decimal represents a rational number.*

The preceding theorems imply that the following is true.

THEOREM 6. *An irrational number has a decimal expression which is neither terminating nor periodic.*

Every real number can be expressed as a decimal, terminating or not, and, conversely, every decimal represents a real number.* Whether a real number is rational or not does not matter to the practical engineer directly, since measurements are never more than approximate and an irrational number can be approximated as closely as we please by terminating decimals. But the concept of irrational numbers is essential to any logical theory of geometric magnitudes, and to the logical derivation of many theorems of algebra and calculus, upon which much practical work is based.

* This will be proved in Chapter XXVII.

Every real number is either positive, negative, or zero according as it lies to the right of, to the left of, or at the origin.

Note that the simple, although tedious, method of successive approximation by trials, which we used above to find the decimal expression for $\sqrt{2}$ can be used not only for square roots, but also for cube roots, fourth roots, etc. It resembles the process of running down a base runner caught trying to steal second base in a baseball game, except that while we continue to pinch the irrational number between narrower and narrower limits, we never tag it exactly.

Note 1. The word "fraction" is often used for any indicated quotient such as $\sqrt{2}/3$ which is not a fraction in the sense of section **7**. To avoid confusion, we shall henceforth refer to the fractions of section 7 as *positive rational numbers*.

Note 2. The meaning of the phrase **correct to the nearest hundredth** is self-explanatory if one thinks of the geometric picture (Fig. 10). For example, 2.374 becomes 2.37 when **corrected** or **rounded off** to the nearest hundredth, while 2.376

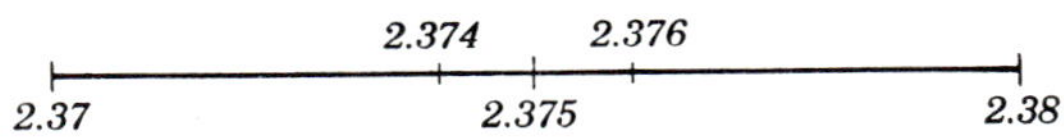

Fig. 10.

becomes 2.38 when corrected to the nearest hundredth. If a number is exactly in the middle of an interval, we agree to assign it to the larger end; thus 2.375 is rounded off to the nearest hundredth as 2.38.

EXERCISES

(*a*) *Approximate as far as three decimal places, that is, locate between successive thousandths;* (*b*) *write the answer rounded off to the nearest hundredth:*

1. $\sqrt{5}$. **2.** $\sqrt{3}$. **3.** $\sqrt[3]{3}$.
4. $\sqrt[3]{4}$. **5.** $\sqrt[3]{5}$. **6.** $\sqrt{10}$.

Show that each of the following numbers is irrational:

7. $5 + \sqrt{2}$. **8.** $\sqrt[3]{2}$. **9.** $\sqrt{2}/3$.
10. $3 - \sqrt{2}$. **11.** $\dfrac{3 + \sqrt{2}}{5}$. **12.** $\dfrac{3 - \sqrt{2}}{5}$.

13. (*a*) Show that 22/7 approximates $\pi = 3.14159\ \ldots$ to the nearest hundredth. (*b*) If we take the decimal expression for 22/7 beyond the hundredths place, will more places yield a closer approximation to π, or not?

Tell whether each of the following is true or false:

14. The decimal expression for $\sqrt{4/9}$ is neither terminating nor periodic.

15. The decimal expression for $\sqrt{5}/3$ is neither terminating nor periodic.

16. Show that $3\frac{10}{71} < \pi < 3\frac{1}{7}$. ($\pi = 3.14159 \ldots$.)

★15. Extraction of Square Roots

While the method of trials used in the preceding section to obtain the decimal expression for $\sqrt{2}$ has the advantage of applying also to cube roots, etc., it is admittedly tedious. The student may recall the following quicker scheme for extracting square roots.

Example 1.

```
        3  5.
      √12'25.
       9
  65 | 3 25
       3 25
       ----
          0.
```

But it is likely that the student memorized this process without ever understanding how or why it works. We shall try to explain the basis for this peculiar procedure, including the "doubling" of the partial answer.

Suppose first that the number N, whose square root we wish to extract, is a perfect square, that is, the square of an integer. Note that $1^2 = 1$, $9^2 = 81$; $10^2 = 100$, $99^2 = 9801$; $100^2 = 10000$ and $999^2 = 998001$; and so on. Hence we see that a one- or two-digit square has a one-digit square root, a three- or four-digit square has a two-digit square root, a five- or six-digit square has a three-digit square root, and so on. Therefore, *the first step in the process is to mark off pairs of digits in the number N to the left of the decimal point, as far as possible:* thus, 12′25. in example 1. Let us separate the answer, which we know will consist of two digits, in the case of example 1, into tens and units, denoting the tens by t and the units by u. That is, we write the answer as $t + u$; in example 1, it turns out that $t = 30$ and $u = 5$, so that $t + u = 35$. Our task is to find t and u. We know that $N = (t + u)^2$. By theorem 2, section 12, this means that $N = t^2 + 2tu + u^2$. The significance of the steps in the process of deter-

★ This section may be omitted without disturbing the continuity of the chapter.

mining t and u will be understood by studying the following analysis of example 1.

$$
\begin{array}{rr|ll}
 & & 3 \quad 5. & = 30 + 5 = t + u \\
 & & \sqrt{12'\ 25.} & = N \\
 & & 9\ (00) & = t^2 \\
2t + u = & 65 & 3\ \ 25 & = N - t^2 \\
 & & 3\ \ 25 & = (2t + u)u = 2tu + u^2 \\
 & & 0 & = N - t^2 - 2tu - u^2.
\end{array}
$$

From the last line we see that $N = t^2 + 2tu + u^2$, or $N = (t + u)^2$. Thus $t + u = 35 = \sqrt{1225}$.

Example 2. Similarly, let h represent the hundreds in the three-digit answer for $\sqrt{54756}$. Then

$$
\begin{array}{rr|ll}
 & & 2 \quad 3 \quad 4. & = 200 + 30 + 4 = h + t + u \\
 & & \sqrt{5'\ 47'\ 56.} & = N \\
 & & 4\ (00\ \ 00) & = h^2 \\
2h + t = & 43(0) & 1\ \ 47\ \ 56 & = N - h^2 \\
 & & 1\ \ 29\ (00) & = (2h + t)t = 2ht + t^2 \\
2h + 2t + u = & 464 & 18\ \ 56 & = N - h^2 - 2ht - t^2 \\
 & & 18\ \ 56 & = (2h + 2t + u)u = 2hu + 2tu + u^2 \\
 & & 0 & = N - h^2 - 2ht - t^2 - 2hu - 2tu - u^2.
\end{array}
$$

From the last line we have

$$N = h^2 + t^2 + u^2 + 2ht + 2hu + 2tu = (h + t + u)^2$$

by (2), section 12. Hence $h + t + u = \sqrt{N}$, or $234 = \sqrt{54756}$.

The zeros written in parentheses, above, are usually omitted.

The student should see for himself that this process can be continued in the usual way to the right of the decimal point, to obtain (approximate) square roots of numbers which may not be perfect squares. Thus

$$
\begin{array}{r|l}
 & 1.\ 4\ \ 1\ \ 4\ \ldots \\
 & \sqrt{2.00\ 00\ 00} \\
 & 1 \\
24 & 1\ 00 \\
 & 96 \\
281 & 4\ 00 \\
 & 2\ 81 \\
2824 & 1\ 19\ 00 \\
 & 1\ 12\ 96 \\
 & 6\ 05 \\
 & \ldots
\end{array}
$$

It will be unnecessary to develop a similar scheme for cube roots, etc., since a general method for performing such computations will be taken up in sections 127 and 130. See also table I.

EXERCISES

Using the usual scheme, calculate the (principal) square root of each of the following:

1. 576.	**2.** 784.	**3.** 841.	**4.** 1521.
5. 2209.	**6.** 54756.	**7.** 61504.	**8.** 108241.
9. 118336.	**10.** 15.1321.	**11.** 27.4576.	**12.** 8.3521.

Approximate the (principal) square root of each of the following to the nearest thousandth:

13. 230.	**14.** 6.36.	**15.** 56.5.	**16.** 84.3.
17. 3.25.	**18.** 3.84.	**19.** 488.	**20.** 577.

16. The System of Complex Numbers

At the beginning of section 14, two motives were presented for the introduction of irrational numbers, one geometric and the other algebraic. Our geometric objective has been accomplished, and hence real numbers are all that are necessary for most geometric purposes. But the system of real numbers is not closed under the operation of taking square roots. This can be seen from the following theorem.

THEOREM 1. *No negative number N has a square root among the real numbers.*

Proof. Let x be any real number. It is either positive, zero, or negative. If x is positive, then x^2 is positive and cannot be equal to N. If x is zero, then x^2 is zero and cannot be equal to N. If x is negative, then x^2 is positive and cannot be equal to N. This completes the proof.

We therefore introduce the symbols $\pm\sqrt{N}$ where N is negative and call them **pure imaginary numbers.** Thus $\sqrt{-4}$ and $\sqrt{-5}$ are pure imaginary numbers. We agree to write $\sqrt{-4} = \sqrt{4}\sqrt{-1}$ and $\sqrt{-5} = \sqrt{5}\sqrt{-1}$, etc. With this agreement we see that every pure imaginary number can be written as the product of a real number and $\sqrt{-1}$. We call $\sqrt{-1}$ the **imaginary unit** and denote it by the letter i; thus

$$i = \sqrt{-1} \text{ or } i^2 = -1. \tag{1}$$

Thus we may write $\sqrt{-4} = 2i$ and $\sqrt{-5} = i\sqrt{5}$. Hence, *every pure imaginary number may be written in the form bi where b is real and* $i = \sqrt{-1}$ $(b \neq 0)$.

The system consisting of all real numbers and pure imaginary numbers together would not be closed under addition, for the sum of a real number and a pure imaginary number would be neither real nor pure imaginary; for example, $2 + 3i$. Symbols of the form $a + bi$, where a and b are real and $i = \sqrt{-1}$, are called **complex numbers.** We call a the **real part** and bi the **imaginary part** of the complex number $a + bi$. We define equality, addition, and multiplication of complex numbers as follows:

DEFINITION 1. $a + bi = c + di$ *if and only if* $a = c$ *and* $b = d$.

DEFINITION 2. $(a + bi) + (c + di) = (a + c) + (b + d)i$.

Example 1. $(3 + 2i) + (5 + 4i) = 8 + 6i$.

DEFINITION 3. $(a + bi)(c + di) = (ac - bd) + (ad + bc)i$.

Example 2. $(3 + 2i)(5 + 4i) = (15 - 8) + (12 + 10)i = 7 + 22i$.

Multiplication of complex numbers may be carried out according to the usual laws, with the additional step that i^2 *is to be replaced by* -1. Thus

$$\begin{aligned}(3 + 2i)(5 + 4i) &= (3 + 2i)5 + (3 + 2i)4i \\ &= 15 + 10i + 12i + 8i^2 \\ &= 15 + 22i + 8(-1) \\ &= 7 + 22i.\end{aligned}$$

Subtraction of $c + di$ can be defined as before (definition 1, section 5), and may be done by addition of $-(c + di)$ as before. Hence

$$(a + bi) - (c + di) = (a - c) + (b - d)i. \tag{2}$$

Example 3. $(5 + 4i) - (3 + 2i) = 2 + 2i$.

We defer further study of the arithmetic of complex numbers until it is needed. See chapter XVIII.

The system of complex numbers includes all previous kinds of numbers. For example, if $b = 0$, the complex number $a + bi$ is real; for example, $2 + 0i = 2$ is real. If $b \neq 0$, the complex number is called **imaginary**; thus $2 + 3i$ is imaginary. If $b \neq 0$ and $a = 0$, the complex number is called pure imaginary; for example, $0 + 3i = 3i$ is pure imaginary. We may classify the entire complex number system as follows:

Complex numbers ($a + bi$ *where* a *and* b *are real numbers and* $i = \sqrt{-1}$).
- *Imaginary numbers* ($b \neq 0$). *Example:* $2 + 3i$.
  - *Pure imaginary numbers* ($b \neq 0$, $a = 0$). *Example:* $0 + 3i$.
- *Real numbers* ($b = 0$). *Example:* $2 + 0i$.
  - *Irrational. Examples:* $\sqrt{2}$, $\sqrt{3}$, $\sqrt[3]{2}$, π, ...
  - *Rational.* (*Expressible as* p/q *where* p *and* q *are integers.*) *Example:* $2/3$.
    - *Integers.* (*Expressible with denominator 1.*) *Example:* $2/1 = 2$.
      - *Negative integers. Examples:* -1, -2, ...
      - *Zero.*
      - *Positive integers or natural numbers.*

Note that we have proceeded a long way from the natural numbers with which we started. At each stage, new kinds of numbers were introduced because the system at hand was not closed under all operations. But it will not be necessary to invent still more numbers now because it can be shown that the system of complex numbers is closed under all the operations studied in this book, such as addition, subtraction, multiplication, division (except for division by zero), and taking nth roots. It can also be shown that the basic laws of algebra are obeyed, such as the associative, commutative, and distributive laws. No useful concept of greater or less can be applied to imaginary numbers. *Hence > and <, and positive and negative will be used only in connection with real numbers.*

The word "imaginary" does not mean that these numbers are somehow not genuine, or impractical or fictitious, although it did have such connotations in the 17th century, when imaginary numbers were not well understood. In fact, a logically satisfactory theory of complex numbers was first given by the Irish mathematician, Sir William Rowan Hamilton, in 1835. In a later chapter we shall give a perfectly "real" graphical interpretation of complex numbers. They are used practically in advanced engineering, the theory of electricity and magnetism, the theory of elasticity and plasticity, the theory of aerodynamics and aeronautical engineering, etc. While other types of numbers exist, the system of complex numbers suffices for much of applied mathematics, and it is with this system of numbers that we shall be concerned.

EXERCISES

Express each of the following in the form bi where b is real:

1. $\sqrt{-16}$. **2.** $\sqrt{-36}$. **3.** $\sqrt{-7}$. **4.** $\sqrt{-13}$.
5. $\sqrt{-x^2}$, $x > 0$. **6.** $3\sqrt{-36}$. **7.** $5\sqrt{-16}$.
8. $4\sqrt{-9}$.

Express each of the following in the form a + bi where a and b are real:

9. $2 + \sqrt{-9}$.
10. $\sqrt{16} + \sqrt{-49}$.
11. $4 + 2\sqrt{-16}$.
12. $3 + 4\sqrt{-9}$.
13. $3 + 2i + 5 + i$.
14. $5 + \sqrt{-9} + 2 + \sqrt{-1}$.
15. $1 - 3\sqrt{-4} - 2 + \sqrt{-9}$.
16. $4 - 3i + 2 + 5\sqrt{-9}$.
17. $(3 - 2i) - (5 - 2i)$.
18. $6 + 5i - (2 - 3i)$.
19. $(-1 + 3i) - (-2 + 4i)$.
20. $(3 + 4i) - (7 + i)$.
21. $3i(4 - 2i)$.
22. $2i(5 + 3i)$.
23. $(2 + 3i)(4 + 2i)$.
24. $(3 - 2i)(4 + 2i)$.
25. $(3 - 2i)(5 - 2i)$.
26. $(4 + i)(5 - 2i)$.
27. $(4 - i)(3 - 2i)$.
28. $(1 - 4i)(3 + 5i)$.
29. $(1 - i)(2 + 3i)(4 + 2i)$.
30. $(5 + 2i)(4 - 3i)(1 - i)$.

List all the adjectives "complex, imaginary, pure imaginary, real, irrational, rational, integral, positive, negative, zero, natural" which apply to each of the following numbers:

31. $2 + 5i$. **32.** $6 - 2\sqrt{-2}$. **33.** $4\sqrt{-3}$.
34. $2i\sqrt{2}$. **35.** $4\sqrt{2}$. **36.** $\sqrt{2}/5$.
37. $-8/7$. **38.** $6/2$. **39.** $(1 + \sqrt{16})/2$.
40. $(1 - \sqrt{5})/2$. **41.** $(2 - \sqrt{4})/4$.

42. What was our motive in introducing (*a*) fractions; (*b*) negative numbers; (*c*) irrational numbers; (*d*) pure imaginary numbers; (*e*) complex numbers?

★17. Conclusion. The Axioms of Algebra

In the course of this chapter we sketched the evolution of the number system and found that some of the familiar algebraic manipulations were logical consequences of a few simple defi-

★ May be omitted without disturbing continuity.

nitions and basic laws. We list these basic laws here for easy reference, *all letters representing arbitrary numbers of the system.*

I. *Law of closure for addition.* Given any pair of numbers a and b, there exists a unique number in the system called the sum of a and b, denoted by $a + b$.

II. *Commutative law for addition.* $a + b = b + a$.

III. *Associative law for addition.* $(a + b) + c = a + (b + c)$.

IV. *Law of closure for multiplication.* Given any pair of numbers a and b, there exists a unique number in the system called the product of a and b, denoted by ab.

V. *Commutative law for multiplication.* $ab = ba$.

VI. *Associative law for multiplication.* $(ab)c = a(bc)$.

VII. *Distributive law.* $a(b + c) = ab + ac$.

VIII. There exists a unique number 0, called *zero*, such that: (a) $0 + a = a$ for any number a; (b) $0 \cdot a = 0$ for any number a; (c) if $a \neq 0$ and $b \neq 0$, then $ab \neq 0$.

IX. *Law of closure for subtraction.* Given any numbers a and b, there exists a unique number x such that $b + x = a$. This number x is denoted by $a - b$.

X. *Law of closure for division.* Given any numbers a and b ($b \neq 0$), there exists a unique number x such that $bx = a$. This number x is denoted by a/b.

XI. $a = a$.

XII. If $a = b$, then $b = a$.

XIII. If $a = b$ and $b = c$, then $a = c$. (*Things equal to the same thing are equal to each other.*)

XIV. If $x = y$ and $u = v$, then (*a*) $x + u = y + v$; (*b*) $x - u = y - v$; (*c*) $xu = yv$; (*d*) $x/u = y/v$ provided u and v are not zero. That is, *if equals are* (*a*) *added to,* (*b*) *subtracted from,* (*c*) *multiplied by,* (*d*) *divided by equals, the results are equal.*

There is a subdivision of the complex number system, called the *real numbers,* which have the following properties:

XV. If a and b are any real numbers, one and only one of the relations $a > b$, $a = b$, or $a < b$ holds. If a real number $x > 0$, it is called *positive:* if $x < 0$, it is called *negative.*

XVI. *Law of signs for multiplication.*

(*a*) If $a > 0$ and $b > 0$, then $ab > 0$.
(*b*) If $a < 0$ and $b < 0$, then $ab > 0$.
(*c*) If $a < 0$ and $b > 0$, then $ab < 0$.

XVII. If $a < b$ and $b < c$, then $a < c$.

XVIII. If $a < b$, then $a + c < b + c$.

XIX. If $a < b$, then $-b < -a$.

XX. If $a < b$ and $c > 0$, then $ac < bc$.

XXI. Every real number can be expressed as a decimal, terminating or non-terminating, and, conversely, every decimal represents a real number.

It would be possible to use these basic laws, plus a few others not stated here, as axioms or postulates in a strictly logical treatment of algebra.* This is not our purpose here, because such a task is best left for more advanced courses. We shall occasionally indicate, however, how these basic laws justify our manipulations.

The natural numbers may be regarded as the basis of all algebra. All the different kinds of numbers can be strictly defined in terms of the natural numbers, although we have not done this here completely. For example, a fraction was defined as a symbol consisting of a pair of natural numbers. L. Kronecker (German, 1823–1891) is said to have remarked that "the whole number was created by God, everything else is man's handiwork." The fundamental character of the natural numbers, alluded to by Kronecker, has led to the extensive development not only of the *theory of numbers,* one of the oldest branches of mathematics, but also of mystical *numerology,* one of the oldest varieties of nonsense.†

* For more detailed indications of the role of axioms in algebra, or, indeed, in any branch of mathematics, see M. Richardson, *Fundamentals of Mathematics,* Macmillan, 1941, and references listed there. The axioms given above are neither independent nor complete.

† See E. T. Bell, *Numerology,* Williams and Wilkins, Baltimore, 1933, for an entertaining account of numerology.

Chapter **II**

What is algebra?

Like a well-cut jewel, algebra presents many facets. It is at the same time a language, a mathematical (logical) science, a collection of techniques of calculation, an important element in the history of human endeavor, and a collection of puzzles.

18. Algebra as a Language

The symbols and technical terms of algebra have been evolved over a long period of history, and have proven themselves most apt for the consideration of quantitative relationships and problems in all fields, such as physical sciences, social sciences, statistics, business, etc. The earliest advances over ordinary prose were largely abbreviations. Thus, in his *Ars Magna de Rebus Algebraicis* (1545), Cardan wrote "cubus p. 6 rebus aequalis 20," which we would express as $x^3 + 6x = 20$. The $=$ sign is found in print for the first time in *The Whetstone of Witte*, by Robert Recorde, published in 1557. He writes that he uses two parallel lines for the equals sign because "noe two thynges could be moare equalle." In the early 17th century, we still find *aaaaa* written for a^5. The ancients had no adequate algebraic symbolism and it is likely that this handicapped their development of algebraic ideas. One has only to try to solve a relatively simple problem without using any algebraic symbolism to realize the necessity for some such compact language. A similar need can be seen in music; imagine an orchestra trying to play a symphony if all directions for playing were in prose instead of the compact musical symbolism. In algebra, as in music, the student should master as soon as possible the symbolism and technical terminology in order to leave his mind free for the interesting things that are expressed in this language. *A definition in this book will be clearly recognizable whether or not it is marked "Definition" because the term being defined will be printed in* **boldfaced (heavy) type.** In algebra, as in music, considerable effort and drill is needed for this purpose. Benjamin Franklin wrote,* "Whatever may have been imputed to some other studies under the notion of insignificancy and loss of time, yet these (mathematics), I believe, never caused repentance in any, except it was for their remissness in the prosecution of them."

19. Algebra as a Logical Science

Any logical science should begin with axioms or assumptions and deduce all further propositions from them by strictly logical argument. The assumptions ought to be as simple, as few, and as plausible as possible. Such a logical organization was first attempted by the ancient Greeks for geometry. A partial set of axioms for algebra is given in section 17, but no attempt to develop algebra completely from these axioms will be made here because such a task presents technical difficulties best left for more advanced courses.†

20. Algebra as a Collection of Techniques of Calculation

This aspect of algebra is doubtless the most familiar to the student, since techniques or processes of calculation are the reasons for much of the usefulness of algebra in applied science and are heavily stressed in elementary teaching. Nevertheless, if one is asked the embarrassing question, "How do you know this technique or process always works?" the only answer is to justify the technique by a reasonable argument. Thus in section 15, the familiar process for extracting square roots has been justified by tracing it back to the simple algebraic fact that

* *On the Usefulness of Mathematics, Complete Works of Benjamin Franklin,* vol. I, p. 421, G. P. Putnam's Sons, The Knickerbocker Press (1887).

† For a more complete discussion of the role of axioms in mathematics, see M. Richardson, *Fundamentals of Mathematics,* Macmillan, 1941.

$(t + u)^2 = t^2 + 2tu + u^2$, which in turn may be derived from the axioms themselves as in section 12. Techniques and processes will not only be explained clearly in this book, but will also be justified by reasonable arguments as far as possible. It is the author's belief that *a process understood will be remembered better and longer than one which is merely memorized.*

21. Algebra as a Branch of Human Endeavor

The search for truth has certainly been, throughout history, one of the most distinctively human of human activities. In it, mathematics has played a central role. The manifold applications of mathematics to astronomy, physics, chemistry, biology, statistics, the social sciences, engineering, etc., would not be possible without algebra, which is basic to trigonometry, analytic geometry, calculus, and higher mathematics. Thus, the industrial revolution in the 18th and 19th centuries, and whatever analogous social changes result from the modern scientific advances in aeronautics, radiation, the release of atomic energy, and automation, are in considerable measure related to progress in mathematics.

From a personal rather than a social point of view, algebra requires and stimulates clear and persistent thinking and clear understanding of concepts and precise definitions.

22. Algebra as a Collection of Puzzles

To the student, algebra presents the appearance of a collection of puzzles, especially in the exercises on which the student must cut his mental teeth. Needless to say, the significance of algebra is far deeper than could possibly be possessed by a mere collection of puzzles. Nevertheless, the puzzle-motive is a universal urge and is surely present in algebra. Solving an algebraic problem yields satisfaction of the same kind as that obtained by solving puzzles of a less useful sort, and at the same time gives the deeper satisfaction of furthering one's mathematical and scientific education. The student, to be successful, should be as reluctant to give up an algebraic problem as he was to give up a useless puzzle when he was very young and full of the pure intellectual curiosity that is evident in almost all children, at least until they are exposed to the educational system. It should also be mentioned that there are many historic puzzles which appear useless on the surface but which have proved to be very fruitful in the advancement of science, if only because the attempts to solve them led to the development of new methods which were useful in other problems.*

* For further details of some puzzles of this sort, see M. Richardson, *Fundamentals of Mathematics,* Chapter VIII, Macmillan, 1941.

Chapter III

Algebraic expressions

23. Positive Integral Exponents and Radicals. Order of Operations

If n is any positive integer, we have defined the **nth power of x** to mean the product of n factors, each equal to x. In symbols, $x^n = xx \ldots x$ where there are n factors on the right. The **exponent** n indicates the number of factors. For example, $x^3 = xxx$. Clearly this definition can apply only to *positive integral* exponents. The following theorems are probably familiar to the student.

THEOREM I. $x^a x^b = x^{a+b}$.

Example: $x^2x^3 = (xx)(xxx) = x^5$.

THEOREM II. $(x^a)^b = x^{ab}$.

Example: $(x^2)^3 = (x^2)(x^2)(x^2) = (xx)(xx)(xx) = x^6$.

THEOREM III. $(xy)^n = x^n y^n$.

Example: $(xy)^3 = (xy)(xy)(xy) = xxxyyy = x^3y^3$.

THEOREM IV. $\left(\frac{x}{y}\right)^n = \frac{x^n}{y^n}$.

Example: $\left(\frac{x}{y}\right)^3 = \frac{x}{y} \cdot \frac{x}{y} \cdot \frac{x}{y} = \frac{x^3}{y^3}$.

THEOREM V. *If $x \neq 0$,*

$$\frac{x^a}{x^b} = \begin{cases} x^{a-b} \text{ if } a > b. \text{ Example: } \dfrac{x^5}{x^2} = \dfrac{xxx\not{x}\not{x}}{\not{x}\not{x}} = x^3. \\ \dfrac{1}{x^{b-a}} \text{ if } a < b. \text{ Example: } \dfrac{x^2}{x^5} = \dfrac{\not{x}\not{x}}{\not{x}\not{x}xxx} = \dfrac{1}{x^3}. \\ 1 \quad \text{ if } a = b. \text{ Example: } \dfrac{x^2}{x^2} = 1. \end{cases}$$

These are the fundamental rules for the manipulation of exponents. Their proofs, for positive integral exponents, are clearly indicated by the examples above.

Proof of Theorem I. $x^a = xx \ldots x$ with a factors on the right, while $x^b = xx \ldots x$ with b factors on the right. Hence $x^a x^b = (xx \ldots x)(xx \ldots x) = xx \ldots x$ where there are $a + b$ factors on the right. Hence, $x^a x^b = x^{a+b}$.

Proof of Theorem II. $(x^a)^b = (x^a)(x^a) \ldots (x^a)$ where there are b pairs of parentheses on the right. But each pair of parentheses on the right encloses the product of a x's. Hence there are ab x's on the right. Hence, $(x^a)^b = x^{ab}$.

Proof of Theorem III. See section 12, Theorem 1.

Proof of Theorem IV. $\left(\dfrac{x}{y}\right)^n = \dfrac{x}{y} \cdot \dfrac{x}{y} \cdot \ldots \cdot \dfrac{x}{y}$. By multiplying fractions, we get $\dfrac{xx \ldots x}{yy \ldots y} = \dfrac{x^n}{y^n}$.

Proof of Theorem V. Case I. Suppose $a > b$. Then

$$\frac{x^a}{x^b} = \frac{\overbrace{(xx \ldots x)}^{b \text{ factors}}\overbrace{(xx \ldots x)}^{a-b \text{ factors}}}{\underbrace{(xx \ldots x)}_{b \text{ factors}}} = \overbrace{(xx \ldots x)}^{a-b \text{ factors}} = x^{a-b}.$$

Case II. Suppose $a < b$. Then

$$\frac{x^a}{x^b} = \frac{\overbrace{(xx \ldots x)}^{a \text{ factors}}}{\underbrace{(xx \ldots x)}_{a \text{ factors}}\underbrace{(xx \ldots x)}_{b-a \text{ factors}}} = \frac{1}{\underbrace{(xx \ldots x)}_{b-a \text{ factors}}} = \frac{1}{x^{b-a}}.$$

Case III. Suppose $a = b$. Then $\dfrac{x^a}{x^b} = \dfrac{x^a}{x^a} = 1$.

Note that in all cases we must suppose $x \neq 0$ because division by zero is excluded (see section 11).

If n is a positive integer greater than 1, and $x^n = a$, then x is called an **nth root of a.** If $x^2 = a$, x is called a **square root** of a; if $x^3 = a$, x is called a **cube root** of a. For example, $+3$ and -3 are both square roots of 9, since $(+3)^2 = 9$ and $(-3)^2 = 9$; 2 is a cube root of 8, since $2^3 = 8$; and -2 is a cube root of -8, since $(-2)^3 = -8$. It will be shown in a later chapter that every number a has n distinct nth roots. To avoid the confusion that would arise if we allowed the radical sign $\sqrt[n]{a}$ to designate any of these nth roots ambiguously, we reserve the radical sign for a particular one of them, called the **principal nth root.** *If a is positive, the principal nth root $\sqrt[n]{a}$ shall be the positive nth root; if a is negative and if n is odd, the principal nth root $\sqrt[n]{a}$ shall be the negative nth root;* no other cases are needed for the present. For example, $\sqrt{9} = 3$, $\sqrt[3]{8} = 2$, $\sqrt[3]{-8} = -2$. If we wish to designate -3 as a square root of 9, we write $-\sqrt{9}$. Obviously, by definition, we have

$$(1) \qquad (\sqrt[n]{a})^n = a.$$

Order of operations. Parentheses ( ), brackets [ ], braces { }, and sometimes the vinculum $\overline{\quad}$, are used to group together the quantities enclosed by them. *Operations within parentheses* (*or other symbols of grouping*) *are to be done first.* Parentheses are thus merely punctuation marks in the written language of algebra, used to prevent ambiguity and misunderstanding, as are punctuation marks in prose. Thus $8 \div 4 \div 2$ might mean either $(8 \div 4) \div 2 = 2 \div 2 = 1$ or $8 \div (4 \div 2) = 8 \div 2 = 4$. Similarly the phrase "beautiful little girls' school" might mean "beautiful (little girls' school)" signifying "beautiful school for little girls," or "(beautiful little girls') school," signifying "school for beautiful little girls," or "(beautiful little) (girls' school)" signifying "girls' school which is beautiful and little."

We agree that *powers take precedence over multiplications and divisions, which in turn precede additions and subtractions, except where otherwise indicated by parentheses.* Thus, $2 \cdot (3 + 4) = 2 \cdot 7 = 14$ while $2 \cdot 3 + 4 = 6 + 4 = 10$. We may also write $2 \cdot (3 + 4)$ as $2[3 + 4]$ or $2\{3 + 4\}$ or $2 \cdot \overline{3 + 4}$. Also $2 \cdot 5^2 = 2 \cdot 25 = 50$, while $(2 \cdot 5)^2 = 10^2 = 100$. Thus, $2 \cdot 5^2 + (2 \cdot 5)^2 = 50 + 100 = 150$.

It is understood that *all operations indicated under a radical sign are to be done before extracting the root.* For example,

$\sqrt{9+16} = \sqrt{25} = 5$ is correct, but $\sqrt{9+16} = 3 + 4 = 7$ is wrong. That is, a radical sign acts as a pair of parentheses, grouping together the quantities under it. In fact, it may be conjectured that the radical sign originates from a script r (for *radix,* Latin word meaning *root*) and the vinculum (Latin word meaning *yoke* or *bond*). Thus $\sqrt{9+16} = r(9+16) = 5$.

EXERCISES

1 to 29. Do exercises 1–29 of section 12.

Simplify each of the following:

30. a^3a^5. **31.** a^8/a^2. **32.** $(a^3)^5$. **33.** a^2/a^6.

34. $\dfrac{a^2b^5c^2}{a^6bc^2}$. **35.** $(3a^3b)(2ab^4)^2$.

36. $\dfrac{(x^3y)^3}{(xy^2)^2}$. **37.** $\dfrac{2(x^2y^4)^2}{(2x^3y^2)^3}$.

38. $(-2x^2y^3)^4$. **39.** $-2(x^2y^3)^4$.

40. $\left(\dfrac{x^2}{y^3}\right)^5$. **41.** $\left(\dfrac{2a^3}{b}\right)^2\left(\dfrac{b^4}{2a^2}\right)^3$.

42. $\dfrac{(xy^2)^3(x^2y^3)^2}{(x^4y^2)^3}$. **43.** $(-3)^2(-2)^3$.

44. $4^9/4^6$. **45.** $\dfrac{(2^3)^4}{(2^42^7)}$. **46.** $\dfrac{8^4}{2^{11}}$.

Assuming that all letters represent positive numbers, simplify:

47. $\sqrt{a^4b^6}$. **48.** $\sqrt{9a^8b^{10}}$. **49.** $\sqrt[3]{x^6y^9}$. **50.** $\sqrt[3]{8a^{12}b^6}$.

Tell whether each of the following is true or false, and, if false, correct the right member:

51. $3^2 \cdot 3^3 = 9^5$. **52.** $3^2 \cdot 3^3 = 3^6$. **53.** $3^2 \cdot 3^3 = 3^8$.

54. $3^2 \cdot 3^3 = 9^6$. **55.** $x^3 + x^3 = 2x^3$. **56.** $3^2 + 3^3 = 3^5$.

57. $5^3 + 5^3 = 5^6$. **58.** $5^3 + 5^3 = 10^3$. **59.** $3^8/3^2 = 3^4$.

60. $(3^2)^3 = 3^5$. **61.** $(4^2)^3 = 4^8$. **62.** $6^3/2^3 = 3^3$.

24. Algebraic Expressions

A **variable** or **unknown** is a letter which is permitted to have different values (that is, to stand for different numbers) during the same discussion. A **constant** is a symbol which is permitted to stand for just one particular number during the discussion, even if we do not specify which number it stands for. It is customary to use the later letters of the alphabet, such as

x, y, z, t, u, for variables, and the early letters of the alphabet, such as a, b, c, d, for constants. Of course, specified numbers like 2, $-\frac{1}{2}$, $\sqrt{2}$, $3 + \sqrt{-2}$, are constants. We shall also use subscripts and primes to distinguish different quantities. For example, a, a' (read "a prime"), a'' (read "a double prime"), a''' (read "a triple prime") may be used to represent four constants with arbitrary values. Similarly, a_0 (read "a sub-nought" or "a sub-zero"), a_1 (read "a sub-one"), a_2 (read "a sub-two") may be used to represent three constants with arbitrary values. Be careful not to confuse subscripts with exponents.

By an **algebraic expression** is meant an expression which can be built up from a limited number of constants and variables connected by a limited number of additions, subtractions, multiplications, divisions, radical signs, and parentheses (or other symbols of grouping).

Examples of algebraic expressions:

(1) $(2x + \sqrt{5})\sqrt[3]{4x - 2} + \dfrac{5a}{x^3}$.

(2) $1 + \sqrt{x}$.

(3) $\sqrt{x + y}$.

(4) $\dfrac{1}{x}$.

(5) $\dfrac{\sqrt{3}\,x^2 - 5x + 1}{4x + 6}$.

(6) $\dfrac{b}{x + 2} + \dfrac{2}{x + 3}$.

(7) $\dfrac{1}{x + y} + \dfrac{1}{x - y}$.

(8) $2x^2y + 3xy^2 + cxy + dx + 4$.

(9) $5x^2 - \frac{1}{3}x - 2$.

(10) $(x + 2)(x + 3)$.

(11) $3x$.

(12) 3.

Note. Powers with positive integral exponents amount to nothing more than multiplication; thus $x^3 = xxx$. Hence x^3 is an algebraic expression. On the other hand, the "exponential" expression 3^x (where x is a variable), and $\log x$, and $\sin x$ are not algebraic expressions. We shall have little to do with non-algebraic expressions in this book.

An algebraic expression which can be written without any variables occurring under radical signs is called a **rational expression.** An algebraic expression which necessarily involves variables occurring under radical signs is called an **irrational expression.**

Thus, examples (1), (2), and (3), above, are irrational expressions, while the rest are rational expressions.

A rational expression which can be written without any divisions by divisors (or denominators) involving variables is called a **rational integral expression,** or simply a **polynomial.**

Thus, examples (8), (9), (10), (11), and (12) are polynomials or rational integral expressions.

A rational expression which is not a polynomial, that is, in which denominators (or divisors) involving variables necessarily occur, is called a **rational fractional expression,** or simply a **fractional expression.**

Thus, examples (4), (5), (6), and (7) are fractional expressions.

If an algebraic expression is written as a succession of quantities, or partial expressions, each separated from its neighbors by a plus or a minus sign, then each quantity or partial expression, taken together with the preceding plus or minus sign, is called a **term.** Such an algebraic expression may be called the **algebraic sum** of its terms. Compare section 10.

Thus, example (9) above has 3 terms, namely $5x^2$, $-\frac{1}{3}x$, and -2. Example (8) has 5 terms. Examples (3), (4), (5), (10), (11), and (12) consist of 1 term each. Examples (1), (2), (6), and (7) consist of 2 terms each.

A term consisting only of the product of positive integral powers of variables and any non-zero constants whatever is called a **monomial.***

Thus, examples (11) and (12) are monomials. Example (9) is an algebraic sum of 3 monomials. Example (8) is an algebraic sum of 5 monomials.

The constant factor in a monomial is called its **coefficient.**

For example, the coefficient of the monomial $-3x^2$ is -3. The coefficients of the terms of example (8) are 2, 3, c, d, and 4. The coefficients of the terms of example (9) are 5, $-\frac{1}{3}$, and -2.

Monomials which differ at most in their coefficients are called **like terms.**

* The word *monomial* is sometimes used to refer to any algebraic expression with one term; in that case, our monomials would be called *rational integral monomials*, since they are special cases of rational integral expressions or polynomials. Since rational integral monomials are the only ones we shall be concerned with, we adopt the usage in the text above. Thus, *an algebraic sum of monomials is a polynomial.* An algebraic sum of terms which are not necessarily (rational, integral) monomials may be called a **multinomial.** An expression with 2 terms may be called a **binomial;** with 3 terms, a **trinomial,** **etc.** Thus, $1 + \sqrt{x}$ is a binomial and a multinomial but not a polynomial in x.

Examples. $2x^3$ and $-5x^3$ are like terms. Also $3x^2y$, $\frac{1}{2}x^2y$, and ax^2y are like terms.

The **degree** of a monomial which involves variables is the sum of the exponents attached to those variables. A monomial which involves no variables, that is, a non-zero constant, is said to be of **degree zero.**

For example, the degree of the monomial $5x^2y$ is $2 + 1$ or 3. The degree of $-2xy^2$ is 3. The degree of $3x^2y^2$ is 4. The degree of $3x$ is 1. The degree of $5x^3$ is 3. The degree of 3 is zero. The degree of $3a$ is also zero, since a is a constant.*

If an algebraic expression, a rational expression, or a polynomial, etc., involves no variables other than x, we shall refer to it as an **algebraic expression in *x*,** a **rational expression in *x*,** or a **polynomial in *x*,** etc., as the case may be. Similarly, a **polynomial in *x* and *y*** shall mean a polynomial involving no variables other than x and y.

Thus, example (8), above, is a polynomial in x and y. Example (9) is a polynomial in x.

EXERCISES

In each of the following algebraic expressions (a) list all the terms (A) irrational expression, (B) rational expression, (C) polynomial, or rational integral expression which apply to it; (b) find the value of it when $x = 3$ and $y = 2$:

1. $2x^3 - 4x^2 + 10$. **2.** $6/x$. **3.** $\sqrt{3x + 7}$.

4. $3x + 2y$. **5.** $\dfrac{2x - 3}{x + 7}$. **6.** $\dfrac{2x^2 - 3x + 1}{5 - x^2}$.

7. $\frac{1}{3}x^3 + x - \sqrt{2}$. **8.** $\sqrt{x^2 + 16}$. **9.** x.

10. $\dfrac{\sqrt{x} + 2}{3x + 1}$.

Name (a) the coefficient, and (b) the degree of each of the following monomials:

11. $-3x^2$. **12.** $6x^3$. **13.** $-x^2y^4$. **14.** $3x^3y$.

15. $-\frac{1}{2}x^2yz^3$. **16.** $-\frac{1}{3}xy^2z$. **17.** 10. **18.** x.

* We could, if we wished, consider also the degree of a monomial with respect to literal constants, but there is nothing to be gained by doing so.

25. Removal and Insertion of Parentheses. Combination of Like Terms

We have the following rule for removal of parentheses (or any other symbol of grouping):

If parentheses immediately follow a minus sign, they may be removed, together with the minus sign, provided the sign of every term within the parentheses is reversed. If parentheses immediately follow a plus sign, they may be removed without altering the signs of terms within the parentheses.

Example 1. $a - (2b - 3c + 4d) = a - 2b + 3c - 4d.$

Example 2. $a + (2b - 3c + 4d) = a + 2b - 3c + 4d.$

Example 3. $a - (-2b + 3c) = a + 2b - 3c.$

Example 4. $a + (-2b + 3c) = a - 2b + 3c.$

Clearly, *parentheses may be inserted according to the same rule.* Thus, the right members of the above examples may be changed to the left members, if desired.

The justification of this rule is found in theorems 1, 2, 3, and 4 of section 10.

In an algebraic sum, like terms may be combined by adding their coefficients, as follows: $2x^2 + 3x^2 = 5x^2$, and $5x^2y - 2x^2y = 3x^2y$.

The justification of this rule is found in the distributive law: $2x^2 + 3x^2 = (2 + 3)x^2 = 5x^2$, and $5x^2y - 2x^2y = (5 - 2)x^2y = 3x^2y$.

Parentheses preceded by a multiplier or factor may be removed by multiplying every term within the parentheses by the multiplier, as in the following examples:

Example 5. $3(2x - 5y) = 6x - 15y.$

Example 6. $$(a + b)(x + y) = (a + b)x + (a + b)y = ax + bx + ay + by.$$

Example 7. $3x^2(2x^3y - 5y) = 6x^5y - 15x^2y.$

The justification of this rule is to be found in the distributive law.

To **simplify** a rational integral expression means to remove all parentheses enclosing variables and to combine like terms, as in the following examples:

Example 8. $(x + 2)(x + 3) = (x + 2)x + (x + 2)3 = x^2 + 2x + 3x + 6 = x^2 + 5x + 6.$

Example 9. Removing the innermost symbols of grouping first, we have:

$$\begin{aligned} &3x^3 + (5x^2 - [3x^2 - 4x\{x - 2\} + 3] - 2) \\ &= 3x^3 + (5x^2 - [3x^2 - 4x^2 + 8x + 3] - 2) \\ &= 3x^3 + (5x^2 - 3x^2 + 4x^2 - 8x - 3 - 2) \\ &= 3x^3 + 5x^2 - 3x^2 + 4x^2 - 8x - 3 - 2 \\ &= 3x^3 + 6x^2 - 8x - 5. \end{aligned}$$

We could also proceed by removing the outermost symbols of grouping first, as follows:

$$\begin{aligned} &3x^3 + (5x^2 - [3x^2 - 4x\{x - 2\} + 3] - 2) \\ &= 3x^3 + 5x^2 - [3x^2 - 4x\{x - 2\} + 3] - 2 \\ &= 3x^3 + 5x^2 - 3x^2 + 4x\{x - 2\} - 3 - 2 \\ &= 3x^3 + 5x^2 - 3x^2 + 4x^2 - 8x - 3 - 2 \\ &= 3x^3 + 6x^2 - 8x - 5. \end{aligned}$$

EXERCISES

Simplify each of the following expressions:

1. $2x + 5y - 3x + 2y.$

2. $3x^2 - 4x + 5 - x^2 + 7x - 2.$

3. $x^2 - 3xy + 2y^2 - 3x^2 + 5xy - y^2.$

4. $x + y + z - 6x + 5y - 6z - y + 5x + 3z.$

5. $(3x - y) - (2x - 3y).$

6. $(5x + 3) - (-2x + 1).$

7. $(2x^2 - 5x + 3) - (3x^2 - 3x - 2).$

8. $(5x^2 + 3x - 5) - (-4x^2 - 3x - 2).$

9. $-[x - x(2x - 3)].$

10. $4 - [3x - 2x(x + 7)].$

11. $x - \{x - [x - (x - y)]\}.$

12. $2x - [3x + (x - 3y) - 2y].$

13. $-3x + x[2 - x(2 - x)].$

14. $5x - 3[3 + x(3 - x)].$

15. $x - [2x - 3x(4x - 1) + (x - 3)].$

16. $5x^2 - [4x + (2x^2 - x) - x(x - 3) - 5].$

17. $3 - \{2 - [x - (4 - x) - 1] + 3[2 - x]\}.$

18. $-\{2 + x[3 - x] - [x - (2x - 3)]\}.$

(*a*) *Insert parentheses preceded by a plus sign about the first two terms and about the second two terms;* (*b*) *insert parentheses preceded by a minus sign about the first two terms and about the second two terms:*

19. $-x + y + 2a - 3b.$

20. $2x - 5y - 4a + 3.$

26. Standard Form of Polynomials

The **standard form** of a polynomial or rational integral expression in x is obtained by removing all parentheses enclosing variables, combining like terms, and arranging terms in order of descending powers of x.

Example 1. To put the polynomial $(x + 2)(x + 3)$ in standard form, we write $(x + 2)(x + 3) = (x + 2)x + (x + 2)3 = x^2 + 2x + 3x + 6 = x^2 + 5x + 6$. The last expression is the standard form of the polynomial.

Example 2. The following polynomials in x are already in standard form:

$$3x^4 + 2x^2 - 5, \tag{1}$$

$$\tfrac{1}{2}x^2 + \sqrt{5}\,x + (3 + \sqrt{-2}), \tag{2}$$

$$ax^2 + (2 + b)x - (c + \sqrt{2}). \tag{3}$$

A polynomial in more than one variable may be put into a **standard form** by removing all parentheses, combining like terms, and arranging terms in order of descending powers of one of the variables.

Example 3. $x^2(2y^3 - xy^2) - (5x^2y^3 - 2x^3y^2 - 4x + 3) = 2x^2y^3 - x^3y^2 - 5x^2y^3 + 2x^3y^2 + 4x - 3 = x^3y^2 - 3x^2y^3 + 4x - 3$. The latter expression is in standard form, arranged according to descending powers of x. If descending powers of y are used, we have the alternative standard form $-3x^2y^3 + x^3y^2 + 4x - 3$.

It is intuitively clear, and it can be proved, that every polynomial can be written in standard form.

The **degree of a polynomial** is the degree of the term or monomial* of highest degree occurring in it when the polynomial is expressed in standard form. When a polynomial is written in standard form, the coefficients of its terms are called the **coefficients of the polynomial.**

For example, the polynomial $(x + 2)(x + 3)$ of example 1 has degree 2. The polynomial (1) above is of degree 4, (2) is of degree 2, and (3) is of degree 2. The polynomial of example 3 has degree 5.

Example 4. The degree of $(x - 2)(x + 5) - x^2$ is 1, since the standard form of this polynomial is $3x - 10$.

Polynomials of degree 1, 2, 3, 4, and 5 are called **linear, quadratic, cubic, quartic,**† and **quintic,** respectively.

* The degree of a monomial is defined in section 24.

† **Biquadratic** is sometimes used synonymously with *quartic.*

By the **general linear polynomial in x** is meant the expression

$$(4) \qquad ax + b$$

where a and b are arbitrary constants, $a \neq 0$. Similarly, the **general quadratic polynomial in x** is the expression

$$(5) \qquad ax^2 + bx + c$$

where a, b, and c are arbitrary constants ($a \neq 0$). The **general cubic polynomial in x** is the expression

$$(6) \qquad ax^3 + bx^2 + cx + d \qquad (a \neq 0).$$

The **general polynomial in x of nth degree,** where n is any positive integer, is the expression

$$ax^n + bx^{n-1} + cx^{n-2} + \dots + kx + l \qquad (a \neq 0)$$

or, in a better notation,

$$(7) \qquad a_0x^n + a_1x^{n-1} + a_2x^{n-2} + \dots + a_{n-1}x + a_n \quad (a_0 \neq 0).$$

For example, using subscripts as in (7), the general polynomials of degrees 1, 2, and 3 may be written $a_0x + a_1$, $a_0x^2 + a_1x + a_2$, and $a_0x^3 + a_1x^2 + a_2x + a_3$, respectively. An arbitrary non-zero constant such as a or a_0 may be considered to be the general polynomial of degree zero.

Note. It can be shown that every rational fractional expression can be expressed as a quotient of two polynomials. For example, we may write

$$\begin{aligned}\frac{1}{x+2} + \frac{2}{x+3} &= \frac{1 \cdot (x+3)}{(x+2)(x+3)} + \frac{2 \cdot (x+2)}{(x+2)(x+3)} \\ &= \frac{3x+7}{x^2+5x+6}\end{aligned}$$

by means of the definition of addition of fractions (definition 2, section 8). Thus *polynomials, or rational integral expressions, and rational expressions are very analogous to integers and rational numbers, respectively.* Many similarities between the properties of integers and the properties of polynomials will be brought out in later chapters.

EXERCISES

Write each of the following polynomials in standard form, and state its degree:

1. $5x^2 - 4x + 5 - x^2 + 2x - 1$.
2. $3 - x^3 + 5x^2 - 4x^3 - x^2 + x - 1$.
3. $(2x^3 + 4x - 1) - (x^3 + x - 2) - x^3 + 3$.
4. $(-x^3 + 2x^2 - 3) - (-x^3 - x^2 + 3x - 2)$.
5. $(x + 2)(x + 3) - (x^2 + 1)$.
6. $(x + 3)(x - 1) - (2x - 5)$.
7. $(x^4y + x^2y^2) - (x^3y - 2x + 3)$.
8. $x^2y + xy - y^2 - (3 - xy)$.

9. If integers are substituted for the variables in a polynomial whose coefficients are all integers, what kind of number must the resulting value of the expression be? Explain and illustrate.

10. If rational numbers are substituted for the variables in a rational expression consisting of the quotient of two polynomials with rational coefficients, what kind of number must the resulting value of the expression be, if it exists at all? Explain and illustrate.

11. Will the value of an irrational expression for any values of the variables always be an irrational number? Explain and illustrate.

27. Equations and Identities

An algebraic expression which contains variables has no definite numerical value unless definite values are assigned to the variables. For example, $2x + 6 = 8$ if $x = 1$, but $2x + 6 = 10$ if $x = 2$, and so on. Hence, the meaning of statements such as

$$(1) \qquad 2(x + 3) = 2x + 6$$

and

$$(2) \qquad 2x + 3 = 11,$$

each of which seems to say that two algebraic expressions are equal, requires explanation.

Equation (1) is true regardless of what value we substitute for x, since, by the distributive law, $2(x + 3) = 2 \cdot x + 2 \cdot 3 = 2x + 6$ for any number x whatever. For example, if we set $x = 1$, (1) becomes $2 \cdot (1 + 3) = 2 \cdot 1 + 6$ or $2 \cdot 4 = 2 + 6$; setting $x = 2$, (1) becomes $2 \cdot (2 + 3) = 2 \cdot 2 + 6$ or $2 \cdot 5 =$

$4 + 6$; and so on. Equation (2) is of a quite different character. It becomes false for $x = 1$, since $5 \neq 11$. In fact, it becomes false if we substitute for x anything but 4; for $x = 4$ it becomes true. We distinguish between these two kinds of equations by means of the following definitions.

An equation which becomes a true statement for all allowable values of the variables is called an **identical equation** or simply an **identity.**

For example, (1) is an identity. By **allowable values** of the variables is meant values for which all the expressions involved have a definite meaning. The reason for the word "allowable" in the definition is that some values of x may have to be excluded from consideration. For example, $\frac{1}{x} = \frac{3}{3x}$ is an identity, but the value $x = 0$ has to be excluded from consideration, since division by zero cannot be permitted (see section 11). Similarly, $\frac{1}{x-2} + \frac{1}{x-3} = \frac{2x-5}{x^2-5x+6}$ is an identity, but the values $x = 2$ and $x = 3$ must be excluded from consideration for the same reason. Both these equations are true, however, for all allowable values of x.

An equation which becomes false for some allowable value(s) of the variable(s) is called a **conditional equation** or simply an **equation.**

For example, (2) is an (conditional) equation.

If an (conditional) equation involves only one variable, a value of the variable for which the equation becomes true is called a **root** or a **solution** of the equation, and is said to **satisfy** the equation. To **solve** an equation in one variable means to find all its roots.

For example, the only root of (2) is 4.

If an (conditional) equation involves more than one variable, a set of values for the variables which makes the equation true is called a **solution** of the equation and is said to **satisfy** the equation.

For example, the equation $x + y = 7$ has the solutions $(x = 1,\ y = 6)$, $(x = 2,\ y = 5)$, $(x = 3,\ y = 4)$, $(x = -1,\ y = 8)$, etc.

The subject of solving (conditional) equations will occupy several later chapters.

28. Verification of Identities. Reversible Steps

The subject of verifying identities is simpler, however. Note that an equation in one variable can be shown to be conditional, and hence not an identity, by merely exhibiting one particular allowable value of the variable for which it becomes a false statement; but a thousand particular values of the variable for which the equation becomes true will not suffice, unassisted by further information, to prove that it is an identity, for they will provide no guarantee that the equation will become true for the 1001th value substituted. (See, however, corollary 2, section 119.)

The simplest way to verify an identity is to reduce one or both members by steps of substitution of equals for equals until they look alike.

Example 1. Verify that $x(x + 3) + 6 = x^2 + 3(x + 2)$ is an identity. Since $x(x + 3) = x^2 + 3x$ and $3(x + 2) = 3x + 6$, we may substitute on both sides, obtaining $x^2 + 3x + 6 = x^2 + 3x + 6$.

If steps other than substitution of equals for equals are used, one must take care that all steps are reversible.

Example 2. Verify that

$$\frac{x + y}{2} + \frac{z}{2} = \frac{x}{2} + \frac{y + z}{2}. \tag{3}$$

We might reason as follows. If (3) is true, then what is obtained by multiplying both sides by 2 will be true (since if equals are multiplied by equals, the results are equal). Thus, if (3) is true, then

$$x - y + z = x + y + z \tag{4}$$

which is clearly true. Does this enable us, without further consideration, to assert that (3) is true? Certainly not, for a true conclusion may be obtained by correct reasoning from a false hypothesis. Thus, "if (3) is true, then (4) is true, and (4) is true" does not guarantee that (3) is true. What we really need is the converse proposition "if (4) is true, then (3) is true," for if the hypothesis of a valid argument is true, then the conclusion must be true. Now we have so far proved only that "if (3) is true, then (4) is true," and we know that the converse (which is now needed) of a valid argument may not be valid. To prove the converse, we may try to deduce (3) from (4). But this may be done by merely reversing the steps of the argument by which we deduced (4) from (3), *provided those steps were all reversible.* In our present example this is the case, for since (4) was derived from (3) by multiplying both sides of (3) by 2, we can reverse the argument by dividing both sides of (4) by 2.

That one must really be careful not to use non-reversible steps will be clearly shown by the next two examples.

Example 3. Let us set out to prove that $3 = 7$. We reason that if $3 = 7$, then $3 \cdot 0$ would be equal to $7 \cdot 0$, because if equals are multiplied by equals the results are equal. But $3 \cdot 0 = 7 \cdot 0$, since $0 = 0$. May we then conclude that $3 = 7$? Of course not! The step of multiplying both sides by 0 cannot be reversed, because to reverse it would require us to divide both sides by 0, which cannot be done (see section 11).

Example 4. Let us set out to prove that

$$x = -x. \tag{5}$$

If equals are squared, the results are equal; that is, *if* $a = b$, *then* $a^2 = b^2$. Hence, squaring both sides of (5), we obtain $(x)^2 = (-x)^2$ or $x^2 = x^2$, which is true. May we conclude that therefore $x = -x$? Certainly not! For the converse of the italicized statement above is not true (compare section 12) and *the step of squaring both sides cannot be reversed.*

The step of substituting equals for equals is always reversible; therefore the method first suggested is always safe, if it is feasible.

EXERCISES

Verify the identities:

1. $x^2 - 2(x - 5) = x(x - 2) + 10.$ **2.** $\frac{2}{x} + \frac{3}{x} = \frac{5}{x}.$

3. $(x + 3)(x + 1) = x^2 + 4x + 3.$

4. $\frac{1}{x - 2} + \frac{1}{x - 4} = \frac{2x - 6}{x^2 - 6x + 8}.$

5. $\frac{1}{x} + \frac{3}{x - 2} = \frac{4x - 2}{x^2 - 2x}.$

Decide whether each of the following is an identity or a conditional equation:

6. $4(x - 2) = 4x - 8.$ **7.** $\frac{2}{x} + \frac{3}{3x} = \frac{5}{4x}.$ **8.** $8 - 3x = 5x.$

9. $\sqrt{x^2 + 16} = x + 4.$ **10.** $\frac{8x}{5} - \frac{3x}{5} = x.$

★11. Do exercise 30, section 12.

29. Elementary Operations with Equations

In the last section we pointed out the desirability of using reversible steps. We shall here indicate several useful operations on equations (either identical or conditional) which are always reversible.

I. *A term may be transferred from one side of an equation to the other, providing its sign is changed.* This operation is called **transposition.**

Thus, $2x + 3 = 11$ becomes $2x = 11 - 3$, or $2x = 8$, by transposition.

The justification for transposing a term is found in the axiom that if equals are added to, or subtracted from, equals, the results are equal. Such a step is always reversible, since, to undo an addition, we have only to subtract, and vice versa.

II. *Both sides of an equation may be multiplied or divided by a non-zero constant.*

Hence $2x = 8$ becomes $x = 4$ by dividing both sides by 2 (or multiplying by $\frac{1}{2}$).

This is clearly reversible, since to undo a multiplication we have only to divide, and vice versa; if we refrain from multiplying by zero, we shall have no trouble with division by zero.

Note that these steps have sufficed to solve the equation $2x + 3 = 11$. For we have proved that if x is such that $2x + 3 = 11$, then $x = 4$. Since all steps in the argument are reversible, the converse is also true, namely, if $x = 4$, then $2x + 3 = 11$. Hence 4 is a root and the only root.

Another useful operation with equations is based on the following proposition (see section 12).

III. *If* $a^2 = b^2$, *then* $a = \pm b$.

This is a reversible step, since the converse is also true, namely, if $a = \pm b$, then $a^2 = b^2$. Note that the $\pm$ sign in III is necessary.

★ May be omitted without disturbing continuity.

Thus $x^2 = 9$ implies $x = \pm 3$, and conversely. Hence ± 3 are the only roots of $x^2 = 9$.

Steps I, II, and III often suffice to solve simple equations.

Two conditional equations are called **equivalent** if they have exactly the same roots or solutions.

The reversible operations above clearly result in equivalent equations. The *following* operations do *not* always result in equivalent equations and *therefore must be used with caution.*

IV. *Multiplying both members by a quantity involving unknowns.*

> *Example 1.* Solve $\dfrac{2x}{x-2} = \dfrac{4}{x-2}\cdot$ Multiplying both sides by $x - 2$, we obtain $2x = 4$. Hence $x = 2$. This proves that if x were a root of the given equation, then x could only be 2. But 2 does not satisfy the given equation, since division by zero is excluded. Hence the given equation has no root at all, and is certainly not equivalent to the derived equation $2x = 4$.

V. *Squaring both members of an equation.* (See the preceding section.)

> *Example 2.* Solve $\sqrt{x-3} = -3$. Squaring both sides yields $x - 3 = 9$, or $x = 12$. Hence if the original equation had a root, it could only be 12. But 12 does not satisfy the original equation, since $\sqrt{12-3} = \sqrt{9}$ stands for $+3$ but not for -3. Hence the given equation is not equivalent to the derived equation $x - 3 = 9$.

VI. *Dividing both members by a quantity involving unknowns*

> *Example 3.* $(x - 1)(x - 2) = 0$ has the roots 1 and 2, as can be seen by substitution. If we divide both sides by $(x - 1)$, we get $x - 2 = 0$, which has only the root 2. Hence $x - 2 = 0$ is not equivalent to the given equation.

Operations IV and V may lead to equations which have roots not possessed by the original equation; such roots are often called **extraneous.** They can always be detected by checking by substitution in the original equation.

Operation VI may lead to equations which do not have roots which are possessed by the original equation. These lost roots are not easily detected.

We give only a few exercises here, since much later work illustrates these considerations.

EXERCISES

Find the roots of the following equations.

1. $2x - 5 = x + 2.$ **2.** $4x + 2 = x + 14.$

3. $\dfrac{3x}{x - 4} = \dfrac{12}{x - 4}.$ **4.** $\dfrac{5x}{x - 3} = \dfrac{15}{x - 3}.$

5. $\sqrt{x + 2} = -5.$ **6.** $\sqrt{x + 2} = 5.$

7. Do exercise 30, section 12.

30. Standard Form of Polynomial Equations. Degree of an Equation

We shall be principally concerned with **polynomial equations,** or **rational integral equations,** that is, with (conditional) equations both of whose members are polynomials or rational integral expressions. For example,

$$(1) \qquad x^3 + 5x^2 - x + 3 = x^3 + x^2 + x + 2$$

is a polynomial equation. By transposing and simplifying, we can write such an equation with zero on the right of the equals sign, and a polynomial in standard form (see section 26) on the left; when this is done, the equation is said to be in **standard form.** For example, (1) may be written in the standard form

$$(2) \qquad 4x^2 - 2x + 1 = 0.$$

When a polynomial equation is written in standard form, the degree of the polynomial on the left of the equals sign is called the **degree** of the equation. Hence equation (1) has the degree 2, since its standard form is (2). Equations of degree 1, 2, 3, 4, and 5 are called **linear, quadratic, cubic, quartic,*** and **quintic** equations, respectively. For example, $2x^3 + x - 5 = 0$ is a cubic equation in x; and $x^2y - 2xy^2 + xy - 5x + 4y - 3 = 0$ is a cubic equation in x and y. It will be seen in later chapters that the solution of polynomial equations increases in difficulty as the degree increases.

By the **general equation in *x* of a given degree** we mean the equation obtained by setting the general polynomial (see section 26) in x of that degree equal to zero. Hence the **general linear equation in *x*** may be written as

$$(3) \qquad ax + b = 0 \ (a \neq 0), \text{ or } a_0x + a_1 = 0 \qquad (a_0 \neq 0)$$

* Or **biquadratic.**

where a, b, a_0, a_1 are arbitrary constants. Similarly, the **general quadratic equation in x** may be written as

$$(4) \quad ax^2 + bx + c = 0 \ (a \neq 0), \qquad \text{or } a_0x^2 + a_1x + a_2 = 0 \quad (a_0 \neq 0).$$

The **general cubic equation in x** may be written as

$$(5) \quad ax^3 + bx^2 + cx + d = 0 \ (a \neq 0), \qquad \text{or } a_0x^3 + a_1x^2 + a_2x + a_3 = 0 \quad (a_0 \neq 0).$$

The **general equation of nth degree in x** may be written as

$$(6) \quad a_0x^n + a_1x^{n-1} + a_2x^{n-2} + \ldots + a_{n-1}x + a_n = 0 \qquad (a_0 \neq 0).*$$

EXERCISES

Write each of the following equations in standard form and state its degree:

1. $3x^4 + 2x^3 + x^2 + 3 = 3x^2(x^2 + x) - x^3 + x - 1$.
2. $x^3 - x + 1 = x^3 - 4x + 3$.
3. $2x^4 + x^2(3x - 1) + 2 = 2x^4 + 3x^2(x - 1) - x - 6$.
4. $x(x^2 - 1) = x(x + 3) - 5$. **5.** $x(2x^2 + 1) = x - 7$.
6. Write the general equations of degree 4 and 5.

* If $a_0 = 0$, the equation would have degree lower than n.

Chapter IV

Functions and graphs

31. Functions

Consider two variables x and y. The variable y is said to be a **function of x** when any scheme or rule (relation, or correspondence) is given whereby to each value of x there correspond one or more definite values of y.

A simple way to define y as a function of x is to let y equal an algebraic expression in x.

Example 1. The rule $y = 2x + 1$ defines y as a function of x. From this rule one can calculate the value of y corresponding to any value of x; thus, if $x = 2$, $y = 5$; and if $x = 3$, $y = 7$, and so on.

Example 2. The rule $y = \dfrac{12}{x - 2}$ defines y as a function of x. If $x = 8$, $y = 2$; if $x = 4$, $y = 6$; if $x = 5$, $y = 4$; if $x = 7$, $y = \frac{12}{5}$ and so on.

Example 3. The rule $y = \pm\sqrt{x}$ defines y as a function of x. If $x = 4$, $y = \pm 2$; if $x = 9$, $y = \pm 3$; if $x = 2$, $y = \pm\sqrt{2}$, and so on.

If the scheme or rule assigns just one value of y to each value of x, the function is called **single-valued;** otherwise, **many-valued.** Thus, the functions given in examples 1 and 2 are single-valued functions, but the function given in example 3 is a many-valued function.

If y is a function of x, we call x the **independent variable,** and y the **dependent variable,** and we say that y **depends on x.** It may sometimes be necessary or desirable to restrict the possible values of the independent variable somewhat; the totality or set of values which x may assume is called the **range** of x. If nothing is said to the contrary, the range of x will be assumed to be all numbers for which there is a corresponding value of y. In example 2, the range of x must not include the number 2, since division by zero is excluded. In example 3, if we wish y to be a real number, then the range of x must be restricted to non-negative real numbers. A function is said to be **defined** for every value of x in the range.

We may also consider functions of two or more independent variables. Thus, the variable z is said to be a function of x and y if a scheme or rule is given whereby to every given pair of values of x and y there corresponds one or more definite values of z. For example, the rule $z = x^2 + y$ defines z as a single-valued function of x and y. If $x = 2$, and $y = 3$, then $z = 7$, and so on.

If a function is defined by letting the values of the dependent variable be obtained by substituting values of the independent variable(s) in a polynomial, or rational integral expression, in the independent variable(s), we say that the function is a **polynomial** or **rational integral function** of the independent variable(s). If a rational expression is used, we call the function a **rational function** of the independent variable(s). If a linear polynomial is used, we call the function a **linear function** of the independent variable(s). If a quadratic polynomial is used, we speak similarly of a **quadratic function.**

Thus, example 1 defines y as a linear function of x. Example 2 defines y as a rational function of x. Also, $z = 2x + 3y - 4$ defines z as a linear function of x and y; and $y = x^2 - 2x + 3$ defines y as a quadratic function of x.

Note that the word "function" is used in mathematics in a technical sense quite different from its everyday meaning as in the statements "pumping blood is the function of the heart," "passing bills is a function of the legislature," or "the Senior Prom was the outstanding function of the school year."

A function may not always be defined as simply as by an algebraic expression. For example, $\log x$ and $\sin x$ are not. Functions may be defined for a limited number of values of the independent variable by means of a table whose entries may be obtained from experimental or observational data. Other ways of defining functions may be discussed elsewhere. In this book most of the functions will be defined by algebraic expressions.

EXERCISES

Find the value of each of the following functions when $x = -2$:

1. $y = 3x^2 - 2x + 11.$ **2.** $y = 2x^3 - x^2.$

3. $y = \dfrac{12}{2 - x}.$ **4.** $y = \dfrac{8}{x - 2}.$

Find the value of each of the following functions when $x = 2$ *and* $y = -3$:

5. $z = xy - 3x^2.$ **6.** $z = x + y + 1.$

7. $z = x^2 + y^2 - xy.$

For what values of x *is each of the following functions not defined?*

8. $y = \dfrac{1}{x}.$ **9.** $y = \dfrac{6}{2 - x}.$ **10.** $y = \dfrac{8}{x^2 - 9}.$

11. $y = \dfrac{1}{(x - 2)(x + 5)}.$ **12.** $y = \dfrac{x - 2}{x - 3} + \dfrac{1}{x + 2}.$

32. Functional Notation

A given function of x, such as $2x + 1$, may be conveniently denoted by a symbol such as $f(x)$, read "f of x," or "the f-function of x." If $f(x) = 2x + 1$, then $f(2)$ stands for the value of the dependent variable, or value of the function, when $x = 2$. Note that $f(x)$ does *not* mean a number f times a number x, but is simply a compact symbol for a function of x. If a is a constant, then $f(a)$ *stands for the value of the function* $f(x)$ *when* x *has the value* a, *or when* a *is substituted for* x. Hence, in the above example, $f(2) = 5$, $f(3) = 7$, $f(a) = 2a + 1$.

If two functions have to be discussed in the same problem, we may use similar symbols, such as $F(x)$, $g(x)$, $D(x)$, $R(x)$, etc., to avoid confusion. Hence, if $F(x) = x^2 + 1$ and $f(x) = 2x + 1$, we have $f(2) = 5$, $F(2) = 5$, $f(3) = 7$, $F(3) = 10$, $f(-2) = -3$, $F(-2) = 5$, and so on.

A function of two variables may be similarly denoted by a symbol of the form $f(x,y)$. Thus, if $f(x,y) = x^2 + 3y$, then $f(2,1) = 7$, $f(3,2) = 15$, $f(1,1) = 4$, $f(-1,0) = 1$, and so on.

If $F(x) = \dfrac{1}{x - 2}$, we shall say that $F(2)$ does not exist, or is not defined, since 2 is not in the range of x.

EXERCISES

Given that $f(x) = 3x + 2$, *find:*

1. $f(2)$. **2.** $f(3)$. **3.** $f(-1)$. **4.** $f(-2)$.
5. $f(-3)$. **6.** $f(0)$. **7.** $f(n)$. **8.** $f(a)$.
9. $f(-r)$. **10.** $f(10k)$.

Given that $d(x) = x^2 - 4$, *find:*

11. $d(2)$. **12.** $d(-2)$. **13.** $d(3)$. **14.** $d(-3)$. **15.** $d(r)$.

Given that $F(x) = 2x^2 - 3x + 2$, *find:*

16. $F(2)$. **17.** $F(-2)$. **18.** $F(r)$. **19.** $F(\frac{1}{2})$.
20. $F(-\frac{1}{2})$. **21.** $F(1.2)$. **22.** $F(-1.2)$.

Given that $G(x) = \dfrac{2x - 1}{x - 4}$, *find:*

23. $G(5)$. **24.** $G(\frac{1}{2})$. **25.** $G(-\frac{1}{2})$. **26.** $G(\frac{3}{4})$. **27.** $G(4)$.

Given that $Q(x) = \dfrac{1}{x - 1} + \dfrac{1}{x - 3}$, *find:*

28. $Q(1)$. **29.** $Q(4)$. **30.** $Q(\frac{1}{2})$.

33. Applications of Functions in Science. Formulas

The study of functions came to be of overwhelming importance in the 17th century, when scientists became greatly interested in the study of varying quantities. If y is a function of x, we say that the value of y **depends** on the value of x. Speaking loosely, we might say that *y varies with x,* or that a change in *x induces* a corresponding change in y. This statement is not strictly true, for we might well have a function which assigns the same value to y no matter what value is given to x, such as the function $y = 3$. Such a function is called a **constant function.** But it will do no great harm if you think of the value of y changing as the value of x changes, provided it is agreed that the change in y may be zero.

Every formula, whether taken from geometry, physics, or other subjects, expresses some variable quantity as a function of other variable quantities.

Example 1. The area A of a square of side x is given by the formula $A = x^2$.

Example 2. The perimeter P of a square of side x is given by the formula $P = 4x$.

Example 3. The distance s, measured in feet, through which a body falls in t seconds is given approximately by the formula $s = 16t^2$, air resistance being neglected.

Example 4. The area of a triangle is given by the formula $A = \frac{1}{2}bh$, where b is the base and h the altitude.

Example 5. The area of a trapezoid is given by the formula $A = \frac{1}{2}h(b + b')$, where h is the altitude and b and b' are the bases (Fig. 11).

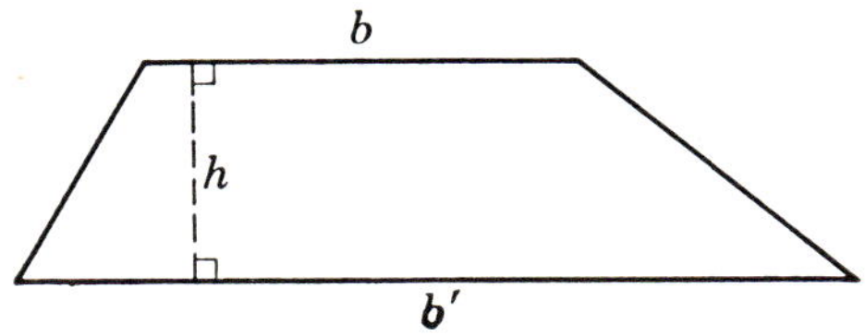

Fig. 11.

Example 6. The simple interest I on P dollars at the rate R (expressed as a decimal) per year for T years is given by the formula $I = PRT$.

Examples 1, 2, 3 are functions of one variable, example 4 is a function of two variables, and examples 5 and 6 are functions of three variables. It is clear that examples of functions abound in many different subjects. They may be exact, as in examples 1, 2, 4, 5, and 6, or approximate, as in example 3. Further exercises involving formulas are found in section 58.

EXERCISES

1. Using the formula of example 3, above, how far will a body fall in (*a*) 2 seconds? (*b*) 5 seconds?

2. Using the formula of example 4, above, what is the area of a triangle with base 5 feet and altitude 4 feet?

3. Using the formula of example 4, above, what is the area of a triangle with base 7 feet and altitude 4 feet?

4. Using the formula of example 5, above, what is the area of a trapezoid with altitude 6 inches and bases 3 inches and 9 inches, respectively?

5. Using the formula of example 6, above, find the simple interest on \$224.00 at the rate of 4% per year for 3 years. (*Hint:* express the rate as a decimal, that is, as 0.04.)

6. The time T measured in seconds, required for a complete oscillation of a simple pendulum of length k feet is given by the formula $T = 2\pi \sqrt{k/g}$ where $\pi = 3.14$ approximately and $g = 32.2$ approximately. If $k = 3.24$ feet, find T correct to the nearest tenth.

7. If F is the temperature in degrees Fahrenheit and C is the temperature in degrees Centigrade, then $F = \frac{9}{5}C + 32$. Find the Fahrenheit temperature corresponding to 35° Centigrade.

8. The volume V of a sphere of radius r is given by the formula $V = \frac{4}{3}\pi r^3$. Find the volume of a sphere of radius 3 feet, correct to the nearest tenth, using $\pi = 3.14$ approximately.

9. The surface area S of a sphere of radius r is given by the formula $S = 4\pi r^2$. Find the surface area of a sphere of radius 3 feet, correct to the nearest tenth, using $\pi = 3.14$ approximately.

10. Using the formulas of exercises 8 and 9, find the volume and area of a sphere of radius 6 inches.

34. Rectangular Coordinates

Choose two perpendicular lines in a plane. Call one the **x-axis** and the other the **y-axis.** The point O where they intersect is called the **origin.** It is customary to consider the x-axis horizontal and the y-axis vertical. Choose a unit of length and lay it off repeatedly on both axes. Choose a positive direction on each axis; it is customary to make the positive side of the x-axis the right side, and the positive side of the y-axis the upper side. Assign real numbers to the points of the axes in the usual way. Then to each point in the plane we may attach a pair of real numbers, called the **x-coordinate** (or **abscissa**) and the **y-coordinate** (or **ordinate**) of the point, respectively. Together these **coordinates** of a point P constitute directions for getting to P from O. The x-coordinate tells us how to proceed from O along the x-axis, and then the y-coordinate tells us how to go in the direction of the y-axis. To reach P from the origin, we go 3 units in the direction of the positive x-axis, and then 2 units in the direction of the positive y-axis. The x and y coordinates are indicated by a symbol of the form (x,y), the x-coordinate always being given first. Thus the point P in Fig. 12 has the coordinates (3,2) while Q has the coordinates (2,3). Similarly, R has the coordinates (0,3), S the coordinates $(-2,3)$, and so on as in Fig. 12. Note that *every point on the x-axis has y-coordinate equal to zero.* Marking a point with given coordinates is called **plotting** the point.

Use of these so-called **rectangular coordinates** was developed by the great French mathematician-philosopher René Descartes, who founded the important subject of analytic geometry* on this basis in 1637.

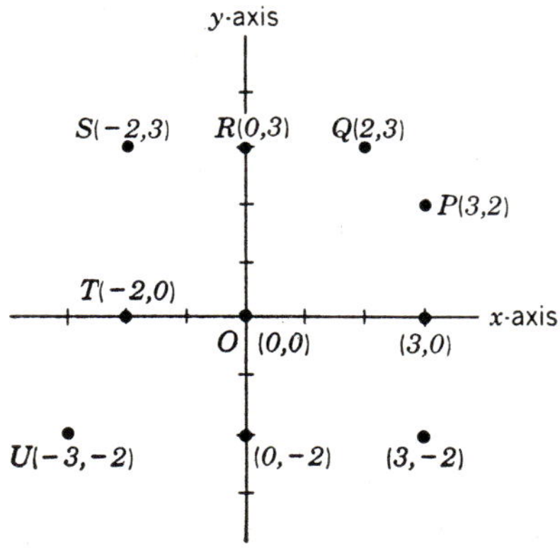

Fig. 12.

EXERCISES

1. Plot the points whose coordinates are (1,2), (2,1), (−1,2), (2,−1), (−2,−1).

2. Plot the points whose coordinates are (2,0), (0,2), (−2,0), (0,−2), (0,0).

3. What is the y-coordinate of any point on the x-axis?

4. What is the x-coordinate of any point on the y-axis?

35. The Graphs of Equations and Functions

By the **graph** of an equation in two variables x and y we mean the set or totality of all those points, and only those points, whose coordinates satisfy the equation. *Only real numbers will be considered.* To plot the graph of an equation we may substitute various real values for one variable in the equation, calculate from the equation the corresponding values of the other variable, and tabulate the results; then we plot the corresponding points and join them with a smooth curve.

* For further indications of how these coordinates enable us to make huge strides in geometry, and for a discussion of other types of coordinates, see M. Richardson, *Fundamentals of Mathematics*, Macmillan, 1941, Chapter IX.

Example 1. Draw the graph of the equation $x + y = 7$. We obtain the table

x	-3	-2	-1	0	1	2	3	4	5	6	7	8	9	10
y	10	9	8	7	6	5	4	3	2	1	0	-1	-2	-3

Plotting these points, we get the straight line in Fig. 13.

It will be shown later that every linear equation in x and y, with real coefficients, represents a straight line. With this knowledge, it would have sufficed to plot two points, with a possible third point as a check. In general, however, many points are necessary to get the shape of the graph.

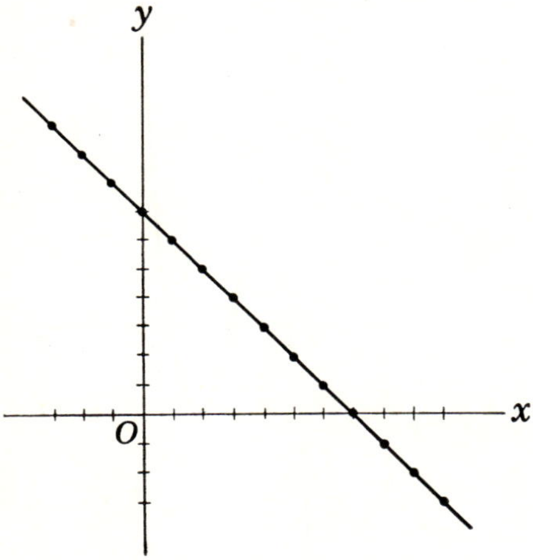

Fig. 13.

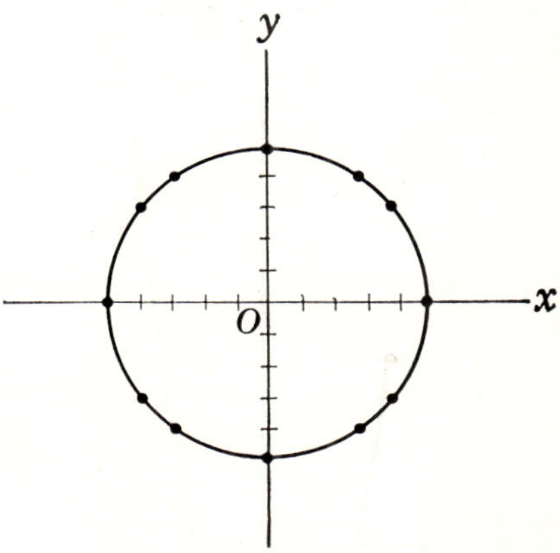

Fig. 14.

Example 2. Draw the graph of the equation $x^2 + y^2 = 25$. We get the table

x	0	± 5	$+3$	-3	$+4$	-4
y	± 5	0	± 4	± 4	± 3	± 3

Plotting these points, we get the circle in Fig. 14.

By the **graph of a function** $f(x)$ we mean the graph of the equation $y = f(x)$.

Example 3. Plot the graph of the function $x^2 + x - 6$. From the equation $y = x^2 + x - 6$ we get the table

x	0	1	-1	2	-2	3	-3	4	-4
y	-6	-4	-6	0	-4	6	0	14	6

Plotting, we get the parabola in Fig. 15.

Example 4. Draw the graph of the function $x^3 - 3x^2 - x + 3$. From the equation $y = x^3 - 3x^2 - x + 3$ we get the table

x	0	1	-1	2	-2	3	4
y	3	0	0	-3	-15	0	15

Plotting these points, we get the graph of Fig. 16.

Much information concerning *how* the quantity y varies as x varies can be obtained at least approximately from the graph. Thus, one might infer from the graph of Fig. 15 that y has a

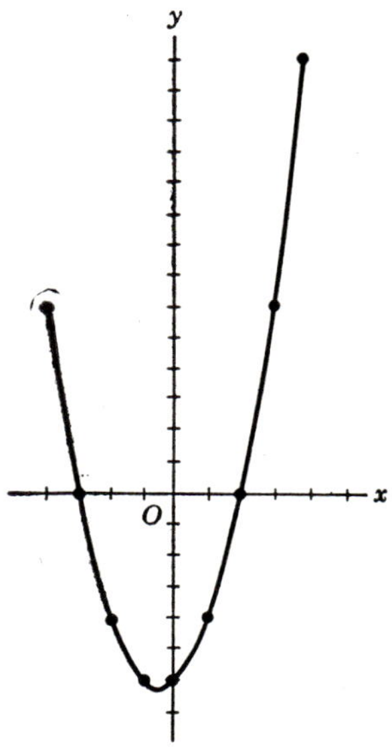

Fig. 15.

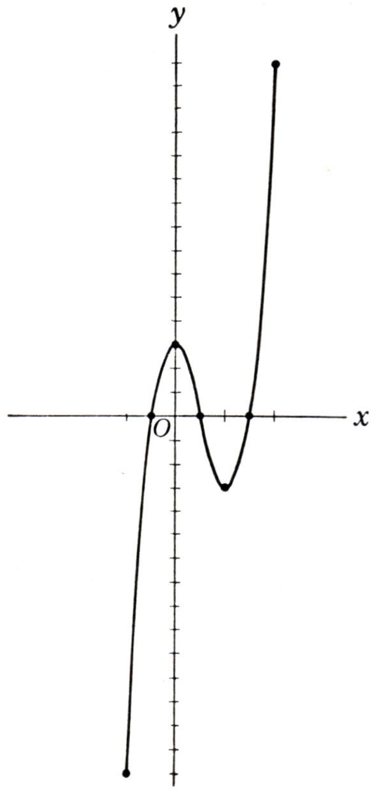

Fig. 16.

minimum value of something less than -6 and assumes this minimum value for a value of x about midway between $x = -1$ and $x = 0$.

EXERCISES

Plot the graph of each of the following functions and estimate from the graph for what (real) values of x the function assumes the value zero:

1. $2x - 8$. **2.** $3x + 12$. **3.** $\frac{1}{3}x + 1$. **4.** $\frac{1}{2}x - 1$.

5. $x^2 - 9$. **6.** $x^2 - 4$. **7.** $x^2 + 1$. **8.** $x^2 + x + 1$.
9. $x^2 + x - 6$. **10.** $x^2 - 5x + 4$.
11. $x^3 - 3x^2 - 6x + 8$. **12.** $x^3 - 3x^2 - 2x + 12$.

Plot the graphs of each of the following equations:

13. $x + 3y = 5$. **14.** $2x - 3y = 2$.
15. $x^2 + y^2 = 36$. **16.** $9x^2 + 4y^2 = 36$. **17.** $xy = 1$.
18. $9x^2 - 4y^2 = 36$. **19.** $x^2 - y^2 = 1$.
20. $y(x - 2) = 1$. **21.** $(y - 3)(x - 2) = 1$.

36. Continuous Functions

In plotting the graph of a function $f(x)$ we join the points actually plotted by a smooth curve when we have enough points to judge the shape of the curve. This involves the assumption, among others, that the curve does not behave wildly between the plotted points, as in Fig. 17. This assumption, although

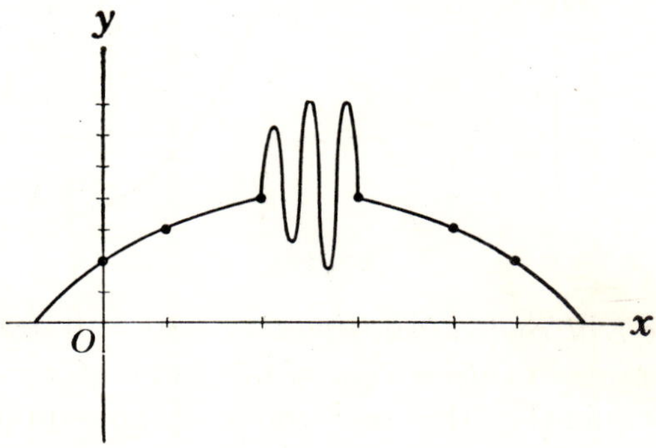

Fig. 17.

rash at this stage of the game, can be justified for polynomial functions by advanced analysis, beyond the scope of this book. By a **continuous function** is meant, speaking very roughly,* one whose graph can be drawn with one continuous stroke of the pencil, that is, without taking pencil from paper. It can be shown that *polynomials in x are continuous at all points of their graphs,* and that *rational functions in x, expressed as* $\frac{f(x)}{d(x)}$ *where*

* This is highly inaccurate, but will suffice for present purposes. For a more precise discussion of continuity, see M. Richardson, *Fundamentals of Mathematics,* Macmillan, 1941.

$f(x)$ and $d(x)$ are polynomials, are continuous except at the roots of the denominator polynomial $d(x)$.

37. Graphical Solution of Equations

Let $f(x)$ be any function of x. If the value of $f(x)$ is zero when the number r is substituted for x, that is, if $f(r) = 0$, then r is a **root** of the equation $f(x) = 0$. If the root r is a real number, then the graph of the function $y = f(x)$ must have a point in common with the x-axis at $x = r$ since $y = 0$ at $x = r$ (Fig. 18).

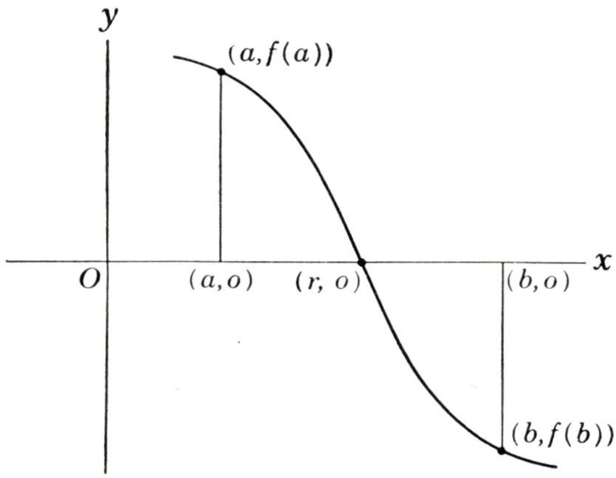

Fig. 18.

Therefore *the real roots of an equation $f(x) = 0$ are the x-coordinates of the points of intersection of the graph of $y = f(x)$ with the x-axis.* Consequently, the real roots of an equation may be estimated from the graph. Imaginary roots cannot be found this way, since the x- and y-coordinates on our graph are real numbers.

An important property of all continuous functions with real values exclusively, and hence, in particular, of all polynomials with only real coefficients, is the following:

If $f(x)$ is continuous, and if $f(a)$ and $f(b)$, (where $a < b$), have opposite signs, then there is at least one real root r of the equation $f(x) = 0$ between a and b (that is, $a < r < b$).

This principle is intuitively clear from Fig. 18, but is too difficult to prove here.

Example 1. The real roots of $x^2 + x - 6 = 0$ are 2 and -3. See Fig. 15.

Example 2. The real roots of $x^3 - 3x^2 - x + 3 = 0$ are ± 1, and 3. See Fig. 16.

Example 3. Estimate graphically the real roots of $x^3 - x^2 - 5x + 5 = 0$. Plotting the function $y = x^3 - x^2 - 5x + 5$, we find the table

x	0	1	2	3	-1	-2	-3
y	5	0	-1	8	8	3	-16

and the graph of Fig. 19. Hence the roots are 1, something between 2 and 3, and something between -2 and -3.

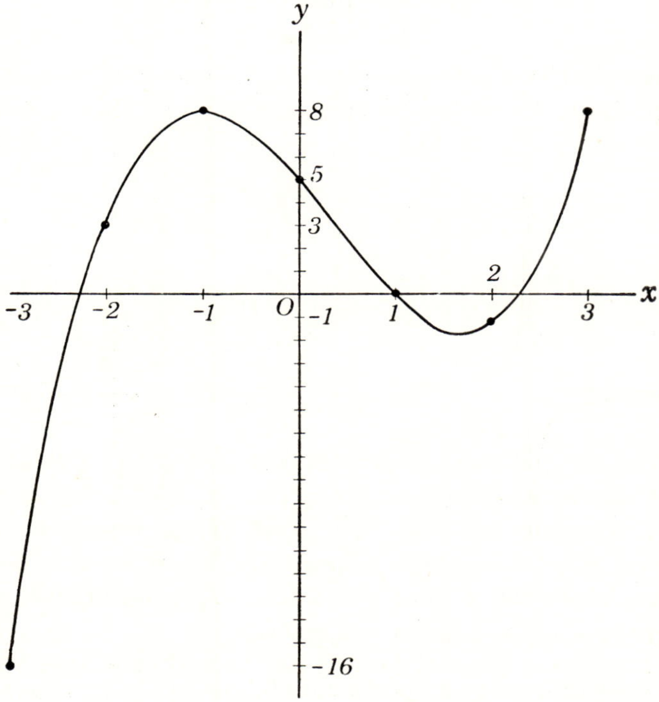

Fig. 19.

This method will be elaborated further in Chapter XIX.

EXERCISES

Estimate from the graph the real roots of each of the following equations:

1. $3x - 6 = 0$. **2.** $2x + 8 = 0$. **3.** $\frac{1}{2}x + 3 = 0$.
4. $\frac{1}{3}x + 1 = 0$. **5.** $x^2 - 1 = 0$. **6.** $x^2 - 16 = 0$.
7. $x^2 - 4x - 5 = 0$. **8.** $x^2 - 6x + 5 = 0$.

9. $x^2 - 4x + 3 = 0.$ **10.** $x^2 + 2x - 3 = 0.$
11. $x^3 - 4x^2 + x + 6 = 0.$ **12.** $x^3 - 7x - 6 = 0.$
13. $6x^2 - 5x - 4 = 0.$ **14.** $6x^2 - 7x - 3 = 0.$
15. $x^2 - 2 = 0.$ **16.** $x^2 - 3 = 0.$

★38. Infinities of a Rational Function

The function $y = 1/x$ is not defined for the value $x = 0$. But the graph of the function exhibits an important property for values of x near zero. Note the following table:

x	± 1	± 0.1	± 0.01	± 0.001	± 0.0001	...	0
y	± 1	± 10	± 100	± 1000	± 10000	...	not defined

It is clear that as x gets nearer to 0, y increases in absolute value without limit. In technical language we say that y **becomes**

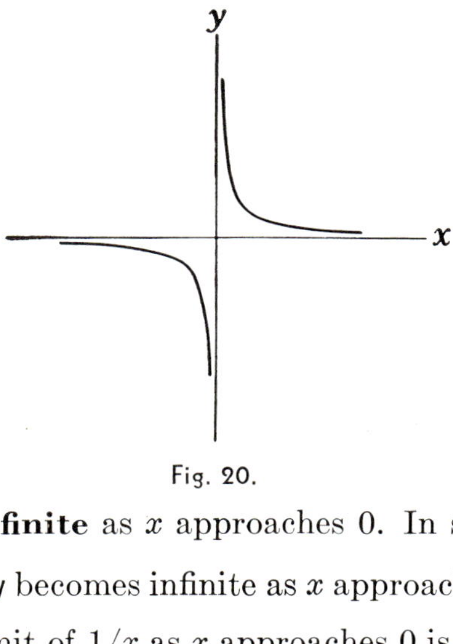

Fig. 20.

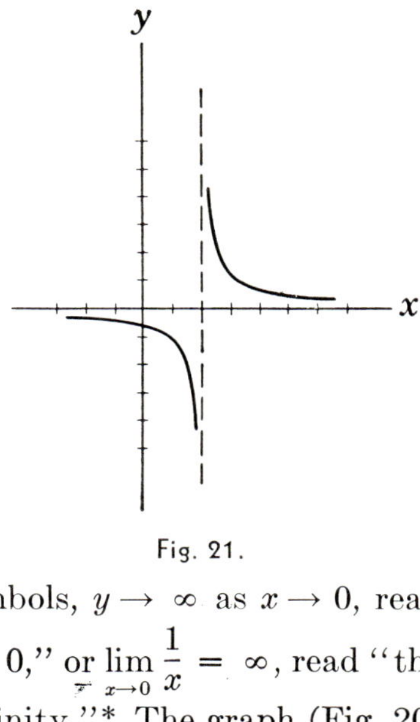

Fig. 21.

infinite as x approaches 0. In symbols, $y \to \infty$ as $x \to 0$, read "y becomes infinite as x approaches 0," or $\lim\limits_{x \to 0} \frac{1}{x} = \infty$, read "the limit of $1/x$ as x approaches 0 is infinity."* The graph (Fig. 20) indicates what this means.

★ This section may be omitted on first reading without disturbing the continuity of the book.

* It is misleading to write $1/0 = \infty$, although this is often done, because it seems to suggest that we get something by dividing 1 by 0. We have seen in section 11 that division by zero is excluded. When the symbol $1/0 = \infty$ is encountered, it should be understood as an abbreviation for the situation described in the text.

Similarly $y = \frac{1}{x - 2}$ becomes infinite at $x = 2$ (Fig. 21). It can be shown that a rational function $y = \frac{f(x)}{d(x)}$, where $f(x)$ and $d(x)$ are polynomials, becomes infinite at every root of $d(x)$ which is not also a root of $f(x)$. A value of x for which a function $y = F(x)$ becomes infinite may be called an **infinity** or **infinite discontinuity*** of the function $F(x)$.

EXERCISES

Plot the graph of each of the following functions and find its infinities:

1. $y = \frac{3}{x}$. **2.** $y = \frac{1}{3x}$. **3.** $y = \frac{1}{x - 2}$. **4.** $y = \frac{1}{3x - 2}$.

5. $y = \frac{1}{2x + 3}$. **6.** $y = \frac{1}{3x - 5}$. **7.** $y = \frac{1}{x^2 - 1}$.

8. $y = \frac{1}{4 - x^2}$. **9.** $y = \frac{1}{(x - 1)(x + 2)}$.

10. $y = \frac{1}{x(x - 3)}$. **11.** $y = \frac{x}{x^2 - 2x}$.

12. $y = \frac{x - 3}{x^2 - 3x}$. **13.** $y = \frac{x^2 - 3x}{x - 3}$.

* For a further discussion of this and other types of discontinuities see M. Richardson, *Fundamentals of Mathematics*, Macmillan, 1941.

Chapter V

Elementary operations with polynomials

39. Integers and Polynomials. Hindu-Arabic Notation

It has already been remarked (section 26) that integers and polynomials (or rational integral expressions) have many similar properties. There are deep-lying reasons for their similarity which are beyond the scope of this book. But even the elementary operations of addition, subtraction, multiplication, and division are much alike for integers and polynomials. These similarities may be traced to the fact that an integer, written in the usual way, is nothing but a particular value of a polynomial in x, namely the value for $x = 10$. Thus 2345 is the value of the polynomial $2x^3 + 3x^2 + 4x + 5$ when $x = 10$, or $2 \cdot 10^3 + 3 \cdot 10^2 + 4 \cdot 10 + 5$. Similarly 5078 is the value of the polynomial $5x^3 + 7x + 8$ when $x = 10$; and -407 is the value of the polynomial $-4x^2 - 7$ when $x = 10$.

Our way of writing integers is called the Hindu-Arabic notation because it was invented by the Hindus about 500 A.D. and brought into Europe by the Arabs when they overran Spain in the 8th century. It represents a great advance over the Roman numerals, as one can see by multiplying 1626 by 62 in the usual way and then trying to multiply the same numbers MDCXXVI by LXII, using only Roman numerals. In practice, the ancients used the abacus, counting-boards with pebbles, and other devices. (The word "calculate" comes from the Latin *calculus* meaning "pebble.") Despite its obvious superiority over previous methods, the Hindu-Arabic notation was not widely used in Europe until the 15th century, after a long struggle against the repressive forces of tradition and resistance to change. In fact the new notation was adopted by the enlightened merchants long before the "learned" authorities became reconciled to it. The importance of the Hindu-Arabic notation for the operation of our complicated modern industrial and commercial civilization is indicated by the fact that it made it possible for the general public, instead of experts only, to learn arithmetic.

The Hindu-Arabic notation is called a *decimal* notation because each digit in a number represents the coefficient of a power of 10. It is called a *positional* notation because the position of each digit tells us the power of 10 of which it is the coefficient. Thus, in school, the student learned to speak of the units column, the tens column, the hundreds column, etc.

40. Addition and Subtraction of Polynomials

In discussing the elementary operations of addition, subtraction, multiplication, and division of polynomials, *it will be understood that all polynomials will first be written in standard form.** This is a convenient practice, although not absolutely essential.

To add two polynomials, we write like terms under each other and then add by columns, as in the following example.

Example 1. To add $2x^2 + 5x + 8$ and $3x^2 + 6x + 5$ we write

$$\begin{array}{r} 2x^2 + 5x + 8 \\ 3x^2 + 6x + 5 \\ \hline 5x^2 + 11x + 13. \end{array}$$

Hence the sum is $5x^2 + 11x + 13$.

Justification of the procedure. The sum $(2x^2 + 5x + 8) + (3x^2 + 6x + 5)$ can be written as $(2x^2 + 3x^2) + (5x + 6x) + (8 + 5)$ by using the associative and commutative laws for addition. This is accomplished in practice by writing like terms under each other in columns. Adding up by columns amounts to using the distributive law which enables us to write $2x^2 + 3x^2 = (2 + 3)x^2 = 5x^2$, and $5x + 6x = (5 + 6)x = 11x$. Thus the sum is $5x^2 + 11x + 13$.

* See section 26 for the definition of standard form of a polynomial.

Comparison with adding integers. To add 258 and 365, we write

$$\begin{array}{l} 258\ (= 2 \cdot 10^2 + \ \ 5 \cdot 10 + \ \ 8) \\ \underline{365\ (= 3 \cdot 10^2 + \ \ 6 \cdot 10 + \ \ 5)} \\ 623\ (= 5 \cdot 10^2 + 11 \cdot 10 + 13). \end{array}$$

The sum 623 is indeed the value of $5x^2 + 11x + 13$ when $x = 10$, or $5 \cdot 10^2 + 11 \cdot 10 + 13$, but must be written 623 because we allow only coefficients from 0 to 9 inclusive. Therefore we use the trick of "carrying." That is, we write $5 \cdot 10^2 + 11 \cdot 10 + 13 = 5 \cdot 10^2 + (10 + 1)10 + (10 + 3) = 5 \cdot 10^2 + 1 \cdot 10^2 + 1 \cdot 10 + 1 \cdot 10 + 3 = 6 \cdot 10^2 + 2 \cdot 10 + 3$.

To subtract a first polynomial from a second, we write the first under the second, placing like terms under each other, change the signs of the terms of the lower polynomial, and add.

Example 2. To subtract $2x^2 - 3x + 5$ from $5x^2 - 2x - 3$, we write

$$\begin{array}{rrr} 5x^2 & - \ 2x & - \ 3 \\ - & + & - \\ \oplus 2x^2 & \ominus \ 3x & \oplus \ 5 \\ \hline 3x^2 & + \ \ x & - \ 8. \end{array}$$

Justification of the procedure. By the rules for removing parentheses (section 25), $(5x^2 - 2x - 3) - (2x^2 - 3x + 5) = 5x^2 - 2x - 3 - 2x^2 + 3x - 5$. The rest of the process is merely addition.

Checking. A convenient way to check is to substitute $x = 10$, say, in the polynomials involved in the process, and perform the corresponding arithmetical operations. Thus, to check example 1, we do what was done above in the paragraph marked "Comparison with adding integers." To check example 2, we similarly substitute $x = 10$ and verify that

$$\begin{array}{rcr} 5 \cdot 10^2 - 2 \cdot 10 - 3 & = & 477 \\ -(2 \cdot 10^2 - 3 \cdot 10 + 5) & = & -175 \\ \hline 3 \cdot 10^2 + 1 \cdot 10 - 8 & = & 302. \end{array}$$

Such a check by substituting a particular value for x cannot, of course, prove that the relation among the polynomials involved is an identity, but it will usually catch any error. Any number may be substituted as a check, but it is advisable to avoid 0 and ± 1 because they are more likely to fail to reveal common errors, such as wrong exponents, since all positive integral powers of 0 are 0 and all positive integral powers of 1 are 1.

Subtraction of course may be checked by applying the definition of subtraction (section 5) and adding the result to the polynomial in the second line, obtaining the polynomial in the first line. Thus example 2 can be checked as follows:

$$\begin{array}{l} 3x^2 + x - 8 \\ \underline{2x^2 - 3x + 5} \\ 5x^2 - 2x - 3. \end{array}$$

The procedures for polynomials in more than one variable are similar.

EXERCISES

Calculate, and check by substitution:

1. Add $4x^2 - 5x + 2$ and $3x^2 + x - 3$.
2. Add $7x^2 - 3x - 5$ and $x^2 - 3x + 8$.
3. Add $2x^2 + 3xy - 2y^2$ and $x^2 - 5xy + 4y^2$.
4. Add $-3x^2 - 5xy + 4y^2$ and $-2x^2 + 3xy - 5y^2$.
5. Add $4x^3 - 3x^2y + 6y^2$ and $2x^3 + 5xy^2 - 2y^2$.
6. Add $3x^3 + 3x^2y - xy$ and $-4x^3 - 5xy + y^3$.
7. Subtract $x^2 - 2xy + 3y^2$ from $4x^2 - xy - 4y^2$.
8. Subtract $3x^2 + 2x^2y - 3y^2$ from $4x^2y - 3xy - y^2$.
9. Subtract $5x^2 - 6x + 4$ from the sum of $-3x^2 + 3x + 1$ and $2x^2 - 5x + 7$.
10. Subtract the sum of $3x^2 - 6x + 2$ and $-4x^2 - x - 5$ from $5x^2 - x + 2$.
11. Subtract the sum of $x^3 + 2x^2y - 4y^3$ and $2x^3 - 5x^2y + 7y^3$ from zero.
12. Subtract the sum of $2x^3 + 3x^2y - 4y^3$ and $x^3 - xy^2 + y - y^3$ from zero.

41. Multiplication of Polynomials

To multiply two monomials, we multiply the coefficients and combine the variable factors according to the rules for exponents (section 23).

Example 1. The product of $2x^3$ and $3x^2$ is $6x^5$.

Justification of the procedure. The product $(2x^3)(3x^2)$ may be written as $(2 \cdot 3)(x^3 \cdot x^2)$ by the associative and commutative laws for multiplication. The rest is obvious.

To multiply a first polynomial by a second polynomial, we multiply each term of the first polynomial by each term of the second and take the algebraic sum of these products. The work may be arranged as in the following example.

Example 2. To multiply $2x^2 - 3x + 4$ by $5x - 6$, we write

$$\begin{array}{rrrr} & 2x^2 & -\ 3x & +\ 4 \\ & & 5x & -\ 6 \\ \hline & -\ 12x^2 & +\ 18x & -\ 24 \\ 10x^3 & -\ 15x^2 & +\ 20x & \\ \hline 10x^3 & -\ 27x^2 & +\ 38x & -\ 24. \end{array}$$

Justification of the procedure. By the distributive law we may write

$$\begin{aligned} (2x^2 - 3x + 4)(5x - 6) & \\ &= (2x^2 - 3x + 4)5x + (2x^2 - 3x + 4)(-6) \\ &= 10x^3 - 15x^2 + 20x - 12x^2 + 18x - 24. \end{aligned}$$

The rest is obvious.

Comparison with multiplication of integers. The comparison is clear from the following example in which the right side is obtained from the left by setting $x = 10$.

Example 3.

$$\begin{array}{rrrrl} & 2x^2 & +\ 5x & +\ 3 & \quad 253 \\ & & 4x & +\ 2 & \quad\ \ 42 \\ \hline & 4x^2 & +\ 10x & +\ 6 & \quad 506 \\ 8x^3 & +\ 20x^2 & +\ 12x & & \quad 1012 \\ \hline 8x^3 & +\ 24x^2 & +\ 22x & +\ 6 & \quad 10626. \end{array}$$

Checking. To check the multiplication of polynomials in x, we may substitute $x = 10$, say, and perform the corresponding arithmetical process. We should obtain for the answer the value of the product polynomial when $x = 10$. Thus the right side of example 3 is a check for the left side.

The rules for polynomials in more than one variable are similar.

EXERCISES

Multiply, and check by substitution:

1. $-2x(5y - 3z)$.

2. $3x(2y - w)$.

3. $2x(x^2 - 4x + 3)$.

4. $3x(-2x^2 + 2x - 5)$.

5. $4x(x^2 + 2xy - 3y^2)$.
6. $-3x^2y(2x^3 - 3y^2)$.
7. $(x - 2)(x - 4)$.
8. $(x - 3)(x + 2)$.
9. $(6x - 3)(x + 5)$.
10. $(2x + 3)(6x + 4)$.
11. $(2x - 1)(3x - 4)$.
12. $(3x + 2)(4x - 5)$.
13. $(x + y)(x + y)$.
14. $(x - y)(x - y)$.
15. $(x + y)(x - y)$.
16. $(x + y)(x^2 - xy + y^2)$.
17. $(x - y)(x^2 + xy + y^2)$.
18. $(x + 2y)(x^2 - xy + y^2)$.
19. $(2x - 3)(2x^2 + 5x - 3)$.
20. $(x - 2y)(x^2 + 2xy + 4y^2)$.
21. $(ax^c + b)(ax^c - b)$.
22. $(ax^c + b)(ax^c + b)$.
23. $(ax^c - b)(ax^c - b)$.
24. $(x + 2)(x^2 - 2x + 4)$.
25. $(2x^2 - 2x + 3)(3x^2 + x - 2)$.
26. $(3x^2 + 2xy - y^2)(2x^2 - xy + 2y^2)$.
27. $(x - y)(x^3 + x^2y + xy^2 + y^3)$.
28. $(2x + 3y)(4x^2 - xy + y^2)$.
29. $(x + y)(x^3 - x^2y + xy^2 - y^3)$.
30. $(3x - 2y)(x^2 - 3xy + 2y^2)$.

42. Division by Monomials

To **divide** a quantity a by a quantity b means to find a quantity c such that $bc = a$. To divide a monomial by a monomial, we divide their coefficients and treat the variable factors according to the laws for exponents.

Example 1. $6x^5/3x^2 = 2x^3$.

To check, we multiply $3x^2 \cdot 2x^3 = 6x^5$.

To divide a polynomial by a monomial, we divide each term of the polynomial by the monomial and take the algebraic sum of the results.

Example 2. $(6x^5 + 15x^4 + 12x^3 + 9x^2) \div 3x^2 = 2x^3 + 5x^2 + 4x + 3$.

To check, we multiply $3x^2(2x^3 + 5x^2 + 4x + 3) = 6x^5 + 15x^4 + 12x^3 + 9x^2$, or else check by substitution.

The justification of this process is obvious, since division has been defined as the inverse of multiplication. However, the quotient may not always be a polynomial or rational integral function, although it is necessarily a rational function; just as the quotient of two integers may fail to be an integer although it is always a rational number. For example, $4x/2x^3 = 2/x^2$ and $5 \div 3 = 5/3$. *Another concept of division for both integers and polynomials will be introduced in the next section.*

EXERCISES

Divide and check:

1. $8x^8/2x^2$. **2.** $9x^9/3x^3$. **3.** $24x^6y^{10}/8x^2y^2$.

4. $28x^4y^9 \div 7xy^4$. **5.** $(18x^3y^2 + 21x^5y^5) \div 3xy^2$.

6. $\dfrac{4x^4 - 16x^3 + 8x^2 + 4x}{4x}$. **7.** $\dfrac{16x^8 - 8x^6 + 4x^2}{-4x^2}$.

8. $\dfrac{3x^7 + 4x^6 - 8x^2}{2x}$. **9.** $\dfrac{15x^3y^4 - 10x^2y^2 - 9xy^2}{-3xy^2}$.

10. $(12x^3y^2 - 8x^2y^3 + 6x^4y^2 - 10x^2y^4) \div (-4x^2y)$.

43. The Division Algorithm*

It is a mildly unpleasant fact that the word "division" is used in two different senses in connection with integers, and in the same two senses in connection with polynomials. In the sense already taken up, to divide an integer f by an integer d ($\neq 0$) means to find a number q such that $f = dq$. In this sense the quotient $q = f/d$ may not always exist in the system of integers but always does in the system of rational numbers. Thus $7 \div 2 = 7/2$. In this sense, division is the inverse of multiplication; thus $2 \cdot \frac{7}{2} = 7$. But there is another sense given to the word *division,* in which division of integers always can be performed within the system of integers, namely, *division with a remainder.* This may be defined as follows:

DEFINITION 1. To **divide** a positive integer f, the dividend, by a positive integer d, the divisor, ($d \leqq f$), means to find a positive integer q, the quotient, and a non-negative integer r, the remainder, such that (a) $f = dq + r$, and (b) $0 \leqq r < d$.

Example 1. Dividing 7 by 2, we obtain the quotient 3 and the remainder 1, so that $7 = 2 \cdot 3 + 1$ and $1 < 2$.

Note that also $7 = 2 \cdot 2 + 3$, but we would not say that in dividing 7 by 2 we get the quotient 2 and the remainder 3, because the remainder is required to be less than the divisor.

This sense of "division" is also a reversal of multiplication.

* The word "algorithm" means "a process of computation" and is derived from the name of al-Khowarizmi, an Arab of the 9th century, who wrote a book on mathematics.

Multiplication by a positive integer is repeated addition, and this kind of division amounts to repeated subtraction as far as it can be done. Thus we can subtract three 2's from 7 and have 1 left over. In general $f = dq + r$ is equivalent to $f - dq = r$ and we subtract as many d's (q of them) as we can until a remainder less than d is left.

It is intuitively clear from Fig. 22 that q and r always exist and are uniquely determined; this could be proved but the proof will be omitted.

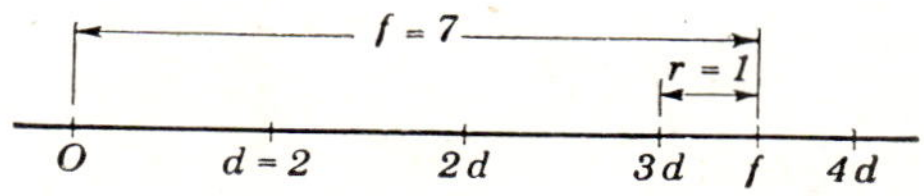

$$f = dq + r$$
$$7 = 2 \cdot 3 + 1$$

Fig. 22.

The process of long division is really repeated subtraction, as can be seen from the following example.

Example 2.

$$\begin{array}{rrl} & 2\ 3 & = q = 20 + 3 \\ d = 12 & \overline{)28\ 1} & = f \\ & 24(0) & = 20 \cdot d \\ & \overline{4\ 1} & = f - 20 \cdot d \\ & 3\ 6 & = 3d \\ & \overline{5} & = r = f - 23d. \end{array}$$

Hence $281 = 12 \cdot 23 + 5$. Thus when 281 is divided by 12, the quotient is 23 and the remainder is 5.

Note that $f = dq + r$ may be written also as $\frac{f}{d} = q + \frac{r}{d}\cdot$ The expression $q + \frac{r}{d}$ is the "quotient" in the earlier sense of division, since if we multiply it by d we get f. The former sense of division is sometimes called *exact division*. If $r = 0$, then d is a **factor** of f, or f is **exactly divisible** by d, and the two senses of division coincide in this case. Fortunately, confusion is unlikely to arise because of this double use of the same word.

As usual, polynomials are similar to integers. As the inverse of multiplication, *exact division* of a polynomial $F(x)$ by a polynomial $D(x)$ would mean finding a function $Q(x)$ such that $F(x) = D(x) \cdot Q(x)$. But exact division of polynomials is not always possible within the system of polynomials, although it is within the system of rational functions, since $Q(x) = F(x)/D(x)$ is always a rational function. Thus

$$(2x^3 + 5x^2 + 7x + 4) \div (x^2 + x + 1) = Q(x)$$

can be written as a rational function

$$\frac{2x^3 + 5x^2 + 7x + 4}{x^2 + x + 1}$$

but is not a polynomial, as will be seen below. As in the case of integers, we define division of polynomials within the system of polynomials as follows:

DEFINITION 2. To **divide** a polynomial $F(x)$, the dividend, by a polynomial $D(x)$, the divisor, means to find a polynomial $Q(x)$, the quotient, and a polynomial $R(x)$, the remainder, such that (a) $F(x) = D(x) \cdot Q(x) + R(x)$, and (b) the degree of $R(x)$ is less than the degree of $D(x)$.

This is also repeated subtraction, since (a) is equivalent to $F(x) - D(x) \cdot Q(x) = R(x)$ and we subtract as "great" a multiple of $D(x)$ from $F(x)$ as we can, so that the remainder has lower degree than $D(x)$. The work may be arranged similarly to long division of integers, as in the following example.

Example 3. To divide $2x^3 + 5x^2 + 7x + 4$ by $x^2 + x + 1$, we write

$$\begin{array}{rl}
 2x \;+ 3 & = Q(x) \\
D(x) = x^2 + x + 1 \,\overline{)\, 2x^3 + 5x^2 + 7x + 4} & = F(x) \\
\underline{2x^3 + 2x^2 + 2x } & = 2x \cdot D(x) \\
3x^2 + 5x + 4 & = F(x) - 2x \cdot D(x) \\
\underline{3x^2 + 3x + 3} & = 3 \cdot D(x) \\
2x + 1 & = R(x) = F(x) \\
& \quad - (2x + 3) \cdot D(x).
\end{array}$$

To check division, we may apply definition 2 and verify that $D(x) \cdot Q(x) + R(x) = F(x)$. Thus in example 2 we find by multiplication and addition that

$$(x^2 + x + 1)(2x + 3) + (2x + 1) = 2x^3 + 5x^2 + 7x + 4.$$

Or we may check by letting $x = 10$, say, and performing the corresponding arithmetical division. Thus, in example 3 we get

$$\begin{array}{rcrcl} & & 23 & = q = & Q(10) \\ D(10) = d = & 111 \,\big|\, & \overline{2574} & = f = & F(10) \\ & & 222 & & \\ & & \overline{\ 354} & & \\ & & 333 & & \\ & & \overline{\ \ 21} & = r = & R(10). \end{array}$$

Note that in example 3 we do not say after the first subtraction that the quotient is $2x$ and the remainder $3x^2 + 5x + 4$, because the remainder is required to have lower degree than the divisor so that further subtraction is impossible. As for integers, we may write $F(x) = D(x) \cdot Q(x) + R(x)$ as

$$\frac{F(x)}{D(x)} = Q(x) + \frac{R(x)}{D(x)}.$$

The function $Q(x) + \dfrac{R(x)}{D(x)}$ would yield $F(x)$ if multiplied by $D(x)$.

If $R(x) = 0$, then $D(x)$ is a **factor** of $F(x)$ or $F(x)$ is **exactly divisible** by $D(x)$, and the two senses of division coincide in this case.

It can be proved that the quotient polynomial $Q(x)$ and remainder polynomial $R(x)$ always exist and are uniquely determined as long as the divisor is not zero. The process of determining them, that is, the division process, is called the **division algorithm.** *Recall that it is intended that all polynomials are to be written in standard form before beginning the division process.* The procedure for polynomials in more than 1 variable is similar.

EXERCISES

Divide and check.

1. $(x^2 - 8x + 12) \div (x - 2)$.
2. $(x^2 - 2x - 15) \div (x + 3)$.
3. $(4x^2 + 4x - 3) \div (2x - 1)$.
4. $(6x^2 - 4 + 5x) \div (3x + 4)$.
5. $(2x^3 - 13x^2 + 27x - 18) \div (2x - 3)$.
6. $(3x^3 + 10x^2 + 9x + 2) \div (3x + 1)$.
7. $(3x^3 - x^2 - 22x + 24) \div (3x - 4)$.
8. $(8x^3 - 10x^2 - x + 3) \div (4x - 3)$.
9. $(5x^3 - x^2 - 5x + 1) \div (x^2 - 1)$.

10. $(5x^3 + 12x^2 - 36x - 16) \div (x^2 + 2x - 8)$.
11. $(x^3 + 8y^3) \div (x + 2y)$. **12.** $(x^3 - 64y^3) \div (x - 4y)$.
13. $(2x^3 - 7x^2y + 12xy^2 - 9y^3) \div (2x - 3y)$.
14. $(x^3 + x^2y - xy^2 + 2y^3) \div (x + 2y)$.
15. $(3x^3 - 5x^2y - 16xy^2 + 12y^3) \div (3x - 2y)$.
16. $(6x^3 - 13x^2y - 19xy^2 + 12y^3) \div (3x + 4y)$.
17. $(x^2 + 3x + 7) \div (x + 2)$.
18. $(x^2 + x - 8) \div (x - 2)$.
19. $(12x^3 + 4x^2 - 13x + 9) \div (2x - 1)$.
20. $(16x^3 + 8x^2 - 39x + 28) \div (4x - 3)$.
21. $(18x^3 - 39x^2 + 9x + 11) \div (9x^2 - 24x + 16)$.
22. $(2x^3 - 13x^2 + 29x - 20) \div (x^2 - 5x + 6)$.
23. $(3x^3 + 10x^2 + 11x + 5) \div (3x^2 + 7x + 2)$.
24. $(3x^3 - x^2 - 15x + 10) \div (x^2 + x - 6)$.
25. $(x^4 - x^2 - 10) \div (x^2 - 4)$.
26. $(x^4 + 2x^2 - 12) \div (x^2 - 3)$.
27. $(x^3 - 6x + 5) \div (x^2 + 3x - 2)$.
28. $(2x^3 + 4x^2 - 5) \div (x^2 + 3)$.

29. Show that $(x + y)$ is a factor of $(x^3 + y^3)$ and find another factor.

30. Show that $(x + y)$ is a factor of $(x^5 + y^5)$ and find another factor.

31. Show that $(x - y)$ is a factor of $(x^3 - y^3)$ and find another factor.

32. Show that $(x - y)$ is a factor of $(x^5 - y^5)$ and find another factor.

33. Show that $(x + y)$ is a factor of $(x^4 - y^4)$ and find another factor.

34. Show that $(x + y)$ is a factor of $(x^6 - y^6)$ and find another factor.

35. Show that $(x^2 + 2x + 4)$ is a factor of $(x^4 + 2x^3 - 8x - 16)$ and find another factor.

36. Show that $x^2 + x + 1$ is a factor of $(x^4 - x^3 - x^2 - 5x - 3)$ and find another factor.

★44. Non-decimal Scales of Notation

The advantages of Hindu-Arabic notation over, say, Roman numerals are due to two features. The first and less important feature is that in Hindu-Arabic notation any natural number, no matter how large, can be expressed by means of a limited number of symbols, namely the digits 0, 1, 2, 3, 4, 5, 6, 7, 8, 9,

★ This section may be omitted without disturbing the continuity.

while in the Roman system it would be impractical to express large numbers without inventing more and more symbols, such as I, V, X, L, C, D, M, ... The second and more significant feature is that the positional character of the Hindu-Arabic notation makes possible the use of the simple algebraic operations with polynomials, with slight modifications, such as carrying, designed to confine all coefficients of the polynomials to the ten digits. Thus, as explained in sections 39 and 40, we add

$$\begin{array}{rl} \overset{1\,1}{2}58 & (= 2 \cdot 10^2 + 5 \cdot 10 + 8) \\ \underline{365} & (= \underline{3 \cdot 10^2 + 6 \cdot 10 + 5)} \\ 623 & (= 5 \cdot 10^2 + 11 \cdot 10 + 13) \end{array}$$

just as the polynomials $2x^2 + 5x + 8$ and $3x^2 + 6x + 5$ except that 10 of the 13 units in the units column are carried as one 10 into the tens column, and 10 of the 12 tens then in the tens column are carried as one 10^2 into the 10^2 column, yielding $6 \cdot 10^2 + 2 \cdot 10 + 3$, or 623.

Neither of these two features makes any essential use of the particular number 10. The prevalence of 10 in number notations is doubtless analogous to the 10 fingers. A similarly convenient positional notation can be built using any other natural number greater than one. Thus, if four were used instead of 10, we would have only four digits 0, 1, 2, and 3, and the columns used in a number would be arranged as powers of 4, the first column on the right being units, the second column being fours, the third being fours-squared, and so on. Thus, the symbol 132 would be interpreted as one 4^2 plus three 4's plus 2 units, or $1 \cdot 4^2 + 3 \cdot 4 + 2$ or the value of the polynomial $1x^2 + 3x + 2$ when $x = 4$. In this so-called 4-scale, the first ten natural numbers would be written as 1, 2, 3, 10, 11, 12, 13, 20, 21, 22. Addition would be done with the help of the addition table

$$\begin{array}{llll} 0 + 0 = 0 & 1 + 0 = 1 & 2 + 0 = 2 & 3 + 0 = 3 \\ 0 + 1 = 1 & 1 + 1 = 2 & 2 + 1 = 3 & 3 + 1 = 10 \\ 0 + 2 = 2 & 1 + 2 = 3 & 2 + 2 = 10 & 3 + 2 = 11 \\ 0 + 3 = 3 & 1 + 3 = 10 & 2 + 3 = 11 & 3 + 3 = 12 \end{array}$$

together with the device of carrying just as in the decimal system. Thus the numbers written in the 4-scale as 132 and 23 are added as follows:

$$\begin{array}{rl}
\overset{1\,1}{132} & (= 1 \cdot 4^2 + 3 \cdot 4 + 2) \\
\underline{23} & (= \qquad\quad \underline{2 \cdot 4 + 3)} \\
221 & (= 1 \cdot 4^2 + 5 \cdot 4 + 5 = 2 \cdot 4^2 + 2 \cdot 4 + 1).
\end{array}$$

The same example rewritten in the familiar 10-scale (decimal system) would appear as:

$$\begin{array}{r}
30 \\
\underline{11} \\
41.
\end{array}$$

The 10-scale appears easier to us only because we were brought up with it. If we had memorized the table for the 4-scale above at an early age and used it consistently ever since, it would be just as easy for us as the decimal system seems to be.

All other operations of arithmetic can be adapted to the 4-scale or any other scale. Thus the multiplication table for the 4-scale is:

$$\begin{array}{llll}
0 \cdot 0 = 0 & 1 \cdot 0 = 0 & 2 \cdot 0 = 0 & 3 \cdot 0 = 0 \\
0 \cdot 1 = 0 & 1 \cdot 1 = 1 & 2 \cdot 1 = 2 & 3 \cdot 1 = 3 \\
0 \cdot 2 = 0 & 1 \cdot 2 = 2 & 2 \cdot 2 = 10 & 3 \cdot 2 = 12 \\
0 \cdot 3 = 0 & 1 \cdot 3 = 3 & 2 \cdot 3 = 12 & 3 \cdot 3 = 21.
\end{array}$$

Example 1. Multiply in the 4-scale the numbers which are written in the 4-scale as 32 and 23. Check by translating both factors and their product into the 10-scale.

Four-scale	*Ten-scale*
32 $(= 3 \cdot 4 + 2)$	14
23 $(= 2 \cdot 4 + 3)$	11
222	14
130	14
2122 $(= 2 \cdot 4^3 + 1 \cdot 4^2 + 2 \cdot 4 + 2)$	154

Example 2. Subtract in the 4-scale the numbers which are already written in the 4-scale: 232 − 23. Check as before.

	Four-scale	*Ten-scale*
(Note the	$2\overset{2\,1}{\not 3}2$ $(= 2 \cdot 4^2 + 3 \cdot 4 + 2)$	46
borrowing	23 $(= \qquad 2 \cdot 4 + 3)$	11
procedure)	203 $(= 2 \cdot 4^2 + 0 \cdot 4 + 3)$	35

Example 3. Divide in the 4-scale the numbers which are already written in the 4-scale: $132 \div 3$. Check as before.

Four-scale

$$\begin{array}{r@{}l}
 & 22 \quad (= 2 \cdot 4 + 2) \\
3) & \overline{132} \quad (= 1 \cdot 4^2 + 3 \cdot 4 + 2) \\
 & \underline{12} \\
 & 12 \\
 & \underline{12}
\end{array}$$

Ten-scale

$$\begin{array}{r@{}l}
 & 10 \\
3) & \overline{30} \\
 & \underline{3}
\end{array}$$

In some parts of higher mathematics, it is convenient to use the 3-scale (ternary notation) and the 2-scale (binary notation). In fact, modern computing machines (electronic brains) frequently use the binary notation because in it we use only two digits, 0 and 1. The first ten natural numbers in the binary system are written as 1, 10, 11, 100, 101, 110, 111, 1000, 1001, 1010. The addition and multiplication tables in the two-scale are particularly simple:

Addition Table		*Multiplication Table*	
$0 + 0 = 0$	$1 + 0 = 1$	$0 \cdot 0 = 0$	$1 \cdot 0 = 0$
$0 + 1 = 1$	$1 + 1 = 10$	$0 \cdot 1 = 0$	$1 \cdot 1 = 1$

Because of the use of only two digits, every number can be recorded as a succession of electric circuits (perhaps with light bulbs) for the columns, a 1 being registered as a closed circuit (bulb lit) and a 0 as an open circuit (bulb dark). Because of the simplicity of the addition and multiplication tables, electric circuits are easily devised to build these tables into the machine, so that arithmetical operations can be performed with the speed with which electricity can pass through wires and operate switching devices.

For scales greater than ten, we would have to invent some new digits. Thus, for the twelve-scale, we would need twelve digits representing natural numbers from zero up to eleven. We might use 0, 1, 2, 3, 4, 5, 6, 7, 8, 9, t, e, where the t stands for ten and e for eleven. Thus $2te$ stands for the value of the polynomial $2x^2 + tx + e$ where $x = 12$ or $2 \cdot 12^2 + t \cdot 12 + e$, which is written in the 10-scale as $288 + 120 + 11 = 419$.

Note: An easy way to convert a number from the ten-scale to the m-scale is to divide repeatedly by m until a quotient less than m is reached, using the remainders and final quotient as the digits from right to left, as in the following example.

Example. Write twenty-five in the 2-scale.

$$\begin{array}{rl} 2\overline{)25} & \text{remainder} = 1 \\ 2\overline{)12} & \text{remainder} = 0 \\ 2\overline{)6} & \text{remainder} = 0 \\ 2\overline{)3} & \text{remainder} = 1 \\ 1 & \end{array}$$

Hence in the 2-scale, twenty-five is 11001.

Sketch of proof. To express the number N in the m-scale, we divide N by m, obtaining

$$(1) \qquad N = mq_1 + r_1, \text{ where } 0 \leqq r_1 < m.$$

Then dividing the quotient q_1 by m we get

$$(2) \qquad q_1 = mq_2 + r_2, \quad 0 \leqq r_2 < m.$$

Dividing q_2 by m we get

$$(3) \qquad q_2 = mq_3 + r_3, \quad 0 \leqq r_3 < m.$$

Suppose that the divisions terminate at this stage because q_3 is less than m. Then combining (1), (2), (3), we have

$$\begin{aligned} N &= mq_1 + r_1 = m(mq_2 + r_2) + r_1 \\ &= m^2q_2 + mr_2 + r_1 = m^2(mq_3 + r_3) + mr_2 + r_1 \\ &= m^3q_3 + m^2r_3 + mr_2 + r_1. \end{aligned}$$

Then N is the value of the polynomial $q_3x^3 + r_3x^2 + r_2x + r_1$ when $x = m$ and the coefficients are all integers between 0 and $m - 1$, inclusive. They are therefore the digits in the expression of N in the m-scale.

EXERCISES

The following numbers are already written in the 4-scale. Rewrite them in the 10-scale:

1. 213. **2.** 321. **3.** 102. **4.** 1023.
5. 1102. **6.** 2031.

The following numbers are already written in the 10-scale. Rewrite them in the 4-scale:

7. 21. **8.** 33. **9.** 40. **10.** 19. **11.** 129. **12.** 73.

The following numbers are already written in the 2-scale. Rewrite them in the 10-scale:

13. 101. **14.** 110. **15.** 111010. **16.** 101010.
17. 1110. **18.** 11111.

The following numbers are already written in the 10-scale. Rewrite them in the 2-scale:

19. 12. **20.** 23. **21.** 98. **22.** 86. **23.** 57. **24.** 46.

Perform the following arithmetical operations in the 4-scale and write the answer in the 4-scale. Check by translating the two given numbers, here written in the 4-scale, and the answer into the 10-scale.

25. $\begin{array}{r} 23 \\ +33 \\ \hline \end{array}$ **26.** $\begin{array}{r} 102 \\ +213 \\ \hline \end{array}$ **27.** $\begin{array}{r} 132 \\ -13 \\ \hline \end{array}$ **28.** $\begin{array}{r} 211 \\ -32 \\ \hline \end{array}$

29. $\begin{array}{r} 32 \\ \times 23 \\ \hline \end{array}$ **30.** $\begin{array}{r} 120 \\ \times 32 \\ \hline \end{array}$ **31.** $2\overline{)320}$. **32.** $3\overline{)231}$.

Perform the following arithmetical operations in the 2-scale and write the answer in the 2-scale. Check by translating the two given numbers, here written in the 2-scale, and the answer into the 10-scale.

33. $\begin{array}{r} 101 \\ +11 \\ \hline \end{array}$ **34.** $\begin{array}{r} 1011 \\ +101 \\ \hline \end{array}$ **35.** $\begin{array}{r} 1101 \\ -111 \\ \hline \end{array}$ **36.** $\begin{array}{r} 11011 \\ -1100 \\ \hline \end{array}$

37. $\begin{array}{r} 101 \\ \times 11 \\ \hline \end{array}$ **38.** $\begin{array}{r} 110 \\ \times 101 \\ \hline \end{array}$ **39.** $11\overline{)1001}$ **40.** $101\overline{)1101}$.

41. (*a*) Express the numbers thirty-two and nineteen in the 5-scale.

(*b*) Make up and use a table of addition for the 5-scale in order to add the two numbers of part (*a*) in the 5-scale. Check by translation into the 10-scale.

(*c*) Make up and use a table of multiplication for the 5-scale in order to multiply the two numbers of part (*a*) in the 5-scale. Check by translation into the 10-scale.

Express each of the following numbers in the (*a*) *10-scale,* (*b*) *7-scale,* (*c*) *3-scale,* (*d*) *12-scale:*

42. Thirty-seven. **43.** Twenty-nine. **44.** Thirty-four.
45. Fifty-eight. **46.** Forty-seven.
47. One-hundred thirty-one. **48.** One-hundred ten.
49. One-hundred forty-three. **50.** One thousand.

Chapter VI

Factoring of polynomials

45. Prime Factors

A polynomial $D(x)$ is said to be a **factor** of a polynomial $F(x)$ if there exists a polynomial $Q(x)$ such that $F(x) = D(x) \cdot Q(x)$; that is, if $F(x)$ is exactly divisible by $D(x)$; or if the remainder is zero when $F(x)$ is divided by $D(x)$.

Thus, if two polynomials are multiplied together, either one is said to be a factor of the product.

To factor a polynomial means to express it as a product of other polynomials.

For example, $x^2 - 1 = (x + 1)(x - 1)$, as we can verify by multiplication. But although $x - 1 = (\sqrt{x} + 1)(\sqrt{x} - 1)$, this is not considered factoring, since the quantities in parentheses are not polynomials, that is, not rational integral functions.

In multiplication, we are given two factors and asked for the product. In (exact) division, we are given the product and one factor and asked for the other factor. In both situations we have a routine procedure for accomplishing our task. In factoring, we are given the product and asked for the factors from which it was compounded. Factoring, or unmultiplying, is consequently more difficult than the other two tasks, just as unscrambling eggs is more difficult than scrambling them, and chemical analysis of a compound is harder than compounding a substance from its components. In fact, we shall have to lean on our experience with multiplication in order to recognize factors, and shall develop no systematic procedure for factoring (except for one theorem to be taken up in Chapter XIX).

In fact, the question whether a polynomial has factors

other than plus and minus itself and ± 1 is not a simple question and depends on what kind of coefficients are to be permitted in the factors. Thus by multiplication we can verify that $(x + \sqrt{-1})(x - \sqrt{-1}) = x^2 + 1$. Hence $x^2 + 1$ has these linear factors if imaginary numbers are permitted in the factors. Similarly $x^2 - 2$ has the factors $(x + \sqrt{2})(x - \sqrt{2})$ if irrational numbers are permitted in the factors. Finally $x + 1$ has the factors $2(\frac{1}{2}x + \frac{1}{2})$, etc., if fractions are allowed in the factors. But neither $x^2 + 1$, $x^2 - 2$, nor $x + 1$ has any factors at all except plus and minus itself and ± 1 if we insist on integral coefficients in the factors.

In the present chapter, we shall usually consider only polynomials with integral coefficients and shall require factors to be polynomials with integral coefficients only.*

A polynomial may be called **prime** if it has no factors other than plus and minus itself and ± 1. To **factor completely** a polynomial means to express it as a product of its prime factors. Under the circumstances mentioned these are uniquely determined just as for integers (compare section 6), except for trivial changes of sign.†

We shall not have any systematic procedure for determining whether or not a polynomial is prime. It is not hard to see that $x^2 + 1, x^2 - 2, x + 1, x^2 + y^2, x + y, x^2 + xy + y^2, x^2 - xy + y^2$, etc., are prime. If we permitted coefficients other than integers, our concept of prime polynomial would be altered, but our procedures would not.

Factoring always should be checked by multiplication.

46. Special Products

Since factoring will be done largely from experience with multiplication, it is advisable to become familiar with the following types of products which are of frequent occurrence. Each can be verified by multiplication.

(1) $a(u - v + w - x) = au - av + aw - ax.$
(2) $(a + b)(a - b) = a^2 - b^2.$
(3) $(a + b)^2 = (a + b)(a + b) = a^2 + 2ab + b^2.$
(4) $(a - b)^2 = (a - b)(a - b) = a^2 - 2ab + b^2.$

* In Chapter XIX we shall permit any complex coefficients.
† Thus, $x + 2 = (-1)(-x - 2)$.

(5) $(x + b)(x + d) = x^2 + (b + d)x + bd.$
(6) $(ax + b)(cx + d) = acx^2 + (ad + bc)x + bd.$

It will be found convenient to memorize formulas (2), (3), and (4). In the case of (5) and (6) one can think mentally of the usual form of the work of multiplication. Thus in the product

$$\begin{array}{c} ax + b \\ \times \\ \underline{cx + d} \end{array}$$

it is seen that the term $(ad + bc)x$ may be called the sum of the "cross-products." Formula (5) is merely the special case of (6) in which $a = c = 1$.

Note. It is important to realize that in formulas such as those listed above, the letters a, b, etc., unlike your instructor, will stand for anything (more precisely, for any quantity). For example,

$$(2x + 3y)(2x - 3y) = 4x^2 - 9y^2$$

is of the type (2) where a stands for $2x$ and b for $3y$. Much of the superiority of algebra over arithmetic is due to the generality arising from the fact that letters often may be used to represent any quantity whatever.

EXERCISES

Multiply each of the following by inspection:

1. $2a(3x + 4y)$.
2. $-3a(x - 2y)$.
3. $-3ab^2(3a^3 - ab)$.
4. $3b(x - 2y + 3z)$.
5. $(x + 5y)(x - 5y)$.
6. $(3x + 2y)(3x - 2y)$.
7. $(1 - 4x^3)(1 + 4x^3)$.
8. $(3 - 4xy)(3 + 4xy)$.
9. $(2x + 3)^2$.
10. $(3x + 5y)^2$.
11. $(2x - 3y)^2$.
12. $(2ax + b)^2$.
13. $(x^3 - 3y^3)^2$.
14. $(3x^3c - ay)^2$.
15. $(4x^2 - 5y^3)^2$.
16. $(x + 2)(x + 4)$.
17. $(x + 3)(x + 5)$.
18. $(x - 2)(x + 4)$.
19. $(x + 2)(x - 4)$.
20. $(x - 2)(x - 4)$.
21. $(2u - 3)(4u - 1)$.
22. $(2y - 1)(y + 4)$.
23. $(3u - 2)(4u + 3)$.
24. $(3x - 2y)(4x + 5y)$.
25. $(2x + 5y)(x - 3y)$.
26. $(4x^2 - 3y^2)(2x^2 + y^2)$.
27. $(2s^2 - 3t^2)(4s^2 + 5t^2)$.
28. $[(x + 3y) - 5][(x + 3y) + 5]$.
29. $[(x + 2y) - z][(x + 2y) + z]$.
30. $[(x + 3) + 3y]^2$.
31. $[(2x + 3) - 4y]^2$.

32. $[(x + y) + z][(x + y) - z]$.
33. $(x + 2y + 3z)(x + 2y - 3z)$.
34. $(x - y - z)^2$. **35.** $(3x - 2y - 4)^2$. **36.** $(2x - 3y + 5)^2$.
37. $[(x + y) + (u - v)][(x + y) - (u - v)]$.
38. $[(x - y) - (u + v)][(x - y) + (u + v)]$.
39. $(x^a + y^b)^2$. **40.** $(x^a - y^b)^2$.
41. $(x^a + y^b)(x^a - y^b)$. **42.** $(ax^b + cy^d)^2$.

47. Simple Types of Factoring

The examples of factoring taken up in this section are done merely by recognizing the type of product which fits the given expression. To check the factoring, we multiply out the factors and see if we get back the original expression.

Case 1. Monomial common factor. Since $a(b - c + d - e) = ab - ac + ad - ae$, we may take out the greatest possible common factor from an algebraic sum of terms.

Example 1. $2x^3y^2 - 6x^2y^3 + 4x^2y^2 = 2x^2y^2(x - 3y + 2)$.

Case 2. Difference of two squares. Since $(a + b)(a - b) = a^2 - b^2$, the difference of the squares of two quantities may be expressed as the product of the sum of the two quantities by their difference.

Example 2. $4x^2 - 9y^2 = (2x)^2 - (3y)^2 = (2x + 3y)(2x - 3y)$.

Case 3. Trinomial perfect squares. Since $(a + b)^2 = a^2 + 2ab + b^2$ and $(a - b)^2 = a^2 - 2ab + b^2$, we may factor these right members by setting them equal to the left members.

Example 3. $4x^2 + 12x + 9 = (2x)^2 + 2(2x) \cdot 3 + 3^2 = (2x + 3)^2$.

Example 4. $9x^2 - 30xy + 25y^2 = (3x)^2 - 2 \cdot (3x)(5y) + (5y)^2 = (3x - 5y)^2$.

Case 4. Sum and difference of two cubes. By multiplication we find that $(a + b)(a^2 - ab + b^2) = a^3 + b^3$ and $(a - b)(a^2 + ab + b^2) = a^3 - b^3$. Hence we may factor the right members by setting them equal to the left members.

Example 5. $8x^3 + 27y^3 = (2x)^3 + (3y)^3 = (2x + 3y)(4x^2 - 6xy + 9y^2)$.

Example 6. $8x^3 - 27y^3 = (2x)^3 - (3y)^3 = (2x - 3y)(4x^2 + 6xy + 9y^2)$.

Case 5. Combination of different types. It is advisable to take out the greatest possible common factor first, and then apply whatever types of factoring seem appropriate.

Example 7.

$$\begin{aligned} 2x^7y^2 - 2x^3y^2 = 2x^3y^2(x^4 - 1) &= 2x^3y^2(x^2 + 1)(x^2 - 1) \\ &= 2x^3y^2(x^2 + 1)(x + 1)(x - 1). \end{aligned}$$

It is important to note that the letters a, b, etc., in the typical formulas above may stand for any quantities. See the note of section 46.

EXERCISES

Factor each of the following completely, and check:

1. $2ax - 8a^3$. **2.** $4x^2y + 10xy^3$.
3. $4x^3y - 6xy^3 + 8x^2y^2$. **4.** $10x^3y^2 - 16x^2y^2 - 4x^2y^3$.
5. $9x^2 - 16y^2$. **6.** $25x^2 - 9$. **7.** $4 - 49x^2y^2$.
8. $1 - 16a^2x^2$. **9.** $81u^2 - 64v^2$. **10.** $36m^2 - 121n^2$.
11. $9x^2 + 12xy + 4y^2$. **12.** $16x^2 + 8x + 1$.
13. $25x^2 - 20xy + 4y^2$. **14.** $25x^2 - 30xy + 9y^2$.
15. $x^2y^2 - 16xy + 64$. **16.** $9x^2 - 30x + 25$.
17. $4a^2x^2 + 4abx + b^2$. **18.** $b^2r^2 + 2bars + a^2s^2$.
19. $x^2 + \frac{b}{a}x + \frac{b^2}{4a^2}$. **20.** $16r^2 - 88rs + 121s^2$.
21. $25x^3 - 4xy^2$. **22.** $50x^2 - 32y^2$. **23.** $18x^4 - 8x^2y^2$.
24. $121x^2y^2 - 49z^2$. **25.** $x^4 - 81$.
26. $16x^4 - y^4$. **27.** $27x^3 + y^3$. **28.** $8p^3q^3 + 27$.
29. $x^3 - 64y^3$. **30.** $1 - 8y^3$. **31.** $64x^6 - y^6$.
32. $x^6 - 729y^6$. **33.** $2a(x + 2y) - b(x + 2y)$.
34. $5(x - y) + 3a(x - y)$. **35.** $(x + 2y)^2 - 16$.
36. $(4x - 3y)^2 - 4z^2$. **37.** $9x^2 - (y - z)^2$.
38. $x^2 - (4y - 3z)^2$. **39.** $(x - 3y)^2 - 4(x - 3y) + 4$.
40. $(3x - y)^2 + 6(3x - y) + 9$.
41. $(2x - y)^2 - (x + 3y)^2$. **42.** $(x - 3y)^2 - (2x - y)^2$.
43. $2x^2y^4 - 54x^5y$. **44.** $3p^6q - 192p^3q^4$.
45. $(x^2 + 4x + 4) - (y - 1)^2$.
46. $(9x^2 - 6x + 1) - (2y - 3)^2$.
47. $(2x + y)^3 - 8y^3$. **48.** $(x - 2y)^3 + 8y^3$.
49. $(x + 2)(x - 3y)^2 - (x + 2)(2x - y)^2$.
50. $(2x + y)(2x - y)^2 - (2x + y)(x - 3y)^2$.
51. $5x^5y - 80xy^5$. **52.** $2x^3y^2 - 12x^2y^2 + 18xy^2$.
53. $3x^5y + 24x^2y^4$. **54.** $2x^5y^2 - 162xy^2$.

48. Factoring by Grouping

Sometimes various terms may be grouped together in such a way that one of the preceding types then applies.

Example 1. $ax + bx - ay - by = (a + b)x - (a + b)y$. Now $(a + b)$ is a common factor, so that the original expression may be factored into $(a + b)(x - y)$.

Example 2. $x^2 + 4x + 4 - y^2 + 6yz - 9z^2 = (x^2 + 4x + 4) - (y^2 - 6yz + 9z^2) = (x + 2)^2 - (y - 3z)^2 = [(x + 2) + (y - 3z)][(x + 2) - (y - 3z)] = [x + 2 + y - 3z][x + 2 - y + 3z]$.

It is sometimes possible to add and subtract the same quantity so that the preceding method may then be applied.

Example 3. To factor $x^4 + x^2y^2 + y^4$ we add and subtract x^2y^2, obtaining $x^4 + 2x^2y^2 + y^4 - x^2y^2 = (x^2 + y^2)^2 - x^2y^2 = (x^2 + y^2 + xy)(x^2 + y^2 - xy)$.

EXERCISES

Factor each of the following, and check:

1. $ax + 5x - ay - 5y$.
2. $3bx + 3by + cx + cy$.
3. $x^2 - 4x - xy + 4y$.
4. $x^3 + 4x^2 - 2xy - 8y$.
5. $2x^3 + 5x^2 - 4x - 10$.
6. $x^3 + 3x^2 + 3x + 9$.
7. $x^2 - 9y^2 + x - 3y$.
8. $x^2 - y^2 - 5x + 5y$.
9. $x^2 + 8x + 16 - y^2$.
10. $3p^3 + 3p^2 - p - 1$.
11. $4x + y - 12x^2 - 3xy$.
12. $5p^2 - 5q^2 + p + q$.
13. $x^2 + 6xy + 9y^2 - z^2 - 2z - 1$.
14. $x^3 - x^2y + 5xy^2 - 5y^3$.
15. $x^4 - 3x^2y^2 + y^4$.
16. $x^4 - 11x^2y^2 + y^4$.
17. $x^4 + x^2 + 1$.
18. $x^4 - 6x^2y^2 + y^4$.
19. $x^4 + 4$.
20. $x^4 + 64$.
21. $x^4 + 2x^2 + 9$.
22. $x^4 + 3x^2 + 4$.
23. $p^4 - 7p^2q^2 + 9q^4$.
24. $u^4 + 64v^4$.
25. $p(x - y) + q(y - x)$.
26. $5a(3x - y) - 3(y - 3x)$.

49. Quadratic Trinomials

If a quadratic trinomial is not prime, it may be factored by trial, as follows.

Example 1. To factor $x^2 - 5x + 6$ we realize that, if it can be factored, it must have factors of the form $(x + b)(x + d)$. Since $(x + b)(x + d) = x^2 + (b + d)x + bd$, we must find integers b and d whose product is 6 and whose sum is -5. Trying various pairs of factors of 6, we arrive at $(x - 2)(x - 3)$.

Example 2. To factor $4x - 11x + 6^2$ we realize that, if it can be factored, it must have factors of the form $(ax + b)(cx + d)$. Since $(ax + b)(cx + d) = acx^2 + (ad + bc)x + bd$, we must find integral values of a, b, c, and d such that $ac = 4$, $bd = 6$, and $ad + bc = -11$. After trying various possibilities, we find the factors $(4x - 3)(x - 2)$.

It is important to check by multiplying out the factors, paying careful attention to the sum of the cross-products.

Note. Examples of this type may be done also by the method of grouping. Thus, $x^2 - 5x + 6$ may be factored by writing $x^2 - 5x + 6 = x^2 - 3x - 2x + 6 = x(x - 3) - 2(x - 3) = (x - 2)(x - 3)$. Similarly, example 2 can be done by observing that $4x^2 - 11x + 6 = 4x^2 - 3x - 8x + 6 = (4x - 3)x - (4x - 3)2 = (4x - 3)(x - 2)$.

EXERCISES

Factor each of the following, if possible, and check:

1. $x^2 - 8x + 12$. **2.** $x^2 + 7x + 12$. **3.** $x^2 - 2x - 15$.
4. $x^2 + x - 12$. **5.** $x^2 + 3x - 18$. **6.** $x^2 - 21x - 22$.
7. $x^2 + x + 3$. **8.** $p^2 + 9p - 36$. **9.** $k^2 - 16k - 36$.
10. $x^2 - x - 5$. **11.** $x^2 - 10xy - 24y^2$.
12. $p^2 - 22pq - 48q^2$. **13.** $m^2 - 16m + 48$.
14. $m^2 - 8mn - 48n^2$. **15.** $2x^2 - 9x + 9$.
16. $3x^2 + 7x + 2$. **17.** $4m^2 + 8mn - 5n^2$.
18. $6p^2 - 19pq + 3q^2$. **19.** $2x^2 + x - 5$.
20. $6x^2 + 5xy - 4y^2$. **21.** $8x^2 - 2x - 3$.
22. $2x^2 + 7x + 3$. **23.** $4x^2 + 5xy - 6y^2$.
24. $12x^2 - 8xy - 15y^2$. **25.** $(x + 2)^2 - 5(x + 2) + 6$.
26. $(x + y)^2 - 7(x + y) + 12$.
27. $(2x - y)^2 + 9(2x - y) - 36$.
28. $(x - 3y)^2 + 7(x - 3y) - 18$.
29. $6(3x - 5)^2 - 5(3x - 5) - 4$.
30. $4(x - 2y)^2 + 5(x - 2y) - 6$.
31. $(x^2 - 8x + 16) - 7(x - 4) + 12$.
32. $(x^2 - 10xy + 25y^2) + 16(x - 5y) - 36$.
33. $12x^3y^2 + 2x^2y^3 - 4xy^4$. **34.** $2x^3 + 4x^2 - 42x$.
35. $4x^4y - 2x^3y^2 - 20x^2y^3$. **36.** $12x^3y - 27x^2y^2 + 6xy^3$.
37. $x^4 + 8x^2 + 12$. **38.** $x^4 - 7x^2 + 12$.
39. $x^6 - 7x^3 - 8$. **40.** $x^6 - 18x^3 - 40$.
41. $2x^4 - 7x^2y^2 - 4y^4$. **42.** $2x^4 - 14x^2y^2 - 36y^4$.

★50. Sum and Difference of Two Like Powers

The following theorems can be proved by methods taken up later; see exercise 25, section 116, and exercise 18, section 101.

THEOREM A. *If n is any positive integer, $x^n - y^n$ has the factor $x - y$.*

THEOREM B. *If n is an even positive integer, $x^n - y^n$ has the factor $x + y$.*

THEOREM C. *If n is an odd positive integer, $x^n + y^n$ has the factor $x + y$.*

THEOREM D. *If n is an even positive integer, neither $x + y$ nor $x - y$ is a factor of $x^n + y^n$.*

Another factor, if one factor is known, may be found by division.

Example 1. $x^5 + 32y^5$ has the factor $x + 2y$ by theorem C. Dividing $x^5 + 32y^5$ by $x + 2y$, we find that $(x^5 + 32y^5) = (x + 2y)(x^4 - 2x^3y + 4x^2y^2 - 8xy^3 + 16y^4)$.

Example 2. $64x^6 - y^6$ has the factor $2x + y$, by theorem B, and the factor $2x - y$ by theorem A. Hence, dividing by $(2x + y)(2x - y) = 4x^2 - y^2$ we obtain the factors $(2x + y)(2x - y)(16x^4 + 4x^2y^2 + y^4)$. The last factor can be factored further as follows: $16x^4 + 4x^2y^2 + y^4 = 16x^4 + 8x^2y^2 + y^4 - 4x^2y^2 = (4x^2 + y^2)^2 - 4x^2y^2 = (4x^2 + y^2 - 2xy)(4x^2 + y^2 + 2xy)$. Hence $64x^6 - y^6 = (2x + y)(2x - y)(4x^2 + y^2 - 2xy)(4x^2 + y^2 + 2xy)$.

This example could be done also by writing $64x^6 - y^6 = (8x^3 + y^3)(8x^3 - y^3)$ and factoring each of these factors by theorems A and C.

Some expressions of the form $x^n + y^n$ with even n can be factored, as in the following examples.

Example 3. $x^6 + y^6 = (x^2)^3 + (y^2)^3 = (x^2 + y^2)(x^4 - x^2y^2 + y^4)$.

Example 4. $x^4 + 4y^4 = x^4 + 4x^2y^2 + 4y^4 - 4x^2y^2 = (x^2 + 2y^2)^2 - 4x^2y^2 = (x^2 + 2y^2 + 2xy)(x^2 + 2y^2 - 2xy)$.

★ This section may be omitted without disturbing the continuity of the chapter.

EXERCISES

Factor each of the following, and check:

1. $x^5 + y^5$. **2.** $x^5 - y^5$. **3.** $x^6 - y^6$.
4. $a^5x^5 + 32y^5$. **5.** $x^4 - 81y^4$. **6.** $x^5 - 32$.
7. $x^5y^5 + 243$. **8.** $128x^7 - 1$. **9.** $64x^6 + 1$.
10. $a^4 + 64b^4$. **11.** $64a^6 - b^6$. **12.** $p^7 + q^7$.
13. $256m^8 - n^8$. **14.** $8x^3 + 27h^3$.

51. H.C.F. and L.C.M.

The **highest common factor** (H.C.F.) of several polynomials (with integral coefficients) is the polynomial of highest degree, with the largest coefficients in absolute value, which is a factor of all the given polynomials.

To find the H.C.F. of several polynomials, we may first express each of them as a product of its prime factors. Then the H.C.F. is the product of all the different prime factors common to all the given polynomials, each taken the least number of times it occurs in any of the given polynomials. See section 6.

Example 1. To find the H.C.F. of $6x^3y^2 - 6xy^4$ and $10x^3y + 20x^2y^2 + 10xy^3$ we write $6x^3y^2 - 6xy^4 = 6xy^2(x^2 - y^2) = 6xy(x + y)(x - y)$ and $10x^3y + 20x^2y^2 + 10xy^3 = 10xy(x^2 + 2xy + y^2) = 10xy(x + y)(x + y)$. Hence the H.C.F. is $2xy(x + y)$.

The **lowest common multiple** (L.C.M.) of several polynomials (with integral coefficients) is the polynomial of lowest degree, with the smallest coefficients in absolute value, of which each of the given polynomials is a factor.

To find the L.C.M. of several polynomials we may first express each of them as a product of its prime factors. Then the L.C.M. is the product of all the different prime factors occurring in the given polynomials, each taken the greatest number of times it occurs in any of the given polynomials.

Example 2. The L.C.M. of the two polynomials in example 1 is $30xy^2(x + y)^2(x - y)$.

Since we are confining ourselves to polynomials with integral coefficients in this chapter, the H.C.F. and L.C.M. are unique except for a factor ± 1. The definitions and procedures here described are analogous to those for integers discussed in section 6. As proved there for integers, we may prove the following theorem:

If L is the L.C.M. of two polynomials A and B and if H is their H.C.F., then $LH = \pm AB$, or $L = \pm AB/H$. This may be used as a check; thus in examples 1 and 2 above we find that $LH = AB = 60x^2y^3(x + y)^3(x - y)$.

EXERCISES

Find the H.C.F. and the L.C.M. of each of the following sets of polynomials, leaving answers in factored form:

1. $8x^3y$, $6xy^2$.
2. $12x^3y^2$, $8xy^3$.
3. $9x^2y^3$, $2x^2y^2$, $3x^4y$.
4. $8x^4y^5$, $4x^2$, y^3, $24x^3y^2$.
5. $5x - 15$, $6x - 18$.
6. $4x + 8$, $6x + 12$.
7. $3x - 6$, $2x^2 - 8$.
8. $4x - 12$, $6x^2 - 54$.
9. $x^2 - y^2$, $x^2 - 2xy + y^2$.
10. $x^2 - 9$, $x^2 + 6x + 9$.
11. $x^2 - 9$, $x^2 - 5x + 6$.
12. $x^2 + x - 2$, $x^2 - x - 6$.
13. $x^2 - 9$, $x^3 + 27$, $x^2 + x - 6$.
14. $x^3 - 8$, $x^2 - 4$, $x^2 - 4x + 4$.
15. $x^2 - 4$, $x^2 - 5x + 6$, $x^3 + 8$.
16. $2x^2 - 12x + 18$, $4x^2 - 36$, $2x^2 + 4x - 30$.
17. $x^3 - y^3$, $x^2 - y^2$, $x^2 - 2xy + y^2$.
18. $x^2 - 5xy + 6y^2$, $x^2 - 4y^2$, $x^2 + 4xy + 4y^2$.
19. $x^4 - y^4$, $x^2 + y^2$, $x^2 + xy - 2y^2$.
20. $x^4 - y^4$, $x^2 - y^2$, $x^2 + 2xy - 3y^2$.
21. $9 - 4x^2$, $2x^2 + x - 6$, $8x^2 - 12x$.
22. $3x - 6$, $4 - x^2$, $x^2 - x - 2$.
23. $3x^2 - 6xy$, $x^2 - 4y^2$, $2x^2 - 8xy + 8y^2$.
24. $8x^3 - 27y^3$, $4x^2 - 9y^2$, $12xy - 8x^2$.

Chapter VII

Elementary operations with fractional expressions

Since a fractional expression or (fractional) rational function becomes a number for any allowable value of the variables, all definitions concerning and operations with fractional expressions are similar to those for numerical fractions. Therefore the student should review carefully sections 7 to 11 inclusive before proceeding with this chapter. Fractional expressions are also referred to as algebraic fractions or simply as fractions.

52. Reduction to Lowest Terms

In Chapter I it was pointed out that *a non-zero common factor may be either introduced into or cancelled from both numerator and denominator of a fraction;* in symbols,

$$(1) \qquad \frac{ac}{bc} = \frac{a}{b}, \text{ provided } c \neq 0.$$

The same statement therefore may be made for fractional expressions. When all common factors except ± 1 have been cancelled from numerator and denominator of a fractional expression, it is said to be **reduced to lowest terms** or **expressed in simplest form.** Hence *to reduce a fractional expression to lowest terms, express numerator and denominator as products of their prime factors and cancel common factors.*

Example 1. $\dfrac{2x^3y^2z}{6x^2yz^3} = \dfrac{xy}{3z^2}.$

Example 2. $\dfrac{x^2 - 7x + 12}{x^2 - 5x + 6} = \dfrac{(x - 4)(x - 3)}{(x - 2)(x - 3)} = \dfrac{x - 4}{x - 2}.$

Note that the expression obtained by reducing to lowest terms is not always quite the same function as that with which we start. Thus in example 1, the reduced fractional expression is defined, and has the value 0 when $x = y = 0$, $z \neq 0$, while the original expression is undefined for such values of the variable. Similarly in example 2, the reduced fractional expression is defined and has the value -1 for $x = 3$, while the original expression is undefined for $x = 3$. This is, however, a subtle difference, which usually may be overlooked without fear by the beginning student.

Recall that $\frac{-a}{b} = \frac{a}{-b} = -\frac{a}{b}$ while $\frac{-a}{-b} = \frac{a}{b}$ (see (4), section 10). It follows that, if we think of three signs attached to every fraction, one before the numerator, one before the denominator, and one before the entire fraction, as $+\frac{+a}{+b}$, *any two of the three signs may be changed without altering the value of the fraction.*

Example 3. $\frac{x - y}{2(y - x)} = -\frac{x - y}{2(x - y)} = -\frac{1}{2}$, since $-(y - x) = x - y$.

EXERCISES

Reduce to lowest terms, leaving the answer in factored form:

1. $\frac{9x}{15x}$.
2. $\frac{14y}{16y}$.
3. $\frac{12x^2y^2}{18x^3y}$.
4. $\frac{16x^2y}{22xy}$.
5. $\frac{21ax^2y^3}{14axy}$.
6. $\frac{20xy^2}{16xy^6}$.
7. $\frac{18(a + b)}{15a(a + b)}$.
8. $\frac{6a^2(2x + y)}{8a(2x + y)}$.
9. $\frac{15xy^3(x - y)}{9xy(x - y)}$.
10. $\frac{12xy^2(2x - y)}{15x^3y^2(2x - y)}$.
11. $\frac{4x + 4xy^2}{3 + 3y^2}$.
12. $\frac{8x^2y - 4xy}{6x - 3}$.
13. $\frac{4x^2 - 4y^2}{3x + 3y}$.
14. $\frac{2x^2 - 2y^2}{5x - 5y}$.
15. $\frac{x^2 - 8x + 16}{x^2 - 16}$.
16. $\frac{x^2 - 36y^2}{x^2 + 12xy + 36y^2}$.
17. $\frac{4x^2 - 20x}{x^2 - 4x - 5}$.
18. $\frac{x^2 + x - 20}{2x^2 + 10x}$.
19. $\frac{2x^2 - 13x + 6}{2x^2 - 7x + 3}$.
20. $\frac{6x^2 - 7x - 3}{12x^2 - 16x - 3}$.
21. $\frac{x^3 - y^3}{x^2 - y^2}$.

22. $\dfrac{x^2 - 9y^2}{x^3 + 27y^3}$.

23. $\dfrac{x^2 - 2xy + y^2}{3y - 3x}$.

24. $\dfrac{x^2 - 2x - 8}{16 - 4x}$.

25. $\dfrac{1 - 27x^3}{9x^2 - 1}$.

26. $\dfrac{x^4 - 16}{(x^2 + 4)(2 - x)}$.

27. $\dfrac{2x^2 - 8y^2}{40y^3 - 5x^3}$.

28. $\dfrac{-x^2 + xy + 6y^2}{5xy - 6y^2 - x^2}$.

29. $\dfrac{15y^2 - 13xy + 2x^2}{2x^2 - 16xy + 30y^2}$.

30. $\dfrac{6x^2 - 35y^2 - xy}{15x^2 + 14y^2 - 41xy}$.

53. Multiplication and Division of Fractional Expressions

By definition 4 (section 7), *the product of two fractions is the fraction whose numerator is the product of the two given numerators and whose denominator is the product of the two given denominators;* in symbols, $\dfrac{a}{b} \cdot \dfrac{c}{d} = \dfrac{ac}{bd}$. Hence to multiply two fractional expressions we have only to follow this procedure, and then, in order to put the answer in simplest form, reduce to lowest terms as in the preceding section.

Example 1. $\dfrac{2}{3} \cdot \dfrac{9}{4} = \dfrac{2 \cdot 3 \cdot 3}{3 \cdot 2 \cdot 2} = \dfrac{3}{2}$.

Example 2. $\dfrac{2x^2y^3z}{3xy^2z^3} \cdot \dfrac{9xyz}{4x^2yz} = \dfrac{3y}{2z^2}$.

Example 3.
$$\frac{x^2 - 5x + 6}{x^2 - 6x + 9} \cdot \frac{x(x - 3)}{x^2 - 4x + 4} = \frac{(x - 2)(x - 3)x(x - 3)}{(x - 3)(x - 3)(x - 2)(x - 2)} = \frac{x}{x - 2}.$$

By theorem 2 (section 7) *the quotient of two fractions may be obtained by inverting the divisor and multiplying.*

Example 4. $\dfrac{2}{3} \div \dfrac{4}{9} = \dfrac{2}{3} \cdot \dfrac{9}{4} = \dfrac{3}{2}$.

Example 5. $\dfrac{3x^2y}{4z^3} \div \dfrac{6x^2y^2}{2z^2} = \dfrac{3x^2y}{4z^3} \cdot \dfrac{2z^2}{6x^2y^2} = \dfrac{1}{4yz}$.

Example 6.
$$\frac{x^2 - 5x + 6}{x^2 - x - 6} \div \frac{4 - x^2}{x^2 + 4x + 4}$$
$$= \frac{(x - 2)(x - 3)}{(x - 3)(x + 2)} \cdot \frac{(x + 2)(x + 2)}{-(x^2 - 4)}$$
$$= -\frac{(x - 2)(x - 3)(x + 2)(x + 2)}{(x - 3)(x + 2)(x + 2)(x - 2)} = -1.$$

EXERCISES

Perform the indicated operations and simplify:

1. $\frac{3}{7} \cdot \frac{35}{12}$. **2.** $\frac{7}{4} \cdot \frac{12}{14}$. **3.** $6 \div \frac{2}{3}$. **4.** $\frac{4}{3} \div 12$.

5. $\dfrac{4xy}{3z} \cdot \dfrac{12z^3}{10x^2y^3}$. **6.** $\dfrac{8x^3yz^2}{12a^2b^3c} \cdot \dfrac{16ab^4c^2}{6xy^3z}$.

7. $\dfrac{15a^3b^2c}{2xyz^3} \div \dfrac{5a^2b^3c^2}{6x^2yz^2}$. **8.** $\dfrac{24a^3b^4c}{15xyz^4} \div \dfrac{18ab^3c^2}{25xy^3z}$.

9. $\dfrac{ab + ac}{2a^3y} \div \dfrac{4b + 4c}{6ay^2}$. **10.** $\dfrac{2x - 6}{6x^2y} \div \dfrac{5x - 10}{30xy}$.

11. $\dfrac{2x^2 - 18}{3x} \div (x - 3)$. **12.** $\dfrac{3x - 6}{x^2 - 5x + 6} \div \dfrac{1}{(x - 3)}$.

13. $\dfrac{x - y}{x + 3y} \cdot \dfrac{x^2 - 9y^2}{x^2 - y^2}$. **14.** $\dfrac{4 - x^2}{x^3 + 5x^2} \cdot \dfrac{5x^2}{x^2 + 5x + 6}$.

15. $-\dfrac{x^2 - x - 6}{4 - x^2} \div \dfrac{x^2 - 8x + 15}{x^3 - 2x^2}$.

16. $\dfrac{x^2 - xy - 6y^2}{x^2 - xy - 12y^2} \div \dfrac{x^2 - xy - 2y^2}{x^2 - 3xy - 4y^2}$.

17. $\dfrac{x^2 + 5x + 6}{x^2 - 1} \div \dfrac{x^2 - 9}{x^2 - 2x - 3}$.

18. $\dfrac{x^2 - 2xy - 3y^2}{x^2 - 9y^2} \div \dfrac{x^2 + 5xy}{x^2 + 8xy + 15y^2}$.

19. $\dfrac{y^2 - 4x^2}{2x^2 - 7xy + 3y^2} \cdot \dfrac{2xy - x^2}{6x^2 + 13xy + 5y^2} \div \dfrac{2x^2 - 4xy}{3x^2 - 4xy - 15y^2}$.

20. $\dfrac{x^2 - x - 20}{x^2 + 2x - 8} \cdot \dfrac{x^2 + 5x}{x^2 - 25} \div \dfrac{x^2 - x - 2}{x + 1}$.

21. $\dfrac{x^2 + 7xy + 10y^2}{x^2 - 5xy + 6y^2} \cdot \dfrac{x^2 + xy - 6y^2}{x^2 - 25y^2} \div \dfrac{x^2 + 5xy + 6y^2}{x^2 - 3xy - 10y^2}$.

22. $\dfrac{x^3 - y^3}{9x^2 - 4y^2} \cdot \dfrac{3x^2 + xy - 2y^2}{x^2 - y^2} \div (x^2 + xy + y^2)$.

23. $\dfrac{4x^2 - 4xy + y^2}{3x^2 + 4xy - 4y^2} \cdot \dfrac{x^2 - 4y^2}{3x^2 + 13xy - 10y^2} \div \dfrac{2x^2 - 5xy + 2y^2}{9x^2 - 12xy + 4y^2}$.

24. $\dfrac{x^2 - 4xy + 4y^2}{x^2 - 7xy + 12y^2} \cdot \dfrac{x^2 - 6xy + 9y^2}{x^2 - 4y^2} \div \dfrac{x^2 - 5xy + 6y^2}{x^2 - 2xy - 8y^2}$.

25. $\dfrac{\dfrac{2x^2 - x - 10}{6x^2 - 19x + 10}}{x^2 - 4}$. **26.** $\dfrac{2x^2 - x - 10}{\dfrac{6x^2 - 19x + 10}{x^2 - 4}}$.

27. $\dfrac{\dfrac{6x^2 - xy - y^2}{4x^2 - y^2}}{6x^2 + 5xy + y^2}$. **28.** $\dfrac{6x^2 - xy - y^2}{\dfrac{4x^2 - y^2}{6x^2 + 5xy + y^2}}$.

29. $\dfrac{\dfrac{15x^2 - 11x - 12}{9x^2 - 16}}{\dfrac{25x^2 + 5x - 6}{15x^2 + 14x - 8}}.$ **30.** $\dfrac{\dfrac{2ax^2 + 2bxy}{3ax^2 - 6abxy + 3ab^2y^2}}{\dfrac{ax^2 + abxy + bxy + b^2y^2}{x^2 - b^2y^2}}.$

54. Addition and Subtraction of Fractional Expressions

By section 8, *to add or subtract two or more fractions having the same denominator, we form the fraction whose numerator is the algebraic sum of the given numerators and whose denominator is the common denominator.*

Example 1. $\dfrac{2}{6} + \dfrac{3}{6} = \dfrac{2 + 3}{6} = \dfrac{5}{6}.$

Example 2. $\dfrac{2}{6} - \dfrac{3}{6} = \dfrac{2 - 3}{6} = -\dfrac{1}{6}.$

Example 3. $\dfrac{3}{x + 3} - \dfrac{3x - 5}{x + 3} + \dfrac{5x - 2}{x + 3}$

$$= \frac{3 - (3x - 5) + (5x - 2)}{x + 3} = \frac{3 - 3x + 5 + 5x - 2}{x + 3}$$

$$= \frac{2x + 6}{x + 3} = \frac{2(x + 3)}{x + 3} = 2.$$

To add or subtract two or more fractions having different denominators, we express the given fractions as equal fractions having a common denominator, by multiplying their numerators and denominators by suitable factors, and then proceed as above.

Example 4. $\dfrac{1}{3} + \dfrac{1}{2} = \dfrac{1 \cdot 2}{3 \cdot 2} + \dfrac{3 \cdot 1}{3 \cdot 2} = \dfrac{2}{6} + \dfrac{3}{6} = \dfrac{5}{6}.$

Example 5. $\dfrac{1}{2} - \dfrac{1}{3} = \dfrac{1 \cdot 3}{2 \cdot 3} - \dfrac{2 \cdot 1}{2 \cdot 3} = \dfrac{3}{6} - \dfrac{2}{6} = \dfrac{1}{6}.$

Example 6. $\dfrac{x}{x - 2} - \dfrac{2}{x + 2} + \dfrac{-8}{x^2 - 4}$

$$= \frac{x}{x - 2} - \frac{2}{x + 2} + \frac{-8}{(x + 2)(x - 2)}$$

$$= \frac{x(x + 2)}{(x - 2)(x + 2)} - \frac{2(x - 2)}{(x - 2)(x + 2)} + \frac{-8}{(x + 2)(x - 2)}$$

$$= \frac{x(x + 2) - 2(x - 2) - 8}{(x + 2)(x - 2)} = \frac{x^2 - 4}{(x + 2)(x - 2)}$$

$$= \frac{(x + 2)(x - 2)}{(x + 2)(x - 2)} = 1.$$

By the **least common denominator** (L.C.D.) of several fractions is meant the L.C.M. of their denominators. It is usually desirable, although not essential, to use the L.C.D. in obtaining the algebraic sum of fractional expressions. In any case answers should be reduced to lowest terms.

EXERCISES

Perform the indicated operations and simplify:

1. $\frac{4}{16} + \frac{7}{16} - \frac{9}{16}$.
2. $\frac{3}{12} - \frac{7}{12} + \frac{8}{12}$.
3. $\frac{1}{4} + \frac{3}{2} - \frac{3}{8}$.
4. $\frac{3}{8} - \frac{1}{6} + \frac{3}{4}$.
5. $\frac{2x}{3} - \frac{x}{6} + \frac{x}{2}$.
6. $\frac{5x}{4} + \frac{3x}{8} - \frac{x}{2}$.
7. $\frac{1}{a} + \frac{3}{2a} + \frac{a}{3}$.
8. $\frac{2a}{5} + \frac{5}{2a} - \frac{3}{a}$.
9. $\frac{1}{a} + \frac{1}{b} - \frac{1}{c}$.
10. $\frac{2}{x} - \frac{3}{y} + \frac{1}{z}$.
11. $\frac{5}{3x} + \frac{2}{3x^2} - \frac{x}{6x^2}$.
12. $\frac{3}{2x} - \frac{x}{3} - \frac{1}{2x^2}$.
13. $\frac{x+1}{3} + \frac{2x}{3} - \frac{2x-3}{2x}$.
14. $\frac{3x-1}{2x^2} + \frac{2}{3x} - \frac{x-3}{3x^2}$.
15. $\frac{x}{x-5} + \frac{5}{x-2} - \frac{5}{x-5}$.
16. $\frac{x}{x-2} - \frac{1}{x+1} + \frac{1}{x-2}$.
17. $\frac{x-3}{2x-1} + \frac{x}{x-2} - \frac{1}{2x-1}$.
18. $\frac{1}{x-2} - \frac{1}{x-3} + \frac{1}{x-4}$.
19. $\frac{a+b}{a-b} - \frac{a-b}{a+b}$.
20. $\frac{x-2y}{x+2y} - \frac{2x-y}{2x+y}$.
21. $\frac{1}{x^2-a^2} - \frac{1}{(x-a)^2}$.
22. $\frac{3}{x+2} + \frac{2}{x-2} - \frac{3x}{x^2-4}$.
23. $\frac{1}{x-2} + \frac{1}{x^2-5x+6}$.
24. $\frac{1}{x-5} - \frac{1}{x^2-9x+20}$.
25. $\frac{2}{x^2-y^2} - \frac{3}{x^2-xy-2y^2} - \frac{1}{x^2-3xy+2y^2}$.
26. $\frac{1}{x^2+2xy-8y^2} - \frac{1}{x^2-4y^2} - \frac{1}{x^2+6xy+8y^2}$.
27. $\frac{1}{x^2+15x+54} - \frac{4}{36-x^2}$.
28. $\frac{2x}{x^2-4y^2} + \frac{1}{2y-x} + \frac{1}{x+2y}$.
29. $\frac{1}{x^2-5xy+6y^2} - \frac{2}{x^2-4xy+3y^2} + \frac{1}{x^2-3xy+2y^2}$.
30. $\frac{x+2y}{2y-x} - \frac{x-2y}{x+2y} - \frac{8xy}{x^2-4y^2}$.

31. $\dfrac{4a^2}{x^2 - a^2} - \dfrac{a - x}{x + a} + \dfrac{x + a}{a - x}.$

32. $\dfrac{x + a}{(a - 1)(a - x)} + \dfrac{a + 1}{(x - 1)(a - 1)} + \dfrac{x + 1}{(x - a)(x - 1)}.$

33. $\dfrac{x - 2y}{x^2 - 2xy - 24y^2} - \dfrac{x - 4y}{x^2 - 2xy - 8y^2} + \dfrac{x + 4y}{x^2 + 8xy + 16y^2}.$

34. $\dfrac{a^2 - x^2}{(x + a)^2} - \dfrac{x^2 - a^2}{x^2 + a^2} - \dfrac{(x - a)^2}{a^2 - x^2}.$

55. Complex Fractions

By a **complex fraction** is meant a fraction whose numerator or denominator or both contains one or more fractions.

Example 1. $\dfrac{2 + \frac{1}{3}}{1 + \frac{2}{3}}$ is a complex fraction.

To evaluate a complex fraction, we may perform the indicated operations in the numerator and in the denominator, and then divide the resulting numerator by the resulting denominator.

Thus in example 1, we may write

$$\frac{\frac{7}{3}}{\frac{5}{3}} = \frac{7}{3} \cdot \frac{3}{5} = \frac{7}{5}.$$

Note that, to avoid ambiguity, we make some fraction lines longer or heavier than others. Thus

$$\cfrac{\frac{a}{b}}{c} = \cfrac{\left(\frac{a}{b}\right)}{\left(\frac{c}{1}\right)} = \frac{a}{b} \cdot \frac{1}{c} = \frac{a}{bc},$$

while

$$\cfrac{a}{\frac{b}{c}} = \cfrac{\left(\frac{a}{1}\right)}{\left(\frac{b}{c}\right)} = \frac{a}{1} \cdot \frac{c}{b} = \frac{ac}{b}.$$

Example 2. $\dfrac{\frac{1}{x} - \frac{1}{y}}{\frac{1}{x} + \frac{1}{y}} = \dfrac{\frac{y - x}{xy}}{\frac{y + x}{xy}} = \dfrac{y - x}{xy} \cdot \dfrac{xy}{y + x} = \dfrac{y - x}{y + x}.$

Another procedure, sometimes expedient, is to multiply both the entire numerator and denominator of the given complex fraction by a suitable quantity. Thus in example 2 we may write

$$\frac{\dfrac{1}{x} - \dfrac{1}{y}}{\dfrac{1}{x} + \dfrac{1}{y}} \cdot \frac{xy}{xy} = \frac{y - x}{y + x}.$$

This amounts to multiplying the given complex fraction by 1.

The algebraic sum of one or more integral rational expressions (or polynomials) and one or more fractional rational expressions is often called a **mixed expression.**

Example 3. $x + 3 + \dfrac{2}{x - 1}$ is a mixed expression.

To evaluate a mixed expression, we may write the integral part over the denominator 1, and proceed in the usual way. Thus in example 3 we write

$$\frac{x + 3}{1} + \frac{2}{x - 1} = \frac{(x + 3)(x - 1)}{x - 1} + \frac{2}{x - 1}$$
$$= \frac{x^2 + 2x - 1}{x - 1}.$$

EXERCISES

Reduce each of the following to a simple fraction in lowest terms:

1. $\dfrac{\dfrac{1}{3} + \dfrac{1}{4}}{\dfrac{1}{3} - \dfrac{1}{4}}.$ **2.** $\dfrac{\dfrac{3}{5} - \dfrac{1}{2}}{\dfrac{3}{5} + \dfrac{1}{2}}.$ **3.** $\dfrac{\dfrac{1}{a} - \dfrac{1}{b}}{\dfrac{1}{a} + \dfrac{1}{b}}.$ **4.** $\dfrac{\dfrac{2}{x} + \dfrac{2}{y}}{\dfrac{3}{x} + \dfrac{3}{y}}.$

5. $\dfrac{\dfrac{x}{y} - \dfrac{y}{x}}{\dfrac{x}{y} + \dfrac{y}{x}}.$ **6.** $\dfrac{3 - \dfrac{4}{5}}{2 + \dfrac{3}{5}}.$ **7.** $\dfrac{\dfrac{2}{3} + \dfrac{5}{12}}{1 - \dfrac{2}{3} \cdot \dfrac{5}{12}}.$

8. $\dfrac{\dfrac{5}{12} - \dfrac{3}{4}}{1 + \dfrac{5}{12} \cdot \dfrac{3}{4}}.$ **9.** $\dfrac{\dfrac{a}{b} + \dfrac{a}{b}}{1 - \dfrac{a}{b} \cdot \dfrac{a}{b}}.$ **10.** $\dfrac{\dfrac{a}{b} - \dfrac{c}{d}}{1 + \dfrac{a}{b} \cdot \dfrac{c}{d}}$

11. $\dfrac{\dfrac{x}{r} \cdot \dfrac{m}{n} - \dfrac{y}{r} \cdot \dfrac{p}{n}}{\dfrac{x}{r} \cdot \dfrac{p}{n} + \dfrac{y}{r} \cdot \dfrac{m}{n}}.$ **12.** $\dfrac{3 - x}{\dfrac{3}{x} - 1}.$ **13.** $\dfrac{\dfrac{x}{y} - 1}{1 - \dfrac{y}{x}}.$

14. $\dfrac{1-\dfrac{x^2}{y^2}}{\dfrac{x}{y}-1}.$ **15.** $\dfrac{x-\dfrac{x^2}{x-y}}{y+\dfrac{y^2}{x-y}}.$ **16.** $\dfrac{\dfrac{4}{y-x}-\dfrac{2}{x^2-y^2}}{\dfrac{1}{y-x}-\dfrac{1}{y+x}}.$

17. $\dfrac{\dfrac{a}{a+b}-\dfrac{b-a}{a}}{\dfrac{a}{a+b}+\dfrac{b-a}{a}}.$ **18.** $\dfrac{\dfrac{a+2}{a-2}-\dfrac{a-2}{a+2}}{\dfrac{a}{a-2}-\dfrac{a}{a+2}}.$

19. $\dfrac{a+1-\dfrac{6}{a}}{a+5+\dfrac{6}{a}}.$ **20.** $\dfrac{\dfrac{2}{3-a}+\dfrac{3}{a^2-9}}{\dfrac{2}{a-3}+3}.$

21. $\dfrac{\dfrac{x-2}{x+2}-\dfrac{x}{2}}{\dfrac{x+2}{x-2}+\dfrac{x}{2}}.$ **22.** $\dfrac{2a-9+\dfrac{4}{a}}{2a+3-\dfrac{2}{a}}.$ **23.** $\dfrac{a}{\dfrac{b}{c}}\div\dfrac{\dfrac{a}{b}}{c}.$

24. $\dfrac{\dfrac{2b^2}{a+b}-a}{a+\dfrac{b^2}{a-2b}}.$ **25.** $\dfrac{m+\dfrac{n^2}{m}+n}{m^2-\dfrac{n^3}{m}}.$ **26.** $\dfrac{\dfrac{p^2}{q^2}-\dfrac{p}{q}+1}{\dfrac{p^3}{q^2}+q}.$

27. Find the value of $\dfrac{x-y}{x+y}$ in terms of a and b if $x=\dfrac{a}{a+b}$ and $y=\dfrac{b-a}{a}.$

28. $1+\dfrac{1}{a-\dfrac{1}{a}}.$ **29.** $1+\dfrac{2}{a+\dfrac{3}{a+\dfrac{4}{a}}}.$ **30.** $1-\dfrac{2}{a-\dfrac{3}{a+\dfrac{4}{a}}}.$

31. Find the value of $\dfrac{x-y}{x+y}$ in terms of a if $x=\dfrac{a}{1-a}$ and $y=\dfrac{1+a}{a}.$

32. $\dfrac{\dfrac{m}{n}}{\dfrac{p}{q}}\div\dfrac{m}{\dfrac{n}{\dfrac{p}{q}}}$

Chapter VIII

Linear equations and linear functions

56. Solution of Linear Equations in One Variable

A **root** of an equation in x is a number which satisfies the equation, or converts it into a true statement when substituted for the variable x. **To solve** an equation means to find all its roots.

Note. Not every equation has a root. Thus $x = x + 1$ and $1/x = 0$ have no root.

A **linear equation** in x is one that can be written in the standard form* $ax + b = 0$, $a \neq 0$, where a and b are constants.

The technique of solving an equation is to manipulate it so that the variable or unknown stands alone on one side of the equation. In the case of a linear equation, this can always be done with the help of axiom XIV, section 17. That is, if equals are (*a*) added to, (*b*) subtracted from, (*c*) multiplied by, (*d*) divided by equals, the results are equal. (See section 29.)

Example 1. Solve

(1) $$3x - 1 = x + 13.$$

We reason as follows. If equation (1) has a root, then this root will also satisfy

(2) $$2x = 14$$

which is obtained by transposing, or by adding one to both sides and subtracting x from both sides. But this root will also satisfy

* See section 30.

the equation obtained by dividing both sides of (2) by 2. Thus

$$(3) \qquad x = 7.$$

We have thus proved that if x satisfies (1), then x must be 7. This means that 7 is the only possible candidate for the position of root. But does it prove that 7 is a root of (1)? This requires the proposition "if $x = 7$, then it satisfies (1)." But this is the converse of what we proved, and it is well known that a correct proposition may have an incorrect converse. For example, "if x is a donkey, then x is a mammal" is true; but the converse "if x is a mammal, then x is a donkey" is false, as every reader can testify personally. Therefore we must prove the converse. This can be done either by verifying that each step of the reasoning above is reversible (see sections 28 and 29), which is the case in example 1, or, more simply, by substituting 7 for x in the original equation (1). Thus, $3 \cdot 7 - 1 \stackrel{?}{=} 7 + 13$, or $20 = 20$. Hence 7 is a root and certainly the only root.

The process of substituting the candidate or candidates for the title of root in the original equation is often called *checking* the solution. This unfortunately suggests that it is a superfluous operation designed only to allay the insecurity of the student. *We have seen that it is an essential part of the work,* unless all steps used are known to be reversible. The student should check every solution.

Example 2. Solve $\frac{5x}{2} - 3 = \frac{x}{3} + 10$. Multiplying both sides by the L.C.D. 6, we obtain $15x - 18 = 2x + 60$. Transposing, we get $13x = 78$. Dividing both sides by 13, we get $x = 6$. Hence 6 is the only possible root. Checking, we find that 6 is a root, since $\frac{5 \cdot 6}{2} - 3 = \frac{6}{3} + 10$, or $12 = 12$.

EXERCISES

Solve and check:

1. $x - 2 = 5.$ **2.** $x + 2 = 5.$ **3.** $3x = 18.$

4. $\frac{x}{3} = 5.$ **5.** $5x + 2 = x + 10.$

6. $3x - 5 = 9 + x.$ **7.** $4x + 6 = 2(x + 7).$

8. $5(x - 2) = 2x + 8.$ **9.** $4 - 2x = 3(5 - 3x).$

10. $4 + 3(x + 2) = 4x - 2.$ **11.** $2x - 5 = 3(-2x + 4) - 7.$

12. $5(x + 3) = 2(2 - 3x) + 7x - 1.$

13. $2x + 3 = 3(2x + 1).$

14. $6y + 4(2y - 1) = 3(y - 3) - 10.$

15. $\frac{z}{2} + \frac{3}{4} = 5\left(z + \frac{1}{4}\right).$ **16.** $\frac{w}{3} + \frac{5}{6} = 4\left(w + \frac{1}{12}\right).$

17. $3[x - (3x - 1) + 5] = \frac{x}{3} + 1.$

18. $2x + [3x - (4x - 5)] = \frac{x}{4} - 2.$

19. $\frac{3x - 1}{2} - \frac{x - 5}{4} = \frac{2x + 1}{2}.$

20. $\frac{u}{3} - \frac{2u - 1}{6} = u - 7.$ **21.** $\frac{w - 2}{2} = \frac{w}{5} + \frac{1}{2}.$

22. $\frac{5x}{6} - \frac{3}{2} = \frac{1}{2} - \frac{x}{6}.$ **23.** $y - \frac{5}{12} = \frac{1}{3}\left(y + \frac{3}{2}\right).$

24. $\frac{3x - 1}{4} - \frac{2x}{6} = \frac{3}{2} + \frac{x - 2}{2}.$

25. $\frac{1}{2}\left(x - \frac{4}{3}\right) = \frac{1}{3}\left(2x - \frac{6}{5}\right).$

26. $\frac{z}{2} - 2z = \frac{3z}{4} - \frac{1}{4}(z - 2).$

27. $x - 2\left[x - \frac{5}{4}(2x - 4) + 1\right] = 3.$

28. $\frac{1}{3}[x - \frac{6}{5}(2x - \frac{1}{2}) + \frac{9}{5}] = 7.$

29. $4x - 0.3 = 0.27 - 6x.$

30. $0.2x - 5.23 = 0.05x - 2.77.$

31. $1.2(x - 4) = 3.6x - 0.24.$

32. $6x - 0.2 = \frac{1}{4}(1 - 12x).$

57. Equations Leading to Linear Equations

Frequently an equation, not linear itself, leads to a linear equation.

Example 1. Solve $\frac{1}{6} + \frac{1}{10} = \frac{1}{x}.$ Multiplying both members by the L.C.D. $30x$, we get $5x + 3x = 30$, or $8x = 30$, or $x = \frac{30}{8} = \frac{15}{4}.$ This proves that $\frac{15}{4}$ is the only possible root. To show that it is a root, we check by substitution:

$$\frac{1}{6} + \frac{1}{10} = \frac{1}{15/4} \qquad \text{or} \qquad \frac{8}{30} = \frac{4}{15}.$$

Example 2. Solve $\frac{2x}{x - 2} = \frac{4}{x - 2}.$ Multiplying both sides by the L.C.D. $x - 2$ we get $2x = 4$, or $x = 2$. This proves that if x

satisfied the original equation, then x could only be 2. But checking by substitution reveals that 2 does not satisfy the original equation, since division by zero is excluded. Hence the given equation has no root at all.

Recall that multiplying both sides by a quantity involving unknowns is not always a reversible step and may lead to non-equivalent equations as in example 2. Compare section 29. Therefore *checking is an essential part of the solution in such a case and not merely a superfluous operation designed to reassure the student.*

EXERCISES

Solve and check:

1. $\dfrac{3}{x} - \dfrac{1}{x} + \dfrac{8}{x} = 2.$ **2.** $\dfrac{3}{x} + 3 = \dfrac{15}{x}.$

3. $\dfrac{3x - 2}{4x} = \dfrac{3}{x} + \dfrac{1}{2}.$ **4.** $\dfrac{3}{x} - \dfrac{12 + x}{7x} = \dfrac{5}{7}.$

5. $\dfrac{3}{x - 3} = \dfrac{4}{x - 4}.$ **6.** $\dfrac{4}{x - 3} = \dfrac{3}{x - 4}.$

7. $x(x - 4) + 3 = x^2 - 5(x + 2).$

8. $(x - 2)(x + 1) = (x - 3)(x + 5) - 3.$

9. $\dfrac{2x}{x - 6} = \dfrac{12}{x - 6}.$ **10.** $\dfrac{5x}{x - 3} = \dfrac{15}{x - 3}.$

11. $\dfrac{2x - 1}{x - 3} = \dfrac{2x - 3}{x - 1}.$ **12.** $\dfrac{6x - 5}{3x - 2} = \dfrac{2x - 3}{x + 1}.$

13. $\dfrac{2}{2y^2 - 3y + 4} = \dfrac{1}{y^2 - 4}.$ **14.** $\dfrac{2}{2y^2 - 5y + 6} = \dfrac{3}{3y^2 - 10}.$

15. $(x - 2)^2 = (x + 5)^2.$

16. $(2x - 3)(8x + 5) = (4x - 3)^2.$

17. $\dfrac{4}{x(x - 2)} - \dfrac{2}{x} = \dfrac{x + 2}{x(x - 2)}.$

18. $\dfrac{2}{x - 1} - \dfrac{1}{2(x - 1)} + \dfrac{1}{2(x - 3)} = \dfrac{2}{(x - 1)(x - 3)}.$

19. $\dfrac{5}{5y^2 - 17y + 6} = \dfrac{2}{2y^2 - 7y + 3}.$

20. $3(x - 2)^2 = (3x - 2)(x + 2).$

21. $\dfrac{x + 3}{x + 4} + \dfrac{x + 4}{x + 3} = \dfrac{2x^2 - 16}{x^2 + 7x + 12}.$

22. $\dfrac{5}{x - 1} = \dfrac{8}{x - 2} - \dfrac{3}{x - 3}.$

23. $\dfrac{3}{x^2 - 1} - \dfrac{4}{x - 1} = \dfrac{3}{2(x - 1)} - \dfrac{3}{2(x + 1)}.$

24. $\dfrac{5y - 3}{y + 2} = \dfrac{6y + 1}{y - 2} - 1.$

25. $\dfrac{x + 2}{2x - 3} + \dfrac{3x - 1}{4x + 5} = \dfrac{10x^2 + x - 3}{8x^2 - 2x - 15}.$

26. $\dfrac{2x - 3}{3x - 2} + \dfrac{2x + 3}{2x + 5} = \dfrac{2(5x^2 - 3)}{6x^2 + 11x - 10}.$

58. Literal Equations and Formulas

The operations of adding, subtracting, multiplying, or dividing equals with equals may be used to solve linear equations with literal* coefficients.

Example 1. Solve $ax - b = cx + d$ for x.

Solution. Adding b to both sides of the equation and subtracting cx from both sides, we obtain $ax - cx = b + d$, or

$$(a - c)x = b + d.$$

Provided $a - c \neq 0$, or $a \neq c$, we may divide both sides by $a - c$, obtaining

$$x = \frac{b + d}{a - c}.$$

The reader should check this solution by substituting in the original equation.

This sort of manipulation will arise frequently in the student's scientific studies when it is desired to rearrange formulas connecting two or more quantities so as to express a certain one of them in terms of the others.

Example 2. Let F be the number of degrees Fahrenheit and C the number of degrees Centigrade. Then $F = \frac{9}{5}C + 32$. Solve this formula for C in terms of F.

Solution. Subtracting 32 from both sides, we have

$$\tfrac{9}{5}C = F - 32.$$

Multiplying both sides by $\frac{5}{9}$, we have

$$C = \tfrac{5}{9}(F - 32).$$

* That is, coefficients which are unspecified quantities.

EXERCISES

Solve the following equations for x, stating any conditions which become necessary to avoid division by zero.

1. $2x - 3ab = 7ab.$

2. $4x - c = x + 5c.$

3. $3a - 2x = x - 9b.$

4. $6x - 3ab = 3x + 6ab.$

5. $3ax - 2b = 10b + 6x.$

6. $3ax - 4a^2 = 12a^2 - ax.$

7. $2(ax - 4b) = b(x - 6).$

8. $ax + d = bx + c.$

9. $2a^2x + b^2 = 4a^2 + bx.$

10. $4ax + b^2 = 16a^2 + bx.$

11. $\dfrac{x+6}{2a} = \dfrac{3}{2} + a.$

12. $\dfrac{4a - x}{5} = \dfrac{x - 4a}{6}.$

13. $\dfrac{x-b}{a} = \dfrac{x-a}{b}.$

14. $a(a - x) = b(x - b) - 2ab.$

15. $\dfrac{1}{a} + \dfrac{1}{x} = \dfrac{1}{b} - \dfrac{1}{x}.$

16. $(x + a)^2 - (x - b)^2 = 2a(a - b).$

17. $\dfrac{1}{x-a} + \dfrac{1}{x-b} = 0.$

18. $\dfrac{x}{ac} - \dfrac{x}{cb} = \dfrac{1}{ab}.$

Solve each of the following for y:

19. $3x - 5y = 2.$

20. $2x + 3y = 4.$

21. $3x + 5y - 3 = 2x - 3y + 6.$

22. $6x - 2y + 1 = 2x + 3y + 11.$

23. $Ax + By = C.$

24. $\dfrac{x}{a} + \dfrac{y}{b} = 1.$

Solve each of the following formulas for the indicated letter, all letters being understood to represent positive quantities:

25. $pv = c$ for v.

26. $I = PRT$ for R.

27. $A = \frac{1}{2}bh$ for b.

28. $A + B + C = 180$ for B.

29. $P = 2l + 2w$ for l.

30. $d = \frac{1}{2}gt^2$ for g.

31. $C = \frac{5}{9}(F - 32)$ for F.

32. $A = \frac{1}{2}h(B + b)$ for B.

33. $V = \frac{1}{3}\pi r^2 h$ for h.

34. $A = P(1 + rt)$ for r.

35. $F = G\dfrac{m_1 m_2}{r^2}$ for m_2.

36. $S = \dfrac{a - rl}{1 - r}$ for r.

37. $S = \dfrac{n}{2}[2a + (n - 1)d]$ for d.

38. $\dfrac{V_1}{V_2} = \dfrac{P_2}{P_1}$ for P_1.

39. $nE = I(R + nr)$ for r.

40. $I = \dfrac{E}{r + \dfrac{R}{n}}$ for n.

41. $d_1 w_1 = d_2 w_2$ for w_2.

42. $S = 2\pi r(r + h)$ for h.

43. $c = \dfrac{P - p}{pt}$ for t.

44. $\dfrac{v^2}{2g} + \dfrac{p}{c} = H$ for c.

45. $p = p_1 - c(h - h_1)$ for h.

46. $T = T_1\left(1 - \dfrac{n-1}{n} \cdot \dfrac{h}{h_0}\right)$ for h.

47. $\dfrac{1}{C} = \dfrac{1}{C_1} + \dfrac{1}{C_2}$ for C.

48. $V = \dfrac{q}{\epsilon_0}\left(\dfrac{1}{r_1} - \dfrac{1}{r_2}\right)$ for r_2.

49. $\dfrac{a - b}{a + b} = \dfrac{c - d}{c + d}$ for a.

50. $A = \dfrac{Bb + Cc}{b + c}$ for b.

59. Graph* of a Linear Function

Consider a linear function $y = ax + b$ $(a \neq 0)$ with real coefficients. It can be proved† that the graph of a linear function in x with real coefficients is an oblique straight line, that is, one which is parallel to neither axis. Let (x_1,y_1) and (x_2,y_2) be any two points on its graph; that is,

$$y_1 = ax_1 + b$$

$$y_2 = ax_2 + b.$$

Then $y_2 - y_1 = a(x_2 - x_1)$, or‡

$$\frac{y_2 - y_1}{x_2 - x_1} = a. \tag{1}$$

The constant a is called the **slope** of the line, since it represents geometrically the ratio of vertical rise over horizontal progress toward the right (Fig. 23), or the vertical rise per unit of horizontal progress toward the right.

Example 1. The graph (Fig. 24) of the function $y = 2x + 4$ rises 2 units per unit of horizontal progress toward the right.

Example 2. The graph (Fig. 25) of the function $y = -\frac{1}{2}x + 4$ sinks 1 unit (rises -1 unit) for every 2 units of horizontal progress toward the right.

* The student should review sections 34 and 35.

† For proof see M. Richardson, *Fundamentals of Mathematics,* Macmillan, 1941, or textbooks on analytic geometry.

‡ Since a linear function is single-valued, two different points on its graph must have different x-coordinates. Thus $x_1 \neq x_2$ or $x_2 - x_1 \neq 0$; hence division by $x_2 - x_1$ is permissible.

These examples indicate the fact that a line with positive slope rises as it goes to the right, while a line with negative slope sinks as it goes to the right.

The constant b is called the **y-intercept** of the line $y = ax + b$, since, by substitution, it is clear that $(0,b)$ are the coordinates of the point where the graph intersects the y-axis. See Figs. 23, 24, 25.

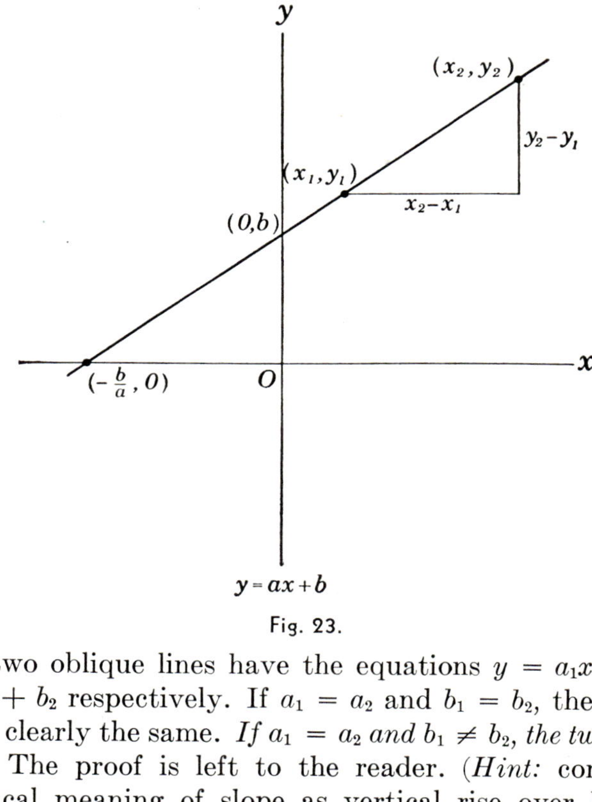

Fig. 23.

Let two oblique lines have the equations $y = a_1x + b_1$ and $y = a_2x + b_2$ respectively. If $a_1 = a_2$ and $b_1 = b_2$, then the two lines are clearly the same. *If* $a_1 = a_2$ *and* $b_1 \neq b_2$, *the two lines are parallel.* The proof is left to the reader. (*Hint:* consider the geometrical meaning of slope as vertical rise over horizontal progress toward the right.)

The solution or root of the equation $ax + b = 0$ *is* $x = -b/a$, *which is the x-coordinate of the point at which the graph of* $y = ax + b$ *intersects the x-axis.* See Figs. 23, 24, 25. The number $-b/a$ is called the **x-intercept** of the line.

Example 3. Solve the equation $2x + 4 = 0$ graphically.

Solution. From the graph (Fig. 24), the solution is $x = -2$, since the line crosses the x-axis at $x = -2$.

Example 4. The solution of the equation $-\frac{1}{2}x + 4 = 0$ is seen similarly to be $x = 8$ (Fig. 25).

It can be seen from the graph that equation (1) above means that the linear function $y = ax + b$ indicates that the variable y

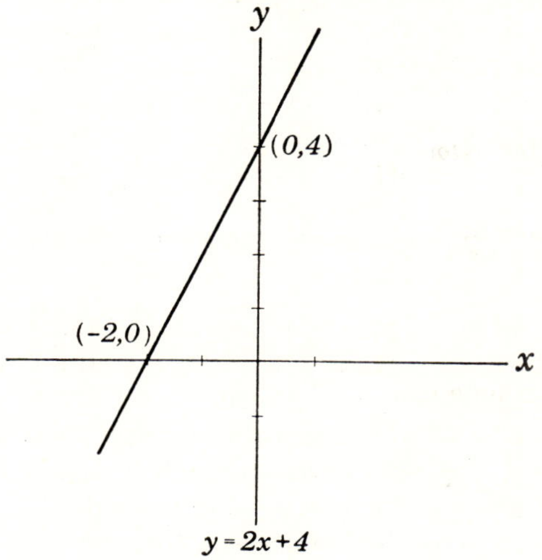

Fig. 24.

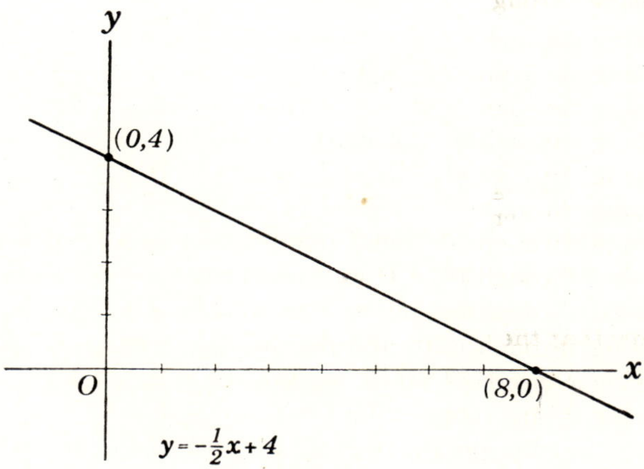

Fig. 25.

has a constant rate of change a with respect to x. That is, the change in y induced by a change in x is always the same constant a times the change in x.

Since a straight line is determined by two points, plotting two points suffices to draw the graph of a linear function. A third point is desirable as a check; if the three points do not lie on a straight line, an error has been made.

EXERCISES

Plot the graph of each of the following functions:

1. $y = 3x + 2$. **2.** $y = 3x - 2$. **3.** $y = \frac{1}{3}x + 2$.
4. $y = \frac{1}{3}x - 2$. **5.** $y = -3x + 2$. **6.** $y = -3x - 2$.
7. $y = -\frac{1}{3}x + 2$. **8.** $y = -\frac{1}{3}x - 2$. **9.** $y = x$.
10. $y = -x$. **11.** $y = 3x$. **12.** $y = -\frac{1}{3}x$.

Solve each of the following equations algebraically and check by plotting the graph:

13. $3x - 12 = 0$. **14.** $2x + 6 = 0$. **15.** $\frac{1}{3}x - 4 = 0$.
16. $\frac{1}{2}x + 4 = 0$. **17.** $5 - 3x = 0$. **18.** $2 + 3x = 0$.
19. $4 - \frac{1}{2}x = 0$. **20.** $\frac{1}{2} - \frac{1}{6}x = 0$.
21. The line $y = 3x + b$ passes through the point (2,4). Find b.
22. The line $y = ax + 3$ passes through the point (3,6). Find a.
23. Find the slope of the line passing through (1,4) and (3,8).
24. Find the slope of the line passing through (−1,3) and (1,−5).

Write an equation of the line satisfying each of the following conditions:

25. Passing through (2,5) with slope 3.
26. Passing through (2,3) with slope 4.
27. Passing through (−2,−3) with slope $\frac{1}{4}$.
28. Passing through (−3,2) with slope −1.
29. Passing through (1,−2) and (−3,8).
30. Passing through (2,−1) and (4,5).
31. Passing through (0,3) and parallel to $y = 4x - 2$.
32. Passing through (0,−4) and parallel to $y = \frac{1}{3}x + 1$.
33. Passing through (−3,0) and parallel to $y = -\frac{1}{2}x + 1$.
34. Passing through (2,0) and parallel to $y = \frac{1}{3}x - 1$.
35. Having x-intercept 3 and y-intercept −9.
36. Having x-intercept −2 and y-intercept 6.

37. Show that the points (1,5), (3,9), and (5,13) are on the same line.

38. Show that the points (−1,7), (1,3), and (5,−5) are on the same line.

39. Decide whether the following pairs of points are on the same line or on parallel lines: (1,5), (5,21) and (8,34), (12,50).

40. Decide whether the following pairs of points are on the same line or on parallel lines: (1,2), (5,22) and (8,37), (12,57).

60. Systems of Two Linear Equations in Two Unknowns

The general linear equation in x and y may be written in the form $Ax + By = C$ where A, B, and C are constants and where A and B are not both zero. A **solution** of such an equation is a pair of values, one for x and one for y, which satisfies the equation. In general, a single equation in more than one unknown may have infinitely many solutions. For example, $x + y = 7$ has the solutions $(x = 1, y = 6)$, $(x = 2, y = 5)$, $(x = 3, y = 4)$, $(x = -1, y = 8)$, etc. No matter what value we choose for x, that value together with the value $y = 7 - x$ will constitute a solution. Therefore there is little point in asking one to "solve" such an equation. But if we have two such equations, we may ask whether they have a common solution, and, if so, what it is. A common solution of two or more equations in two or more unknowns is called a **solution of the system of equations.** Thus the system of equations

$$\left.\begin{aligned} &(1) \qquad x + y = 7 \\ &(2) \qquad x - y = 1 \end{aligned}\right\}$$

has the solution $(x = 4, y = 3)$ and no other solution.

The technique of solving a system of two linear equations in two unknowns is to reduce it to one equation in one unknown by eliminating one of the unknowns. This may be done by substitution, as follows.

Example 1. Solve the system of equations (1) and (2). We reason as follows. If x and y satisfy (1), then they also satisfy $y = 7 - x$. Substituting this in (2), we have $x - (7 - x) = 1$, or $x - 7 + x = 1$, or $2x = 8$, or $x = 4$. Then $y = 7 - x = 7 - 4 = 3$. This proves that if (x,y) satisfy the system, then x can only be 4 and y can only be 3. The converse is also true, since all steps taken are reversible. Nevertheless, it is desirable to check the converse by substitution in both equations. Thus $4 + 3 = 7$ and $4 - 3 = 1$. This illustrates the first method for solving systems of two linear equations in two unknowns x and y.

Method of elimination by substitution. (*a*) *Solve one equation for one unknown in terms of the other.* (*b*) *Substitute this expression in the second equation, thus obtaining an equation in the other unknown only.* (*c*) *Solve this equation for the other unknown.* (*d*) *Substitute this value in the expression obtained in step* (*a*), *thus obtaining the value for the first unknown.* (*e*) *Check by substitution in both given equations.*

Example 2. Solve the system

$$\left.\begin{aligned} 2x + 3y &= 13 \quad (3)\\ 5x - 2y &= 4. \quad (4)\end{aligned}\right\}$$

Solution. (*a*) From (3), we obtain $y = \frac{13 - 2x}{3}$. (*b*) Substituting this in (4), we obtain $5x - 2\left(\frac{13 - 2x}{3}\right) = 4$. (*c*) Then $15x - 26 + 4x = 12$, or $19x = 38$, and $x = 2$. (*d*) $y = \frac{13 - 4}{3} = 3$. (*e*) Checking, $2 \cdot 2 + 3 \cdot 3 = 13$ and $5 \cdot 2 - 2 \cdot 3 = 4$.

A second method, often more convenient, will be illustrated by using it to solve example 2.

We obtain equations equivalent to (3) and (4) by multiplying them by any constant except 0. Let us multiply both sides of (3) by 2 and both sides of (4) by 3, obtaining

$$\left.\begin{aligned} 4x + 6y &= 26\\ \text{and} \qquad 15x - 6y &= 12.\end{aligned}\right\}$$

Since if equals are added to equals the results are equal, we may add the left members and right members of these equations and obtain the equation

$$19x = 38.$$

Hence $x = 2$.

Similarly, multiplying (3) and (4) by 5 and -2 respectively, we obtain

$$\left.\begin{aligned} 10x + 15y &= 65\\ -10x + 4y &= -8.\end{aligned}\right\}$$

Adding these equations, we get

$$19y = 57.$$

Hence $y = 3$. The check is as above.

This illustrates the second method for solving systems of two linear equations in two unknowns.

Method of elimination by multiplying and adding. (*a*) *Multiply both equations by constants chosen so that the coefficients of one unknown become negatives of each other.* (*b*) *Add the resulting equations, thus eliminating one unknown.* (*c*) *Solve the resulting equation for the other unknown.* (*d*) *Repeat the process with the roles of the unknowns interchanged, solving for the first unknown.* (*e*) *Check by substituting in both equations.*

Variants of these methods are easily devised.

Another method of solving systems of linear equations will be taken up in Chapter XX.

EXERCISES

Solve by any method, unless otherwise directed, and check:

1. $\begin{cases} 3x + y = 9, \\ 2x - 2y = 14. \end{cases}$

2. $\begin{cases} 2x - 3y = 13, \\ 3x + y = 3. \end{cases}$

3. $\begin{cases} 2x + 6y = 0, \\ x + 4y = 1. \end{cases}$

4. $\begin{cases} 2x - 3y = 5, \\ 5x - 4y = 2. \end{cases}$

5. $\begin{cases} 3x + 4y = 5, \\ 6x - y = 1. \end{cases}$

6. $\begin{cases} 4x - 2y = 3, \\ 3x + 3y = 4. \end{cases}$

7. $\begin{cases} 2x - 3y = 8, \\ 3x + 5y = 12. \end{cases}$

8. $\begin{cases} 3x - 5y = 10, \\ 4x - 3y = 6. \end{cases}$

9. $\begin{cases} 0.03x = 0.02y - 0.13, \\ 0.02x + 0.05y - 0.04 = 0. \end{cases}$

10. $\begin{cases} 0.5 - 0.4x = 0.3y, \\ 0.2x + 0.1y - 0.2 = 0. \end{cases}$

11. $\begin{cases} 0.02x - 0.01y - 0.07 = 0, \\ 0.03x - 0.02y - 0.13 = 0. \end{cases}$

12. $\begin{cases} 0.3x - 0.4y + 0.5 = 0, \\ 0.2x + 0.2y - 0.3 = 0. \end{cases}$

13. $\begin{cases} x + \frac{5}{2}y = 2, \\ \frac{1}{2}x + y = \frac{1}{2}. \end{cases}$

14. $\begin{cases} \frac{2}{3}x - y = \frac{1}{3}, \\ x + \frac{1}{3}y = \frac{7}{3}. \end{cases}$

15. $\begin{cases} \frac{1}{4}x + \frac{1}{3}y = \frac{5}{12}, \\ \frac{1}{2}x + y = 1. \end{cases}$

16. $\begin{cases} \frac{1}{3}x - \frac{1}{2}y = \frac{23}{6}, \\ \frac{1}{2}x + \frac{1}{3}y = \frac{1}{3}. \end{cases}$

17. $\begin{cases} \dfrac{x}{2} - \dfrac{y}{3} + \dfrac{7}{6} = 0, \\ \dfrac{x}{6} - \dfrac{y}{3} + \dfrac{1}{6} = 0. \end{cases}$

18. $\begin{cases} \dfrac{x}{3} + \dfrac{y}{6} = \dfrac{1}{2}, \\ \dfrac{x}{2} + \dfrac{3y}{10} = 1. \end{cases}$

19. $\begin{cases} \dfrac{x-2}{2} + \dfrac{y+3}{3} = 5, \\ \dfrac{x+3}{3} - \dfrac{y-2}{2} = 7. \end{cases}$

20. $\begin{cases} \dfrac{x+3}{3} - \dfrac{y+2}{2} = 4, \\ \dfrac{x-2}{2} + \dfrac{y-3}{3} = 6. \end{cases}$

21. $\begin{cases} \dfrac{x+y}{2} - \dfrac{x-y}{3} = 2, \\ \dfrac{x-y}{4} + \dfrac{x+y}{3} = 1. \end{cases}$

22. $\begin{cases} \dfrac{2x+3y}{2} + \dfrac{2x-3y}{3} = 1, \\ \dfrac{3x+2y}{3} - \dfrac{3x-2y}{2} = -1. \end{cases}$

23. $\begin{cases} x + y = 7a, \\ x - y = a. \end{cases}$

24. $\begin{cases} x + y = 4a, \\ x - y = 2b. \end{cases}$

25. $\begin{cases} 3x - y + 7b = 0, \\ x + 2y + 7a = 0. \end{cases}$

26. $\begin{cases} 5x + 2y = a, \\ 2x - 3y = 8a. \end{cases}$

27. $\begin{cases} x + 3y - 5a = 0, \\ 3x - 2y = 4a - 11. \end{cases}$

28. $\begin{cases} 2x + y = 9a, \\ 7x - 2y = 4a. \end{cases}$

29. $\begin{cases} \dfrac{x}{4a} + \dfrac{y}{3a} = \dfrac{1}{12}, \ (a \neq 0), \\[2mm] \dfrac{x}{3b} + \dfrac{y}{5b} = \dfrac{1}{15}, \ (b \neq 0). \end{cases}$

30. $\begin{cases} \dfrac{x}{a} + \dfrac{y}{b} = 1, \\[2mm] \dfrac{x}{a} - \dfrac{y}{b} = 2. \end{cases}$

31. $\begin{cases} ax + by = a^2 + b^2, \\ bx - ay = a^2 + b^2, \ (a^2 + b^2 \neq 0). \end{cases}$

32. $\begin{cases} ax + by = a^2 - b^2, \\ bx + ay = a^2 - b^2, \ (a^2 - b^2 \neq 0). \end{cases}$

33. $\begin{cases} a_1x + b_1y = k_1, \\ a_2x + b_2y = k_2, \ (a_1b_2 - a_2b_1 \neq 0). \end{cases}$

34. $\begin{cases} ax - by = c, \\ dx + ey = f, \ (ae + bd \neq 0). \end{cases}$

35. $\begin{cases} (2b - a)y = 2cx - 2ac, \\ (b - 2a)y = cx - 2ac \ (a \neq 0, \ c \neq 0). \end{cases}$

36. $\begin{cases} c(y - c) = (a - b)x - a^2 + b^2, \\ c(y - c) = b^2 - bx \ (a \neq 0, \ c \neq 0). \end{cases}$

37. $\begin{cases} (a - b)x - cy = 0, \\ bx + cy = 2ab \ (a \neq 0, \ c \neq 0). \end{cases}$

61. Graphical Interpretation. Inconsistent and Dependent Equations

It can be proved* that every linear equation

$$(1) \qquad ax + by = c$$

with real coefficients a, b, and c, not both a and b equal to zero, has a straight line for its graph; and conversely every straight line in the plane has such a linear equation. If $a = 0$, $b \neq 0$, we may write the equation (1) as $y = \dfrac{c}{b}$, which clearly represents a

* For a proof, see M. Richardson, *Fundamentals of Mathematics*, Macmillan, 1941, or textbooks on analytic geometry.

horizontal straight line through the point $\left(0, \frac{c}{b}\right)$. If $b = 0$, $a \neq 0$, then equation (1) becomes $x = \frac{c}{a}$, which clearly represents a vertical line through the point $\left(\frac{c}{a}, 0\right)$. If neither a nor b is zero, then equation (1) may be written as $y = -\frac{a}{b}x + \frac{c}{b}$, which represents an oblique straight line with slope $-\frac{a}{b}$ and y-intercept $\frac{c}{b}$. To plot the graph of a linear equation, therefore, requires plotting only two points; a third point may be desirable as a check.

Consider two linear equations, as

(2) $$x + y = 7,$$

(3) $$x - y = 1.$$

Every point in the plane whose coordinates satisfy equation (2) lies on the graph of equation (2) (Fig. 26). Every point whose

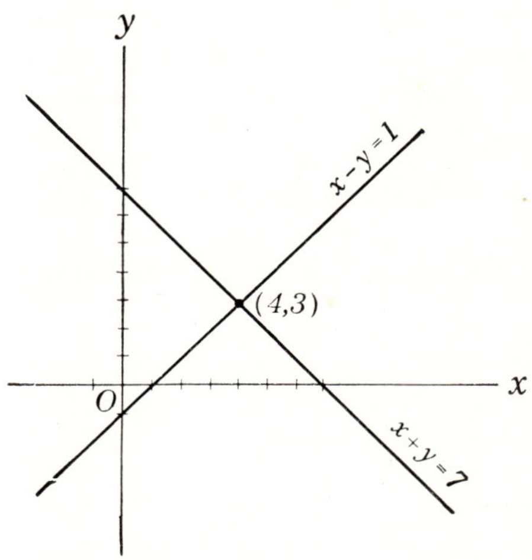

Fig. 26.

coordinates satisfy (3) lies on the graph of equation (3). Hence *a common solution of the system of equations (2) and (3) must be the coordinates of a point of intersection of these two lines, and conversely*. Usually two linear equations have one and only one common solution, but not always. Two equations which have a unique common solution are often called **simultaneous.**

Not every pair of linear equations in x and y have a common solution. For example, it is clear that the equations

$$x + y = 7 \tag{4}$$

and

$$x + y = 5 \tag{5}$$

can have no common solution, since the sum of two numbers x and y cannot be both 7 and 5. Two equations which have no common solution are called **inconsistent** or **incompatible.** The graphs of inconsistent equations are parallel lines, since they have no point of intersection.

On the other hand, the system consisting of equation (5) and equation

$$2x + 2y = 10 \tag{6}$$

have all their solutions in common. That is, they represent the same line, since any pair of numbers satisfying (5) also satisfies (6), and conversely, because $2x + 2y = 2(x + y)$. Two linear equations in x and y which represent the same line are called **dependent** on each other.

EXERCISES

Decide which of the following systems are simultaneous, inconsistent, or dependent; solve if possible, and check by graphing each pair of equations:

1. $\begin{cases} 2x - 3y = 4, \\ 4x - 9y = 8. \end{cases}$ **2.** $\begin{cases} 4x - 3y = 1, \\ 8x + 6y = 14. \end{cases}$

3. $\begin{cases} 4x + 2y = 5, \\ 6x + 3y = 8. \end{cases}$ **4.** $\begin{cases} 3x - y = 6, \\ 9x - 3y = 12. \end{cases}$

5. $\begin{cases} 2x - 4y = 6, \\ 3x - 6y = 9. \end{cases}$ **6.** $\begin{cases} 3x - y = 5, \\ 6x - 2y = 10. \end{cases}$

7. $\begin{cases} 5x + 2y = 1, \\ 5x - 2y = 9. \end{cases}$ **8.** $\begin{cases} 4x - 2 = 6y, \\ 2x = 3y + 1. \end{cases}$

9. $\begin{cases} 4x = 8 - 6y, \\ 2x + 3y = 2. \end{cases}$ **10.** $\begin{cases} x - 4y + 3 = 0, \\ 4x = 8y - 6. \end{cases}$

11. $\begin{cases} 9x - 6y = 3, \\ 12x = 8y + 4. \end{cases}$ **12.** $\begin{cases} x - 3y = 5, \\ 6y = 2x - 10. \end{cases}$

★**13.** Show that the equations

$$\begin{cases} a_1x + b_1y = c_1, \\ a_2x + b_2y = c_2, \end{cases}$$

are dependent if $\frac{a_1}{a_2} = \frac{b_1}{b_2} = \frac{c_1}{c_2}$. (*Hint:* Let the value of this common ratio be k.)

★**14.** Show that the equations in exercise 13 are inconsistent if

$$\frac{a_1}{a_2} = \frac{b_1}{b_2} \neq \frac{c_1}{c_2}.$$

15. Solve each of the systems in exercises 1–10, section 60, graphically, and check by solving algebraically.

62. Systems of Three Linear Equations in Three Unknowns

If the system of three linear equations

$$\begin{cases} a_1x + b_1y + c_1z = k_1, \\ a_2x + b_2y + c_2z = k_2, \\ a_3x + b_3y + c_3z = k_3, \end{cases}$$

in x, y, and z has a unique common solution, as is usually the case, it may be found by an extension of the methods used in section 60. To solve a system of three linear equations in three unknowns, we may proceed as follows:

(*a*) Between any two of the three equations, one unknown may be eliminated, resulting in an equation in the other two unknowns.

(*b*) Between any other pair of the three equations, the same unknown may be eliminated, resulting in a second equation in the other two unknowns.

(*c*) The two equations in the same two unknowns obtained in steps (*a*) and (*b*) may be solved.

(*d*) The values obtained in step (*c*) may be substituted in any one (the simplest) of the three given equations, which then may be solved for the third unknown.

(*e*) The solution should be checked by substituting in the three given equations.

★ May be omitted without disturbing continuity.

Example. Solve the system

$$\begin{aligned} &(1) && \begin{cases} x + 2y + 3z = 9, \\ 2x - y + 2z = 11, \\ 3x + 4y - 2z = -4. \end{cases} \\ &(2) && \\ &(3) && \end{aligned}$$

(*a*) Multiplying equation (1) by 2 and equation (2) by -3, we get

(4) $$2x + 4y + 6z = 18,$$

(5) $$-6x + 3y - 6z = -33.$$

Adding (4) and (5), we eliminate z, obtaining

(6) $$-4x + 7y = -15.$$

(*b*) Adding (2) and (3), we eliminate z, obtaining

(7) $$5x + 3y = 7.$$

(*c*) Solving (6) and (7) by the methods of section 60, we get

(8) $$x = 2 \text{ and } y = -1.$$

(*d*) Substituting in (1), we get $2 - 2 + 3z = 9$, or

(9) $$z = 3.$$

(*e*) Substituting (8) and (9) in the three given equations, we find that they check.

The same procedure can be used to solve a system of 4 linear equations in 4 unknowns, and so on, by successively reducing the number of equations and the number of unknowns by one. Thus a system of 4 equations in 4 unknowns may be reduced to a system of 3 equations in 3 unknowns by eliminating the same unknown between 3 different pairs of the 4 given equations. The resulting system of 3 equations may be solved for the 3 unknowns as in the above example.

Variants of this procedure are easily devised.

Another method for dealing with systems of linear equations will be taken up in Chapter XX, where the theory of such systems will be discussed in greater detail.

EXERCISES

Solve and check:

1. $\begin{cases} 2x - 2y + 3z = 1, \\ x - 3y - 2z = -9 \\ x + y + z = 6. \end{cases}$

2. $\begin{cases} x - 2y = 0 \\ z - 2y = 0 \\ x + y + z = 1. \end{cases}$

3. $\begin{cases} x + y = -4 \\ x + z - 1 = 0 \\ 2z - y + 3x = 4. \end{cases}$

4. $\begin{cases} x + 4y - z = -3 \\ 2y - x + 2z = 2 \\ 2z - 3x + 2y = 0. \end{cases}$

5. $\begin{cases} 3x + y + 4z = 0 \\ 5x + y + 3z = 1 \\ x - 3y - 4z = 5. \end{cases}$

6. $\begin{cases} u - 3v + 2w = 8 \\ 2u - v + 2w = 2 \\ 6u + w + 3v + 3 = 0. \end{cases}$

7. $\begin{cases} 2A + B + C = 0 \\ 6A - 2B + 3C = 1 \\ A + 2B - 7C = 1. \end{cases}$

8. $\begin{cases} r + 3s + t = 4 \\ 2s - 3r + 3 = t \\ 2r - 3t = 2s + 5. \end{cases}$

9. $\begin{cases} 2x + 3y + z = 2a \\ 6x + 6y + 2z = 5a \\ 3x - 3y - \dfrac{z}{4} = 0. \end{cases}$

10. $\begin{cases} x + y - z = 0 \\ x - y + z = c - b \\ x - y - z = a - c. \end{cases}$

11. $\begin{cases} x + y + z = 2 \\ 2x - z + t = 3 \\ 3x + 2y + t = 4 \\ 3y + 2z + 2t = 7. \end{cases}$

12. $\begin{cases} 2t - 2x + y + 4 = 0 \\ 4t = y - z - 3 \\ x + 2y = 3z - 14 \\ 6t + 13 = 3x - 2y. \end{cases}$

13. $\begin{cases} x - z = t \\ 2x + y + 2u - 3t = 2 \\ 3z + u + t = 2 \\ x + z = y \\ x + y + z + t + u = 5. \end{cases}$

14. $\begin{cases} x + y + z + t + u = 3 \\ 2x + t = 0 \\ 2x - u = 0 \\ x - y - z - t = 2 \\ y - z + u - t = 2. \end{cases}$

63. Systems of Equations Reducible to Linear Equations

Equations which are not linear themselves may lead to linear equations.

Example 1. Solve the system

$$\begin{cases} \dfrac{y - x}{2x - 4} = \dfrac{1}{2} \\[2ex] \dfrac{x + 3}{xy + 4x} = \dfrac{1}{y}. \end{cases}$$

Solution. Clearing fractions in each equation by multiplying both sides by the L.C.D., we get

$$\begin{cases} 2y - 2x = 2x - 4 \\ xy + 3y = xy + 4x, \end{cases}$$

or,

$$\begin{cases} 4x - 2y = 4 \\ 4x - 3y = 0. \end{cases}$$

Solving this system, we get $x = 3$, $y = 4$, which checks in both given equations.

If the given equations are linear in the reciprocals of the unknowns, it is better not to clear fractions but rather to regard the reciprocals as the unknowns temporarily.

Example 2. Solve the system

$$\begin{cases} \dfrac{1}{x} + \dfrac{2}{y} = \dfrac{5}{6} & (1) \\[2ex] \dfrac{3}{x} - \dfrac{1}{y} = \dfrac{3}{4}. & (2) \end{cases}$$

Solution. Multiplying (2) by 2 and adding to (1), we get $\dfrac{7}{x} = \dfrac{28}{12}$, or $x = 3$. Multiplying (1) by 3 and subtracting (2), we get $\dfrac{7}{y} = \dfrac{7}{4}$, or $y = 4$. The reader should check by substituting in both equations.

Note. If desired, the solution of example 2 may be obtained by letting $u = \dfrac{1}{x}$ and $v = \dfrac{1}{y}\cdot$ Then (1) and (2) become

$$\begin{cases} u + 2v = \dfrac{5}{6} \\[2ex] 3u - v = \dfrac{3}{4} \end{cases}$$

respectively. Solving this system for u and v, we get $u = 1/3$, $v = 1/4$. Hence $x = 3$ and $y = 4$.

EXERCISES

Solve and check:

1. $\begin{cases} \dfrac{4x - 7}{2x + 2} = \dfrac{12y + 3}{6y - 4}, \\[2ex] \dfrac{4x + 7}{8y + 1} = \dfrac{2x + 3}{4y + 5}. \end{cases}$

2. $\begin{cases} \dfrac{4x + 2y}{3x - 3y} = \dfrac{5}{6}, \\[2ex] \dfrac{2x - 3y}{5x + 3y} = \dfrac{3}{4}. \end{cases}$

3. $\begin{cases} \dfrac{x - 2}{x - 1} = \dfrac{y + 1}{y}, \\[2ex] \dfrac{x + 4}{x - 2} = \dfrac{y - 5}{y + 1}. \end{cases}$

4. $\begin{cases} \dfrac{y - 4}{x + 2} = \dfrac{y}{x - 2}, \\[2ex] \dfrac{x - 2}{y - 3} = \dfrac{x - 4}{y - 4}. \end{cases}$

5. $\begin{cases} \dfrac{3x + 2}{12y - 4} = \dfrac{2x + 3}{8y - 6}, \\[2ex] \dfrac{2y}{3x + 12} = \dfrac{2y + 1}{3x + 15}. \end{cases}$

6. $\begin{cases} \dfrac{2x + y + 3}{5x - y - 1} = \dfrac{2}{3}, \\[2ex] 4x - 3y = 17. \end{cases}$

7. $\begin{cases} \dfrac{3}{x} - \dfrac{4}{y} = 3, \\[2ex] \dfrac{6}{x} + \dfrac{2}{y} = 1. \end{cases}$

8. $\begin{cases} \dfrac{6}{x} + \dfrac{1}{y} = 1, \\[2ex] \dfrac{9}{x} - \dfrac{2}{y} = 5. \end{cases}$

9. $\begin{cases} \dfrac{1}{x} + \dfrac{2}{y} = 4, \\ \dfrac{5}{x} - \dfrac{3}{y} = -19. \end{cases}$

10. $\begin{cases} \dfrac{3}{x} - \dfrac{4}{y} = -1, \\ \dfrac{9}{x} + \dfrac{12}{y} = 9. \end{cases}$

11. $\begin{cases} \dfrac{2}{x} + \dfrac{3}{y} = 7, \\ \dfrac{6}{x} - \dfrac{6}{y} = 1. \end{cases}$

12. $\begin{cases} \dfrac{3}{x} + \dfrac{2}{y} = 8, \\ \dfrac{2}{x} - \dfrac{1}{y} = 10. \end{cases}$

13. $\begin{cases} \dfrac{5}{x} - \dfrac{5}{y} = 7, \\ \dfrac{10}{x} + \dfrac{5}{y} = 2. \end{cases}$

14. $\begin{cases} \dfrac{1}{y} - \dfrac{1}{x} = 1, \\ \dfrac{7}{y} + \dfrac{3}{x} = 5. \end{cases}$

15. $\begin{cases} \dfrac{1}{x} + \dfrac{3}{y} + \dfrac{2}{z} = 9, \\ \dfrac{2}{x} + \dfrac{2}{y} - \dfrac{1}{z} = 11, \\ \dfrac{3}{x} - \dfrac{2}{y} + \dfrac{4}{z} = -4. \end{cases}$

16. $\begin{cases} \dfrac{1}{x} + \dfrac{1}{y} + \dfrac{1}{z} = 6, \\ \dfrac{3}{x} - \dfrac{2}{y} + \dfrac{2}{z} = 1, \\ \dfrac{2}{x} + \dfrac{3}{y} - \dfrac{1}{z} = 9. \end{cases}$

17. $\begin{cases} \dfrac{1}{x} + \dfrac{1}{y} + \dfrac{1}{z} = 3, \\ \dfrac{3}{x} + \dfrac{4}{y} + \dfrac{2}{z} = 4, \\ \dfrac{2}{x} + \dfrac{3}{y} - \dfrac{1}{z} = -5. \end{cases}$

18. $\begin{cases} \dfrac{1}{x} + \dfrac{1}{y} + \dfrac{1}{z} = 0, \\ \dfrac{2}{x} - \dfrac{1}{y} + \dfrac{5}{z} = 3, \\ -\dfrac{1}{x} - \dfrac{2}{y} + \dfrac{1}{z} = 0. \end{cases}$

19. $\begin{cases} \dfrac{3}{x} + \dfrac{1}{y} + \dfrac{1}{z} = 4, \\ \dfrac{2}{x} + \dfrac{3}{y} - \dfrac{2}{z} = -5, \\ \dfrac{2}{x} - \dfrac{1}{y} - \dfrac{3}{z} = -3. \end{cases}$

20. $\begin{cases} \dfrac{2}{x} + \dfrac{5}{y} + \dfrac{3}{z} = 7, \\ \dfrac{5}{x} + \dfrac{9}{y} - \dfrac{7}{z} = 5, \\ \dfrac{3}{x} + \dfrac{2}{y} - \dfrac{4}{z} = -2. \end{cases}$

21. $\begin{cases} \dfrac{1}{3(x-1)} - \dfrac{1}{2(y+1)} = \dfrac{7}{6}, \\ \dfrac{1}{3(x-1)} - \dfrac{1}{6(y+1)} = \dfrac{1}{6}. \end{cases}$

22. $\begin{cases} \dfrac{3}{x-1} + \dfrac{2}{y-1} = 4, \\ \dfrac{1}{x-1} + \dfrac{3}{y-1} = 13. \end{cases}$

23. $\begin{cases} \dfrac{a}{x} - \dfrac{b}{y} = 1, \\ \dfrac{b}{x} + \dfrac{a}{y} = 1, \ (a \neq \pm b, \ a^2 + b^2 \neq 0). \end{cases}$

24. $$\begin{cases} \dfrac{3}{4(x-2)} + \dfrac{1}{y-3} = \dfrac{1}{4}, \\ \dfrac{1}{2(x-2)} + \dfrac{1}{4(y-3)} = 1. \end{cases}$$

64. Verbal Problems

Many problems written in prose, practical and otherwise, lead to linear equations or systems of linear equations. The most important part of the work of solving such a problem is that of translating it from the clumsy language of everyday prose into the more convenient language of algebraic symbolism. Hence for each problem the student should write out a careful list or vocabulary of all symbols to be used and express all quantities to be considered in terms of those symbols.* If you fail to write down what your symbols stand for, you are writing in a secret code, thus defeating the purpose of written language, which is to communicate ideas clearly. In particular, be careful to make note of the units in which the quantities mentioned are expressed and to convert them to similar units before working out the problem.

The following procedure should be used in solving all problems:

(*a*) Read the problem, to get the general idea.

(*b*) Read it again, paying attention to the relationships among the various quantities involved.

(*c*) Choose symbols for what appear to be the basic quantities which have to be found (the unknowns).

(*d*) Read the problem again, phrase by phrase, and write down the list or vocabulary of all quantities involved together with their expressions in terms of the unknowns chosen in step (*c*).

(*e*) Read the problem again, paying careful attention to the parts which express relationships among the quantities which can be translated into equations.

* Various schemes of arrangement of this vocabulary on the page, such as "boxes," or other bookkeeping devices, seem to be popular. While they may save a few seconds in certain "types" of problems, they are restricted to certain types and tend in general to obscure the reasoning even for the types to which they apply. It is better, in the author's opinion, for the student to learn to translate problems into algebra apart from classifying them into types, so that he will become unafraid to attack an unfamiliar problem.

(*f*) When you have as many equations as unknowns, solve the equations.

(*g*) Check the answers against the language of the original verbal problem, *not in the equations obtained in step* (*e*).

Step (*f*) may take the most room on paper, but steps (*a*)–(*e*) merit at least as much attention.

Example 1. A man is 24 years older than his son. Eight years ago, he was twice as old as his son was. What are their present ages?

First solution. Let x be the age of the son now, expressed in years. Then $x + 24$ is the father's age now, in years. Eight years ago the son was $x - 8$ years old, and the father was $x + 24 - 8 = x + 16$ years old. The second sentence of the problem may be translated into the equation

$$x + 16 = 2(x - 8).$$

Solving this equation, we get $x = 32$, the son's present age. Hence $x + 24 = 56$, the father's present age. Checking, we find that 8 years ago the son was 24 and the father 48, or twice 24.

Second solution. Let x be the father's age now, in years, and let y be the son's age now, in years. Eight years ago the father was $x - 8$ years old and the son was $y - 8$ years old. Then the first sentence of the problem may be translated into the equation

$$x = y + 24. \tag{1}$$

The second sentence of the problem may be translated into the equation

$$x - 8 = 2(y - 8). \tag{2}$$

Solving the system of equations (1) and (2), we get $x = 56$ and $y = 32$. The check is as before.

Example 2. A can do a job in 6 days alone, and B can do it alone in 10 days. How long will it take them to do the job together?

Solution. Let x be the number of days required for both to do the job together. In one day, A can do 1/6 of the job, B can do 1/10 of the job, and both together can do $1/x$ of the job. Therefore, we have the equation

$$\frac{1}{6} + \frac{1}{10} = \frac{1}{x}.$$

To solve this equation, we multiply both sides by the L.C.D. $30x$, obtaining $5x + 3x = 30$ or $x = 30/8 = 15/4 = 3\frac{3}{4}$ days. The check is left to the student.

Example 3. How much acid must be added to 48 grams of a solution 75% pure in order to obtain a new solution 76% pure?

Solution. Let x be the number of grams of acid to be added. The total weight of the new solution therefore will be $48 + x$ grams. The number of grams of acid in the original solution is $0.75 \times 48 = 36$. Hence the number of grams of acid in the new solution is $36 + x$. Since the new solution is to be 76% pure, we have the equation

$$36 + x = 0.76(48 + x).$$

Solving the equation, we get $36 + x = 36.48 + 0.76x$, or $0.24x = 0.48$, or $x = 2$ grams. The check is left to the student.

Example 4. A lever, or teeterboard, or seesaw, is a rigid bar supported at one point called the fulcrum. If a weight w is

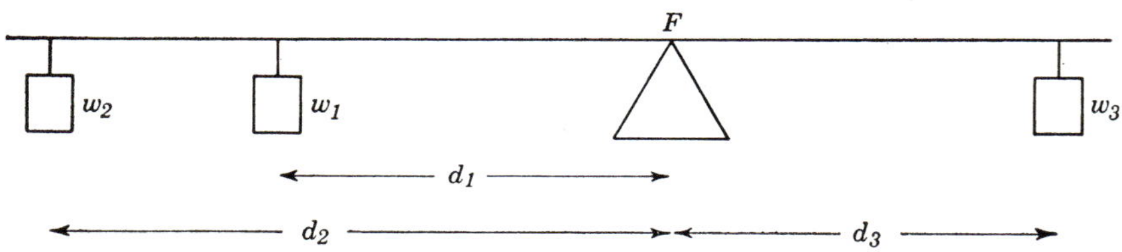

Fig. 27.

attached to the bar at a point at distance d from the fulcrum, the *moment* of this weight with respect to the fulcrum is defined as the product wd. A principle of physics states that if several weights are so attached, the system will be in balance or equilibrium if and only if the sum of the moments of the weights on one side of the fulcrum equals the sum of the moments on the other side. In Fig. 27, we have $w_1d_1 + w_2d_2 = w_3d_3$. If $d_1 = 3$ feet, $d_2 = 5$ feet, $w_1 = 2$ pounds, $w_2 = 3$ pounds, $w_3 = 7$ pounds, find the value of d_3 which will produce equilibrium.

Solution. Substituting in the above formula, $2 \cdot 3 + 3 \cdot 5 = 7d_3$, or $7d_3 = 21$, or $d_3 = 3$ feet.

Example 5. A boat, operating uniformly at full power, goes 5 miles downstream in 60 minutes and returns in 90 minutes. What would be the speed of the boat in still water, and what is the rate of the current?

Solution. Let x be the speed of the boat in still water, and let y be the speed of the current, measured in miles per hour. Then the speed of the boat downstream is $x + y$ and the speed upstream is $x - y$.* Since rate times time is distance, for uniform motion, we have the equations

$$1 \cdot (x + y) = 5$$
$$1.5 \cdot (x - y) = 5.$$

Solving this system of equations, we find that $x = 4\frac{1}{6}$ miles per hour and $y = 5/6$ miles per hour. The check is left to the student.

One reason for checking against the wording of the problem and not against the equation (or equations) is that the student may have solved the equation(s) correctly but may have formed the wrong equation(s). Another reason, apart from mistakes, is that a number may satisfy the equation but not the problem; for when we form the equation we assert that *if* the unknowns satisfy the conditions of the problem, *then* they must satisfy the equation. But this does *not* imply the converse proposition, that *if* a number satisfies the equation, *then* it must satisfy the problem. (The converse of a correct proposition may be incorrect.) Thus the solutions of the equation are the only *possible* solutions of the problem, but whether they are in fact *solutions* of the problem remains to be seen by checking. Therefore checking the problem is an essential step.

Note. In practice, it can even happen that the problem asked is insoluble, although the resulting equation may have a solution. The information that the problem is insoluble may well be valuable.

Example 6. An absent-minded professor has a certain number of students in his class, but has forgotten how many. However, he does remember clearly that if he had three times as many, he would have seven more than he now has. How many students has he?

Solution. Let x be the number of students he has. Then $3x = x + 7$. Solving this equation, we get $x = 7/2$, or $3\frac{1}{2}$ students. While 7/2 does check in the equation, it cannot possibly be an answer to the problem since the answer must be an integer, of necessity. But our reasoning has proved that 7/2 is the only possible answer. Hence the problem has no solution, and we can

* This sentence idealizes the actual situation and is in fact only a crude first approximation to the truth. For a more accurate discussion, one would have to take account of the shape of the boat and the advanced theory of the dynamics of fluids. Therefore we shall always make the above idealization here in all similar problems; it is good enough for our purpose.

only conclude that the professor's memory is worse than we thought.

A more significant example of this phenomenon will be given in section 82.

To demonstrate the superiority of algebraic methods over those of arithmetic, one has only to try to solve the problems in the exercises below without using algebraic symbols. Until the 16th century no adequate algebraic symbolism was used.

EXERCISES

1. The sum of two numbers is 30. Five times the smaller exceeds twice the larger by 17. Find the numbers.

2. The sum of two numbers is 53 and their difference is 19. Find the numbers.

3. A man's age is now five ninths of what it will be 18 years from now. How old is he now?

4. One fifth of a man's age 6 years ago equals one seventh of his age 10 years hence. How old is he now?

5. A collection of nickels and quarters totals $3.40. The total number of coins is 24. How many of each kind are there?

6. A collection of nickels and quarters totals $2.60. The total number of coins is 24. How many of each kind are there?

7. A collection of nickels, dimes, and quarters totals $3.40. The total number of coins is 32, and there are twice as many dimes as quarters. How many of each kind are there?

8. A collection of nickels, dimes, and quarters totals $2.70. The total number of coins is 26, and there are twice as many nickels as quarters. How many of each kind are there?

9. A rectangular field is four times as long as it is wide. If it were 6 feet shorter and 3 feet wider, its area would be increased by 54 square feet. Find its length and width.

10. A rectangle is 30 feet longer than it is wide. If its length were increased by 50 feet and its width were diminished by 8 feet, its area would be increased by 200 square feet. Find its dimensions.

11. A man invests part of $6000 at 4% interest and the rest at 3%. His total annual income is $225. How much has he invested at each rate?

12. A student has quiz grades of 69 and 77. What grade must he achieve on a third quiz in order to have an average of 80?

13. A student has an average grade of 77 on three quizzes. What grade must he achieve on a fourth quiz in order to have an average of 80?

14. A shopper spends half her money in the first shop she visits. In the second shop she spends $\frac{1}{3}$ of what remains. In the third shop she spends \$25. She then has \$15 left. How much had she at the start?

15. Find three consecutive integers whose sum is 114.

16. Find three consecutive even integers whose sum is 132.

17. Find three consecutive odd integers whose sum is 87.

18. Three years from now a man will be three times as old as his son will be. Two years ago he was four times as old as his son was. How old are they now?

19. A can do a job alone in 8 days, and B can do it alone in 4 days. How long will it take them to do the job together?

20. A can do a job alone in 8 days, and B can do it alone in 6 days. How long will it take them to do the job together?

21. A can do a job in 6 days, and A and B can do the job together in 2 days. How long would it take B to do the job alone?

22. A can do a job alone in 8 days, B can do it alone in 6 days, and C can do it alone in 10 days. How long will it take them to do the job together?

23. One input pipe can fill a tank alone in 8 hours, another input pipe can fill it alone in 6 hours, and a drain pipe can empty the full tank in 10 hours. If the tank is empty, and all three pipes are wide open, how long will it take to fill the tank?

24. Achilles races with a tortoise, giving the tortoise a handicap of 990 yards. Achilles runs at the rate of 500 yards per minute while the tortoise runs at the rate of 5 yards per minute. How long will it take Achilles to catch the tortoise?

25. A freight train, running at the rate of 20 miles per hour, leaves a station 3 hours before an express, which travels in the same direction at the rate of 50 miles per hour. How long after the express leaves will it catch the freight, and how far from the station will they be?

26. A man walks 31 miles, partly at the rate of 2 miles per hour and the rest at the rate of 5 miles per hour. If he had walked at the rate of 5 miles per hour for the same time as he actually walked at the rate of 2 miles per hour, and vice versa, he would have covered 15 miles more than he did. How long did it take him to walk the 31 miles?

27. An airplane travels 990 miles in 3 hours with the wind and makes the return trip in 3 hours and 40 minutes. Find the speed of the airplane in still air and the speed of the wind.

28. A boat travels for 4 hours with a current of 2 miles per hour, and returns the same distance against the current in 5 hours. What is the boat's speed in still water, and how far did it travel each way?

29. A boat travels 5 miles downstream in 15 minutes and makes the return trip in 20 minutes. Find the speed of the boat in still water and the speed of the current.

30. An airplane bucks a headwind of 40 miles per hour for 4 hours, and returns the same distance assisted by a tailwind of 40 miles per hour in 3 hours. What is its speed in still air and how far did it fly each way?

31. At what time between 4 and 5 o'clock do the hands of a clock coincide?

32. At what time between 3 and 4 o'clock do the hands of a clock coincide?

33. A 60-pound boy sits on one side of a teeterboard 4 feet from the fulcrum. A second boy sits on the other side 5 feet from the fulcrum and balances the first boy. Find the weight of the second boy.

34. How far from the fulcrum of a teeterboard must an 80-pound boy sit to balance a 50-pound boy who is 4 feet from the fulcrum?

35. A 50-pound boy and a 30-pound boy sit at opposite ends of a 16-foot teeterboard. How far from the first boy should the fulcrum be placed in order to make them balance the board?

36. How many pounds of force must be applied to one end of a lever 6 feet long in order to move a 100-pound boulder if the fulcrum is one foot from the boulder?

37. A 5-pound weight is hung 4 feet from the fulcrum of a lever, and a 2-pound weight is hung on the same side 2 feet from the fulcrum. Where must an 8-pound weight be hung in order to balance them?

38. How much weight can be balanced by a force of 40 pounds exerted at the end of a 4-foot crowbar if the fulcrum is 6 inches from the opposite end of the bar?

39. How many gallons of cottonseed oil must be added to 45 gallons of a solution of olive oil which is 90% olive oil in order to make the resulting solution 80% olive oil?

40. A mixture contains 30 pounds of sugar and 50 pounds of flour. How much flour must be added to make a mixture that is 75% flour?

41. How much water must be evaporated from 80 pounds of a 12% solution of salt in order to obtain a 20% solution?

42. How many ounces of each of a 50% solution and a 10% solution must be mixed in order to obtain 24 ounces of a 25% solution?

43. A confectioner wishes to mix candy worth 70 cents per pound with candy worth 40 cents per pound to make 42 pounds of an assortment worth 50 cents per pound. How much of each kind should he use?

44. A tobacconist wishes to mix tobacco worth 13 cents per pound with tobacco worth 18 cents per pound in order to make 80 pounds of a mixture worth 15 cents per pound. How many pounds of each kind should he use?

45. If the length of a rectangle were increased by 2 feet and the width decreased by 2 feet, the area would be decreased by 20 square feet. If the length were decreased by 2 feet and the width increased by 1 foot, the area would be increased by 1 square foot. Find the dimensions of the rectangle.

46. Three pounds of salt and 5 pounds of sugar cost 94 cents. Six pounds of salt and 2 pounds of sugar cost 76 cents. Find the cost per pound of salt and sugar.

47. A certain two-digit number is equal to five times the sum of its digits. If 9 were added to the number, its digits would be reversed. Find the number. (*Hint:* If u is the units digit and t is the tens digit, the number is $10t + u$.)

48. The units digit of a two-digit number is 2 less than 3 times the tens digit. The sum of the digits is 10. Find the number.

49. Find the angles of a triangle if one angle is half the second, and the third is twice the sum of the first two. (*Hint:* The sum of all the angles of a triangle is 180 degrees.)

50. The age of Diophantus, a brilliant Greek mathematician of about 250 A.D., may be calculated from an epitaph which reads as follows: Diophantus passed one sixth of his life in childhood, one twelfth in youth, and one seventh more as a bachelor; five years after his marriage was born a son who died four years before his father at half his father's final age.

Chapter IX

Integral and fractional exponents

65. Negative Integers and Zero as Exponents

In section 12 and section 23, we defined positive integral exponents as follows:

DEFINITION 1. *If* n *is any positive integer,* $x^n = xx \ldots x$ *where there are* n *factors on the right of the equals sign.*

With this definition, we proved, in section 23, that positive integral exponents obey the following laws:

I. $x^a x^b = x^{a+b}$. *Example.* $x^2x^3 = xx \cdot xxx = x^5$.

II. $(x^a)^b = x^{ab}$. *Example.* $(x^2)^3 = (x^2)(x^2)(x^2) = x^6$.

III. $(xy)^a = x^a y^a$. *Example.* $(xy)^3 = xyxyxy = xxxyyy = x^3y^3$.

IV. *If* $y \neq 0$,

$$\left(\frac{x}{y}\right)^a = \frac{x^a}{y^a}. \quad \textit{Example.} \left(\frac{x}{y}\right)^3 = \frac{x}{y} \cdot \frac{x}{y} \cdot \frac{x}{y} = \frac{x^3}{y^3}.$$

V. *If* $x \neq 0$,

(*i*) $\dfrac{x^a}{x^b} = x^{a-b}$ if $a > b$. *Example.* $\dfrac{x^5}{x^2} = \dfrac{xxx\not{x}\not{x}}{\not{x}\not{x}} = x^3$.

(*ii*) $\dfrac{x^a}{x^b} = \dfrac{1}{x^{b-a}}$ if $a < b$. *Example.* $\dfrac{x^2}{x^5} = \dfrac{\not{x}\not{x}}{\not{x}\not{x}xxx} = \dfrac{1}{x^3}$.

(*iii*) $\dfrac{x^a}{x^b} = 1$ if $a = b$. *Example.* $\dfrac{x^2}{x^2} = 1$.

The occurrence of three different cases in *V* is a source of inconvenience, especially if we wish to use literal exponents rather than particular numbers, and it is natural to wish to unify them.

Let us boldly use $V(i)$ mechanically even when $a < b$. For example, $\frac{x^2}{x^5} = x^{2-5} = x^{-3}$. But a negative exponent has no meaning under definition 1. The symbol x^{-3} cannot possibly be interpreted as the product of -3 x's, except possibly at a séance; but x^{-3} will not stand for the product of the ghosts of 3 departed x's. The symbol x^{-3} does not come under the jurisdiction of definition 1 at all and is at present a completely undefined symbol. This is fortunate because, since x^{-3} has no preassigned meaning, we are free to give it any meaning that suits our convenience without fear of contradicting previous results. It will be convenient to assign to the symbol x^{-3} as its meaning the right answer to our problem, that is $1/x^3$, for then we can dispense with $V(ii)$ altogether. Hence we are led to make the following definition.

Definition 2. *If $-n$ is a negative integer,*

$$x^{-n} = \frac{1}{x^n} \qquad (x \neq 0).$$

Example 1. $\frac{x^2}{x^5} = x^{-3} = \frac{1}{x^3}.$

Example 2. $2^{-3} = \frac{1}{2^3} = \frac{1}{8}.$

Similarly let us apply $V(i)$ when $a = b$. For example, we may write $\frac{x^2}{x^2} = x^{2-2} = x^0$. But the symbol x^0 has no sense under either definition 1 or definition 2, since 0 is neither positive nor negative. This is again fortunate, since we are free to assign to the symbol x^0 any meaning that suits our convenience. Again we choose the right answer, 1, since this allows us to use $V(i)$ in all cases. Hence we make the following definition.

Definition 3. *If $x \neq 0$, $x^0 = 1$.*

Example 3. $5^0 = 1$.

We can now replace $V(i)(ii)(iii)$ above by the single law:

$V.$ $\frac{x^a}{x^b} = x^{a-b}$ $(x \neq 0)$.

It can be proved that the laws *I*, *II*, *III*, *IV*, and the new *V* hold for all integral exponents, positive, negative, or zero.*

EXERCISES

Simplify each of the following by removing non-positive exponents:

1. $3x^{-2}$. **2.** $(3x)^{-2}$. **3.** $3x^{0}$. **4.** $(3x)^{0}$. **5.** $3x^{-2}y^{3}$.

6. $3a^{-2}b^{0}$. **7.** a^{-1}/b^{-1}. **8.** $a^{-1}b^{-1}$. **9.** $\dfrac{a^{-1}-b^{-1}}{a^{-1}+b^{-1}}$.

10. $\dfrac{x^{-1}}{x^{-1}+2y^{-1}}$. **11.** $\dfrac{2b^{-1}}{a^{-1}-2b^{-1}}$. **12.** $\dfrac{x}{y^{-1}}+\dfrac{y}{x^{-1}}$.

13. $\dfrac{x^{-1}+y^{-1}}{x^{-2}-y^{-2}}$. **14.** $\dfrac{1}{x^{-1}+y^{-1}}$.

15. $(x^{-1}+y^{-1})(x+y)^{-1}$. **16.** $\dfrac{a^{-1}-b^{-1}}{a^{2}-b^{2}}$.

17. $\dfrac{1}{x^{-2}-y^{-2}}$. **18.** $\dfrac{x^{-2}+y^{-2}}{x^{-2}-y^{-2}}$.

Evaluate each of the following:

19. $(-2)^{-3}$. **20.** $(-2)^{-4}$. **21.** $(\frac{1}{2})^{-3}$.
22. 5^{0}. **23.** $(\frac{1}{2})^{0}$. **24.** $2\cdot 3^{-2}$.
25. $(2\cdot 3)^{-2}$. **26.** $2\cdot 10^{0}$. **27.** $(2\cdot 10)^{0}$.
28. $2\cdot 10^{2}$. **29.** $2\cdot 10^{-2}$. **30.** $(2\cdot 10)^{-2}$.

Simplify, using the laws I–V for exponents:

31. $(2x^{3}y^{2})(4x^{-2}y^{-3})$. **32.** $\dfrac{6x^{4}y^{3}}{4x^{-3}y^{-3}}$.

33. $(2x^{-2})^{-3}$. **34.** $\dfrac{(xy^{3})^{-2}}{(x^{2}y)^{-3}}$.

35. $x^{-2}(x^{7}+2x^{3}+5x^{2})$. **36.** $\dfrac{(ab^{2})^{-3}(a^{2}b^{-1})^{2}}{(a^{3}b^{4})^{-1}}$.

37. $(2x^{0}y^{2})^{3}(y^{-2})^{2}(x^{3}y^{2})^{0}$. **38.** $\dfrac{(3x^{2}y)^{-2}(xy)^{4}}{(x^{-1}y^{3})^{3}}$.

39. $10^{3}\cdot 10^{0}\cdot 10^{-2}$. **40.** $\dfrac{10^{5}\cdot 10^{-6}}{10^{-2}}$.

41. $\dfrac{10^{7}\cdot 10^{-4}}{10^{3}}$. **42.** $\dfrac{(3\cdot 10^{4})(6\cdot 10^{-1})}{2\cdot 10^{2}}$.

43. $\dfrac{(2\cdot 10^{3})^{2}(6\cdot 10^{-1})^{2}}{2\cdot 10^{3}}$. **44.** $\dfrac{(2\cdot 10^{2})^{3}(6\cdot 10^{-2})^{2}}{(4\cdot 10^{-1})^{2}(2\cdot 10^{3})^{2}}$.

* For indications of the proof, see M. Richardson, *Fundamentals of Mathematics,* Macmillan, 1941.

66. A Use of Powers of Ten in Scientific Writing

When very large or very small numbers are used in scientific writing, it is customary to express them in terms of powers of ten, as follows. We write $1{,}000{,}000 = 10^6$, $256{,}000{,}000 = 256 \times 10^6$ or 25.6×10^7 or 2.56×10^8; we use the $\times$ sign for multiplication to avoid confusion with the decimal point. Similarly, we write $0.0001 = 10^{-4}$, $0.000256 = 256 \times 10^{-6}$ or 25.6×10^{-5} or 2.56×10^{-4}. Multiplying a number by 10^4 moves the decimal point four places to the right, while multiplying by 10^{-4} moves the decimal point four places to the left.

It can be proved that every positive real number can be expressed as a number between 1 and 10 multiplied by a suitable integral power of 10; when this is done, the number is said to be expressed in **standard form.**

For example, 256,000,000 has the standard form 2.56×10^8; 0.000256 has the standard form 2.56×10^{-4}; and 2.56 has the standard form 2.56×10^0.

This standard form is more compact and comprehensible than the everyday way of writing such numbers with large numbers of zeros before or after the decimal point. It also lends itself to calculations with cumbersome numbers, large or small, which are made easier by the use of the laws of exponents.

Example. Calculate $\dfrac{256{,}000{,}000 \times 0.0004}{2000}$. We write

$$\frac{2.56 \times 10^8 \times 4 \times 10^{-4}}{2 \times 10^3} = 5.12 \times 10^{8-4-3} = 5.12 \times 10^1 = 51.2.$$

While the powers of ten are combined according to the laws of exponents, their coefficients are multiplied and divided in the ordinary way. How this may be avoided by expressing them too as (fractional) powers of ten will be discussed in the chapter on logarithms.

EXERCISES

Express each of the following in standard form:

1. 500,000. **2.** 32,000. **3.** 3,470,000.
4. 2,340,000,000. **5.** 0.427. **6.** 0.0268.
7. 4.16. **8.** 41.8. **9.** 0.00043. **10.** 0.00000567.

Express each of the following in ordinary decimal (positional) notation:

11. 10^5. **12.** 10^7. **13.** 10^{-5}. **14.** 10^{-7}.
15. 3.78×10^7. **16.** 3.67×10^9. **17.** 5.76×10^{-5}.
18. 9.13×10^{-4}. **19.** 6.27×10^{-6}. **20.** 4.64×10^0.

Calculate, using the laws for exponents, and express the result in ordinary positional (decimal) notation:

21. $\dfrac{(8 \times 10^7) \times (3 \times 10^{-2})}{2 \times 10^4}$. **22.** $\dfrac{(5.2 \times 10^6) \times (4 \times 10^{-8})}{13 \times 10^{-4}}$.

23. If c represents the velocity of radiant energy in a vacuum, L its wave length, and f its frequency, then $f = c/L$. Suppose $c = 3 \times 10^{10}$ centimeters per second and $L = 6 \times 10^{-5}$ centimeters. Find f. (*Hint:* write $3 \times 10^{10} = 30 \times 10^9$.)

24. An angstrom is a unit of length equal to 10^{-8} centimeters. If one centimeter equals 0.3937 inch, express in inches the wave length of red light with a wave length of 8000 angstroms.

25. A light-year is a unit of length equal to 5.88×10^{12} miles. The distance of the cluster of stars called the Pleiades is 1.2936×10^{15} miles. How many light-years is this?

26. The Great Nebula in Andromeda is approximately 5.292×10^{18} miles away. How many light-years is this? Use the data of exercise 25.

27. If a gram equals 0.002205 pound and the mass of the earth is 5.97×10^{27} grams, find the mass of the earth in pounds.

28. The mass of the sun is 1.98×10^{33} grams. How many pounds is this? Use the data of exercise 27.

29. One coulomb equals 3.00×10^9 statcoulombs. The charge on an electron is 4.80×10^{-10} statcoulombs. How many coulombs is this?

30. One parsec equals 3.084×10^{13} kilometers. One astronomical unit (A.U.) equals 1.495×10^8 kilometers. How many astronomical units are there in 1.5 parsecs?

67. Fractional Exponents

In section 65 we extended the concept of exponent to include all integers as exponents. It is natural to try to extend it still further, to include fractions. In making such an extension (since, if we do it at all, we do it for our own convenience) we wish to preserve the validity of the basic laws *I–V* of section 65; just as in extending the concept of number itself, in chapter I, from positive integers to other kinds of numbers, we wished to preserve the validity of the basic laws such as the associative, distributive, and commutative laws.

The symbol $x^{\frac{1}{2}}$ has no meaning under any of the previous definitions of exponent, since $\frac{1}{2}$ is not an integer. Therefore we are free to give it any meaning that suits our convenience. If law II (section 65) is to operate, then $(x^{\frac{1}{2}})^2$ will have to yield $x^{\frac{1}{2}\cdot 2} = x^1 = x$. Hence $x^{\frac{1}{2}}$ will have to stand for some quantity which when squared yields x; but this can only be $+\sqrt{x}$ or $-\sqrt{x}$. Hence we choose the definition: $x^{\frac{1}{2}} = \sqrt{x}$. This fits also with law *I* (section 65), for $x^{\frac{1}{2}}x^{\frac{1}{2}} = x^{\frac{1}{2}+\frac{1}{2}} = x^1 = x$ represents the fact that $\sqrt{x}\sqrt{x} = x$. More generally we define $x^{\frac{1}{q}} = \sqrt[q]{x}$, so that law *II* will operate correctly; thus, $(x^{\frac{1}{q}})^q = x^{\frac{1}{q}\cdot q} = x^1 = x$. Finally, if law *II* is to operate, we must have

$$x^{p/q} = x^{\frac{1}{q}\cdot p} = (x^{\frac{1}{q}})^p = (\sqrt[q]{x})^p$$

and

$$x^{p/q} = x^{p\cdot\frac{1}{q}} = (x^p)^{\frac{1}{q}} = \sqrt[q]{x^p}.$$

Thus we are led to make the following definition.

DEFINITION 4. *If p/q is any rational number, expressed with positive denominator q, we define $x^{p/q} = (\sqrt[q]{x})^p = \sqrt[q]{x^p}$.*

Example 1. $x^{-\frac{2}{3}} = \sqrt[3]{x^{-2}} = (\sqrt[3]{x})^{-2}$. In particular, $8^{-2/3} = \sqrt[3]{8^{-2}} = \sqrt[3]{\dfrac{1}{64}} = \dfrac{1}{4}$ or $8^{-2/3} = (\sqrt[3]{8})^{-2} = 2^{-2} = \dfrac{1}{2^2} = \dfrac{1}{4}$.

Note that any rational number can be expressed with positive denominator, as $\dfrac{3}{-2} = \dfrac{-3}{2}$, so that this is no restriction.

It can now be proved* that fractional exponents obey laws *I–V*, and also that any fractional exponent may be replaced by an equal fraction. Thus $x^{4/6} = x^{2/3}$.

It is desirable for the present to confine ourselves to non-negative values of x in fractional powers of x in order to avoid

* For indications of the proof, see M. Richardson, *Fundamentals of Mathematics*, Macmillan, 1941, section 52.

difficulties involving imaginary numbers and other difficulties whose explanation is best left for more advanced courses. For example, from $(-1)^{1/3} = (-1)^{2/6}$ we might be tempted to conclude that $\sqrt[3]{-1} = \sqrt[6]{(-1)^2}$ or $\sqrt[3]{-1} = \sqrt[6]{+1}$, or $-1 = +1$.

Fractional exponents may be used to simplify radical expressions, as in the following example, but their real importance lies in the fact that they form a basis for the theory of logarithms, discussed in chapter XXIII, which may be taken up next if desired.

Example 2. $\sqrt[5]{x^3} \cdot \sqrt[3]{x^2} = x^{\frac{3}{5}}x^{\frac{2}{3}} = x^{\frac{3}{5}+\frac{2}{3}} = x^{19/15} = \sqrt[15]{x^{19}}$ or $(\sqrt[15]{x})^{19}$.

While positive integral exponents were used by René Descartes (French, 1596–1650), negative and fractional exponents were not used extensively until the time of Isaac Newton (English, 1642–1727). The concept of exponent can be extended further to include all real and even all complex numbers as exponents, but this is beyond the scope of this book.

EXERCISES

Evaluate each of the following:

1. $16^{1/4}$. **2.** $16^{-1/4}$. **3.** $16^{-1/2}$. **4.** $16^{1/2}$.
5. $8^{-1/3}$. **6.** $8^{2/3}$. **7.** $36^{1/2}$. **8.** $27^{2/3}$.
9. $16^{-3/4}$. **10.** $27^{-4/3}$. **11.** $100^{-3/2}$. **12.** $8^{4/3}$.
13. $16^{3/2}$. **14.** $16^{-6/4}$. **15.** $10^{1/3} \times 10^{5/3}$.
16. $10^{5/2} \times 10^{1/2}$. **17.** $\dfrac{10^{1/2} \times 10^{3/2}}{10^{-1}}$.
18. $\dfrac{10^{-4/5} \times 10^{9/5}}{10^{-2}}$. **19.** $(10^6)^{1/3}$.
20. $(2^{16})^{1/4}$. **21.** $(5^4)^{-1/4}$. **22.** $(5^{-3})^{-1/3}$.

Simplify each of the following, all letters representing positive numbers:

23. $(8a^6)^{1/3}$. **24.** $(16a^4)^{-1/2}$. **25.** $(x^3x^{1/2})^{-2}$.
26. $(x^{1/2}x^{2/3})^6$. **27.** $\left(\dfrac{a^{1/2}b^{4/3}}{a^{-2}b}\right)^3$. **28.** $(9x^{-4}y^2)^{3/2}$.
29. $\left(\dfrac{9x^3y}{4x^{-1}y^{-3}}\right)^{1/2}$. **30.** $\left(\dfrac{64a^{-5}y^7}{a^{-1}y^4}\right)^{2/3}$. **31.** $\sqrt[4]{x^3}\,(\sqrt[3]{x})^2$.
32. $\sqrt[5]{x^2}\,(\sqrt{x})^3$. **33.** $\dfrac{(\sqrt[3]{x})^5}{\sqrt[3]{x^2}}$. **34.** $\dfrac{\sqrt[3]{x^5}}{\sqrt[4]{x^3}}$.
35. $(a^{1/2} + b^{1/2})^2$. **36.** $(a^{1/2} - b^{1/2})^2$.
37. $(a^{1/3} + b^{1/3})^3$. **38.** $(a^{1/3} - b^{1/3})^3$.

Chapter X

Radicals

68. Basic Laws for Radicals

If $x^n = a$, where n is an integer greater than 1, then x is called an ***n*th root of *a*.** It will be proved in chapter XVIII that every complex number has n distinct nth roots. To avoid confusion, we single one out, to be called the **principal *n*th root,** as follows.

If a is a positive (real) number, the principal nth root will be the positive nth root. If n is odd and a is a negative (real) number, the principal nth root will be the negative nth root. For the present no other cases will be considered. It can be proved that the principal nth root is uniquely determined.

The **radical** $\sqrt[n]{a}$ shall stand for the principal nth root. The quantity a is called the **radicand,** n is called the **index** or **order** of the radical. The index 2 is usually omitted from the symbol.

Examples. $\sqrt{9} = 3$, not -3. To designate -3 we would write $-\sqrt{9}$. Thus $\pm\sqrt{9} = \pm 3$. Also, $\sqrt[3]{-8} = -2$.

Radicals obey the following laws:

I. $(\sqrt[n]{a})^n = a$. *Example:* $(\sqrt[3]{6})^3 = 6$.

II. $\sqrt[n]{ab} = \sqrt[n]{a}\,\sqrt[n]{b}$. *Example:* $\sqrt[3]{24} = \sqrt[3]{8 \cdot 3} = \sqrt[3]{8}\,\sqrt[3]{3} = 2\sqrt[3]{3}$.

III. $\sqrt[n]{\dfrac{a}{b}} = \dfrac{\sqrt[n]{a}}{\sqrt[n]{b}}$. *Example:* $\sqrt[3]{\dfrac{9}{8}} = \dfrac{\sqrt[3]{9}}{\sqrt[3]{8}} = \dfrac{\sqrt[3]{9}}{2} = \dfrac{1}{2}\sqrt[3]{9}$.

IV. $\sqrt[m]{\sqrt[n]{a}} = \sqrt[mn]{a}$. *Example:* $\sqrt[4]{\sqrt[3]{a}} = \sqrt[12]{a}$.

V. $\sqrt[n]{a^m} = (\sqrt[n]{a})^m = a^{m/n}$. *Example:* $\sqrt[4]{a^3} = (\sqrt[4]{a})^3 = a^{3/4}$.

To avoid difficulties with imaginary numbers, we shall hereafter in this chapter suppose that all letters represent positive (real) numbers.

Proof of I. By definition.

Proof of II. If $x = \sqrt[n]{a}$ and $y = \sqrt[n]{b}$, then $x^n = a$ and $y^n = b$. Thus $(xy)^n = x^n y^n = ab$. But x and y are positive; hence xy is positive. Therefore xy is the principal nth root of ab, or $xy = \sqrt[n]{ab}$. This completes the proof. If a or b or both are negative and n is odd, a similar proof can be given.

Proof of III. Similar to proof of *II*. We leave this to the student as an exercise.

Proof of IV. Let $x = \sqrt[n]{a}$. Then $x^n = a$. Let $y = \sqrt[m]{x}$. Then $y^m = x$. Raising both sides to the nth power, we have $(y^m)^n = x^n$, or $y^{mn} = a$. But y is positive and hence is the principal mnth root of a, or $y = \sqrt[mn]{a}$. By substitution, $\sqrt[m]{\sqrt[n]{a}} = \sqrt[mn]{a}$.

Proof of V. This is left to the student as an exercise.

69. The Simplest Form of a Radical

A rational number is called a **perfect nth power** if it is the nth power of some rational number. A rational expression is called a **perfect nth power** if it is the nth power of some rational expression.

Examples. $\frac{4}{9}$ is a perfect square because $\frac{4}{9} = (\frac{2}{3})^2$. Similarly, $8a^6b^9$ is a perfect cube because $8a^6b^9 = (2a^2b^3)^3$.

An irrational nth root of a rational number is sometimes called a **surd.** If the index of the root is 2, the surd is called a **quadratic surd.** Thus, $\sqrt{3}$ and $\sqrt{\frac{1}{3}}$ are quadratic surds.

The following are the usual ways in which radicals are simplified.

(*A*) *Removal of factors from the radicand.* To remove factors from a radicand, express the radicand as a product of factors which are, as far as possible, perfect nth powers. Then use law *II* and extract the nth roots of all perfect nth powers.

Example 1. $\sqrt{18a^3b^5} = \sqrt{9a^2b^4 \cdot 2ab} = \sqrt{9a^2b^4}\,\sqrt{2ab} = 3ab^2\sqrt{2ab}$.

Example 2. $\sqrt[3]{-5x^7} = \sqrt[3]{(-1)x^6 \cdot 5x} = \sqrt[3]{(-1)x^6}\,\sqrt[3]{5x} = -x^2\,\sqrt[3]{5x}$.

It is sometimes possible to compute a root approximately by removing factors as above, and then using table I.

Example 3. $\sqrt{125} = \sqrt{25 \cdot 5} = 5\sqrt{5}$. From table I we find $\sqrt{5} = 2.236$ approximately. Hence $\sqrt{125} = 5(2.236) = 11.180$ approximately.

(*B*) *Rationalizing the denominator.* It is usually convenient to remove denominators of radicands, as follows:

Example 4.

$$\sqrt[3]{\frac{5}{2x^2}} = \sqrt[3]{\frac{5}{2x^2}\cdot\frac{4x}{4x}} = \sqrt[3]{\frac{20x}{8x^3}} = \frac{\sqrt[3]{20x}}{\sqrt[3]{8x^3}} = \frac{\sqrt[3]{20x}}{2x} = \frac{1}{2x}\sqrt[3]{20x}.$$

This process is called **rationalizing the denominator.** To rationalize the denominator of a radicand of index n, multiply numerator and denominator of the radicand by the simplest expression that will make the denominator a perfect nth power. Then use law *III*, as in the above example.

Example 5. Compute $\sqrt{\frac{2}{3}}$ by use of table I.

Solution. $\sqrt{\frac{2}{3}} = \sqrt{\frac{2}{3}\cdot\frac{3}{3}} = \frac{\sqrt{6}}{\sqrt{9}} = \frac{1}{3}\sqrt{6} = \frac{1}{3}(2.449) = 0.816$ approximately.

(*C*) *Lowering the index of a radical.* This can sometimes be done by expressing the radicand as a power of some quantity and using law *V*.

Example 6. $\sqrt[6]{27} = \sqrt[6]{3^3} = 3^{3/6} = 3^{1/2} = \sqrt{3}$.

A radical will be said to be in **simplest form** *if we have*

(*a*) *reduced the radicand to a single fraction in lowest terms;*
(*b*) *rationalized the denominator;*
(*c*) *removed all factors which are perfect nth powers, where n is the index of the radical;*
(*d*) *lowered the index of the resulting radical as far as possible.*

Example 7.

$$\sqrt{\frac{1}{a} + (2 + a)} = \sqrt{\frac{1 + 2a + a^2}{a}} = \sqrt{\frac{(1 + a)^2}{a}\cdot\frac{a}{a}} = \frac{1 + a}{a}\sqrt{a}.$$

EXERCISES

Reduce to simplest form:

1. $(\sqrt{5})^2$. 2. $(\sqrt[3]{5})^3$. 3. $(\sqrt[5]{a})^5$. 4. $(\sqrt[3]{a^2})^3$.
5. $\sqrt{3}\cdot\sqrt{12}$. 6. $\sqrt{2}\cdot\sqrt{18}$. 7. $\sqrt[3]{3}\cdot\sqrt[3]{9}$.
8. $\sqrt[3]{2}\cdot\sqrt[3]{4}$. 9. $\sqrt[5]{4}\cdot\sqrt[5]{8}$. 10. $\dfrac{\sqrt[3]{32}}{\sqrt[3]{4}}$.
11. $\dfrac{\sqrt[3]{54}}{\sqrt[3]{2}}$. 12. $\dfrac{\sqrt{48}}{\sqrt{12}}$. 13. $\sqrt{18}$. 14. $\sqrt{48}$.
15. $\sqrt{72}$. 16. $\sqrt{75}$. 17. $\sqrt{80}$. 18. $\sqrt{98}$.
19. $\sqrt[3]{32}$. 20. $\sqrt[3]{81}$. 21. $\sqrt{\frac{1}{8}}$. 22. $\sqrt{\frac{3}{2}}$.
23. $\sqrt{\frac{5}{8}}$. 24. $\sqrt{\frac{7}{12}}$. 25. $\sqrt[4]{\sqrt[3]{5}}$. 26. $\sqrt[3]{\sqrt{2}}$.
27. $\sqrt[8]{2^4}$. 28. $\sqrt[6]{7^3}$. 29. $\sqrt[6]{64}$. 30. $\sqrt[9]{27}$.
31. $\sqrt{\dfrac{1-\frac{1}{2}}{2}}$. 32. $\sqrt{\dfrac{1+\frac{1}{2}}{2}}$. 33. $\sqrt{1-\left(\dfrac{\sqrt{3}}{2}\right)^2}$.
34. $\sqrt{1-(\frac{4}{5})^2}$. 35. $\sqrt[3]{-64}$. 36. $\sqrt[3]{-27}$.
37. $\sqrt[5]{-32}$. 38. $\sqrt[3]{-125}$. 39. $\sqrt{16x^8y^{10}}$.
40. $\sqrt{9a^{16}x^2}$. 41. $\sqrt{4x^{36}y^9}$. 42. $\sqrt[3]{-8a^6x^9}$.
43. $\sqrt[3]{-27a^8y^6}$. 44. $\sqrt[3]{16a^6b^8}$. 45. $\sqrt[3]{54x^{10}y^8}$.
46. $\sqrt{81x^9y^{25}}$. 47. $\sqrt{\dfrac{a^6}{9b^8}}$. 48. $\sqrt{\dfrac{12x^9}{y^8}}$.
49. $\sqrt{\dfrac{16}{25x^8y^5}}$. 50. $\sqrt{\dfrac{4x^7}{9y^3}}$. 51. $\sqrt{\dfrac{18x^5}{25y^7}}$.
52. $\sqrt{\dfrac{50x^5y^6}{48z^3}}$. 53. $\sqrt[3]{\dfrac{8x^8}{y^6}}$. 54. $\sqrt[3]{\dfrac{54x^7}{y^8}}$.
55. $\sqrt{a^2x^2+b^2x^2}$. 56. $\sqrt{x^2y^2-x^2z^2}$.
57. $\sqrt{16x^2+16}$. 58. $\sqrt{9x^2-36}$.
59. $\sqrt{\dfrac{1}{x}+(6+9x)}$. 60. $\sqrt{\dfrac{4}{x}-4+x}$.
61. $\sqrt{\dfrac{9}{2x}+6+2x}$. 62. $\sqrt{\dfrac{9}{x}-12+4x}$.

Compute by use of table I:

63. $\sqrt{\frac{2}{5}}$. 64. $\sqrt{\frac{7}{2}}$. 65. $\sqrt{\frac{7}{3}}$. 66. $\sqrt{\frac{5}{7}}$.
67. $\sqrt{\frac{13}{6}}$. 68. $\sqrt{\frac{13}{8}}$. 69. $\sqrt[3]{\frac{5}{3}}$. 70. $\sqrt[3]{\frac{3}{5}}$.

70. Addition and Subtraction of Radicals

Two or more radicals are called **similar** if they have the same index and radicand. An algebraic sum of terms involving radicals may be simplified by reducing each radical to simplest form and collecting terms with similar radicals, as follows.

Example. $\sqrt{8} + \sqrt{18} - \sqrt{32} = 2\sqrt{2} + 3\sqrt{2} - 4\sqrt{2} = (2 + 3 - 4)\sqrt{2} = \sqrt{2}$.

EXERCISES

Simplify and collect terms:

1. $3\sqrt{2} + 5\sqrt{2} - 2\sqrt{2}$.

2. $5\sqrt{3} - 3\sqrt{3} - 4\sqrt{3}$.

3. $4\sqrt[3]{5} + 5\sqrt[3]{5} - 2\sqrt[3]{5}$.

4. $3\sqrt{2} - 7\sqrt{2} + \sqrt{2}$.

5. $\sqrt{8} + \sqrt{18} - \sqrt{32}$.

6. $\sqrt{50} - \sqrt{8} + \sqrt{72}$.

7. $\sqrt{12} - \sqrt{48} + \sqrt{75}$.

8. $\sqrt{98} - \sqrt{18} + \sqrt{128}$.

9. $\sqrt{72} - \sqrt{\frac{1}{2}} + \sqrt{\frac{9}{2}}$.

10. $\sqrt{75} - \sqrt{27} - \sqrt{\frac{4}{3}}$.

11. $3\sqrt{48} - 2\sqrt{12} - \sqrt{\frac{1}{3}}$.

12. $3\sqrt{\frac{25}{2}} - \sqrt{\frac{1}{8}} - 2\sqrt{50}$.

13. $\sqrt{16a} - \sqrt{36a} - \sqrt{9a}$.

14. $\sqrt{72x^3} + \sqrt{8x^3} - \sqrt{18x^3}$.

15. $\sqrt{32a^2x} - \sqrt{8b^2x} + \sqrt{50x}$.

16. $\sqrt{72b^2x} + \sqrt{98x} - \sqrt{18a^2x}$.

17. $\sqrt[3]{16a^3x} + \sqrt[3]{54b^3x}$.

18. $\sqrt{36x^3y} + \sqrt{16xy^3} + \sqrt{18xy} - \sqrt{8x^3y}$.

19. $\sqrt{18} + \sqrt{12} + 2\sqrt{\frac{1}{2}} + 6\sqrt{\frac{1}{3}}$.

20. $\sqrt{\dfrac{25}{3x}} - 3\sqrt{\dfrac{1}{12x}}$.

21. $a\sqrt{\dfrac{b}{a}} - \sqrt{\dfrac{a^3b}{4}}$.

22. $\sqrt{\dfrac{x+1}{x-1}} + 2\sqrt{1 - \dfrac{1}{x^2}} - \sqrt{\left(x - \dfrac{1}{x}\right)\dfrac{1}{x}}$.

23. $\sqrt{\dfrac{x+y}{x-y}} - \sqrt{\dfrac{x-y}{x+y}}$.

24. $\sqrt{1 - \dfrac{x}{y}} + 2\sqrt{\dfrac{y^2 - xy}{y^2}}$.

25. If $r_1 = \dfrac{-b + \sqrt{b^2 - 4ac}}{2a}$ and $r_2 = \dfrac{-b - \sqrt{b^2 - 4ac}}{2a}$, find $r_1 + r_2$.

26. $\sqrt{\dfrac{a-b}{a+b}} + \sqrt{\dfrac{a+b}{a-b}} - \sqrt{c^2a^2 - c^2b^2}$.

71. Multiplication with Radicals of the Same Index

This is done by means of law *II*, section 68, together with the usual rules of algebra.

Example 1. $\sqrt[3]{4x}\sqrt[3]{2x^2y^4} = \sqrt[3]{8x^3y^4} = 2xy\sqrt[3]{y}$.

Example 2. Multiply $(2\sqrt{3} + \sqrt{2})(\sqrt{3} - 3\sqrt{2})$.

Solution:

$$\begin{array}{r} 2\sqrt{3} + \sqrt{2} \\ \sqrt{3} - 3\sqrt{2} \\ \hline -6\sqrt{6} - 3\sqrt{4} \\ 2\sqrt{9} + \sqrt{6} \qquad \\ \hline 6 - 5\sqrt{6} - 6 \qquad \end{array} = -5\sqrt{6}.$$

EXERCISES

Multiply and simplify:

1. $\sqrt{5} \cdot \sqrt{3}$. **2.** $\sqrt{10} \cdot \sqrt{6}$. **3.** $\sqrt{6} \cdot \sqrt{8}$.
4. $\sqrt{18}\,\sqrt{12}$. **5.** $(3\sqrt{2})(4\sqrt{3})$.
6. $(2\sqrt{3})(7\sqrt{5})$. **7.** $(2\sqrt{6})(3\sqrt{8})$.
8. $(3\sqrt{15})(2\sqrt{5})$. **9.** $(\sqrt[3]{4})(\sqrt[3]{12})$.
10. $(4\sqrt[3]{6})(2\sqrt[3]{9})$. **11.** $\sqrt{12x} \cdot \sqrt{3xy}$.
12. $(6\sqrt{x})(\sqrt{12x^7})$. **13.** $(4\sqrt{6x})^2$.
14. $(3\sqrt{2x^3})^2$. **15.** $(4\sqrt{3x^2y})^2$. **16.** $(5\sqrt{2x^3y^2})^2$.
17. $(2\sqrt[3]{3x})^3$. **18.** $(3\sqrt[3]{2x})^3$. **19.** $(3\sqrt{x^2+y^2})^2$.
20. $(5\sqrt{x^2-y^2})^2$. **21.** $[(x+y)\sqrt{x+y}]^2$.
22. $\sqrt{a+b} \cdot \sqrt{a-b}$. **23.** $\sqrt{2}\,(\sqrt{3}+\sqrt{6})$.
24. $\sqrt{3}\,(\sqrt{2}-\sqrt{6})$. **25.** $\sqrt{6}\,(\sqrt{3}-\sqrt{2})$.
26. $3\sqrt{2}\,(\sqrt{6}-2\sqrt{3})$. **27.** $(2+\sqrt{3})(2-\sqrt{3})$.
28. $(3-\sqrt{2})(2-\sqrt{3})$. **29.** $(3\sqrt{2}+2\sqrt{3})^2$.
30. $(2\sqrt{5}-\sqrt{2})^2$. **31.** $(3\sqrt{5}-2)(\sqrt{5}+3)$.
32. $(3\sqrt{8}-2\sqrt{2})^2$.
33. $(\sqrt{2}-2\sqrt{3})(3\sqrt{2}+\sqrt{3})$.
34. $(3\sqrt{2}-2)(2\sqrt{3}+3)$. **35.** $(3\sqrt{2}-\sqrt{6})(\sqrt{3}+3)$.
36. $(2\sqrt{5}-\sqrt{3})(3\sqrt{3}+\sqrt{5})$.

Find the value of:

37. x^2+2x+4 when $x = 3+\sqrt{2}$.
38. x^2-2x-3 when $x = 2-\sqrt{2}$.
39. x^2-4x-1 when $x = 2-\sqrt{5}$.
40. $2x^2-5x-3$ when $x = \sqrt{3}+2$.
41. $2x^2-3xy$ when $x = \sqrt{2}+3$, $y = \sqrt{2}-3$.
42. x^2+y^2-3xy when $x = 2-\sqrt{3}$, $y = 2+\sqrt{3}$.
43. r_1r_2 if $r_1 = \dfrac{-b+\sqrt{b^2-4ac}}{2a}$ and $r_2 = \dfrac{-b-\sqrt{b^2-4ac}}{2a}$.

Simplify:

44. $(2-\sqrt{x-2})^2$. **45.** $(3+\sqrt{x-2})^2$.
46. $(3+2\sqrt{x+4})^2$. **47.** $(\sqrt{x+3}-\sqrt{x-3})^2$.
48. $(2\sqrt{x+1}-\sqrt{x-2})^2$.
49. $(3\sqrt{2x+1}-2x)^2$.
50. $(2\sqrt{2x+3}-3\sqrt{2x-3})^2$.

72. Division with Radicals. Rationalizing Denominators

Division of two radicals of the same index can be done by law *III*, section 68, and simplifying.

Example 1. $\dfrac{\sqrt{3}}{\sqrt{2}} = \sqrt{\dfrac{3}{2}} = \sqrt{\dfrac{3}{2}\cdot\dfrac{2}{2}} = \dfrac{1}{2}\sqrt{6} = \dfrac{\sqrt{6}}{2}.$

An alternative procedure is to rationalize the denominator directly by removing radicals from the denominator by means of multiplying by whatever is needed to make the radicand in the denominator a perfect power. Thus in example 1, we could write

$$\frac{\sqrt{3}}{\sqrt{2}} = \frac{\sqrt{3}}{\sqrt{2}}\cdot\frac{\sqrt{2}}{\sqrt{2}} = \frac{\sqrt{6}}{2}.$$

Example 2. $\dfrac{1}{\sqrt{8}} = \dfrac{1}{\sqrt{8}}\cdot\dfrac{\sqrt{2}}{\sqrt{2}} = \dfrac{\sqrt{2}}{4}.$

The rationalized form, having no radicals in the denominator, is considered simpler, because it is more convenient for computing decimal values. Thus to compute $1/\sqrt{8}$ directly requires long division of 1 by $\sqrt{8} = 2.818$ approximately, while $\sqrt{2}/4$ requires only mental division of $\sqrt{2} = 1.414$ approximately, by 4. Thus the answer is $\frac{1}{4}(1.414) = 0.354$ approximately.

A denominator consisting of a binomial quadratic surd $\sqrt{a} + \sqrt{b}$ can be rationalized by multiplying by the **conjugate surd** $\sqrt{a} - \sqrt{b}$, and vice versa.

Example 3.

$$\frac{1}{\sqrt{5}-\sqrt{2}} = \frac{1}{\sqrt{5}-\sqrt{2}}\cdot\frac{\sqrt{5}+\sqrt{2}}{\sqrt{5}+\sqrt{2}} = \frac{\sqrt{5}+\sqrt{2}}{3}.$$

Example 4. $\dfrac{1}{7+\sqrt{3}} = \dfrac{1}{7+\sqrt{3}}\cdot\dfrac{7-\sqrt{3}}{7-\sqrt{3}} = \dfrac{7-\sqrt{3}}{46}.$

EXERCISES

Rationalize denominators and simplify, leaving answers in radical form unless directed otherwise:

1. $\dfrac{\sqrt{5}}{\sqrt{2}}.$ **2.** $\dfrac{2\sqrt{8}}{\sqrt{3}}.$ **3.** $\dfrac{\sqrt{15}}{\sqrt{3}}.$ **4.** $\dfrac{2\sqrt{3}}{\sqrt{6}}.$

5. $\dfrac{\sqrt{15x^4}}{\sqrt{3x}}.$ **6.** $\dfrac{\sqrt{5x^2y}}{\sqrt{12x^5}}.$ **7.** $\dfrac{\sqrt{5x^3y}}{\sqrt{18xy^3}}.$ **8.** $\dfrac{\sqrt{5x}}{\sqrt{8x^3y}}.$

9. $\dfrac{\sqrt[3]{5}}{\sqrt[3]{4}}$. **10.** $\dfrac{3\sqrt[3]{16x^2}}{\sqrt[3]{9x}}$. **11.** $\dfrac{1}{2-\sqrt{3}}$.

12. $\dfrac{1}{4+\sqrt{3}}$. **13.** $\dfrac{4}{\sqrt{3}-\sqrt{2}}$. **14.** $\dfrac{2}{\sqrt{6}-\sqrt{2}}$.

15. $\dfrac{\sqrt{2}}{\sqrt{6}+\sqrt{2}}$. **16.** $\dfrac{\sqrt{6}-\sqrt{2}}{\sqrt{3}}$. **17.** $\dfrac{\sqrt{5}+\sqrt{6}}{\sqrt{3}}$.

18. $\dfrac{2\sqrt{3}}{3\sqrt{2}-2\sqrt{3}}$. **19.** $\dfrac{\sqrt{3}-1}{\sqrt{3}+1}$.

20. $\dfrac{\sqrt{5}-3}{2-\sqrt{5}}$. **21.** $\dfrac{2\sqrt{5}-\sqrt{2}}{3\sqrt{5}+2\sqrt{2}}$.

22. $\dfrac{\sqrt{6}+2\sqrt{3}}{2\sqrt{6}-\sqrt{3}}$. **23.** $\dfrac{4\sqrt{5}+\sqrt{3}}{2\sqrt{3}-\sqrt{5}}$.

24. $\dfrac{3\sqrt{7}-2\sqrt{2}}{2\sqrt{7}+3\sqrt{2}}$. **25.** $\dfrac{1+\dfrac{1}{\sqrt{3}}}{1-\dfrac{1}{\sqrt{3}}}$.

26. $\dfrac{1-\dfrac{\sqrt{2}}{2}}{1+\dfrac{\sqrt{2}}{2}}$. **27.** $\dfrac{\sqrt{x-1}}{1-\sqrt{x-1}}$. **28.** $\dfrac{\sqrt{x-a}}{b+\sqrt{x-a}}$.

29. Compute, by use of table I, the answers to the first four exercises.

73. A Use of Fractional Exponents

Certain operations with radicals are best done by means of fractional exponents. For example, to multiply or divide two radicals of different orders, we may convert these radicals to fractional powers of the radicand, change the exponents of those fractional powers to equivalent fractions with their least common denominator, convert back to radical form, and simplify.

Example 1. Simplify $\sqrt[3]{x^2}\,\sqrt[5]{y^3}$. We write $x^{2/3}y^{3/5} = x^{10/15}y^{9/15} = \sqrt[15]{x^{10}}\,\sqrt[15]{y^9} = \sqrt[15]{x^{10}y^9}$.

Example 2. Simplify $\dfrac{\sqrt{6}}{\sqrt[3]{2}}$. We write $\dfrac{6^{1/2}}{2^{1/3}} = \dfrac{6^{3/6}}{2^{2/6}} = \dfrac{\sqrt[6]{6^3}}{\sqrt[6]{2^2}} = \sqrt[6]{\dfrac{6^3}{2^2}} = \sqrt[6]{54}$.

EXERCISES

Simplify:

1. $\sqrt{x}\cdot\sqrt[5]{x}$. **2.** $\sqrt{a}\cdot\sqrt[3]{a}$. **3.** $\sqrt[3]{x^2}\cdot\sqrt[5]{x^3}$.

4. $\sqrt{x} \cdot \sqrt[4]{x^3}$. **5.** $\dfrac{\sqrt{10}}{\sqrt[3]{2}}$. **6.** $\sqrt{2} \cdot \sqrt[3]{4}$.

7. $\sqrt{8}/\sqrt[3]{4}$. **8.** $\sqrt[3]{\sqrt[4]{5}}$. **9.** $\sqrt{\sqrt[3]{7}}$.

10. $\dfrac{\sqrt[4]{8}}{\sqrt[3]{4}}$. **11.** $\dfrac{\sqrt[6]{32}}{\sqrt[4]{8}}$. **12.** $\dfrac{\sqrt[3]{x^2}}{\sqrt{3xy^2}}$.

13. $\sqrt[6]{x^2y}\,\sqrt{y^3}$. **14.** $\dfrac{\sqrt[3]{4a}}{\sqrt{2a}}$.

Chapter XI*

Quadratic equations and quadratic functions

74. Solution of Quadratic Equations by Factoring

Every quadratic equation in x can be written in the form

$$(1) \qquad ax^2 + bx + c = 0 \qquad (a \neq 0)$$

or, dividing both members by a, as

$$(2) \qquad x^2 + px + q = 0$$

where $p = b/a$ and $q = c/a$.

If the first-degree term is missing, the equation is called a **pure** quadratic. To solve such an equation is easy, for the roots of $x^2 = N$ are obviously $x = \pm\sqrt{N}$. Thus the roots of $x^2 - 9 = 0$ are ± 3. The roots of $x^2 - 10 = 0$ are $\pm\sqrt{10}$. The roots of $x^2 + 4 = 0$ are $\pm\sqrt{-4}$ or† $\pm 2i$.

If the left member of (1) or (2) can be expressed as a product of two factors of the first degree, the equation may be solved by setting each factor equal to zero separately, as in the following example.

Example 1. Solve $x^2 - 5x + 4 = -2$.

Solution. We write the equation as $x^2 - 5x + 6 = 0$. Factoring, we get $(x - 2)(x - 3) = 0$. Setting each factor equal to zero, we get $x - 2 = 0$ and $x - 3 = 0$. Hence $x = 2$ and $x = 3$. Checking, by substituting, we get: for $x = 2$, $2^2 - 5 \cdot 2 + 4 = -2$, or $-2 = -2$; for $x = 3$, $3^2 - 5 \cdot 3 + 4 = -2$, or $-2 = -2$.

* The student should review sections 12–16 at this point.

† The student should review section 16.

Justification of the procedure. By theorem 1, section 11, the product of several factors can be zero when and only when at least one of the factors is zero. To solve the equation $x^2 - 5x + 6 = 0$ (to which the original equation is equivalent*) means to find all values of x for which the left member has the value zero. But the left member can be expressed as the product of $(x - 2)$ and $(x - 3)$. This product therefore is zero when and only when $x - 2 = 0$ or $x - 3 = 0$, that is, when $x = 2$ or 3. Therefore if x satisfies the original equation, then x can only be either 2 or 3. The converse "if x is either 2 or 3, then x satisfies the original equation" is certainly correct because all steps taken are reversible, but it is advisable to check by substitution in any case.

A quadratic equation has two roots, one arising from each factor; if the roots arising from both factors are the same, we still say there are two roots, and their common value is called a **double root** *of the equation.*

Example 2. The equation $x^2 - 6x + 9 = 0$ has the double root 3, since the left member is $(x - 3)(x - 3)$.

EXERCISES

Solve and check:

1. $x^2 + 2x - 8 = 0.$ **2.** $x^2 + 6x + 5 = 0.$
3. $x^2 - x - 4 = 2.$ **4.** $x^2 - 2x - 18 = x.$
5. $2x^2 - 98 = 0.$ **6.** $3x^2 - 48 = 0.$
7. $4x^2 - 25 = 0.$ **8.** $9x^2 - 4 = 0.$
9. $x^2 + 10x - 16 = 3x + 2.$
10. $x^2 + 6x - 31 = 5 - 3x.$
11. $x^2 = 16x + 36.$ **12.** $x^2 - 19x - 53 = 3x - 5.$
13. $6x^2 + 7x - 5 = 0.$ **14.** $4x^2 + 4x - 15 = 0.$
15. $6x^2 + 11x - 10 = 0.$ **16.** $6x^2 + 5x - 4 = 0.$
17. $8x^2 + 5x - 3 = 0.$ **18.** $6x^2 - 19x + 3 = 0.$
19. $12x^2 + 11x - 15 = 0.$ **20.** $12x^2 - 25x + 2 = 0.$
21. $x^2 - 5 = 0.$ **22.** $9x^2 - 5 = 0.$
23. $4x^2 - 10 = 0.$ **24.** $3x^2 - 1 = 0.$
25. $x^2 - 2kx + k^2 = 0.$ **26.** $x^2 - kx - 6k^2 = 0.$
27. $21x^2 + 5kx - 4k^2 = 0.$ **28.** $3x^2 - 19kx + 20k^2 = 0.$
29. $x^4 - 12x^2 + 32 = 0.$ (*Hint:* Let $u = x^2$; solve for u, then for x.).
30. $x^4 - 29x^2 + 100 = 0.$ **31.** $x^4 - 8x^2 + 15 = 0.$

* See section 29 for the definition of *equivalent* equations.

32. $x^4 - 4x^2 + 3 = 0.$ **33.** $x^2 + 9 = 0.$

34. $x^2 + 25 = 0.$ **35.** $4x^2 + 9 = 0.$ **36.** $16x^2 + 1 = 0.$

37. $(x - 7)^2 = 81.$ **38.** $(x + 5)^2 = 49.$

39. $(x - 2)^2 - 5(x - 2) + 6 = 0.$

40. $(x + 3)^2 + 4(x + 3) - 21 = 0.$

41. $x^2 - 7x = 0.$ **42.** $3x^2 - 15x = 0.$

43. $3x^2 + 5x = 0.$ **44.** $2x^2 + 9x = 0.$

45. $4x^2 + 20x + 25 = 0.$ **46.** $9x^2 - 12x + 4 = 0.$

47. Solve $s = \frac{1}{2}gt^2$ for t. **48.** Solve $T = \frac{1}{2}mv^2$ for v.

49. Solve $A = \pi r^2$ for r. **50.** Solve $V = \pi r^2 h$ for r.

75. Completing the Square

Since $x^2 + 2hx + h^2 = (x + h)^2$, it follows that if the square of half the coefficient of x be added to an expression of the form* $x^2 + px$, it will become a perfect square. That is, $x^2 + px + \left(\frac{p}{2}\right)^2 = \left(x + \frac{p}{2}\right)^2\cdot$ The geometric meaning of the phrase "completing the square" will be obvious in terms of the areas in the jigsaw puzzle of Fig. 28.

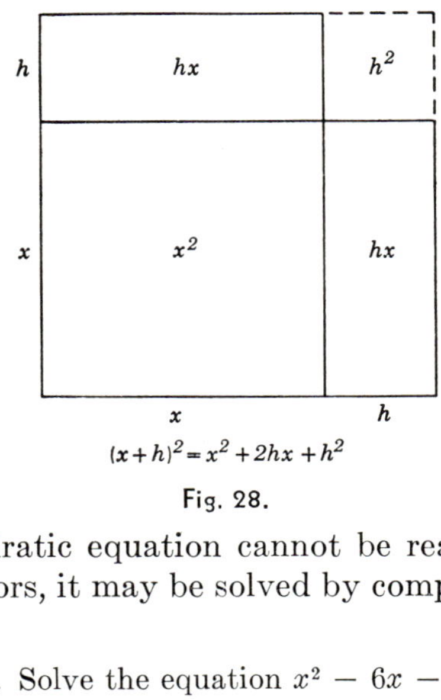

Fig. 28.

When a quadratic equation cannot be readily expressed as a product of factors, it may be solved by completing the square, as follows.

Example 1. Solve the equation $x^2 - 6x - 1 = 0.$

Solution. Transpose the constant term, obtaining $x^2 - 6x = 1.$ We may complete the square on the left side of the equation by adding the square of half the coefficient of x, that is, 3^2 or 9. But

* Writing p for $2h$, or $p/2$ for h.

if we add 9 to the left member, we must of course also add 9 to the right member. Thus we get $x^2 - 6x + 9 = 1 + 9$ or $(x - 3)^2 = 10$. By *III*, section 29, we may write $x - 3 = \pm\sqrt{10}$, or $x = 3 \pm \sqrt{10}$. Checking, we get: for $x = 3 + \sqrt{10}$, $(3 + \sqrt{10})^2 - 6(3 + \sqrt{10}) - 1 = 0$, or $9 + 6\sqrt{10} + 10 - 18 - 6\sqrt{10} - 1 = 0$, or $0 = 0$; for $x = 3 - \sqrt{10}$, $(3 - \sqrt{10})^2 - 6(3 - \sqrt{10}) - 1 = 9 - 6\sqrt{10} + 10 - 18 + 6\sqrt{10} - 1$, or $0 = 0$.

The roots $3 \pm \sqrt{10}$ of the equation of example 1 are irrational.* If a decimal approximation is desired, it may be obtained by working out $\sqrt{10}$ (or from table I), obtaining 3 ± 3.162, or 6.162 and -0.162 approximately.

It can also happen that the roots may be imaginary† as in the following example.

Example 2. Solve the equation $x^2 + 2x + 4 = 0$.

Solution. We write $x^2 + 2x = -4$. Completing the square we get $x^2 + 2x + 1 = -4 + 1$, or $(x + 1)^2 = -3$. Hence $x + 1 = \pm\sqrt{-3}$, or $x + 1 = \pm i\sqrt{3}$. Hence $x = -1 \pm i\sqrt{3}$. Checking we get: for $x = -1 + i\sqrt{3}$, $(-1 + i\sqrt{3})^2 + 2(-1 + i\sqrt{3}) + 4 = 0$, or $1 - 2i\sqrt{3} - 3 - 2 + 2i\sqrt{3} + 4 = 0$, or $0 = 0$; for $x = -1 - i\sqrt{3}$ we get $(-1 - i\sqrt{3})^2 + 2(-1 - i\sqrt{3}) + 4 = 0$, or $1 + 2i\sqrt{3} - 3 - 2 - 2i\sqrt{3} + 4 = 0$, or $0 = 0$.

If the coefficient of x^2 is different from 1, divide through by this coefficient before completing the square, as follows.

Example 3. Solve $2x^2 - 7x + 3 = 0$.

Solution. Transposing the 3, and dividing by 2, we get $x^2 - \frac{7}{2}x = -\frac{3}{2}$. Adding $(\frac{7}{4})^2 = \frac{49}{16}$ to both sides, we get $(x - \frac{7}{4})^2 = \frac{25}{16}$ or $x - \frac{7}{4} = \pm\frac{5}{4}$. Hence $x = 3$ and $x = \frac{1}{2}$. The check is left to the student.

EXERCISES

Solve by completing the square, and check:

1. $x^2 - 16x = 36$.
2. $x^2 + 6x = 27$.
3. $x^2 + 14x + 24 = 0$.
4. $x^2 - 10x + 21 = 0$.
5. $6x^2 + 7x - 5 = 0$.
6. $4x^2 = 5 - 19x$.
7. $18x^2 = 9x + 20$.
8. $12x^2 = 5 - 7x$.
9. $2x^2 + 7x + 3 = 0$.
10. $8x^2 + 15x - 2 = 0$.

Solve by completing the square, and check, (a) leaving answers in

* See sections 13, 14, and 15.

† The student should review section 16.

simplest radical form if irrational, or in terms of i if imaginary; (*b*) *if the roots are irrational, approximate them to the nearest thousandth:*

11. $x^2 - 4x + 1 = 0.$
12. $x^2 - 7x + 2 = 0.$
13. $2x^2 - 10x + 11 = 0.$
14. $2x^2 - 6x + 1 = 0.$
15. $x^2 + 2x - 4 = 0.$
16. $3x^2 + 3x - 2 = 0.$
17. $x^2 - 2x + 5 = 0.$
18. $x^2 - 6x + 34 = 0.$
19. $x^2 + x + 1 = 0.$
20. $x^2 + x + 3 = 0.$
21. $x^2 - 5x + 1 = 0.$
22. $2x^2 - 5x - 5 = 0.$
23. $3x^2 - 11x - 70 = 0.$
24. $3x^2 + 7x - 3 = 0.$
25. $(x + 1)(x + 5) = 13.$
26. $2x^2 + x + 5 = 0.$
27. $x^2 + px + q = 0.$
28. $ax^2 + bx + c = 0$, if $a \neq 0$.

76. The Quadratic Formula

By completing the square, we can prove that the roots of the general quadratic equation

$$(1) \qquad ax^2 + bx + c = 0 \qquad (a \neq 0)$$

are given by the formula

$$(2) \qquad x = \frac{-b \pm \sqrt{b^2 - 4ac}}{2a}.$$

Proof. Dividing both members of (1) by a, and transposing the constant term, we get

$$x^2 + \frac{b}{a}x = -\frac{c}{a}.$$

To complete the square, we add $\left(\frac{b}{2a}\right)^2$ to both sides, obtaining

$$x^2 + \frac{b}{a}x + \left(\frac{b}{2a}\right)^2 = -\frac{c}{a} + \left(\frac{b}{2a}\right)^2$$

or

$$\left(x + \frac{b}{2a}\right)^2 = -\frac{c}{a} + \frac{b^2}{4a^2}$$

or

$$\left(x + \frac{b}{2a}\right)^2 = \frac{b^2 - 4ac}{4a^2}.$$

Therefore,

$$x + \frac{b}{2a} = \frac{\pm\sqrt{b^2 - 4ac}}{2a}$$

or

$$x = \frac{-b \pm \sqrt{b^2 - 4ac}}{2a}.$$

The student should check by substituting both answers (one

with the + sign and the other with the − sign before the radical) in (1); both answers check.

To solve any quadratic equation whatever, we have only to substitute the values of its coefficients a, b, and c into the formula (2) and simplify. As before, we may check by substitution.

Example 1. Solve $x^2 - 5x + 6 = 0$. Here $a = 1$, $b = -5$, and $c = 6$. Hence the roots are

$$\frac{-(-5) \pm \sqrt{(-5)^2 - 4 \cdot 1 \cdot 6}}{2 \cdot 1} = \frac{5 \pm 1}{2} = \begin{cases} 3 \\ 2. \end{cases}$$

Check as before.

Example 2. Solve $x^2 - 6x - 1 = 0$. Here $a = 1$, $b = -6$, and $c = -1$. The roots are

$$\frac{-(-6) \pm \sqrt{(-6)^2 - 4 \cdot 1 \cdot (-1)}}{2 \cdot 1} = \frac{6 \pm \sqrt{40}}{2} = \frac{6 \pm 2\sqrt{10}}{2}$$
$$= 3 \pm \sqrt{10}.$$

Check by substituting as before.

Note that the roots $3 \pm \sqrt{10}$ are irrational.*

If a decimal approximation is desired, we may work out $\sqrt{10}$ (or use table I), obtaining 3 ± 3.162, or 6.162 and −0.162, approximately.

Example 3. Solve $x^2 + 2x + 4 = 0$. Here $a = 1$, $b = 2$, $c = 4$. The roots† are

$$\frac{-2 \pm \sqrt{2^2 - 4 \cdot 1 \cdot 4}}{2 \cdot 1} = \frac{-2 \pm \sqrt{-12}}{2} = \frac{-2 \pm 2\sqrt{-3}}{2}$$
$$= \frac{2(-1 \pm \sqrt{-3})}{2} = -1 \pm \sqrt{-3} = -1 \pm i\sqrt{3}.$$

Check by substituting as before.

Note. A pure quadratic equation should be solved by taking square roots, with the ± sign. If a trinomial quadratic equation can be factored easily, it should be solved by factoring. In all cases it may be solved by formula.

EXERCISES

Solve by use of the quadratic formula, and check:

1–26. The equations of exercises 1–26, section 75, inclusive.

* The student should review sections 13, 14, and 15.

† The student should review section 16.

27. $x^2 = 3x + 10.$

28. $11x = 4 - 3x^2.$

29. $4x^2 = 12x - 9.$

30. $9 + 24x + 16x^2 = 0.$

31. $2x^2 - \sqrt{3}\,x + 2 = 0.$

32. $2x^2 - \sqrt{2}\,x + 3 = 0.$

33. $\sqrt{2}\,x^2 + x + \sqrt{8} = 0.$

34. $\sqrt{2}\,x^2 - 5x + \sqrt{18} = 0.$

35. $\sqrt{3}\,x^2 = 2x + \sqrt{12}.$

36. $\sqrt{27}\,x^2 = \sqrt{48} - 5x.$

37. $x^2 - 2hx + k = 0.$

38. $x^2 + 2mx + n = 0.$

39. $ax^2 + 2dx + c = 0.$

40. $ax^2 - 2dx - k = 0.$

41. $x^2 - 2ax + (a^2 - b) = 0.$

42. $x^2 - 2ax + (a^2 + b^2) = 0.$

43. $x^2 - 4ix - 7 = 0.$

44. $4ix^2 + 3x - i = 0.$

45. $12ix^2 - 5x + 3i = 0.$

46. $3ix^2 + 8x - 3i = 0.$

77. Character of the Roots

Up to this point in the present chapter, no restriction has been imposed on the coefficients a, b, and c of the quadratic equation $ax^2 + bx + c = 0$ $(a \neq 0)$. That is, everything said so far is true no matter what complex* numbers a, b, and c represent. *Let us now consider only equations with real coefficients.*

The two roots, which we shall denote by r_1 and r_2 respectively, are given by the formulas

$$(1) \qquad r_1 = \frac{-b + \sqrt{b^2 - 4ac}}{2a} \text{ and } r_2 = \frac{-b - \sqrt{b^2 - 4ac}}{2a}.$$

From (1) and the fact that the sum, product, difference, or quotient of two real numbers (division by zero excluded) is a real number, it follows that the roots are both imaginary if $b^2 - 4ac$ is negative, both real if $b^2 - 4ac$ is not negative, and both equal if $b^2 - 4ac$ is zero. The quantity $b^2 - 4ac$ is called the **discriminant** of the equation $ax^2 + bx + c = 0$. Hence we have the following summary:

Quadratic Equations with Real Coefficients

Discriminant	Positive	Zero	Negative
Character of the roots	Real and unequal	Real and equal	Imaginary and unequal

If we restrict ourselves further to equations with rational coefficients, we may draw further conclusions. Since the sum, product,

* The student should review sections 11–16, inclusive.

difference, or quotient of two rational numbers is a rational number (division by zero excluded), it follows that the discriminant is a rational number. The roots will be rational if and only if the discriminant is a perfect square. If the discriminant is not a perfect square, its square roots are irrational. Hence we have the following summary:

Quadratic Equations with Rational Coefficients

Discriminant	Positive and a Perfect Square	Positive but Not a Perfect Square	Zero	Negative
Character of the roots	Real, rational, and unequal	Real, irrational, and unequal	Real, rational, and equal	Imaginary and unequal

Example 1. Without solving the equation, describe the character of the roots of $x^2 - 6x - 1 = 0$.

Solution. The coefficients are rational. The discriminant $b^2 - 4ac = (-6)^2 - 4 \cdot 1 \cdot (-1) = 40$. This is positive but not a perfect square. Hence the roots are real, irrational, and unequal.

Example 2. Find the values of k for which the equation $x^2 + kx + 9 = 0$ has equal roots.

Solution. The roots can be equal if and only if the discriminant is zero, that is, when $k^2 - 36 = 0$ or $k = \pm 6$. *Check:* when $k = 6$, the given equation is $x^2 + 6x + 9 = 0$, which has the roots -3, -3; when $k = -6$, the given equation is $x^2 - 6x + 9 = 0$ which has the roots 3, 3.

Note. From (1), it follows that the roots r_1 and r_2 are equal if and only if $b^2 - 4ac = 0$, regardless of what kind of numbers the coefficients are.

EXERCISES

Calculate the discriminant and describe the character of the roots without solving, if possible; check by solving:

1. $x^2 - 4x - 12 = 0$.
2. $x^2 - x - 42 = 0$.
3. $6x^2 - 5 + 7x = 0$.
4. $6x^2 + 15 - 19x = 0$.
5. $x^2 - 2x - 1 = 0$.
6. $64 - 16x + x^2 = 0$.
7. $81 + 16x^2 - 72x = 0$.
8. $x^2 + 2x + 6 = 0$.
9. $8x^2 + 2x + 3 = 0$.
10. $18x^2 + 6x + 1 = 0$.
11. $x^2 - 7x = 0$.
12. $3x^2 + 5x = 0$.

13. $\frac{13}{4}x^2 + 5x + 13 = 0.$

14. $\frac{1}{2}x^2 - x - \frac{1}{3} = 0.$

★**15.** $x^2 + \sqrt{5}\,x - 5 = 0.$

★**16.** $\sqrt{2}\,x^2 + \sqrt{5}\,x - \sqrt{18} = 0.$

★**17.** $x^2 - \sqrt{3}\,x + 7 = 0.$

★**18.** $\sqrt{8}\,x^2 - x + \sqrt{2} = 0.$

★**19.** $x^2 + 4ix - 4 = 0.$

★**20.** $x^2 + 2ix - 3 = 0.$

★**21.** $x^2 + 2ix - 5 = 0.$

★**22.** $3ix^2 + 5x - i = 0.$

★**23.** $2x^2 - \sqrt{8}\,x + 1 = 0.$

★**24.** $x^2 - \sqrt{12}\,x + 3 = 0.$

Find the value or values of k for which the equation will have equal roots:

25. $kx^2 - 24x + 9 = 0.$

26. $25x^2 + 60x + k = 0.$

27. $9x^2 + kx + 4 = 0.$

28. $9x^2 + 3kx + 64 = 0.$

29. $2kx^2 + 6kx + 9 = 0.$

30. $kx^2 + 24x + 9k = 0.$

31. $x^2 - 4kx + 16k = 0.$

32. $16x^2 - 12kx + 9k = 0.$

Find the value or values of k for which the function is a perfect square:

33. $4x^2 - 4kx + 5k.$

34. $25x^2 + 6kx + 9.$

35. $16x^2 + kx + 9.$

36. $kx^2 - 3kx + 9.$

Find the real values of k for which the roots of the equation are (a) real and unequal, (b) imaginary and unequal:

37. $x^2 + 4x + k = 0.$

38. $x^2 + 8x + k = 0.$

39. $x^2 - kx + 9 = 0.$

40. $x^2 + kx + 36 = 0.$

78. The Graph of a Quadratic Function. Graphical Solution of Quadratic Equations*

It can be proved that the graph of every quadratic function $y = ax^2 + bx + c$ $(a \neq 0)$ with real coefficients is a parabola. (See Fig. 29.) To plot the graph we substitute various values for x, calculate the corresponding values of y, plot the points (x,y) thus tabulated, and join them by a smooth curve.

Example 1. Plot the graph of $y = x^2 + x - 6$.
Solution. We get the table

x	0	1	−1	2	−2	3	−3	−4
y	−6	−4	−6	0	−4	6	0	6

Plotting these points, we get the graph of Fig. 29.

The real roots of the quadratic equation $ax^2 + bx + c = 0$ with

★ May be omitted without disturbing continuity.
* The student should review sections 34–37.

real coefficients are the real values of x for which $y = ax^2 + bx + c = 0$. *Thus, they are the x-coordinates of the points where the graph crosses the x-axis.*

For example, the real roots of $x^2 + x - 6 = 0$ are 2 and -3. (See Fig. 29.)

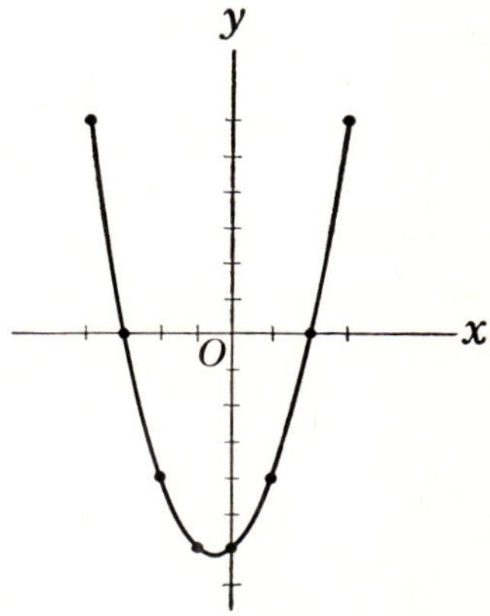

Fig. 29.

Note. When both roots are real and equal, the parabola is tangent to the x-axis. When the roots are imaginary, the graph has no point in common with the x-axis. If the roots are real and unequal, the graph crosses the x-axis in two distinct points. The student should explain how one can tell which of these three cases will arise by means of calculating the discriminant $b^2 - 4ac$. See section 77.

EXERCISES

Draw the graph of each of the following functions, plotting at least eight points for each having x-coordinates which are successive integers:

1. $x^2 - 3$. **2.** $\frac{1}{3}x^2 + 2$. **3.** $3 - x^2$. **4.** $2 - \frac{1}{3}x^2$.
5. $x^2 - 2x - 3$. **6.** $3 - x - x^2$. **7.** $x^2 + 2x + 4$.
8. $x^2 - 4x + 4$. **9.** $9 + 6x + x^2$. **10.** $1 + 5x - 2x^2$.

Estimate the real roots, if any, of each equation graphically:

11. $x^2 - 3x - 4 = 0$. **12.** $x^2 + 2x - 3 = 0$.
13. $3 + 2x - x^2 = 0$. **14.** $6 + x - x^2 = 0$.
15. $4x^2 + 4x + 1 = 0$. **16.** $9x^2 - 16 = 0$.
17. $4x^2 - 9 = 0$. **18.** $4x^2 - 12x + 9 = 0$.
19. $x^2 + x + 1 = 0$. **20.** $\frac{1}{4}x^2 + 2 = 0$.
21. $x^2 - 4x = 0$. **22.** $x^2 + 2x = 0$.

Plot the graphs of both functions on the same set of axes:

23. $x^2 - 2x - 3$ and $-x^2 + 2x + 3$.
24. $x^2 - 3x - 4$ and $-x^2 + 3x + 4$.
25. $x^2 - 3$ and $-x^2 + 3$. **26.** $x^2 + 4$ and $-x^2 - 4$.
27. $x^2 - 6x + 9$ and $-x^2 + 6x - 9$.
28. $x^2 - 2x$ and $-x^2 + 2x$.

Without plotting the graph, decide by means of the discriminant whether the graph will cross the x-axis in two distinct points, will be tangent to the x-axis, or will not meet the x-axis at all:

29. $4x^2 - 12x + 9$.
30. $x^2 - 7x + 12$.
31. $x^2 - 8x + 15$.
32. $x^2 - 4x + 4$.
33. $x^2 + 2x + 4$.
34. $9x^2 + 4$.
35. $x^2 - 3x - 5$.
36. $3x^2 + 2x - 5$.

Find the values of k for which the graph of each function will (a) be tangent to the x-axis; (b) cross the x-axis in two distinct points; (c) not meet the x-axis at all:

37. $kx^2 - 2x + 1$.
38. $2x^2 - 4x + k$.
39. $4x^2 - 3kx + 9$.
40. $4x^2 + 4kx + 1$.
41. $9x^2 + 6kx + 25$.
42. $kx^2 - 3kx + 9$.

79. The Sum and Product of the Roots

The roots r_1 and r_2 of $ax^2 + bx + c = 0$ $(a \neq 0)$ are given by the formulas

$$r_1 = \frac{-b + \sqrt{b^2 - 4ac}}{2a} \text{ and } r_2 = \frac{-b - \sqrt{b^2 - 4ac}}{2a}.$$

By direct calculation, which should be checked by the student we find

$$(1) \qquad r_1 + r_2 = -\frac{b}{a}$$

and

$$(2) \qquad r_1 r_2 = \frac{c}{a}.$$

Example 1. The sum of the roots of $x^2 + x - 6 = 0$ is $-1/1 = -1$. The product of the roots is $-6/1 = -6$.

Example 2. If one root of $x^2 + x + c = 0$ is 2, find the other root.

First solution. The sum of the roots is -1. Let $r_1 = 2$. Then $2 + r_2 = -1$. Hence $r_2 = -3$.

Second solution. Since $x = 2$ must satisfy the equation, we have $2^2 + 2 + c = 0$. Hence $c = -6$. Solving $x^2 + x - 6 = 0$, we find that the second root is -3.

Example 3. If one root of $x^2 + bx - 6 = 0$ is 2, find the other root.

First solution. The product of the roots is -6. Let $r_1 = 2$. Then $2 \cdot r_2 = -6$. Hence $r_2 = -3$.

Second solution. Since $x = 2$ must satisfy the equation, we have $2^2 + 2b - 6 = 0$. Hence $b = 1$. Solving $x^2 + x - 6 = 0$, we find that the second root is -3.

By means of (1) and (2) we can prove the following theorem:

THEOREM 1. *If r_1 and r_2 are the roots of $ax^2 + bx + c = 0$, then $ax^2 + bx + c = a(x - r_1)(x - r_2)$.*

Proof. We may write $ax^2 + bx + c = a\left(x^2 + \frac{b}{a}x + \frac{c}{a}\right) = a(x^2 + [-r_1 - r_2]x + r_1r_2)$. By (5), section 46, the latter expression is equal to $a(x - r_1)(x - r_2)$.

Example 4. Form a quadratic equation with integral coefficients having the roots $\frac{1}{2}$ and $-\frac{1}{3}$.

Solution. By theorem 1, we may write $a(x - \frac{1}{2})(x + \frac{1}{3}) = 0$ as an equation having the desired roots. To get integral coefficients, we may let $a = 6 = 2 \cdot 3$, and write $2 \cdot 3(x - \frac{1}{2})(x + \frac{1}{3}) = 0$, or $2(x - \frac{1}{2}) \cdot 3(x + \frac{1}{3}) = 0$, or $(2x - 1)(3x + 1) = 0$, or, finally, $6x^2 - x - 1 = 0$.

EXERCISES

Without solving, find the sum and product of the roots:

1. $x^2 - 8x + 15 = 0$.
2. $x^2 - 4x + 7 = 0$.
3. $3x^2 + 5x - 12 = 0$.
4. $2x^2 - 6x + 15 = 0$.
5. $x = 5 - 3x^2$.
6. $8 - 3x = 2x^2$.
7. $2x^2 - 7 = 0$.
8. $3x^2 + 11 = 0$.
9. $2x^2 + 11x = 0$.
10. $3x^2 - 4x = 0$.

11. If one root of $x^2 - 4x + k = 0$ is 6, find the other root and the value of k.

12. If one root of $x^2 - 3x + h = 0$ is 8, find the other root and the value of h.

13. If one root of $8x^2 + kx - 10 = 0$ is -2, find the other root and the value of k.

14. If one root of $12x^2 + hx + 12 = 0$ is $\frac{3}{4}$, find the other root and the value of h.

15. If the sum of the roots of $2hx^2 - 24x + 15h + 5 = 0$ is 6, find the value of h and the roots.

16. If the product of the roots of $(k - 5)x^2 - 68x + 3k = 0$ is 4, find the value of k and the roots.

17. If one root of $9x^2 - 15x + h = 0$ exceeds the other by 3, find the roots and the value of h.

18. If one root of $2x^2 - 13x + 5k - 4 = 0$ is 12 times the other, find the roots and the value of k.

19. If one root of $2hx^2 + (3h - 6)x - 9 = 0$ is the negative of the other, find the value of h and the roots.

20. If the sum of the roots of $2kx^2 - 8x + 3k - 1 = 0$ is 4, find the value of k and the roots.

Write a quadratic equation with integral coefficients having the given numbers as roots:

21. $6, -2$. **22.** $2, -5$. **23.** $\frac{1}{2}, -\frac{2}{3}$.

24. $\frac{4}{3}, -\frac{1}{4}$. **25.** $-\frac{5}{3}, 2$. **26.** $\frac{6}{5}, -\frac{4}{7}$.

27. $\pm 2\sqrt{3}$. **28.** $\pm 3\sqrt{5}$. **29.** $\pm i\sqrt{5}$.

30. $\pm 2i\sqrt{3}$. **31.** $2 \pm \sqrt{3}$. **32.** $4 \pm 2\sqrt{5}$.

33. $3 \pm 2i$. **34.** $\dfrac{5 \pm \sqrt{3}}{2}$.

35. $\dfrac{2 \pm \sqrt{3}}{4}$. **36.** $\dfrac{3 \pm i\sqrt{2}}{2}$.

37. $a \pm \sqrt{b}$, where a and b are integers.

38. $a \pm bi$, where a and b are integers.

80. Equations Leading to Quadratics

(*A*) *Fractional equations often lead to quadratics.*

Example 1. Solve the equation

$$\frac{x^2}{(x-2)(x-3)} = \frac{2}{x-2} + \frac{6}{(x-2)(x-3)}.$$

Solution. Multiplying both members of the equation by the L.C.D. $(x - 2)(x - 3)$, we get

$$x^2 = 2(x - 3) + 6,$$

or

$$x^2 - 2x = 0,$$

whose roots are 2 and 0. This proves that if x satisfies the original equation, it can only be either 2 or 0. That is, 2 and 0 are the only eligible candidates for the position of root. To decide whether they are really roots, we must investigate the correctness of the converse* propositions: "If $x = 2$, then x satisfies the original equation," and "If $x = 0$, then x satisfies the original equation." This is most easily done by substituting the candidates for the rootship in the original equation. We find that 0 is a root, since

$$\frac{0^2}{(0-2)(0-3)} = \frac{2}{0-2} + \frac{6}{(0-2)(0-3)},$$

* The student should review section 29.

or $$\frac{0}{6} = -1 + \frac{6}{6}, \quad \text{or} \quad 0 = -1 + 1.$$

But 2 is not a root, since division by 0 is excluded.

The unsuccessful candidates for the rootship, if any, are often called **extraneous roots,** but this is merely a euphemistic way of saying that they are not roots at all. The occurrence of extraneous roots shows clearly that checking is not merely a superfluous step but is an essential part of the work.

(B) Sometimes an equation may become quadratic upon making a change of variable, as in the following example.

Example 2. Solve $\frac{1}{x^4} - \frac{5}{x^2} + 6 = 0.$

Solution. Let $u = \frac{1}{x^2}$. Then the equation becomes $u^2 - 5u + 6 = 0$, which has the solutions $u = 2$ and $u = 3$. Hence $\frac{1}{x^2} = 2$ and $\frac{1}{x^2} = 3$. From $\frac{1}{x^2} = 2$ we obtain $x = \pm \frac{1}{\sqrt{2}} = \pm \frac{\sqrt{2}}{2}$. From $\frac{1}{x^2} = 3$ we obtain $x = \pm \frac{1}{\sqrt{3}} = \pm \frac{\sqrt{3}}{3}$. The student should check all four roots in the original equation.

EXERCISES

Solve and check:

1. $x = \frac{10}{x - 9}$.

2. $6x = \frac{7x - 4}{x - 3}$.

3. $\frac{x}{4 - x} = \frac{2}{x - 4}$.

4. $\frac{x}{6 - x} = \frac{2}{x - 6}$.

5. $\frac{x^2}{(x - 3)(x + 1)} = \frac{3}{x - 3} - \frac{3}{(x - 3)(x + 1)}$.

6. $\frac{2x^2}{(x + 2)(x - 1)} + \frac{4}{x + 2} + \frac{4}{(x + 2)(x - 1)} = 0.$

7. $\frac{x - 1}{x + 2} = \frac{4x + 1}{2(x + 1)}$.

8. $\frac{4}{x + 7} + \frac{9}{x + 3} = 1.$

9. $\frac{3}{3x + 1} + \frac{2}{5x - 5} = \frac{1}{3x - 7}$.

10. $\frac{(x - 3)^2}{(x - 2)^2} - \frac{4(x - 1)}{x - 2} = \frac{-8}{x - 2} + 5.$

11. $\frac{1}{x} + \frac{1}{x + 1} = \frac{1}{x + 2} + \frac{1}{x + 3}$.

12. $\dfrac{x+3}{x+2}+1=\dfrac{x+2}{x+1}$. **13.** $\dfrac{1}{x}+\dfrac{1}{x-a}+\dfrac{1}{x+a}=0$.

14. $\dfrac{x}{(x-1)(x-2)}-\dfrac{1}{2(x-2)}+\dfrac{1}{2(x-1)}=0$.

15. $x^4-15x^2+36=0$. **16.** $x^4-20x^2+64=0$.

17. $\dfrac{100}{x^4}-\dfrac{29}{x^2}+1=0$. **18.** $\dfrac{96}{x^4}-\dfrac{22}{x^2}+1=0$.

19. $(x-4)^4-11(x-4)^2+18=0$. (*Hint:* Let $u=[x-4]^2$.)

20. $(x+3)^4-2(x+3)^2-24=0$.

21. $(x^2-2x)^2-7(x^2-2x)-8=0$. (*Hint:* Let $u=x^2-2x$.)

22. $(x^2+4x)^2-(x^2+4x)-20=0$.

23. $(x^3+8)+(2x^2+x-6)=0$. (*Hint:* Factor each expression in parentheses and extract common factor.)

24. $(x^3-1)+(x^2+5x-6)=0$.

25. $x^6+7x^3-8=0$. **26.** $27x^6+26x^3-1=0$.

81. Irrational Equations

An **irrational equation** is one in which an irrational expression* is involved. In solving equations of this type, we shall again find it necessary to check the converse proposition by substitution in the original equation.†

Example 1. Solve the equation $x-7=\sqrt{x-5}$.

Solution. Squaring both sides we get $x^2-14x+49=x-5$, or $x^2-15x+54=0$. The roots of the latter equation are 6 and 9. We have so far proved that if x satisfies the original equation, then x can only be 6 or 9. That is, 6 and 9 are the only candidates eligible for the position of root. To see whether or not they are roots, we must check the correctness of the converse propositions: "If $x=6$, then x satisfies the original equation," and "If $x=9$, then x satisfies the original equation." This is done by substituting in the original equation. For $x=6$, we get $6-7=\sqrt{6-5}$, or‡ $-1=+1$. Thus 6 is not a root. But for $x=9$, we get $9-7=\sqrt{9-5}$, or $2=2$. Thus, 9 is a root.

Example 2. Solve $\sqrt{x-2}=\sqrt{x}+2$.

Solution. Squaring both sides, we get $x-2=x+4\sqrt{x}+4$. Since we have not yet gotten rid of the radicals completely, we transpose, obtaining $-6=4\sqrt{x}$, or $-3=2\sqrt{x}$, and square again. Thus $9=4x$, or $x=9/4$. Therefore 9/4 is the only

* See section 24 for definition.

† Compare the preceding section, and section 29.

‡ Recall that the radical sign represents the principal root only.

possible root. But substitution reveals that it is not a root, since $\sqrt{\frac{9}{4}} - 2 = \sqrt{\frac{9}{4}} + 2$ yields $\sqrt{\frac{1}{4}} = \sqrt{\frac{9}{4}} + 2$, or $\frac{1}{2} = \frac{3}{2} + 2$, which is false. Hence the given equation has no root at all.

Since extraneous roots occur, some step in the argument must fail to be reversible.* Which step is it?

EXERCISES

Solve and check:

1. $\sqrt{x + 9} = 5.$
2. $\sqrt{x + 9} = -5.$
3. $\sqrt{x - 1} = x - 3.$
4. $\sqrt{x - 1} = 3 - x.$
5. $\sqrt{3x} = 3x - 2.$
6. $\sqrt{3x} = 2 - 3x.$
7. $\sqrt{3x + 4} = 2 + \sqrt{2x - 4}.$
8. $\sqrt{x - 4} = 9 - \sqrt{x + 5}.$
9. $\sqrt{x - 4} = \sqrt{x + 5} - 9.$
10. $\sqrt{3x - 2} = \sqrt{x - 1} - 1.$
11. $\sqrt{x + 3} + \sqrt{2x + 7} = 1.$
12. $\sqrt{6x + 1} = 1 + \sqrt{4x - 1}.$
13. $\sqrt{6x + 4} = 1 + \sqrt{6x}.$
14. $\sqrt{3x - 7} + \sqrt{3x} = 7.$
15. $\sqrt{3 - 2x} = 3 + \sqrt{2 + 2x}.$
16. $\sqrt{2x + 5} + \sqrt{4x + 3} = 3.$
17. $\sqrt{x + 1} + \sqrt{x - 1} = 4.$
18. $\sqrt{2x + 5} - \sqrt{x + 3} = 2.$
19. $\sqrt{6x - 2} - \sqrt{2x} + \sqrt{2x - 1} = 0.$
20. $\sqrt{2x + 9} - \sqrt{x + 4} = \sqrt{x + 1}.$
21. $\sqrt{3x + 9} - \sqrt{x + 5} = \sqrt{2x + 8}.$
22. $\dfrac{1}{\sqrt{x - 3}} - \sqrt{\dfrac{2x - 5}{x - 3}} = 2.$

82. Verbal Problems

Many verbal problems lead to the solution of quadratics. The student should review carefully the seven steps in solving a verbal problem given in section 64.

Example. Within a rectangular garden 6 yards wide and 12 yards long, we are to pave a path of uniform width around the boundary so as to leave an area of 40 square yards for flowers. How wide should we make the path?

Solution. Let x = width of path, in yards. Then the width of the actual flower plot is $6 - 2x$, and the length is $12 - 2x$. The

* See section 29.

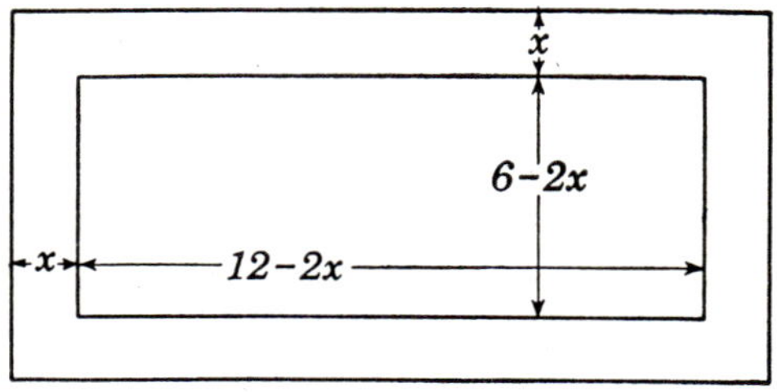

Fig. 30.

area therefore is $(6 - 2x)(12 - 2x)$. Hence the equation is

$$(6 - 2x)(12 - 2x) = 40. \tag{1}$$

The solutions of (1) are 1 and 8.

Clearly 1 satisfies the conditions of the verbal problem, since the flower plot will then have the dimensions 4 yards by 10 yards. But while 8 is a root of equation (1), it cannot satisfy the conditions of the problem, since you cannot pave a path 8 yards wide in a garden only 6 yards wide. How can this happen? See section 64.

EXERCISES

1. The area of a rectangle is 51 square feet. Its perimeter is 40 feet. Find its length and width.

2. The area of a rectangle is 504 square yards. Its perimeter is 100 yards. Find its length and width.

3. The two legs of a right triangle differ by 7 feet. The hypotenuse is 17 feet long. Find the legs. (*Hint:* Use the Pythagorean Theorem.)

4. One leg of a right triangle is 4 feet longer than twice the other leg. The hypotenuse is 26 ft. long. Find the legs.

5. The number of inches in the perimeter of a certain square is equal to the number of square inches in its area. Find the length of the side of the square.

6. The length of a rectangle is 6 feet greater than its width. Its area is 832 square feet. Find its dimensions.

7. Within a rectangular garden 10 yards wide and 20 yards long, we wish to pave a walk around the borders of uniform width so as to leave an area of 96 square yards for flowers. How wide should the walk be?

8. Find two consecutive positive integers whose product is 306.

9. Find two consecutive even positive integers whose product is 528.

10. Find two consecutive odd positive integers whose product is 323.

11. Find a positive number which when diminished by 14 is equal to 51 times the reciprocal of the number.

12. Find a positive number which when increased by 10 is 56 times the reciprocal of the number.

13. The sum S of the first n consecutive positive integers 1, 2, 3, . . ., n is given by the formula $S = \frac{1}{2}n(n + 1)$. How many of these consecutive positive integers must be added to obtain the sum 153?

14. The sum S of the first n consecutive even positive integers 2, 4, 6, . . ., $2n$ is given by the formula $S = n(n + 1)$. How many of these consecutive even positive integers must be added to obtain the sum 272?

15. The sum of the digits of a two-digit number is 10. The tens digit is 2 less than the square of the units digit. Find the number.

16. If the price of eggs rises 10 cents per dozen, one will be able to get 2 dozen fewer eggs with $6.00 than was possible at the lower price. What was the lower price?

17. A car travels 120 miles. A second car travels 10 miles per hour faster than the first and makes the same trip in 2 hours less. Find the speed of each.

18. A train travels 300 miles at a uniform speed. If its speed had been 15 miles per hour less, the trip would have taken 1 hour and 40 minutes more. Find the actual speed of the train.

19. A boat travels 30 miles downstream and 30 miles back upstream in 4 hours. If the speed of the current is 4 miles per hour, what is the boat's speed in still water, and how long did the trip take each way?

20. The speed of a stream is 3 miles per hour. A boat takes $\frac{3}{4}$ hour longer to travel 35 miles upstream than it does for the return trip downstream. What is the speed of the boat in still water and how long did the trip take each way?

21. A stone is dropped from the top of a vertical cliff 100 feet high. Using the formula $s = 16t^2$ for the distance s (measured in feet) through which a body will fall in t seconds, find the number of seconds it takes for the stone to reach bottom.

22. A stone is dropped from the top of a vertical wall 36 feet high. Using the formula of exercise 21, find the number of seconds it takes for the stone to reach bottom.

23. A stone is dropped from a cliff overlooking a lake, and the sound of the splash is heard 9 seconds after the stone was dropped. If the velocity of sound were 1024 feet per second, how high is the cliff above the lake? (*Hint:* use the formula of exercise 21 together with other considerations.)

24. If a body is thrown vertically upward with an initial velocity of v feet per second, its height h above the starting level, measured in feet, after t seconds, is given approximately by the formula $h = vt - 16t^2$, neglecting air resistance. (*a*) Solve this formula for t; (*b*) find t if $v = 128$ ft. per sec. and $h = 192$ ft.

25. Using the formula of exercise 24, how long will it take for a ball thrown vertically upward with an initial velocity of 60 feet per second to reach the point 54 feet above starting level? Explain the physical significance of the fact that there are two answers.

26. A alone can do a job in 2 days less than it takes B alone to do the same job. Working together they complete the job in $3\frac{3}{7}$ days. How long would it take each to do the job alone?

27. One pipe alone can fill a tub in 6 minutes less than it takes a second pipe alone to fill the same tub. Both pipes together can fill the tub in 4 minutes. How long does it take each pipe alone to fill the tub?

28. A pet shop bought a litter of puppies for $80. All but 3 of them were sold, at a profit of $6 each, for a total of $80. How many puppies were in the litter?

29. If $2000 invested at interest compounded annually amounts to $2121.80 in two years, what is the rate of interest?

30. Two straight roads cross at right angles. Two men start simultaneously from the point of intersection, A walking at the rate of 2.5 miles per hour on one road, B walking at the rate of 6 miles per hour on the other road. How long will it take for them to be 26 miles apart?

31. In calculating the current in a simple series electrical circuit with inductance L henrys, resistance R ohms, and capacitance C farads, it becomes necessary to solve the equation $Lx^2 + Rx + \frac{1}{C} = 0$. Find x in terms of L, R, and C.

★83. Maxima and Minima of Quadratic Functions

The graph of a quadratic function $y = ax^2 + bx + c$ with real coefficients ($a \neq 0$) is a parabola and obviously has a highest point if it opens downward, or a lowest point if it opens upward. The coordinates of this **maximum** or **minimum** point may be found by completing the square as follows.

Example 1. Find the coordinates of the maximum or minimum point of $y = x^2 - 6x - 1$.

Solution. We add and subtract the square of half the coefficient of x, obtaining $y = x^2 - 6x + 9 - 1 - 9$ or

$$y = (x - 3)^2 - 10. \tag{1}$$

★ This section may be omitted without disturbing the continuity of the book.

Since $(x - 3)^2$ is never negative for real values of x, the minimum y is clearly -10 and is attained when $(x - 3)^2 = 0$, or $x = 3$. Hence the minimum point has the coordinates $(3, -10)$.

Example 2. Find the coordinates of the maximum or minimum point of $y = 5 - 12x - 3x^2$.

Solution. We write $y = -3(x^2 + 4x) + 5$. To complete the square within the parentheses we must insert 4. Hence we must add 12 outside the parentheses, obtaining $y = -3(x^2 + 4x + 4) + 5 + 12$, or

$$(2) \qquad y = -3(x + 2)^2 + 17.$$

Since $(x + 2)^2$ is never negative for real values of x, the maximum value of y is clearly 17, and this is attained when $(x + 2)^2 = 0$, or $x = -2$. Hence the maximum point has the coordinates $(-2,17)$.

Finding the maximum or minimum values of a function $y = f(x)$ and the value of x for which it is attained is clearly a problem of paramount practical importance. In engineering, business, etc., we want to get the best result, expend the least energy, and so on. Whenever the function involved is quadratic, we may solve such a problem as above. A general method (for quadratic or non-quadratic functions) is provided in the subject known as *differential calculus.**

Example 3. Find the dimensions of the rectangle with maximum area whose perimeter is 40 feet, and find the maximum area.

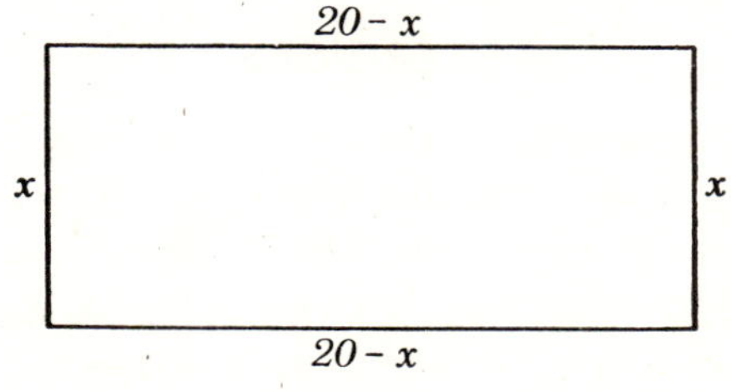

Fig. 31.

Solution. Let x be the width of the rectangle, measured in feet. Then $20 - x$ is the length and $y = x(20 - x)$ is the area (Fig. 31). Thus the function whose maximum we wish is the function

$$(3) \qquad y = 20x - x^2,$$

or

$$y = -(x^2 - 20x).$$

* For a brief introduction to this subject, see M. Richardson, *Fundamentals of Mathematics,* Macmillan, 1941, Chapter XI.

Completing the square, we get $y = -(x^2 - 20x + 100) + 100$ or

$$y = -(x - 10)^2 + 100.$$

Hence the maximum area is 100 square feet and is attained when $x = 10$. The required rectangle therefore is a square. The graph

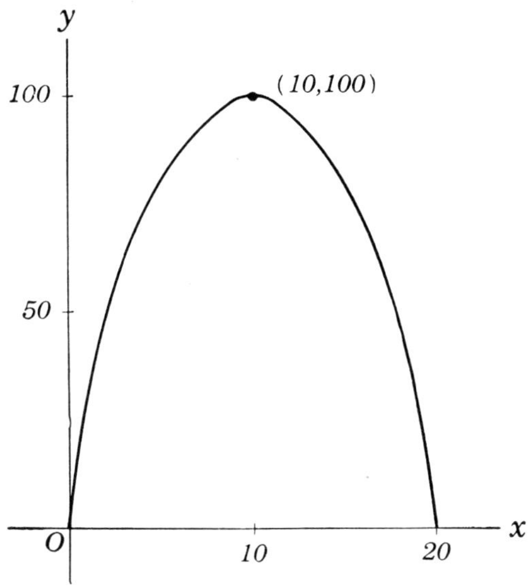

Fig. 32.

of the function (3) given in Fig. 32 indicates the relationship between this example and the preceding ones.

EXERCISES

Find the coordinates of the maximum or minimum points, as the case may be, state whether the point determined is a maximum or minimum point, and sketch the graph of each of the following:

1. $y = x^2 - 8x + 3$. **2.** $y = x^2 + 6x + 1$.
3. $y = 1 + 4x - x^2$. **4.** $y = 4 + 6x - x^2$.
5. $y = 2x^2 - 10x + 7$. **6.** $y = 4 + 6x - 3x^2$.
7. $y = 2 + 4x + 4x^2$. **8.** $y = 4 + 2x - 4x^2$.
9. $y = 15 - 6x - 6x^2$. **10.** $y = 3 - 2x + 3x^2$.

11. Find two numbers the sum of which is 30 and such that their product is a maximum.

12. Find the dimensions of the rectangle with maximum area the perimeter of which is 48 feet. What is the maximum area?

13. Find two numbers the sum of which is 28 and such that the sum of their squares is minimum.

14. Find the dimensions of the rectangle the perimeter of which is 40 feet and such that the square of its diagonal is a minimum.

15. A body is thrown straight up from the ground with an initial velocity of 96 feet per second. Neglecting air resistance, its height h, measured in feet, after t seconds, is given by the formula $h = 96t - 16t^2$. After how many seconds does the body reach its maximum height, and what is the maximum height?

16. A projectile is thrown straight up from a height of 6 feet, with an initial velocity of 192 feet per second. Neglecting air resistance, its height h, measured in feet, after t seconds is given by the formula $h = 6 + 192t - 16t^2$. After how many seconds does the projectile reach its maximum height, and what is the maximum height?

17. If 400 people will attend a moving picture theater when the admission price is 30 cents, and if the attendance decreases by 40 for each 10 cents added to the price, then what price of admission will yield the greatest gross receipts?

18. A telephone company can get 1000 subscribers at a monthly rate of $5.00 each. It will get 100 more subscribers for each 10-cent decrease in the rate. What rate will yield the maximum gross monthly income and what will this income be?

19. It is desired to enclose a rectangular field adjacent to the straight bank of a river. No fence is needed along the river bank. If 160 yards of fence is to be used, what are the dimensions of the rectangle which will enclose the maximum area, and what is the maximum area?

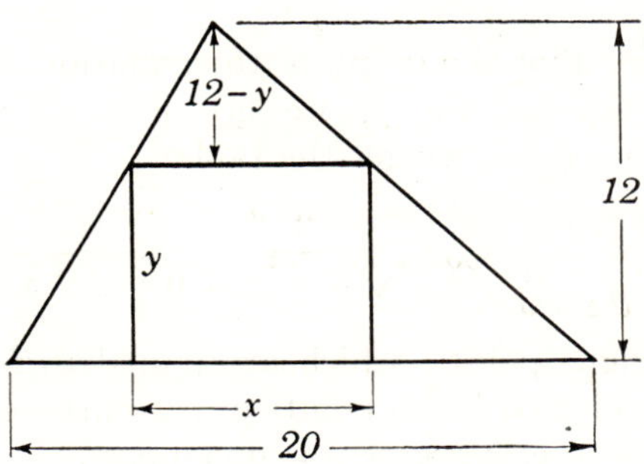

Fig. 33.

20. The base of a triangle is 20 feet and its altitude is 12 feet. Find the dimensions of the rectangle with maximum area that can be inscribed in the triangle with one side along the base (Fig. 33).

21. Show that if $y = ax^2 + bx + c$ $(a \neq 0)$ has real coefficients, then the maximum or minimum point of the graph has the coordinates $x = -b/2a$, $y = (4ac - b^2)/4a$, and that the point is a maximum or a minimum according as a is negative or positive.

Chapter XII

Systems of equations in two unknowns involving quadratics

84. The Graph of a Quadratic Equation in x and y

To plot the graph of an equation of the second degree in x and y with real coefficients, it is usually desirable to solve the equation for y in terms of x, or for x in terms of y, and calculate a table, as in the following example. Square roots may be worked out approximately, or read from table I. The table should always include the values $x = 0$ and $y = 0$ if possible; these will yield the points where the curve crosses the y-axis and the x-axis, respectively.

Example 1. Plot the graph of the equation $9x^2 + 4y^2 = 36$.

Solution. Solving for y, we get $y = \pm\frac{1}{2}\sqrt{36 - 9x^2}$. Substituting values for x, we get the table:

x	0	± 2	$+1$	-1
y	± 3	0	$\pm\frac{3}{2}\sqrt{3} = \pm 2.6$	$\pm\frac{3}{2}\sqrt{3} = \pm 2.6$

This yields eight points, which are joined to form the curve in Fig. 34. It is clear that no points of the curve exist for $x > 2$, or $x < -2$, since in either case y is imaginary.

Example 2. Plot the graph of $xy = 1$.

Solution. Solving for y in terms of x, we get $y = 1/x$. Clearly no points of the curve exist with $x = 0$ or $y = 0$. We get the table

x	4	3	2	1	$\frac{1}{2}$	$\frac{1}{3}$	$\frac{1}{4}$	-4	-3	$\dots$
y	$\frac{1}{4}$	$\frac{1}{3}$	$\frac{1}{2}$	1	2	3	4	$-\frac{1}{4}$	$-\frac{1}{3}$	$\dots$

This yields the graph of Fig. 35.

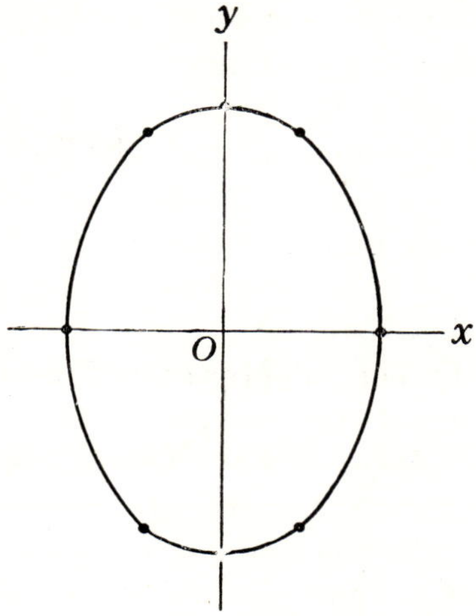

Fig. 34.

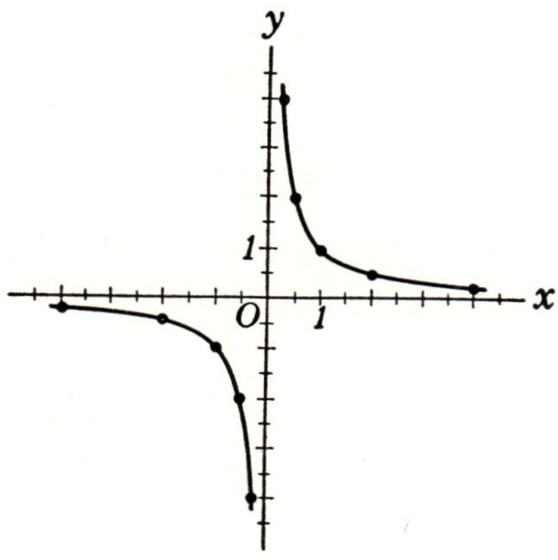

Fig. 35.

The general equation of second degree in x and y may be written as

$$(1) \qquad Ax^2 + Bxy + Cy^2 + Dx + Ey + F = 0.$$

The following facts, proved in textbooks on analytic geometry, will be helpful in plotting graphs.

If the coefficients are all real, the graph is either an ellipse (Fig. 34), a hyperbola (Fig. 35), a circle, a parabola (Fig. 29), or, in special cases, a pair of straight lines (distinct or coincident), a single point, or non-existent. In particular:

If $B^2 - 4AC$ is positive, the graph is a hyperbola, or, in special cases, a pair of intersecting straight lines;

If $B^2 - 4AC$ is zero, the graph is a parabola, or, in special cases, two parallel or coincident straight lines, or non-existent;

If $B^2 - 4AC$ is negative, the graph is an ellipse, or, in special cases, a circle, a single point, or non-existent;

If $B = 0$ and $A = C$, the graph is a circle, or, in special cases, a single point or non-existent.

All the graphs of second-degree equations in x and y are called **conic sections** because they are the intersections of a right circular cone and a plane. These curves were studied by the ancient Greeks without thought of practical applications. A great book about them was written by Apollonius of Perga (3rd century B.C.). Their equations were first used to study them in the 17th century by Descartes, Fermat, Wallis, and others. They have many practical applications in physics. For example, the path of a projectile is, neglecting air resistance, a parabola; the orbits of the planets about the sun are, neglecting slight perturbations, ellipses.*

EXERCISES

Plot the graph of each of the following and identify the curve:

1. $x^2 + y^2 = 25.$ **2.** $x^2 + y^2 = 100.$

3. $y = 2x^2 + 4x - 3.$ **4.** $x^2 + y^2 - 4x + 2y - 4 = 0.$

5. $4x^2 + 9y^2 = 36.$ **6.** $4x^2 + y^2 = 16.$

7. $x^2 - y^2 = 1.$ **8.** $xy - 2y = 1.$ **9.** $x^2 + y^2 = -1.$

10. $x^2 + y^2 = 1.$ **11.** $x^2 + y^2 = 0.$

12. $9x^2 + 4y^2 - 18x + 16y - 11 = 0.$

13. $x^2 - y^2 = 0.$ (*Hint:* The equation is equivalent to the system $x + y = 0$, $x - y = 0$.)

14. $x^2 - y^2 - 4x - 6y - 6 = 0.$

15. $x^2 - 2xy + y^2 = 0.$ (*Hint:* The equation is equivalent to $[x - y]^2 = 0$.)

16. $9x^2 - 24xy + 16y^2 - 200x - 150y = 0.$

17. $(x + 3)(y - 1) = 0.$ **18.** $36x^2 + 24xy + 29y^2 = 5.$

85. Graphical Solution of Simultaneous Equations

A common (real) solution of two equations in x and y are the coordinates of a point belonging to both graphs, that is, a point

* For other applications, see M. Richardson, *Fundamentals of Mathematics*, Macmillan, 1941, and textbooks on analytic geometry.

of intersection of the two curves. Hence the common real solutions can be estimated from the graph.

Example 1. Find graphically the common solutions of

$$3x^2 + 4y^2 = 192$$

and

$$3x^2 - y^2 = 12.$$

Plotting the graphs of the two equations, we get Fig. 36, from which we read (approximately) the common solutions:

x	4	4	−4	−4
y	6	−6	6	−6

The graphical solution yields only the *real* common solutions,

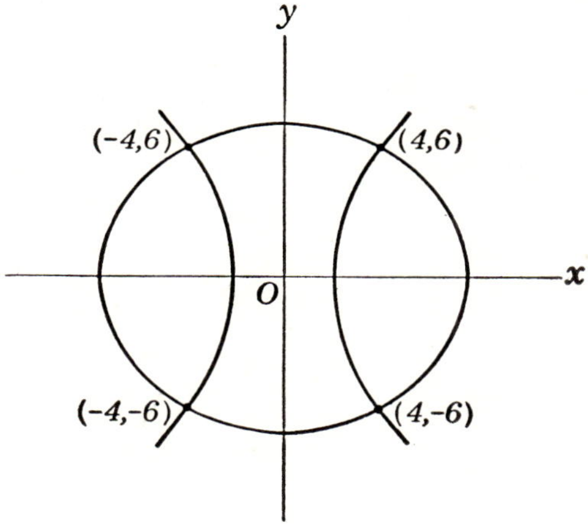

Fig. 36.

since the x- and y-coordinates on the graph are always real numbers, and even these solutions are usually found only approximately, since they are based on drawing and eyesight. To find accurate solutions in all cases we must use algebraic methods, as in the following sections. However, it is desirable to plot the graphs in any case, since they serve as a rough check on the algebraic solution. For many practical problems, a graphical solution is sufficiently accurate. Two quadratic equations in x and y can have no more than four points of intersection, and they may have fewer.

EXERCISES

Estimate graphically the real solutions of each of the following systems of equations:

1. $\begin{cases} x^2 + y^2 = 25 \\ 4y - 3x = 0. \end{cases}$

2. $\begin{cases} x^2 + y^2 = 34 \\ y - x = 2. \end{cases}$

3. $\begin{cases} x^2 + y^2 = 25 \\ x + 2y = 10. \end{cases}$

4. $\begin{cases} x^2 + y^2 = 13 \\ 2y - x = 4. \end{cases}$

5. $\begin{cases} 3x^2 + y^2 = 12 \\ 3x + y = 6. \end{cases}$

6. $\begin{cases} 4x^2 + 25y^2 = 100 \\ 5y - 2x = 10. \end{cases}$

7. $\begin{cases} xy = 12 \\ x + y = 7. \end{cases}$

8. $\begin{cases} xy = 1 \\ 3y - 5x = 2. \end{cases}$

9. $\begin{cases} x^2 = 4y \\ x + 2y = 4. \end{cases}$

10. $\begin{cases} y^2 = 4x \\ x + y = 3. \end{cases}$

11. $\begin{cases} x^2 - y^2 = 5 \\ 3x + y = 11. \end{cases}$

12. $\begin{cases} 7y^2 - 6xy = 8 \\ 2y - 3x = 5. \end{cases}$

13. $\begin{cases} y^2 = 2x \\ x^2 = 2y. \end{cases}$

14. $\begin{cases} x^2 = y \\ x = 4y - 3. \end{cases}$

15. $\begin{cases} x^2 + y^2 = 25 \\ 4x^2 = 9y. \end{cases}$

16. $\begin{cases} (x - 2)^2 + y^2 = 5 \\ x^2 + (y - 1)^2 = 10. \end{cases}$

17. $\begin{cases} x^2 + y^2 = 10 \\ 9x^2 + y^2 = 18. \end{cases}$

18. $\begin{cases} 2x^2 + y^2 = 18 \\ x^2 + 2y^2 = 33. \end{cases}$

19. $\begin{cases} x^2 - y^2 = 5 \\ 2x^2 + y^2 = 22. \end{cases}$

20. $\begin{cases} 2x^2 + y^2 = 33. \\ x^2 - y^2 = 15. \end{cases}$

21. $\begin{cases} x^2 = 9y \\ x^2 - y^2 = 8. \end{cases}$

22. $\begin{cases} 3x^2 = 4y \\ 9x^2 + 4y^2 = 72. \end{cases}$

23. $\begin{cases} x^2 - y^2 = 7 \\ x^2 + y^2 = 25. \end{cases}$

24. $\begin{cases} x^2 = 2y + 10 \\ x^2 + y^2 = 25. \end{cases}$

86. Algebraic Solution When One Equation Is Linear and One Quadratic

In this case, the linear equation can be solved for one variable in terms of the other and this expression substituted in the quadratic equation, thus yielding a quadratic in the other variable only. The solution is completed as in the following example.

Example 1. Solve algebraically the system:

(1) $$xy = 1$$

(2) $$3y - 5x = 2.$$

Solution. From the linear equation (2) we get

(3) $$y = \frac{5x + 2}{3}.$$

Substituting this in the quadratic equation (1), we get

(4) $$\frac{5x^2 + 2x}{3} = 1.$$

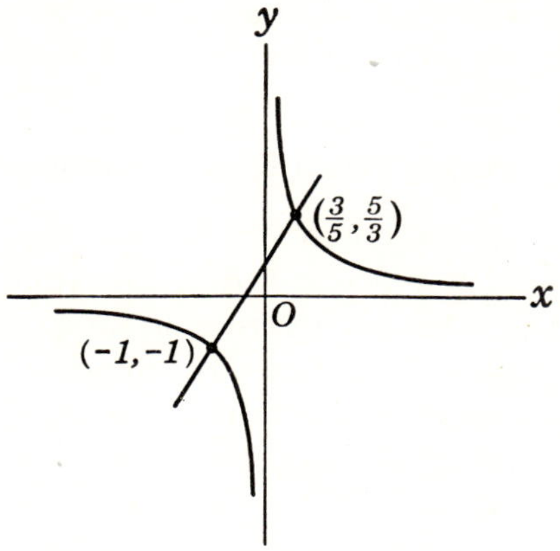

Fig. 37.

Solving (4), we get $x = 3/5$ and $x = -1$. Substituting these results in (3), we get

$$\text{for } x = \tfrac{3}{5},\ y = \tfrac{5}{3}$$

and

$$\text{for } x = -1,\ y = -1.$$

Both solutions should be checked by substituting in both equations.

EXERCISES

Solve algebraically and check:

1–12. Solve exercises 1–12 of section 85.

13. $\begin{cases} (x-3)^2 + y^2 = 5 \\ 2x - y = 6. \end{cases}$ **14.** $\begin{cases} (y+1)^2 = x - 3 \\ y = 4x - 16. \end{cases}$

15. $\begin{cases} (x-4)^2 + y = 5 \\ y = x - 1. \end{cases}$ **16.** $\begin{cases} (x-2)^2 + (y+1)^2 = 5 \\ y = 2x - 5. \end{cases}$

17. $\begin{cases} x^2 + y^2 - 2y = 9 \\ 2x + y + 4 = 0. \end{cases}$

18. $\begin{cases} x^2 - 8x + y^2 + 2y + 7 = 0 \\ y = 2x - 4. \end{cases}$

19. $\begin{cases} x^2 + xy - y^2 = 1 \\ 3y - 2x = 5. \end{cases}$ **20.** $\begin{cases} x^2 + y^2 - 2xy = 9 \\ 3x - 2y = 7. \end{cases}$

21. Prove algebraically that the graphs of $x^2 + y^2 = 1$ and $x + y = 7$ do not intersect.

22. Prove algebraically that the graphs of $2x^2 + y^2 = 2$ and $x - y = 10$ do not intersect.

23. Find the positive value of k for which the graph of $x + y = k$ is tangent to the graph of $x^2 + y^2 = 1$, and find the coordinates of the point of tangency.

24. Find the positive value of k for which the graph of $3x + 5y = k$ is tangent to the graph of $16x^2 + 25y^2 = 400$, and find the coordinates of the point of tangency.

87. Algebraic Solution When Both Equations Are Quadratic

The task of eliminating one variable was easily accomplished in the preceding section because one equation was linear, and from it one could get a simple expression for y in terms of x or for x in terms of y. If both equations are quadratic, the same procedure could be followed but one would in general get from one equation by means of the quadratic formula a cumbersome expression for y in terms of x which, when substituted in the other equation, would yield a complicated equation involving radicals. A better method of eliminating one variable from two polynomial equations of any degree in x and y will be discussed in section 146. For the present, we shall present a few special devices for the solution of special cases only. It can be proved that a system consisting of two quadratics in x and y can have at most four different solutions.

Case A. If the equation $Ax^2 + Bxy + Cy^2 + Dx + Ey + F = 0$ is such that $B = 0$ and $A = C \neq 0$, then it can be shown that its graph is a circle (except for trivial exceptional cases). If both equations of the system are of this type, the common solutions may be found as follows. By multiplication and addition we may eliminate the second-degree terms; then we may solve the resulting linear equation with either one of the given quadratic equations.

Example 1. Solve the system

$$\begin{cases} x^2 - 4x + y^2 = 1 & (1) \\ x^2 + y^2 - 2y = 9. & (2) \end{cases}$$

Solution. Multiplying the first equation by -1 and adding the result to the second equation, we eliminate the second-degree terms obtaining

$$4x - 2y = 8$$

or

$$2x - y = 4. \quad (3)$$

Solving (3) and (1) simultaneously as in the preceding section, we obtain the solutions $x = 3$, $y = 2$ and $x = 1$, $y = -2$. These solutions should be checked in both original equations.

EXERCISES

Solve and check:

1. $\begin{cases} x^2 + y^2 + 4x = 1 \\ x^2 + y^2 - 2y = 9. \end{cases}$ **2.** $\begin{cases} x^2 + y^2 + 2x + 2y = 3 \\ (x + 3)^2 + y^2 = 10. \end{cases}$

3. $\begin{cases} (x + 3)^2 + (y - 1)^2 = 5. \\ (x + 1)^2 + y^2 = 10. \end{cases}$

4. $\begin{cases} (x + 2)^2 + (y - 3)^2 = 5 \\ x^2 + (y - 2)^2 = 10. \end{cases}$

5. $\begin{cases} x^2 + y^2 + 6x - 4y = 0 \\ (x - 5)^2 + (y - 2)^2 = 29. \end{cases}$

★6. Show that if two circles intersect in two distinct points, then the linear equation obtained by eliminating the second-degree terms must be the equation of the straight line joining the points of intersection.

Case B. When both equations have the form $Ax^2 + Cy^2 + F = 0$, the system is linear in the variables x^2 and y^2 and can be solved by multiplication and addition, as in the following example.

Example 2. Solve the system:

$$\begin{cases} 3x^2 + 4y^2 = 192 \\ 3x^2 - y^2 = 12. \end{cases}$$

Solution. Multiplying the second equation by -1 and adding to the first, we get $5y^2 = 180$, or $y^2 = 36$, or $y = \pm 6$. Multiplying the second by 4 and adding to the first, we get $15x^2 = 240$, or $x^2 = 16$, or $x = \pm 4$. Hence there are four solutions:

x	4	4	-4	-4
y	6	-6	6	-6

See Fig. 36, section 85.

EXERCISES

Solve and check:

7. $\begin{cases} 2x^2 - 3y^2 = 6 \\ 3x^2 - 5y^2 = 7. \end{cases}$ **8.** $\begin{cases} y^2 - 8x^2 = 8 \\ x^2 + 2y^2 = 19. \end{cases}$

9. $\begin{cases} x^2 + 2y^2 = 13 \\ 2x^2 - 5y^2 = 8. \end{cases}$ **10.** $\begin{cases} 8x^2 - y^2 + 9 = 0 \\ 3x^2 + y^2 = 31. \end{cases}$

11. $\begin{cases} 2x^2 + y^2 = 5 \\ 2x^2 - 3y^2 = 3. \end{cases}$ **12.** $\begin{cases} 2x^2 - 3y^2 = 4 \\ 3x^2 + 2y^2 = 19. \end{cases}$

13. $\begin{cases} 3x^2 + 2y^2 = 1 \\ 12x^2 - y^2 = 1. \end{cases}$ **14.** $\begin{cases} 3x^2 + 8y^2 = 21 \\ x^2 + 4y^2 = 10. \end{cases}$

★ May be omitted without disturbing continuity.

Case C. When one equation can be factored, the solution may be accomplished as in the following examples.

Example 3. Solve

$$\begin{cases} x^2 + y^2 = 25 & (4) \\ 12x^2 + 7xy - 12y^2 = 0. & (5) \end{cases}$$

Solution. The second equation can be factored, obtaining $(4x - 3y)(3x + 4y) = 0$. This is equivalent to the two linear equations

$$4x - 3y = 0 \tag{6}$$

and

$$3x + 4y = 0. \tag{7}$$

Hence the given system is equivalent to two systems of the type studied in section 86, namely (4) and (6) together, and (4) and (7) together. Solving (4) and (6), we get $x = 3$, $y = 4$, and $x = -3$, $y = -4$. Solving (4) and (7), we get $x = 4$, $y = -3$, and $x = -4$, $y = 3$. Hence the given system has all four solutions.

If no first-degree terms are present, the system can sometimes be reduced to the preceding case by eliminating the constants.

Example 4. Solve

$$\begin{cases} 3x^2 + xy = 3 & (8) \\ 4y^2 - xy = 4. & (9) \end{cases}$$

Solution. To eliminate the constants, we multiply (8) by 4 and (9) by -3 and add, obtaining

$$12x^2 + 7xy - 12y^2 = 0. \tag{10}$$

This may be factored into $(4x - 3y)(3x + 4y) = 0$. We may then solve either of the given equations together with both the equations obtained by setting these factors equal to zero. Thus we may solve the systems

$$\begin{cases} 3x^2 + xy = 3 \\ 4x - 3y = 0 \end{cases} \tag{11}$$

and

$$\begin{cases} 3x^2 + xy = 3 \\ 3x + 4y = 0. \end{cases} \tag{12}$$

Solving (11), we get

$$3x^2 + \frac{4x^2}{3} = 3$$

$$9x^2 + 4x^2 = 9$$

$$13x^2 = 9$$

$$x = \pm \frac{3}{\sqrt{13}} = \pm \frac{3\sqrt{13}}{13}$$

$$y = \pm \frac{12\sqrt{13}}{39} = \pm \frac{4\sqrt{13}}{13}.$$

Solving (12), we get

$$3x^2 - \frac{3x^2}{4} = 3$$

$$12x^2 - 3x^2 = 12$$

$$9x^2 = 12$$

$$x = \pm \frac{\sqrt{12}}{3} = \pm \frac{2\sqrt{3}}{3}$$

$$y = \mp \frac{\sqrt{3}}{2}.$$

Hence the given system has the four solutions

x	$\frac{3\sqrt{13}}{13}$	$-\frac{3\sqrt{13}}{13}$	$+\frac{2\sqrt{3}}{3}$	$-\frac{2\sqrt{3}}{3}$
y	$\frac{4\sqrt{13}}{13}$	$-\frac{4\sqrt{13}}{13}$	$-\frac{\sqrt{3}}{2}$	$+\frac{\sqrt{3}}{2}$

Other cases will be left to the ingenuity of the student, or they may be solved by means of the general method taken up in section 146.

EXERCISES

Solve the following systems, leaving irrational answers in radical form and check:

15. $\begin{cases} x^2 + y^2 = 25 \\ 12x^2 - 25xy + 12y^2 = 0. \end{cases}$

16. $\begin{cases} x^2 + y^2 = 16 \\ 2x^2 - 3xy + y^2 = 0. \end{cases}$

17. $\begin{cases} x^2 + y^2 = 25 \\ (x - y + 5)(3x - 4y) = 0. \end{cases}$

18. $\begin{cases} 16x^2 + 25y^2 = 400 \\ (4x + 5y - 20)(8x + 5y - 40) = 0. \end{cases}$

19. $\begin{cases} x^2 - y^2 = 9 \\ 4x^2 - (5 + 4\sqrt{2})xy + 5\sqrt{2}\,y^2 = 0. \end{cases}$

20. $\begin{cases} 9x^2 + 5y^2 = 161 \\ 20x^2 - 23xy + 6y^2 = 0. \end{cases}$

21. $\begin{cases} 3x^2 - 5y^2 = 7 \\ 2x^2 - 7xy + 6y^2 = 0. \end{cases}$

22. $\begin{cases} 5x^2 + 12y^2 = 128 \\ 3x^2 + 4xy - 4y^2 = 0. \end{cases}$

23. $\begin{cases} x^2 + 2xy + y^2 = 1 \\ xy + 2x^2 = 2. \end{cases}$

24. $\begin{cases} 3x^2 - 5xy = 2 \\ 2y^2 + xy = 4. \end{cases}$

25. $\begin{cases} 2x^2 + xy + y^2 = 8 \\ 3xy + y^2 = 10. \end{cases}$

26. $\begin{cases} 3x^2 + 4xy + y^2 = 5 \\ 3xy + 10y^2 = 4. \end{cases}$

27. $\begin{cases} x^2 + y^2 = 5 \\ xy = 2. \end{cases}$

28. $\begin{cases} x^2 + y^2 = 10 \\ xy = 3. \end{cases}$

88. Verbal Problems

The following verbal problems may lead to systems of equations involving quadratics. The student should follow carefully the seven steps outlined in section 64.

Note. The problem of finding the intersection of two curves is one of considerable practical importance beyond the problems below. Thus in navigation a point is located usually as the intersection of two arcs of circles, or of two straight lines, or even of two hyperbolas. In practical engineering problems, it is often necessary to find the coordinates of points of intersection of two graphs.

EXERCISES

1. Find two positive numbers which differ by 3 and whose product is 208.

2. Find two positive numbers whose sum is 21 and the sum of whose squares is 245.

3. The area of a right triangle is 210 square feet, and its hypotenuse is 37 feet long. Find the legs.

4. The area of a rectangle is 420 square inches, and a diagonal is 29 inches long. Find the length and width of the rectangle.

5. The perimeter of a rectangle is 70 feet and its area is 304 square feet. Find its length and width.

6. The perimeter of a rectangle is 92 inches and a diagonal is 34 inches long. Find its length and width.

7. If a bus travelling between two towns were to increase its usual speed by 10 miles per hour, it would make the trip in 30 minutes less than its usual time. But if its usual speed were decreased by 10 miles per hour, the trip would take 1 hour more. Find the usual speed of the bus and the length of the trip.

8. A man walks 16 miles in 40 minutes less time than his son.

The son's speed is 4/5 mile per hour less than the father's speed. Find the speed of each.

9. The hypotenuse of a right triangle is 19.5 feet long. If each leg were increased by 4.5 feet, the hypotenuse would be increased by 6 feet. Find the legs of the original triangle.

10. The diagonal of a rectangle is 37 inches long. If the length of the rectangle were decreased by 5 inches and its width increased by 4 inches, the diagonal would be decreased by 3 inches. Find the length and width of the original rectangle.

11. A rectangular flower plot has area 504 square feet, and is surrounded by a path 3 feet wide. The area of the path is 312 square feet. Find the length and width of the flower plot.

12. A and B can do a job in 4 days working together. A alone would take twice as long to do the job as B alone would take. How long would each alone require to do the job?

13. A man spent $4.80 for a certain number of items. If the price per item were decreased by 5 cents, he could have bought 16 more items for the same amount of money. How many items were bought and at what price per item?

14. A rectangular piece of cardboard has an area of 120 square inches. By cutting a square 2 inches on each side from each of the four corners and folding up the sides, an open box is formed having a volume of 96 cubic inches. Find the length and width of the original cardboard.

15. The diagonal of a rectangle is 16 inches longer than one of its sides and 8 inches longer than the other side. Find the sides of the rectangle.

16. The hypotenuse of a right triangle is 45 feet. If each leg were 6 feet shorter, the area would be decreased by 171 square feet. Find the legs of the given triangle.

Chapter XIII

Ratio, proportion, and variation

89. Ratio and Proportion

The **ratio** of one number a to another number b ($\neq 0$) is the quotient a/b. The ratio a/b is sometimes written $a:b$. Four numbers a, b, c, d are said to be **in proportion** if $a/b = c/d$ (or $a:b = c:d$). A **proportion** is a statement that two ratios are equal. The proportion $a:b = c:d$ is often read "a is to b as c is to d." In the proportion $\frac{a}{b} = \frac{c}{d}$ the numbers a and d are called the **extremes** and the numbers b and c are called the **means.**

If $a:b = c:x$, then x is called the **fourth proportional** to a, b, and c. If $a:b = b:x$, then x is called the **third proportional** to a and b. If $a:x = x:b$, then x is called a **mean proportional** to a and b.

Proportions were studied by the ancient Greek geometers. If two triangles are similar, then corresponding sides are in proportion. This may be written as follows: If the lengths of the sides of one triangle, all measured in terms of the same unit, are denoted by a, b, and c, and the lengths of the corresponding sides of the other (similar) triangle are denoted by a', b', and c', measured in terms of the same unit, then $a:a' = b:b' = c:c'$. The latter equality means that $\frac{a}{a'} = \frac{b}{b'} = \frac{c}{c'}$.

The following theorems are found in Euclid's *Elements*, the great work on geometry written about 320 B.C. upon which geometry textbooks are based. The student should carry out the proofs. It is assumed throughout that values of the letters which would make any denominator zero are excluded.

THEOREM I. *If* $\frac{a}{b} = \frac{c}{d}$, *then* $ad = bc$. *Or, in any proportion the product of the means equals the product of the extremes.*

Hint for proof: Multiply both members by the L.C.D. bd.

THEOREM II. *If* $\frac{a}{b} = \frac{c}{d}$, *then* $\frac{a}{c} = \frac{b}{d}$. *That is, the means of any proportion may be interchanged.*

Hint for proof: Divide both members of $ad = bc$ by cd.

In II, the second proportion is said to be obtained from the first by **alternation.**

THEOREM III. *If* $\frac{a}{b} = \frac{c}{d}$, *then* $\frac{b}{a} = \frac{d}{c}$.

Hint for proof: Divide both members of $ad = bc$ by ac.

In III, the second proportion is said to be obtained from the first by **inversion.**

THEOREM IV. *If* $\frac{a}{b} = \frac{c}{d}$, *then* $\frac{a+b}{b} = \frac{c+d}{d}$.

Hint for proof: Add 1 to both members of the given proportion and simplify.

In IV, the second proportion is said to be obtained from the first by **composition.**

THEOREM V. *If* $\frac{a}{b} = \frac{c}{d}$, *then* $\frac{a-b}{b} = \frac{c-d}{d}$.

Hint for proof: Subtract 1 from both members and simplify.

In V, the second proportion is said to be obtained from the first by **division.**

THEOREM VI. *If* $\frac{a}{b} = \frac{c}{d}$, *then* $\frac{a+b}{a-b} = \frac{c+d}{c-d}$.

Hint for proof: Divide the second proportion of IV by the second proportion of V and simplify.

In VI, the second proportion is said to be obtained from the first by **composition and division.**

Example 1. To find the fourth proportional to 2, 3, and 4 we write $\frac{2}{3} = \frac{4}{x}$. Solving, we get $2x = 12$, or $x = 6$. *Check:* $\frac{2}{3} = \frac{4}{6}$ is correct.

Example 2. To find a mean proportional to 2 and 8 we write $\frac{2}{x} = \frac{x}{8}$ or $x^2 = 16$, or $x = \pm 4$. Hence either $+4$ or -4 is a mean proportional to 2 and 8.

Note 1. Obviously, the fourth proportional to a, b, c satisfies the equation $a/b = c/x$ or $x = bc/a$. Thus, to find the fourth proportional, we multiply the second number by the third and divide by the first.*

Note 2. In practical applications, care must be taken with the units in which measurements are given. In particular, in a proportion the two ratios must be expressed in *consistent units.* Thus a velocity of 3 feet per 2 seconds is a ratio 3/2 feet per second or 1.5 ft./sec. It may also be expressed as 1 yard per 2 seconds or $\frac{1}{2}$ yard per second or 0.5 yd./sec. But it would be incorrect to write $\frac{3}{2} = \frac{1}{2}$. If the numerator and denominator of a ratio represent measurements expressed in the same units, the quotient or ratio is sometimes called a **pure** or **dimensionless number,** since it remains the same if the units are changed in a consistent way and therefore may be written without any unit. Thus 12 inches:1 inch = 12/1; if both numerator and denominator are expressed in yards, we get

$$\frac{1}{3}\text{ yd.}:\frac{1}{36}\text{ yd.} = \frac{\frac{1}{3}\text{ yd.}}{\frac{1}{36}\text{ yd.}} = \frac{36}{3} = \frac{12}{1}.$$

The ratio of two measurements of the same kind is most conveniently expressed as such a pure or dimensionless number.

Example 3. Express the ratio of 3 inches to 1 foot as a pure number.

Solution. Using the same units, the desired ratio is that of 3 inches to 12 inches, or 3:12, or 1/4.

A symbol of the form $a:b:c = 2:3:6$ is understood to be equivalent to the proportions $a:b = 2:3$ and $b:c = 3:6$, or to the equalities $\dfrac{a}{2} = \dfrac{b}{3} = \dfrac{c}{6}$. A symbol of the form $a:b:c$ is called a **continued ratio.**

EXERCISES

Express each ratio as a fraction in simplest form:

1. 3:12. **2.** 5.6:1.4. **3.** $\frac{1}{3}:\frac{1}{4}$. **4.** $\frac{3}{4}:\frac{9}{20}$.

* This rule is known as the *Rule of Three* and was considered difficult in the Middle Ages, as indicated by the Mother Goose rhyme:

"Multiplication is vexation,
Division is as bad;
The Rule of Three doth puzzle me,
And Practice drives me mad."

5. $3.2x^3:4.8x$.
6. $5.7xy^3:9.5x^3y^2$.
7. $(x^2 - 9):(2x + 6)$.
8. $\dfrac{x^3 - 8}{x^2 - 4}$.

Express each ratio as a pure or dimensionless number in simplest form:

9. 4 feet to 80 inches.
10. 6 inches to $\frac{1}{4}$ yard.
11. 36 hours to 2 days.
12. 80 minutes to 2 hours.
13. 528 feet per second to 720 miles per hour.
14. 40 pounds per square inch to 1440 pounds per square foot.

Express each equation in fractional form and solve for x:

15. $10:x = 2$.
16. $(2 + x):x = 6:x$.
17. $1:(x - 1) = 2:(x - 2)$.
18. $(x - 1):(x - 2) = (x + 3):(x - 4)$.
19. $(2x - 1):(2x + 2) = (2x + 3):(2x - 4)$.
20. $(3x - 1):(2x + 3) = (3x - 4):(2x + 2)$.

Find the fourth proportional to the three given quantities:

21. 3, 5, 57.
22. 2, 3, 34.
23. 2, 5, 14.
24. $2x$, $5x^3$, $3x^3$.
25. a, ab, b.
26. a, b, ab.
27. ab, a, b.
28. $x + 1$, $x - 1$, $x^2 - 1$.

Find the mean proportionals to each pair of given quantities:

29. 2, 8.
30. 2, 18.
31. 1/3, 1/12.
32. 3/4, 3/64.
33. a, b.
34. $9/a^5$, a.
35. $\dfrac{x - 2}{x + 2}$, $x^2 - 4$.
36. $8/xy^5$, $2xy$.

Find the third proportional to each pair of given quantities:

37. 3, 6.
38. 2, 12.
39. 1/2, 1/8.
40. 2/3, 3/8.
41. a, b.
42. $a^2 - b^2$, $a - b$.

43. A line segment 30 inches long is divided into two parts, the lengths of which are in the ratio 2:3. Find the lengths of the parts.

44. A line segment 133 inches long is divided into two parts, the lengths of which are in the ratio 2:5. Find the lengths of the parts.

45. A line segment 143 inches long is divided into three parts, the lengths of which are in the ratio 2:5:6. Find the lengths of the parts.

46. One hundred and thirty-five battleships are divided among three nations in the ratio 3:5:7. How many does each nation get?

47. The sides of a triangle are 6, 10, and 12 inches long, respectively. In a similar triangle, the shortest side is 8 inches long. Find the other sides.

48. The sides of a triangle are 10, 14, and 20 inches long, respectively. In a similar triangle, the longest side is 34 inches long. Find the other sides.

49. A man 5 feet 9 inches tall casts a shadow 11.5 feet long. At the same time of day, a near-by flagpole casts a shadow 76 feet long. How high is the flagpole?

50. A boy 5 feet tall stands 26 feet from the foot of a lamppost. The light of the lamppost casts the boy's horizontal shadow 10 feet long. How high is the lamppost?

51. The area A of a circle is given by the formula $A = \pi r^2$ where r is the radius. Show that if A_1 and A_2 are the areas of two circles of radii r_1 and r_2, respectively, then $A_1:A_2 = r_1^2:r_2^2$.

52. The volume V of a sphere is given by the formula $V = \frac{4}{3}\pi r^3$ where r is the radius. Show that the volumes of two spheres are to each other as the cubes of their radii, that is, $V_1:V_2 = r_1^3:r_2^3$.

53. Carry out the proofs of theorems I–VI inclusive in the text above.

90. The Terminology of Variation

The terms discussed in this section are often used in scientific writing. When y is a linear function of x of the special form

$$y = kx, \tag{1}$$

where k is a constant, we say that y **varies directly as** x, or y **is directly proportional to** x. We call k the **constant of proportionality** or **factor of proportionality.** The statement "y varies directly as x" is sometimes written symbolically as

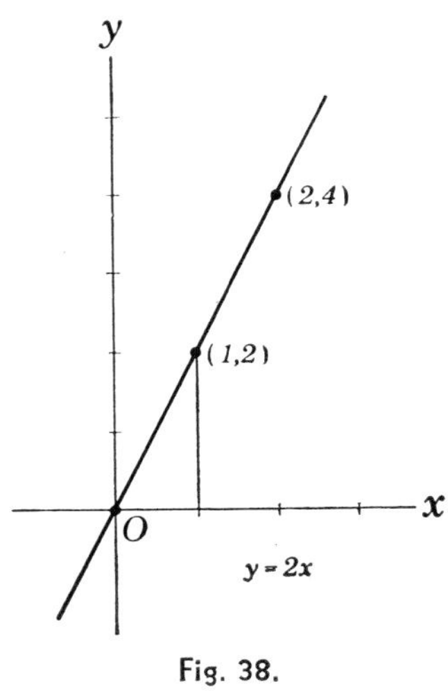

Fig. 38.

$y \propto x$, but for practical use the form (1) is preferable. The graph of such a function is an oblique straight line passing through the origin with slope k (Fig. 38).

Example 1. The circumference of a circle varies directly as the radius since $C = 2\pi r$. In this example the factor of proportionality is 2π.

If sufficient information is given, it is often possible to determine the value of the factor of proportionality and write y as a definite function of x.

Example 2. Suppose y varies directly as x and $y = 6$ when $x = 3$. (*a*) Write y as a function of x. (*b*) Find the value of y when $x = 4$.

Solution. $y = kx$. Hence $6 = 3k$, or $k = 2$. Therefore (*a*) $y = 2x$; (*b*) $y = 8$ when $x = 4$.

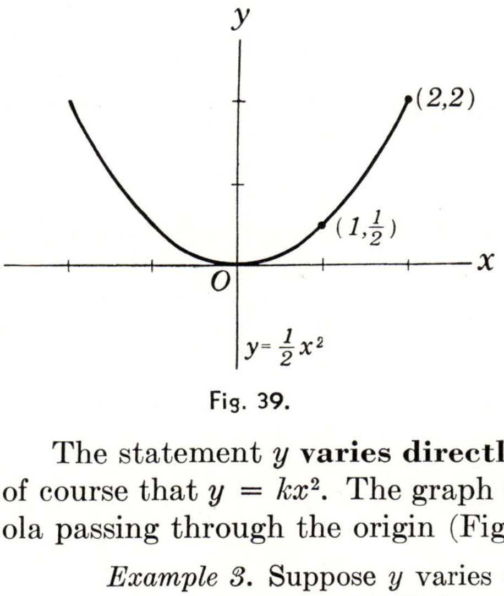

Fig. 39.

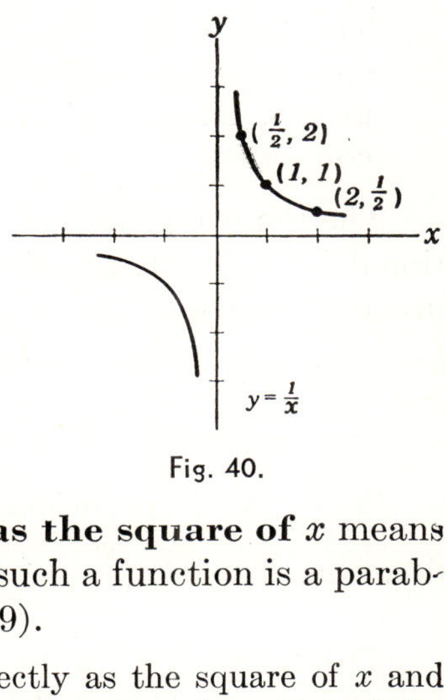

Fig. 40.

The statement y **varies directly as the square of** x means of course that $y = kx^2$. The graph of such a function is a parabola passing through the origin (Fig. 39).

Example 3. Suppose y varies directly as the square of x and $y = 12$ when $x = 2$. (*a*) Write y as a function of x; (*b*) find the value of y when $x = 4$.

Solution. $y = kx^2$. Hence $12 = 4k$, or $k = 3$. Therefore (*a*) $y = 3x^2$; (*b*) $y = 48$ when $x = 4$.

When $y = \frac{k}{x}$, k being a constant, we say that y **varies inversely as** x or y is **inversely proportional to** x. The graph of such a function is a hyperbola (Fig. 40).

Example 4. If y varies inversely as x and $y = 4$ when $x = 3$,

(a) write y as a function of x; (b) find the value of y when $x = 6$.

Solution. $y = k/x$. Hence $4 = k/3$, or $k = 12$. Therefore (a) $y = 12/x$; (b) $y = 2$ when $x = 6$.

The statement y **varies inversely as the square of** x means, of course, that $y = k/x^2$.

When $y = kuv$, k being a constant, we say that y **varies jointly as** u and v. Similarly, we say that y **varies jointly as** u and v **and inversely as** w if $y = \dfrac{kuv}{w}$.

Example 5. Newton's law of gravitation says that the force F exerted by two particles on each other varies jointly as their masses m_1 and m_2 and inversely as the square of the distance d between them. In symbols,

$$F = \frac{km_1m_2}{d^2}.$$

The factor of proportionality k in this example is found to be approximately 6.66×10^{-8} when the units employed are centimeters, grams, and seconds; it is known as the **gravitational constant.**

The following theorems explain the use of the word "proportional" in direct variation, and "inversely proportional" in inverse variation.

THEOREM 1. *If y varies directly as x, and $y = y_1$ when $x = x_1$, and $y = y_2$ when $x = x_2$, then* $\dfrac{y_1}{y_2} = \dfrac{x_1}{x_2}$.

Proof. By hypothesis, $y_1 = kx_1$ and $y_2 = kx_2$. Dividing the first equation by the second, we get $\dfrac{y_1}{y_2} = \dfrac{kx_1}{kx_2}$, or $\dfrac{y_1}{y_2} = \dfrac{x_1}{x_2}$.

THEOREM 2. *If y varies inversely as x, and $y = y_1$ when $x = x_1$, and $y = y_2$ when $x = x_2$, then* $\dfrac{y_1}{y_2} = \dfrac{x_2}{x_1}$.

Proof. By hypothesis, $y_1 = k/x_1$ and $y_2 = k/x_2$. Dividing the first equation by the second, we have $\dfrac{y_1}{y_2} = \dfrac{k/x_1}{k/x_2}$, or $\dfrac{y_1}{y_2} = \dfrac{x_2}{x_1}$.

EXERCISES

1. If y varies directly as x, and $y = 28$ when $x = 7$, (a) express y as a function of x; (b) find the value of y when $x = 6$; (c) how is y affected if x is tripled?

2. If y varies inversely as x and $y = 5$ when $x = 3$, (*a*) express y as a function of x; (*b*) find the value of y when $x = 6$; (*c*) how is y affected when x is doubled?

3. If y varies jointly as u and v and $y = 12$ when $u = 4$ and $v = 6$, (*a*) express y as a function of u and v; (*b*) find the value of y when $u = 2$ and $v = 7$.

4. If y varies directly as the square of u and inversely as the cube of v, and $y = 2$ when $u = 2$ and $v = 2$, (*a*) express y as a function of u and v; (*b*) find the value of y when $u = 4$ and $v = 4$.

5. Suppose that the maximum range of a projectile varies directly as the square of the initial velocity. If the maximum range is 15,000 feet when the initial velocity is 300 feet per second, (*a*) write the maximum range r as a function of the initial velocity v; (*b*) find the maximum range when $v = 500$ feet per second; (*c*) what is v when $r = 60{,}000$ feet?

6. Boyle's law says that, for an enclosed gas at a constant temperature, the pressure p varies inversely as the volume v. If $v = 600$ cubic inches when $p = 24$ pounds per square inch, (*a*) express p as a function of v; (*b*) find p when $v = 720$ cubic inches; (*c*) find v when $p = 100$ pounds per square inch.

7. The weight of a body is essentially the gravitational force exerted on it by the earth. Hence, as a special case of Newton's law of gravitation (example 5 above), we deduce that the weight w of a body varies inversely as the square of its distance d from the center of gravity of the earth. Assuming this to be true, and taking the earth to be a sphere of radius 4000 miles, how much would a girl weigh at a height of 100 miles above the surface of the earth if she weighs 100 pounds at the surface?

8. If a ball rolls down an inclined plane, the distance traversed varies as the square of the time, neglecting friction. If the ball rolls 8 feet in the first second, how far will it go in the first 4 seconds?

9. Neglecting air resistance, the distance a body falls from rest varies directly as the square of the time. If a body falls 64 feet in the first 2 seconds, how far will it have fallen in the first 5 seconds?

10. The surface area S of a sphere varies directly as the square of the radius r. If $S = 16\pi$ square inches when $r = 2$ inches, (*a*) find the formula for S in terms of r; (*b*) find S when $r = 4$ inches; (*c*) find r when $S = 100\pi$ square inches.

11. The volume V of a right circular cone varies jointly as the altitude h and the square of the radius r of the base. If $V = \pi$ cubic inches when $r = 1$ inch and $h = 3$ inches, (*a*) find the formula for V in terms of r and h; (*b*) find V when $r = 2$ inches and $h = 6$ inches.

12. The illumination I from a source of light varies inversely as the square of the distance d from the source. If $I = I_0$ when $d = 16$ feet, how far from the source will the illumination be 4 times as much?

13. The period T of a simple pendulum (that is, the time required for a complete oscillation) varies directly as the square root of the length L of the pendulum. If $T = 1.5$ seconds when $L = 2$ feet, find T when $L = 8$ feet.

14. According to the astronomer Kepler (1571–1630), the squares of the periods of the planets vary directly as the cubes of their mean distances from the sun. The period of the earth is 1 year. Taking the mean distance of the earth from the sun as the unit of distance, the mean distance of Jupiter is 5.2 units. Find the period of Jupiter to the nearest tenth of a year.

15. The volume V of a gas varies directly as the absolute temperature T and inversely as the pressure p. If a certain quantity of gas occupies 500 cubic feet at a pressure of 53 pounds per square foot and an absolute temperature of 500 degrees, what volume will it occupy at 600 degrees absolute temperature and pressure of 106 pounds per square foot?

16. The cost of labor varies jointly as the number of workers and the number of days they work. If 8 men working 9 days each are paid \$576, how many days will it take 6 men to earn \$624?

17. The current I in a wire varies directly as the electromotive force E and inversely as the resistance R. If $I = 22$ amperes when $E = 110$ volts and $R = 5$ ohms, find I when $E = 195$ volts and $R = 15$ ohms.

18. The lift force on a monoplane wing varies jointly as the area of the wing and the square of the velocity (other factors being kept fixed). If the area is increased by 20% and the velocity decreased by 10%, show that the lift is decreased by approximately 3%.

19. Other conditions remaining the same, the thrust T of a propeller varies jointly as the fourth power of its diameter and the square of the number n of revolutions per second. Show that if the number of revolutions per second is doubled and the diameter is decreased by 50%, the thrust will be decreased by 75%.

Chapter XIV

Permutations and combinations

91. Fundamental Principles

The following principles will provide the basis for the work of this chapter.

(A) *If one event can occur in h different ways, and if, after it has happened in one of these ways, another event can occur in k different ways, then both events can occur, in the stated order, in* $h \cdot k$ *different ways.*

Since any one of the h ways in which the first event can occur can be coupled with any one of the k ways in which the second event can occur, there are obviously h groups of k pairs or $h \cdot k$ pairs of possible occurrences.

Example 1. A game consists of drawing 2 cards in succession from a pack of 4 different cards marked a, b, c, and d, respectively, the first card not being replaced before the second is drawn. How many different results can there be?

Solution. The first drawing can produce any one of 4 different results. The second drawing can produce any one of 3 different results. Both drawings in the stated order can produce $4 \cdot 3 = 12$ different results, namely *ab, ac, ad; ba, bc, bd; ca, cb, cd; da, db, dc*.

Example 2. The same as example 1 except that the first card is to be replaced before the second card is drawn.

Solution. The first drawing can produce 4 different results. The second can then produce 4 different results again. Both therefore can occur in the stated order in $4 \cdot 4 = 16$ different ways, namely *aa, ab, ac, ad; ba, bb, bc, bd; ca, cb, cc, cd; da, db, dc, dd*.

If the two events are **independent** of each other, that is, if neither is influenced by the outcome of the other, then we may state the following principle:

(B) *If one of two independent events can occur in h different ways, and the other in k different ways, then both can occur, disregarding their order, in $h \cdot k$ ways.*

Example 3. A luncheon tray is to contain a sandwich and a beverage. There are 4 different kinds of sandwich and 3 different kinds of beverage. How many different lunches can be formed?

Solution. The order is immaterial; we do not care whether the sandwich or the beverage is put on the tray first. Hence, according to principle B, there are $4 \cdot 3 = 12$ different lunches.

Two events are called **mutually exclusive** if they cannot both occur together.

(C) *If two events are mutually exclusive, and one can occur in h different ways while the other can occur in k different ways, then either one or the other (but not both) can occur in $h + k$ different ways.*

Example 4. The same as example 3, except that, because of financial embarrassment, a lunch is now to consist of either a sandwich or a beverage but not both.

Solution. According to principle **C**, $4 + 3 = 7$ different lunches are possible.

In connection with principle C, care must be taken that the events are mutually exclusive.

Example 5. We are to choose either a heart or an ace from a deck of 52 playing cards. How many choices are there?

Solution. There are 13 hearts and 4 aces. But it is wrong to say that there are 17 choices, because these events are not mutually exclusive; there is one heart which is also an ace. Actually there are only 16 different choices.

Example 6. How many different symbols, each consisting of 6 different letters in succession, can be formed from the letters of the word "algorithm," if each symbol is to begin and end with a vowel?

Solution. There are 3 vowels and 6 consonants, and 6 places to be filled: _ _ _ _ _ _. For the first place we may choose any one of the 3 vowels. For the last place we may choose any one of the 2 remaining vowels. The first of the 4 remaining places may be filled by any one of the 7 remaining letters, the next by any one of the 6 remaining letters, the next by any one of the 5 remaining letters, and the last by any one of the 4 remaining letters. We write the numbers of choices over the appropriate blank spaces, thus: $\underline{3} \cdot \underline{7} \cdot \underline{6} \cdot \underline{5} \cdot \underline{4} \cdot \underline{2}$ and multiply these numbers, by virtue of principle A. Hence there are 5040 different symbols.

Example 7. (*a*) How many different arrangements of 6 distinct books each can be made on a shelf with space for 6 books? (*b*) If 3 of the books are English and 3 are French, how many arrangements of the 6 books can be made on the same shelf if books in the same language are to be kept together?

Solution. (*a*) In the first space we may put any one of the 6 books, in the next space any one of the 5 remaining books, and so on. Hence there are $\underline{6} \cdot \underline{5} \cdot \underline{4} \cdot \underline{3} \cdot \underline{2} \cdot \underline{1} = 720$ arrangements. (*b*) If the English books are at the left, we may put any one of the 3 English books in the first position, any one of the 2 remaining English books in the next place, and so on; hence there are $\underline{3} \cdot \underline{2} \cdot \underline{1} \cdot \underline{3} \cdot \underline{2} \cdot \underline{1} = 36$ such arrangements. But there are 36 others having the French books at the left. Hence there are $36 + 36 = 72$ arrangements in all.

EXERCISES

1. In how many ways can 5 students be seated in a row of 5 seats?

2. How many different signals can be made from 5 different flags if each signal is to consist of 3 flags hung in a horizontal row?

3. How many different numbers of 2 different digits each can be formed using the digits 1, 3, 5, 7, 9?

4. How many different numbers of 3 different digits each can be formed using the digits 0, 2, 4, 6, 8?

5. How many different symbols, each symbol consisting of 3 letters in succession, can be formed using the letters a, b, x, u, r, t if (*a*) no letter may be repeated in the same symbol, (*b*) repetitions are allowed?

6. How many different numbers of 3 different digits each can be formed from the digits 2, 3, 5, 6, 7, 9, if each number is to be (*a*) odd, (*b*) even?

7. In how many ways can the positions of president, vice-president, and secretary be filled in a club of 12 members if no person is to hold more than one position?

8. Two cubical dice, one red and one white, are thrown. In how many ways can they fall?

9. Three different cubical dice are thrown. In how many different ways can they fall?

10. If a penny, nickel, dime, quarter, and half dollar are tossed together, in how many ways may they fall?

11. How many numbers of 4 different digits, each greater than 5000, can be formed from the digits 2, 4, 5, 7, 8, 0?

12. In how many ways can a baseball team of 9 men be arranged in batting orders, if a certain 4 men must occupy the first 4 positions in some order?

13. In how many ways can a party of 6 people be seated in a row of 6 seats (*a*) if a certain 2 insist on sitting next to each other, (*b*) if the same 2 refuse to sit next to each other?

14. In how many ways can a set of 5 different mathematics books and 3 different physics books be placed on a shelf with space for 8 books if all books on the same subject are to be kept together?

15. A telephone dial has 10 holes. How many different signals, each consisting of 7 impulses in succession, can be formed (*a*) if no impulse is to be repeated in any given signal, (*b*) if repetitions are permitted?

16. In how many different ways can 5 people be seated in a row of 7 seats?

17. (*a*) How many different symbols, each consisting of 4 different letters in succession, can be formed from the letters of the word *certain?* (*b*) How many of these begin and end with a vowel? (*c*) How many begin with a consonant and end with a vowel?

18. In how many ways can 4 boys and 5 girls be seated in a row of 9 seats if boys and girls are to occupy alternate seats?

19. In how many different ways can the letters of the name *Naomi* be arranged if all vowels are to be kept together?

20. In how many different ways can the letters of the name *Leonard* be arranged if consonants and vowels are to alternate?

21. (*a*) How many numbers can be formed using some or all of the digits 1, 3, 4, 7, 8 if no number is to contain repeated digits? (*b*) How many of these numbers will be even? (*c*) How many will be greater than 350?

22. How many arrangements of 3 different English books and 6 different French books can be made on a shelf with space for 9 books if all the English books are to be kept together?

23. From 3 English, 4 French, and 6 German books, all different, how many different sets can be chosen, each set consisting of 2 books in different languages?

24. How many 4-digit numbers begin with an odd digit and end with an even digit? (*Hint:* Zero is to be counted among the even digits.)

92. Permutations of *n* Distinct Objects Taken *r* at a Time

An arrangement of a set of objects in some order in a straight line is called a **permutation** of these objects. More precisely, if we have n distinct objects, any arrangement of r of them in some order in a straight line is called a **permutation of the n distinct objects taken r at a time.** The number of permutations of n distinct objects taken r at a time is denoted by $P(n,r)$; other symbols in common use are ${}^nP_r$, ${}_nP_r$, P_r^n, $P_{n,r}$.

Example 1. Find the number of permutations of the four letters a, b, c, d taken two at a time.

Solution. Any one of the four letters may be put in the first position, and any one of the three remaining letters in the second position. Hence there are $4 \cdot 3 = 12$ permutations, which are listed in example 1, section 91. Thus $P(4,2) = 12$.

Example 2. The number of permutations of the four letters a, b, c, d taken all at a time is similarly $\underline{4} \cdot \underline{3} \cdot \underline{2} \cdot \underline{1} = 24$. Thus $P(4,4) = 24$.

These permutations are:

abcd	*bacd*	*bcad*	*bcda*
abdc	*badc*	*bdac*	*bdca*
acbd	*cabd*	*cbad*	*cbda*
acdb	*cadb*	*cdab*	*cdba*
adbc	*dabc*	*dbac*	*dbca*
adcb	*dacb*	*dcab*	*dcba.*

Example 3. The number of permutations of the four letters a, b, c, d taken three at a time is $\underline{4} \cdot \underline{3} \cdot \underline{2} = 24$. Thus, $P(4,3) = 24$.

These permutations are:

abc	*abd*	*acd*	*bcd*
acb	*adb*	*adc*	*bdc*
bac	*bad*	*cad*	*cbd*
bca	*bda*	*cda*	*cdb*
cab	*dab*	*dac*	*dbc*
cba	*dba*	*dca*	*dcb.*

We have

$$P(n,n) = n \cdot (n-1) \cdot (n-2) \dots 2 \cdot 1, \tag{1}$$

since the first position can be filled by any one of the n distinct objects, the next by any one of the remaining $(n-1)$, and so on.

Introducing the symbol

(2) $$n! = n(n-1)(n-2)\ \ldots\ 2\cdot 1,$$

read **"factorial *n*,"** we have

(3) $$P(n,n) = n!$$

For example, $1! = 1$, $2! = 2\cdot 1 = 2$, $3! = 3\cdot 2\cdot 1 = 6$, $4! = 4\cdot 3\cdot 2\cdot 1 = 24$, $5! = 5\cdot 4\cdot 3\cdot 2\cdot 1 = 120$, and so on.

If r is less than n, then

(4) $$P(n,r) = n(n-1)(n-2)\ \ldots\ (n-r+1)$$

where there are r factors in the right member, since there are r places to be filled by n distinct objects. Multiplying numerator and denominator of the right-hand member of (4) by

$$(n-r)(n-r-1)(n-r-2)\ \ldots\ 2\cdot 1,$$

we get

$$P(n,r) = \frac{n(n-1)(n-2)\ \ldots\ (n-r+1)\cdot(n-r)(n-r-1)\ \ldots\ 2\cdot 1}{(n-r)(n-r-1)\ \ldots\ 2\cdot 1}$$

or

(5) $$P(n,r) = \frac{n!}{(n-r)!} \qquad \text{if } r < n.$$

In practice, it will sometimes be desirable to work problems directly by use of the fundamental principles, and sometimes by use of these formulas.

EXERCISES

Evaluate each of the following symbols:

1. $7!$ **2.** $\frac{6!}{3!}$. **3.** $\frac{8!}{6!}$. **4.** $P(5,2)$.

5. $P(9,2)$. **6.** $P(5,5)$. **7.** $P(7,3)$. **8.** $P(7,4)$.

9. $\frac{P(6,3)}{3!}$. **10.** $\frac{P(8,2)}{2!}$.

11. How many different arrangements can be formed on a shelf with space for 3 books if there are 6 different books available?

12. A bookshelf has space for 4 books. If there are 7 different books available, how many different arrangements can be made on the shelf?

13. A bookshelf has space for 6 books. There are 5 different English books and 6 different French books available. If we wish to put 3 English books and 3 French books on the shelf, how many arrangements can be made keeping all books in the same language together?

14. In how many ways can a coach assign positions on a baseball team of 9 men if only 3 men are qualified to be pitcher and only 2 others are qualified to be catcher but all are able to play any other position?

15. (*a*) In how many ways can 7 people sit in a row of 7 seats if a certain 4 people must sit side by side? (*b*) In a row of 8 seats?

16. How many numbers, each of 5 different digits, can be formed from the digits 3, 4, 5, 6, 7 if (*a*) the numbers 4 and 5 are to be next to each other, (*b*) the numbers 4 and 5 are not to be next to each other?

17. How many different signals each consisting of one flag or more hung in a horizontal line, can be formed from a set of 5 different flags?

18. How many positive whole numbers can be formed from some or all of the digits 2, 3, 4, 5, 6 if no number is to have repeated digits?

19. How many positive even numbers can be formed from some or all of the digits 2, 3, 4, 5, 6 if no number is to have repeated digits?

20. How many positive whole numbers can be formed from some or all of the digits 0, 1, 2, 3, 4 if no number is to have repeated digits?

★21. How many positive even numbers can be formed from some or all of the digits 0, 1, 2, 3, 4 if no number is to have repeated digits?

93. Permutations Where Some Things Are Alike

The number of permutations of the distinct letters a, b, c, d taken all at a time is 24. The permutations are listed in example 2, section 92. If three of the letters were indistinguishable, for example, if instead of a, b, c, d we had a, x, x, x, then there would be only four permutations, namely

$$axxx, \qquad xaxx, \qquad xxax, \qquad xxxa.$$

Clearly each column of $3! = 6$ permutations of a, b, c, d listed in example 2, section 92, gives rise to only one permutation of $axxx$, since the letters bcd may be permuted among themselves in 3! ways which are indistinguishable for xxx. Hence we may arrive at the answer 4 by dividing the answer of example 2, section 92, by 3!; thus

$$\frac{4!}{3!} = \frac{4 \cdot 3 \cdot 2 \cdot 1}{3 \cdot 2 \cdot 1} = 4.$$

In general, if a set of n objects is such that a set of p of them are indistinguishable, another set of q of them are indistinguish-

★ May be omitted without disturbing continuity.

able, and so on, then the number of permutations of the n objects taken all at a time is

$$(1) \qquad \frac{n!}{p!\,q!\,\ldots}.$$

Example 1. How many permutations can be made of the letters in the word *Mississippi*, taken all at a time?

Solution. There are 11 letters of which 4 i's, 4 s's, and 2 p's are indistinguishable. Hence the answer is

$$\frac{11!}{4!\,4!\,2!} = \frac{11 \cdot 10 \cdot 9 \cdot \not{8} \cdot 7 \cdot \not{6} \cdot 5 \cdot \not{4} \cdot \not{3} \cdot \not{2} \cdot 1}{\not{4} \cdot \not{3} \cdot \not{2} \cdot \not{4} \cdot \not{3} \cdot \not{2} \cdot \not{2}} = 11 \cdot 10 \cdot 9 \cdot 7 \cdot 5 = 34650.$$

EXERCISES

1. How many permutations can be formed from the letters of the word *reiterate* taken all at a time?

2. How many permutations can be formed from the letters of the word *institution* taken all at a time?

3. How many permutations can be formed from the letters a, a, a, b, b, c, c, c, c, d taken all at a time?

4. How many different eight-digit numbers can be formed from the digits 2, 2, 2, 5, 5, 5, 5, 3?

5. How many different signals, each consisting of 7 flags hung in a horizontal line, can be formed from a set of 4 indistinguishable blue flags and 3 indistinguishable yellow flags?

6. How many different eight-digit numbers can be formed from the digits 5, 5, 5, 5, 3, 3, 7, 7?

7. A modern town is laid out in square blocks. By how many different routes can we go from the corner of East First Street and North First Avenue to the corner of East Seventh Street and North Fifth Avenue, each route being of the smallest possible length, that is, 10 blocks?

94. Circular Permutations

We agree that when objects are arranged in a circle, only their relative order as we traverse the circle in a definite sense is to be considered. Hence the arrangements

(1)

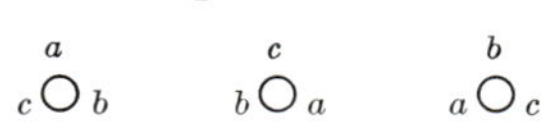

are regarded as the same. Similarly the arrangements

(2)

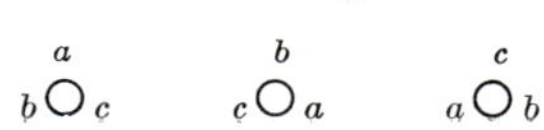

are regarded as the same. Thus, there are only two possible arrangements of three distinct objects in a circle. Therefore, to calculate the number of arrangements in circular order, we must first fix the position of one object arbitrarily, and then we may calculate the number of permutations of the remaining objects as if they were in a straight line. Hence *n distinct objects can be arranged in a circle in* $(n - 1)!$ *ways.*

Example. A group of 6 people can be arranged at a circular table in $5! = 120$ ways.

Note. In arranging keys on a ring, it is usually agreed that two arrangements are the same if one can be gotten from the other by turning the ring over like a flapjack. Thus the arrangements (1) and (2) above are the same with this agreement. In this case we must divide the answer by 2. Thus, there is only one possible arrangement of three different keys on a ring. In general, *there are* $\frac{(n-1)!}{2}$ *different arrangements of n keys on a ring.*

EXERCISES

The following questions are to be answered in accordance with the conventions set forth above:

1. In how many different orders can 5 people take seats at a round table?

2. In how many different orders can 8 children join hands in a ring?

3. In how many different orders can 6 keys be placed on a key ring?

4. How many different necklaces can be formed from 9 differently colored beads?

5. In how many orders can a party of 3 girls and 3 boys be placed at a round table so that boys and girls alternate?

6. In how many orders can a party of 5 girls and 5 boys be placed at a round table so that boys and girls alternate?

7. In how many orders can 6 people be seated at a round table (*a*) if a certain 2 insist on sitting next to each other, (*b*) if they refuse to sit next to each other?

95. Combinations of *n* Distinct Objects Taken *r* at a Time

A set of *r* objects chosen from a given set of *n* distinct objects, without regard to the order in which they are chosen or arranged, is called a **combination of *n* distinct objects taken *r* at a**

time. The number of combinations of n distinct objects taken r at a time is denoted by $C(n,r)$. Other symbols in common use are ${}^nC_r$, ${}_nC_r$, C_r^n, $C_{n,r}$ and $\binom{n}{r}$.

Example 1. The number of combinations of the four letters a, b, c, d taken two at a time is 6. The combinations are ab, ac, ad, bc, bd, cd. Clearly each of these combinations gives rise to $2! = 2$ permutations which are listed in example 1, section 91. Hence $C(4,2) = \dfrac{P(4,2)}{2!} = 6$.

Example 2. The number of combinations of the four letters a, b, c, d taken three at a time is 4. The combinations are abc, abd, acd, bcd. Each of these gives rise to $3! = 6$ permutations which are listed in columns in example 3, section 92. Hence $C(4,3) \cdot 3! = P(4,3)$, or $C(4,3) = \dfrac{P(4,3)}{3!} = \dfrac{4 \cdot 3 \cdot 2}{3 \cdot 2} = 4$.

Clearly, any combination of n distinct things taken r at a time gives rise to $r!$ permutations since each set of r objects can be permuted among themselves in $r!$ ways. Thus $C(n,r) \cdot r! = P(n,r)$, or

$$(1) \qquad C(n,r) = \frac{P(n,r)}{r!} \qquad (0 < r < n).$$

Using (5), section 92, we have

$$(2) \qquad C(n,r) = \frac{n!}{r!(n-r)!} \qquad (0 < r < n).$$

Using (4), section 92, we have

$$(3) \qquad C(n,r) = \frac{n(n-1)(n-2)\ \ldots\ (n-r+1)}{r!} \qquad (0 < r < n).$$

We write $C(n,0) = C(n,n) = 1$, since there is only one way to select no things from the given n things, and only one way to select n things from the given n things, namely to take them all.

Note 1. Whenever we select a set of r things from a set of n distinct things, we also automatically select a set of $n - r$ things to be left behind. Hence

$$(4) \qquad C(n,r) = C(n, n-r).$$

This can also be proved by using (2).

Note 2. Notice that $9! = 9 \cdot 8! = 9 \cdot 8 \cdot 7!$, and so on. Hence, by (2), we may write $C(9,7) = C(9,2) = \dfrac{9!}{2!7!} = \dfrac{9 \cdot 8 \cdot 7!}{2 \cdot 7!} = 36$.

Example 3. How many committees of 5 can be formed from a group of 6 Democrats and 3 Republicans if each committee is to have at least 3 Democrats?

Solution. The number of committees with:

exactly 3 Democrats is $C(6,3) \cdot C(3,2) = 20 \cdot 3$	$= 60;$
exactly 4 Democrats is $C(6,4) \cdot C(3,1) = 15 \cdot 3$	$= 45;$
exactly 5 Democrats is $C(6,5)$	$= 6;$
either 3, 4, or 5 Democrats is the total	111.

Hence the total number of committees with at least 3 Democrats is 111.

EXERCISES

Evaluate each symbol:

1. $C(6,2)$. **2.** $C(5,3)$. **3.** $C(8,3)$. **4.** $C(8,5)$.
5. $C(10,8)$. **6.** $C(7,3) \cdot C(5,2)$.

In how many ways can we select:

7. A committee of 4 from a group of 10 people?

8. A committee of 5 from a group of 7 people?

9. A set of 3 books from a set of 9 different books?

10. A set of 6 books from a set of 9 different books?

11. How many different sums of money, each composed of 3 coins, can be formed from a cent, a nickel, a dime, and a quarter?

12. (*a*) How many straight lines are determined by the vertices of a regular hexagon? (*b*) How many of these lines go through a given vertex?

13. How many triangles are determined by the vertices of a regular hexagon?

14. In how many ways can we select a committee of 3 Democrats and 2 Republicans from a group of 7 Democrats and 6 Republicans?

15. In how many ways can we select a set of 3 mathematics books and 2 physics books from a set of 7 mathematics books and 5 physics books, all different?

16. From a group of 10 Democrats and 8 Republicans, how many different committees of 7 can be chosen which contain (*a*) exactly 4 Democrats? (*b*) at least 4 Democrats? (*c*) at most 4 Democrats?

17. (*a*) How many committees of 5 can be chosen from a group of

15 men? (*b*) How many of these will include a specified man A? (*c*) From how many will A be excluded?

18. How many committees of 5 can be chosen from a group of 8 men if each committee must include 2 certain men A and B?

19. In how many ways can we select a committee of 5 from a group of 10 men (*a*) if a certain 2 men insist on serving together or not at all? (*b*) if a certain 2 men refuse to serve together?

20. How many different committees of 7 can be chosen from a group of 8 Democrats and 5 Republicans if each committee is to contain (*a*) exactly 4 Democrats? (*b*) at least 4 Democrats? (*c*) at most 4 Democrats?

21. In how many different orders can we shelve sets of 5 books, each set consisting of 3 mathematics books and 2 physics books, if the books are to be chosen from a set of 7 mathematics books and 6 physics books, all different?

22. If $P(n,4) = 11{,}880$, find $C(n,4)$.

23. If $C(n,4) = 210$, find $P(n,4)$.

24. Prove that $C(n - 1, r) + C(n - 1, r - 1) = C(n,r)$.

25. (*a*) How many straight lines are determined by 8 points, no 3 of which are in the same straight line? (*b*) How many of these lines pass through any given point?

26. From an urn containing 5 black and 6 white balls, in how many ways can we draw a set of 6 balls of which 4 are white and 2 black?

27. In how many ways can 10 different books be divided among A, B, and C so that 5 are given to A, 3 to B, and 2 to C?

28. In a league of 8 baseball teams, how many games will be played in a season if each team plays 22 games with every other team?

29. How many numbers of 5 different digits, each number to contain 3 odd and 2 even digits, can be formed from the digits 1, 2, 3, 4, 5, 6, 7, 8, 9?

30. How many different sums of money can be formed from a cent, a nickel, a dime, a quarter, and a half dollar, using any number of coins at a time?

31. A cent, a nickel, a dime, a quarter, and a half dollar are tossed simultaneously. In how many ways can it happen that 2 coins fall heads and the remaining 3 tails?

32. (*a*) How many different hands of 13 cards each can be dealt from a pack of 52 different cards? (*b*) How many different sets of 4 hands can be dealt, ignoring the order in which the 4 hands are distributed around the bridge table?

Chapter XV

Probability

96. Definition of Probability

In everyday speech, the word "probably" is used in different vague senses, difficult to unify, as in the statements "It will probably rain tomorrow," "I shall probably pass mathematics," or "King Arthur probably existed." Some writers assert that probability is a measure of subjective belief. Since different people believe opposite statements with equal firmness, and the same person often believes opposite statements at different times, subjective theories of probability involve severe difficulties. We shall here present a more objective theory and shall restrict our discussion to experiments which have a limited number of mutually exclusive possible outcomes, which we shall call the **outcomes** or **possibilities** of the experiment.

Example 1. If a coin is tossed, there are two possible outcomes: head (H) or tail (T).

Example 2. If a single cubical die is thrown, there are six possible outcomes: 1, 2, 3, 4, 5, 6 facing up, respectively.

Example 3. If a nickel and a dime are tossed, there are four possible outcomes: HH, HT, TH, TT.

In the experiment of tossing a well-balanced coin, no reliable prediction can be made as to whether it will fall head or tail on any particular trial. But in a long sequence of trials, great regularity is observed in the fractional part or "percentage" of the trials in which heads turn up. If $f(n)$ is the number of trials in which heads turns up in the first n trials, then this fractional part $f(n)/n$ is termed the **relative frequency** of heads in n trials. By experiment we might find that in tossing a particular coin, we get

n = number of trials	10	100	1000	10000	100000	...
$f(n)$ = number of heads	2	45	512	4982	50137	...
$f(n)/n$ = relative frequency	0.20	0.45	0.512	0.4982	0.50137	...

Here it appears that the relative frequency is approaching the value 1/2 or 50% as n increases. Hence it appears plausible to *assign* the value 1/2 as the **probability** of the event "heads," and to interpret this number as the limiting value* towards which the relative frequency of heads tends as the number of trials increases. Practically, this is interpreted to mean that in a sufficiently long sequence (run) of trials, heads may be expected about half the time, and hence tails also about half the time.

Similarly, in example 2, we would assign the value 1/6 as the probability of each of the six possible outcomes. And in example 3, we would assign 1/4 as the probability of each of the four outcomes.

By an **event** E, we shall mean any set (or collection) of possible outcomes of our experiment. Thus in example 1, heads is an event, tails is another event, heads or tails is an event. In example 2, throwing a six is an event, throwing an even number is an event, throwing anything other than a six (abbreviated not-six) is an event, throwing an odd number is an event. In example 3, throwing exactly one head is an event, throwing at least one head is an event, throwing exactly two heads is an event, throwing no head is an event.

DEFINITION. *By the* ***probability*** $P(E)$ *of an event* E *is meant the sum of the probabilities assigned to the possible outcomes belonging to the set* E*. The "impossible event" containing no possible outcomes is assigned probability zero.*

Thus in example 1, the probability $P(H)$ of heads is 1/2; the probability $P(T)$ of tails is 1/2. The probability $P(H \text{ or } T)$ of obtaining either heads or tails is 1; we must get either heads or tails 100% or all of the time. The probability of obtaining neither H nor T is 0; this happens 0% or none of the time.

In example 2, the probability $P(6)$ of throwing a 6 is 1/6.

* For an accurate definition of *limit* see section 176.

The probability P(even) of throwing an even number is 3/6. The probability P(not-six) of throwing anything other than a six (i.e., a 1, 2, 3, 4, or 5) is 5/6. The probability P(even or odd) of throwing either an even number or an odd number is 1; this must happen 100% of the time. The probability of throwing neither an even nor an odd number is 0; this must happen none of the time.

In example 3, the probability of exactly one head is 1/2; of at least one head is 3/4; of exactly two heads is 1/4, of no head is 1/4. The probability of either at least one head or no head is 1.

A certain event is assigned probability 1, an impossible event is assigned probability 0, and the probability of any event must be a number between 0 and 1 inclusive.

In the case of a biased coin, or a loaded die, experience might well lead us to assign other values as the probabilities of the possible outcomes, but in any case they are intended to represent approximately the relative frequencies or percentages of occurrence of these outcomes in a sufficiently long run. Since one of the mutually exclusive possible outcomes must occur, the probability that "one of the possible outcomes will occur" or "either the first outcome, or the second outcome, or, . . . , or the last outcome will occur" is 1. *Usually, in the absence of evidence to the contrary, we will assign equal probabilities to the possible outcomes. In this case we say that we assume that the possible outcomes are equally likely. As far as the mathematical theory is concerned, however, we may regard the probabilities of the possible outcomes as assigned arbitrarily, so long as they are numbers between 0 and 1 inclusive subject to the requirement that the sum of all of them must be 1.* These assigned numbers may be regarded as a hypothesis for the particular problem and, like any scientific hypothesis, will be judged by whether or not its consequences work out well in practice. The proper concern of the mathematical theory is to calculate the probabilities of more complicated events, given the probabilities of the possible outcomes.

Example 4. Suppose a die is loaded so that the numbers 1, 2, 3, 4, 5, 6 turn up with relative frequencies $\frac{1}{3}$, $\frac{1}{12}$, $\frac{1}{12}$, $\frac{1}{6}$, $\frac{1}{6}$, $\frac{1}{6}$ respectively. Taking these as the assigned probabilities of the possible outcomes, the probability of throwing an even number is $\frac{1}{12} + \frac{1}{6} + \frac{1}{6} = \frac{5}{12}$, and the probability of an odd number is $\frac{1}{3} + \frac{1}{12} + \frac{1}{6} = \frac{7}{12}$.

Two events E and F are termed **mutually exclusive** if no possible outcome belongs to both the sets E and F.

THEOREM 1. *If E and F are mutually exclusive events, then $P(E \text{ or } F) = P(E) + P(F)$.*

Proof. By definition, the probability of the event "E or F" is the sum of the probabilities of the outcomes belonging to either E or F. Since E and F have no outcome in common, by hypothesis, this is precisely the sum of the probabilities of the outcomes in event E plus the sum of the probabilities of the outcomes in event F.

Theorem 1 can be extended to any number of mutually exclusive events $E_1, E_2, \ldots, E_n$, obtaining $P(E_1 \text{ or } E_2 \text{ or } \ldots \text{ or } E_n) = P(E_1) + P(E_2) + \ldots + P(E_n)$.

THEOREM 2. *If E is any event, $P(E) + P(\text{not-}E) = 1$, or $P(\text{not-}E) = 1 - P(E)$.*

Proof. The events E and not-E are mutually exclusive. Every possible outcome belongs to E or not-E. Since $P(E$ or not-$E)$ is therefore 1, the theorem follows from Theorem 1.

THEOREM 3. *If the possible outcomes are supposed to be equally likely, and if there are a total number t of possible outcomes, and an event E contains s of these possible outcomes then $P(E) = s/t$. Briefly, with equally likely outcomes, the probability of an event is the number s of successful cases over the total number t of cases.*

Proof. Assuming equally likely outcomes, the probability of any particular outcome is $1/t$ since the sum of all their probabilities must be 1. Hence

$$P(E) = \frac{1}{t} + \frac{1}{t} + \ldots + \frac{1}{t}$$

where there are s terms on the right. Hence $P(E) = s/t$.

> *Example 5.* The probability that an ace will be drawn in a single drawing from a pack of 52 playing cards is $\frac{4}{52} = \frac{1}{13}$ by theorem 3. The probability of drawing a king is also $\frac{4}{52} = \frac{1}{13}$. By theorem 1, the probability of drawing either an ace or a king in a single drawing is $\frac{4}{52} + \frac{4}{52} = \frac{2}{13}$.

The following theorem generalizes theorem 1 for cases where the events E and F are not mutually exclusive.

THEOREM 4. *$P(E \text{ or } F) = P(E) + P(F) - P(E \text{ and } F)$.*

Proof. $P(E) + P(F)$ yields the sum of the probabilities of all outcomes in E plus the sum of the probabilities of all outcomes in F. If E and F have any outcomes in common then their contributions have now been added in twice. Hence we must subtract their sum $P(E \text{ and } F)$ once to get the correct sum for $P(E \text{ or } F)$.

Example 6. The probability of drawing a heart in a single drawing from a pack of 52 playing cards is 13/52. The probability of drawing an ace is 4/52. The probability of drawing either an ace or a heart is $\frac{13}{52} + \frac{4}{52} - \frac{1}{52} = \frac{16}{52} = \frac{4}{13}$ since $\frac{1}{52}$ is the probability of drawing both an ace and a heart (i.e., the ace of hearts).

The student should now review carefully the fundamental principles of section 91, chapter XIV, which will be useful in working problems on probability.

Example 7. Three different English books and 3 different French books are to be arranged on a shelf with space for 6 books. If they are stacked at random, what is the probability that they will be arranged with all books in the same language together?

Solution. According to the fundamental principles of section 91, chapter XIV, there are $\underline{6} \cdot \underline{5} \cdot \underline{4} \cdot \underline{3} \cdot \underline{2} \cdot \underline{1}$ ways in which the books can be stacked. Of these only twice $\underline{3} \cdot \underline{2} \cdot \underline{1} \cdot \underline{3} \cdot \underline{2} \cdot \underline{1} = 36$ or 72 will satisfy the condition that all books in the same language are together. (See example 7, section 91, chapter XIV.) Hence the probability is $72/720 = 1/10$.

Example 8. An urn contains 10 balls, the probabilities of obtaining which are assumed to be equal, and of which 3 are red, 4 are blue, 2 are green, and 1 is yellow. Then the probability of obtaining a red ball in a single draw is 3/10, the probability of drawing a blue ball is 4/10, green 2/10, yellow 1/10. The probability of drawing a white ball is 0, while the probability of drawing a colored ball is $10/10 = 1$.

The theory of probability, properly used, has been found to be of great value in such diverse subjects as physical chemistry, statistical mechanics, heredity, biometrics, econometrics, the theory of gases, quality control in mass production industries, insurance, social sciences, games of chance, experimental method in general, etc.

Note. The theory of probability was founded by two great French mathematicians Blaise Pascal and Pierre de Fermat in the 17th century. Pascal was a child prodigy who did valuable research in mathematics at the age of 16.

EXERCISES

Assume equally likely outcomes unless otherwise specified.

1. Find the probability of throwing a three with a single die.

2. Throw a single die 100 times and record the number of threes obtained in 10, 20, 30, . . ., 100 throws.

3. Find the probability of throwing a total of two with a pair of dice.

4. Find the probability of throwing a total of three with a pair of dice; a total of four; a total of seven. Which is more probable, a total of three or a total of seven?

5. (*a*) Find the probability of drawing a spade from a pack of 52 playing cards; (*b*) an ace; (*c*) the ace of spades.

6. Toss a coin 100 times and record the number of heads obtained in 10, 20, 30, . . ., 100 tosses.

7. Find the probability of obtaining (*a*) no heads; (*b*) 1 head; (*c*) 2 heads in tossing a pair of coins.

8. Toss a pair of coins 100 times and record the relative frequencies of 0, 1, and 2 heads.

9. Five different books are to be placed at random on a bookshelf with space for 5 books. What is the probability that a certain 2 books will be next to each other?

10. An urn contains 5 white, 3 black, and 2 green balls. (*a*) Find the probability of drawing a black ball in a single draw; (*b*) a white ball; (*c*) a white or green ball.

11. In a single throw of a single die, what is the probability of obtaining a three or anything larger?

12. In a single throw of a pair of dice, what is the probability of obtaining a total of nine or more?

13. If 3 balls are drawn simultaneously from the urn of exercise 10, what is the probability that 2 will be white and 1 black?

14. If 3 balls are drawn simultaneously from an urn containing 5 white and 3 black balls, what is the probability that all 3 will be white?

15. If 9 people are seated at random in a row of 9 seats, what is the probability that a certain 4 people will occupy adjacent seats?

16. If 5 cards are drawn from a deck of 52 playing cards, what is the probability that they will be, regardless of order, the ace, king, queen, jack, and ten of spades?

17. A number of 3 different digits is formed from the digits 1, 2, 3, 4, 5. (*a*) What is the probability that it will be even? (*b*) What is the probability that it will be greater than 325?

18. If 7 people are seated at random at a round table, what is the probability that a certain 2 will be neighbors?

19. Find the probability of drawing (*a*) either an ace or a king in a single draw from a deck of 52 playing cards; (*b*) either an ace or a spade.

20. If 5 cards are drawn from a deck of 52 playing cards, what is the probability that they will be all of the same suit?

21. If the letters *f*, *f*, *f*, *l*, *u* are arranged in a line at random, find the probability that they will spell the word *fluff*.

22. A committee of five is chosen by lot from a group of 5 Democrats and 4 Republicans. Find the probability that the committee will be composed of (*a*) exactly 3 Democrats and 2 Republicans; (*b*) at least 3 Democrats; (*c*) all Democrats.

23. If four boys and four girls are seated in a row of 8 seats by lot, find the probability that: (*a*) all members of the same sex are seated next to each other; (*b*) boys and girls alternate.

24. If a nickel, dime, and quarter are tossed, find the probability of obtaining: (*a*) no heads; (*b*) at least one head; (*c*) exactly two heads; (*d*) at least two heads.

25. Four different letters are dropped on the floor and then inserted at random into the four differently addressed envelopes. Find the probability that: (*a*) every letter is inserted into the proper envelope; (*b*) at least one letter is inserted into a wrong envelope; (*c*) exactly two letters are inserted into the proper envelopes.

26. A committee of 5 is chosen by lot from a group of ten men. Find the probability that a certain two men, Mr. A and Mr. B, will: (*a*) both be on the committee; (*b*) both be off the committee; (*c*) not both be on the committee.

27. A loaded die is such that the probabilities of the numbers 1, 2, 3, 4, 5, 6 turning up are respectively $\frac{1}{3}$, $\frac{1}{6}$, $\frac{1}{6}$, $\frac{1}{9}$, $\frac{1}{9}$, $\frac{1}{9}$. Find the probability of throwing (*a*) an even number; (*b*) an odd number; (*c*) either a one or a six; (*d*) neither a one nor a six.

28. A wheel of fortune is divided into sectors by means of radii. The area of the red sector is 40% of the disk, the green sector is 25%, the blue 15%, the yellow 20%. Find the probability that it will come to rest on (*a*) either green or blue; (*b*) not yellow; (*c*) a primary color.

29. In a single throw of a single die, find the probability of obtaining either a 5 or a 6 or an even number.

30. In drawing a single card from a deck of 52 playing cards, find the probability of drawing either a spade or a face-card (Jack, Queen, King).

97. Statistical Probability

When the probabilities of the possible outcomes are assigned before or without experiment or observation, we use the term

a priori probability. A priori probability is of great importance in both theoretical and practical work, and we shall develop it further in the next section. But it would hardly do for such things as life expectancy in insurance, or other statistical questions. For such considerations, we use relative frequency as the definition of probability. That is, if $f(n)$ is the number of favorable events in n trials, the relative frequency $f(n)/n$ is called the **statistical probability,** or **a posteriori probability,** or **empirical probability** of success. It may be expected to change if n is changed, that is, if the basic data or number of cases examined is changed.

For life insurance, the Commissioner's Standard Ordinary Mortality Table (table VIII) is based on many observations of holders of life insurance policies.

Example. What is the probability that a man aged 32 will live to 50?

Solution. From the table, of 917880 men alive at age 32, 810900 are alive at age 50. Hence the probability is $810900/917880 = 0.883$ or 88.3%.

Needless to say, this means nothing to any particular man since it takes no account of his health, occupation, etc. But it has statistical significance.

Note. Statistical probabilities may, in many practical situations, be used to suggest the assignment of the probabilities of the possible outcomes to be used, instead of merely supposing equally likely outcomes.

EXERCISES

Using table VIII, estimate the statistical probability of:

1. A 10-year-old reaching the age of 35.
2. A 10-year-old reaching the age of 65.
3. A 10-year-old reaching the age of 70.
4. A 30-year-old reaching the age of 65.
5. A 35-year-old reaching the age of 70.
6. A 31-year-old reaching the age of 85.

98. Conditional Probabilities. Independent Events. Repeated Trials

In this section we sketch briefly some more complicated but useful notions.

Example 1. Consider 3 white balls and 1 black ball in an urn. The experiment is to draw two balls in succession from the urn, the first ball not being replaced before the second drawing. What is the probability that both balls will be white?

Solution: Let A be the event that a white ball is obtained on the first draw. Let B be the event that a white ball is obtained on the second draw. Assuming equally likely outcomes, $P(A) = \frac{3}{4}$. We denote by $P(B|A)$ the probability of B assuming that A has already happened; in this case, the probability of a white ball on the second drawing assuming that a white ball has been already drawn. Then $P(B|A) = \frac{2}{3}$. In other words a first white ball occurs 75% of the time, and about $66\frac{2}{3}$% *of these times,* a second white ball occurs. Hence both A and B, i.e., two white balls in succession, will occur $\frac{2}{3}$ of $\frac{3}{4}$ of the times; i.e., $P(A \text{ and } B) = \frac{2}{3} \cdot \frac{3}{4} = \frac{1}{2}$.

This experiment has twelve outcomes; labelling the four balls w_1, w_2, w_3, b, the possible outcomes are:

w_1w_2	w_2w_1	w_3w_1	bw_1
w_1w_3	w_2w_3	w_3w_2	bw_2
w_1b	w_2b	w_3b	bw_3

of which 6 are clearly successful, so that the probability $P(A \text{ and } B) = 6/12 = 1/2$. Nine of the twelve are successful for A, while 6 of these 9 are then successful for B.

The probability $P(B|A)$ of B, assuming A has already happened, is called the **conditional probability of *B* if *A*.**

Example 1 makes it plausible that we should have in general the relation

$$P(A \text{ and } B) = P(A)P(B|A) \tag{1}$$

or

$$P(B|A) = \frac{P(A \text{ and } B)}{P(A)}. \tag{2}$$

Two events are called **independent** if the occurrence or non-occurrence of the first has no effect on the occurrence or non-occurrence of the second; in symbols, $P(B) = P(B|A)$. In the case of independent events, (1) becomes

$$P(A \text{ and } B) = P(A)P(B). \tag{3}$$

The result can be extended to any number of independent events $E_1, E_2, \ldots, E_n$, obtaining

$$P(E_1 \text{ and } E_2 \text{ and } \ldots \text{ and } E_n) = P(E_1)P(E_2)\ldots P(E_n).$$

Example 2. The same as example 1 except that the first ball is replaced before the second drawing. In this case, $P(A) = 3/4$, and $P(B|A) = P(B) = 3/4$. Hence $P(A \text{ and } B) = P(A)P(B) = \frac{3}{4} \cdot \frac{3}{4} = \frac{9}{16}$. In this experiment there are 16 outcomes:

w_1w_1	w_2w_1	w_3w_1	bw_1
w_1w_2	w_2w_2	w_3w_2	bw_2
w_1w_3	w_2w_3	w_3w_3	bw_3
w_1b	w_2b	w_3b	bb

of which 9 constitute the event "A and B."

When an experiment is given repeated trials "under similar conditions" this means that the successive trials are understood to be independent of each other. For such trials we have the following theorem.

THEOREM 1. *If p is the probability that an event will occur in a single trial, then the probability that it will occur exactly r times out of n trials is* $C(n,r) \cdot p^r q^{n-r}$, *where* $q = 1 - p$ *is the probability of failure in a single trial.*

Proof. The probability that the event will occur in any particular set of r trials and fail in the remaining $n - r$ trials is $p^r q^{n-r}$, by (3). But the particular set of r trials may be selected from the set of n trials in $C(n,r)$ ways. Hence, by theorem 1 of section 96, we must add $C(n,r)$ terms each of which has the value $p^r q^{n-r}$. Therefore the desired probability is $C(n,r)p^r q^{n-r}$.

Example 3. Find the probability of throwing a two with a single die exactly 3 times out of 5 trials.

Solution. The probability of throwing a two in a single trial is 1/6. Hence the answer is

$$C(5,3)\left(\frac{1}{6}\right)^3\left(\frac{5}{6}\right)^2 = 10 \cdot \frac{1}{216} \cdot \frac{25}{36} = \frac{125}{3888}.$$

Example 4. A mass production factory makes rods which are to be 1 inch in diameter, with a tolerance of 0.1 inch. That is, they are acceptable if their diameters lie between 0.9 inch and 1.1 inches. In normal production it is found that 1 out of 10 will fail to meet this requirement. Suppose, on a given occasion, a sample of 20 shows 5 unsatisfactory rods. Can this be attributed to chance? That is, what is the likelihood that 5 unsatisfactory rods should occur out of 20? By theorem 1, the probability is $C(20,5)(\frac{1}{10})^5(\frac{9}{10})^{15} = 0.03$ approximately. Since this is a very small probability, careful investigation of the production process is indicated to discover the cause of the excessive failures.

If an experiment has n outcomes, the probabilities of which are $p_1, p_2, \ldots, p_n$ respectively, and if a person wins an amount R_1 if the first outcome occurs, an amount R_2 if the second outcome occurs, and so on, then the amount

$$p_1R_1 + p_2R_2 + \ldots + p_nR_n$$

is called his **mathematical expectation.** It represents approximately his average winnings per trial in a sufficiently long sequence of trials.

Example 5. A man will win \$1 if he throws a 1, 2, 3, or 4 and will lose \$3 (win −\$3) if he throws a 5 or 6 with a single die. His mathematical expectation is

$$\tfrac{1}{6}(1) + \tfrac{1}{6}(1) + \tfrac{1}{6}(1) + \tfrac{1}{6}(1) + \tfrac{1}{6}(-3) + \tfrac{1}{6}(-3) = -\tfrac{2}{6} = -\tfrac{1}{3} \text{ dollar.}$$

If the probability of an event is s/t, the **odds in favor** of the event are s to $t - s$, and the **odds against** the event are $t - s$ to s.

Example 6. The odds against throwing a two with a single throw of a single die are 5 to 1.

EXERCISES

Assume equally likely outcomes unless otherwise specified.

1. What is the probability of throwing a six twice in succession with a single die?

2. What is the probability of drawing an ace twice in succession from a pack of 52 playing cards (*a*) if each drawn card is replaced in the pack before drawing again; (*b*) if drawn cards are not replaced?

3. Find the probability of drawing an ace 4 times in succession from a pack of 52 playing cards (*a*) if each drawn card is replaced in the pack before drawing again; (*b*) if drawn cards are not replaced.

4. An urn contains 4 white and 5 black balls. If drawn balls are to be replaced in the urn before the next drawing, what is the probability of drawing (*a*) 2 white balls in succession; (*b*) 3 black balls in succession?

5. The same as exercise 4 except that drawn balls are not to be replaced.

6. What is the probability that a six will turn up exactly once in 4 successive throws of a single die?

7. What is the probability that a six will turn up exactly twice in 4 successive throws of a single die?

8. Find the probability that a six will turn up at least once in 5 successive throws of a single die. (*Hint:* Consider the probability of failure.)

9. If 10 coins are tossed in succession, what is the probability that (*a*) exactly 3 will be heads; (*b*) at least 3 will be heads?

10. If 9 coins are tossed, find the probability that (*a*) exactly 7 will be heads; (*b*) at least 7 will be heads; (*c*) at most 7 will be heads.

11. If 5 dice are thrown, find the probability that (*a*) exactly 3 of them will turn up a six; (*b*) at least 3 of them will turn up a six; (*c*) at most 3 of them will turn up a six.

12. If a man will win \$60 if he throws a total of seven in a single throw of a pair of dice, and will lose \$15 if he doesn't, what is his mathematical expectation?

13. What is the probability of throwing a total of seven with a pair of dice (*a*) 5 times in succession? (*b*) 10 times in succession? (*c*) Is the probability in (*b*) half as great as that in (*a*)?

14. If 13 cards are drawn from a pack of 52 playing cards, what is the probability that they will all be spades?

15. The probability that A will be alive next Christmas is 3/4. The probability that B will be alive next Christmas is 2/3. What is the probability that at least one of them will be alive next Christmas? (*Hint:* Consider the probability of failure.)

16. Four different red and 4 different blue books are shelved at random on a shelf with space for 8 books. Find the probability that they will be shelved with colors alternating.

17. In a tennis tournament, the odds that A will win the tournament are 4 to 3 and the odds that B will win are 1 to 4. Find the odds that either A or B will win.

18. The probability that A will solve a problem is 2/3. The probability that B will solve the problem is 3/4. If both work on it independently, what is the probability that the problem will be solved?

19. Two people A and B are to draw alternately one ball at a time from an urn containing 3 white and 2 black balls, drawn balls not being replaced. If A takes the first turn, what is the probability that A will be the first to draw white?

20. A machine has two component parts, A and B. If either one breaks down, the machine breaks down. The probability that A will break down in 1 year is 1/4. The probability that B will break down in 1 year is 1/6. Find the probability that the machine will break down in 1 year.

21. A machine has three parts A, B, C. If any one of them breaks down, the machine breaks down. Suppose that the probability that

A breaks down in 1 year is 1/5; similarly for B and for C. Find the probability that the machine will break down in 1 year.

22. (*a*) Two men are to be chosen at random from a group of 7. What is the probability that a certain 2, A and B, will be those chosen? (*b*) The 7 men are to be seated in a row of 7 seats. What is the probability that a certain 2, A and B, will be seated next to each other?

23. (*a*) Find the probability of throwing a total of seven at least once in 3 successive throws of a pair of dice. (*b*) What are the odds against this happening?

24. Five students are assigned seats in a row of 5 seats. (*a*) If they sit down at random, find the probability that every one will sit in his own seat. (*b*) Find the probability that at least 3 will sit in their own seats.

25. Under normal conditions 10% of a certain mass production item will be defective. In a certain sample of 10 items, 3 of them are found to be defective. What is the probability that this happened by chance?

26. One urn contains 4 black balls and 2 white balls. A second urn contains 1 black ball and 5 white balls. If we are to draw just 1 ball, what is the probability that it will be white?

27. In a "multiple choice" test, each question is to be answered by selecting 1 of 5 different proposed answers, of which only 1 is right. If there are 10 questions on the test, what is the probability of (*a*) getting exactly 6 right by pure guesswork; (*b*) getting at least 6 right by pure guesswork?

28. A gambler bets \$72 that he will obtain a total of either eight or five in a single throw of a pair of dice. He loses \$72 if he fails. What are the odds in his favor and what is his mathematical expectation?

29. A coin is biased so that the probabilities of heads and tails are 2/3 and 1/3 respectively. A second coin is biased so that the probabilities of heads and tails are 3/5 and 2/5 respectively. If both coins are tossed, find the probability of (*a*) two heads; (*b*) exactly one head; (*c*) no head; (*d*) at least one head.

30. A loaded die is such that the probabilities of throwing 1, 2, 3, 4, 5, 6 are respectively $\frac{1}{3}$, $\frac{1}{6}$, $\frac{1}{6}$, $\frac{1}{9}$, $\frac{1}{9}$, $\frac{1}{9}$. A second die is such that the probabilities of throwing 1, 2, 3, 4, 5, 6 are respectively $\frac{1}{4}$, $\frac{1}{8}$, $\frac{1}{8}$, $\frac{1}{6}$, $\frac{1}{6}$, $\frac{1}{6}$. In a single throw of the pair of dice, find the probability of throwing (*a*) a total of two; (*b*) a total of three; (*c*) a total of seven; (*d*) a total of twelve.

31. If the first coin mentioned in exercise 29 is tossed 5 times, what is the probability of obtaining exactly two heads?

32. If the first die mentioned in exercise 30 is thrown 6 times, what is the probability of obtaining (*a*) exactly 3 sixes; (*b*) at least 5 sixes; (*c*) at most two sixes?

33. Prove formula (1) of this section, assuming equally likely outcomes. *Hint:* let a be the number of outcomes in which both A and B occur, b the number of outcomes in which A but not B occurs, c the number of outcomes in which B but not A occurs, d the number of outcomes in which neither A nor B occurs. Let $n = a + b + c + d$ be the total number of outcomes. Then

$$P(A \text{ and } B) = \frac{a}{n},\ P(B|A) = \frac{a}{a+b},\ P(A) = \frac{a+b}{n}.$$

34. Using the notations of the hint in exercise 33, prove that if equally likely outcomes are assumed then $P(A \text{ and } B) = P(A|B)P(B)$.

35. Prove formula (3) of this section, assuming equally likely outcomes. *Hint:* using the notations of the hint of exercise 33, the hypothesis of independence, namely $P(B|A) = P(B)$, implies that

$$\frac{a}{a+b} = \frac{a+c}{n}; \quad \text{hence} \quad \frac{a+c}{n} \cdot \frac{a+b}{n} = \frac{a}{n}.$$

Chapter XVI

Mathematical induction

99. The *n*th Term of a Sequence

By a **sequence** of numbers is meant a succession $a_1, a_2, a_3, \ldots, a_n, \ldots$. The subscript n indicates the position of the term a_n in the sequence. Thus a_1 is the first term, a_2 the second term, and in general, a_n is the **nth term,** or **general term,** of the sequence. A sequence may be specified by stating its nth term as a function of n, as follows.

Example 1. The sequence whose nth term is $2n$ has the numbers $2 \cdot 1 = 2$, $2 \cdot 2 = 4$, $2 \cdot 3 = 6$, $2 \cdot 4 = 8$, $2 \cdot 5 = 10$ as the first five terms.

Example 2. The sequence whose nth term is $1/n^2$ has the numbers $1, \frac{1}{4}, \frac{1}{9}, \frac{1}{16}, \frac{1}{25}$ as the first five terms.

However, there is no possibility of conversely determining the nth term as a function of n from a knowledge of the first five (or first thousand) terms alone. For example, the nth term of the sequence of example 1, namely

$$(1) \qquad 2, 4, 6, 8, 10, \ldots$$

is $2n$. But the first five terms given in (1) are also the first five terms of the sequence whose nth term is

$$(2) \qquad 2n + (n-1)(n-2)(n-3)(n-4)(n-5),$$

and also the first five terms of the sequence whose nth term is

$$(3) \qquad 2n + (n-1)(n-2)(n-3)(n-4)(n-5)f(n),$$

where $f(n)$ may be almost any function of n at all.

Note. It follows that the question, "What is the next term of the sequence 2, 4, 6, 8, 10, . . . ?" which is often asked on intelligence tests, cannot be answered logically. For if the nth term were $2n$, the next term would be $2 \cdot 6 = 12$; but if the nth term were (2), then the next term would be $2 \cdot 6 + 5 \cdot 4 \cdot 3 \cdot 2 \cdot 1 = 132$; indeed, the next term may be any number at all if the function $f(n)$ in (3) is properly chosen and (3) then taken as the nth term.

EXERCISES

Write (a) the first 5 terms, (b) the 10th term, (c) the kth term, (d) the $(k + 1)$th term, of the sequence whose nth term is:

1. $2n$ **2.** $4n$. **3.** n^2. **4.** $3n - 1$. **5.** $n(n + 1)$.

6. $\dfrac{(-1)^n}{n^3}$. **7.** $\dfrac{(-1)^{n+1}}{n(n + 1)}$. **8.** $5n + 3$.

100. The Axiom of Mathematical Induction

The following so-called **axiom of mathematical induction** is one of the basic axioms of algebra.

If *a proposition involving the positive integer n can be proved to have the following properties:*

(A) The proposition is correct for $n = 1$;

(B) If k is any value of n for which the proposition is true, then the proposition is also true for the next value $n = k + 1$,

then *the proposition is true for all positive integral values of n.*

This axiom is extremely plausible. For if (A) is established, the proposition is true for $n = 1$. But if (B) is established, then it is true for $1 + 1$ or 2. But then (B) also implies that it is true for $2 + 1$ or 3. And so on. But we cannot go on so for more than a limited number of cases in a lifetime. Hence we assume the axiom above.

The axiom of mathematical induction may be made intui-

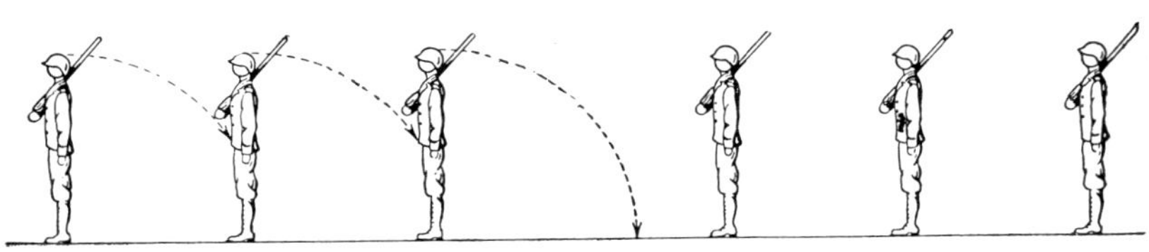

Condition (b) not satisfied

Fig. 41.

tively plausible by considering an endless single file of toy soldiers (Fig. 41). Suppose we wish to knock over all the soldiers. We cannot knock over more than a limited number of them if we knock over one at a time. But we can be sure they will all fall if we can make sure of two things, namely that:

(*a*) The first soldier is knocked over;

(*b*) The soldiers are spaced so that if any one is knocked over, it automatically knocks over the next. Clearly, (*a*) corresponds to (*A*) and (*b*) to (*B*) in the axiom of mathematical induction.

101. Proofs by Mathematical Induction

Let us prove the following theorem:

THEOREM 1. *If n is any positive integer, then*

$$3 + 6 + 9 + \ldots + 3n = \frac{3n(n+1)}{2}. \tag{1}$$

Note that (1) asserts that

for $n = 1$, $$3 = \frac{3 \cdot 1(1+1)}{2},$$

for $n = 2$, $$3 + 6 = \frac{3 \cdot 2(2+1)}{2},$$

for $n = 3$, $$3 + 6 + 9 = \frac{3 \cdot 3(3+1)}{2},$$

for $n = 4$, $$3 + 6 + 9 + 12 = \frac{3 \cdot 4(4+1)}{2},$$

and so on. But we cannot prove the infinite number of propositions contained in theorem 1 by checking one case at a time in this way. Hence we prove it by means of the axiom of mathematical induction, as follows:

Proof. (*A*) For $n = 1$, we verify that (1) is true by direct substitution of 1 for n; that is, we verify that

$$3 = \frac{3 \cdot 1(1+1)}{2}.$$

(*B*) We prove the so-called auxiliary proposition, that *if the proposition is true for $n = k$, then it must be true for the next value $n = k + 1$.*

The hypothesis of this proposition is obtained by substituting $n = k$ in the proposed formula (1). The conclusion is obtained by substituting $n = k + 1$ in (1). Thus,

Hypothesis. $3 + 6 + 9 + \ldots + 3k = \frac{3k(k+1)}{2} = \frac{3k^2 + 3k}{2}.$

Conclusion. $3 + 6 + 9 + \ldots + 3(k+1)$

$$= \frac{3(k+1)(k+1+1)}{2} = \frac{3k^2 + 9k + 6}{2}.$$

Proof. By hypothesis,

$$3 + 6 + 9 + \ldots + 3k = \frac{3k^2 + 3k}{2}.$$

Adding the $(k+1)$th term, $3(k+1)$, to both sides, we obtain

$$(2) \quad 3 + 6 + 9 + \ldots + 3k + 3(k+1) = \frac{3k^2 + 3k}{2} + 3(k+1),$$

which is then true, because if equals are added to equals the results are equal. The left member of (2) is the same as the left member of the conclusion. It remains only to verify that the right member of (2) is the same as the right member of the conclusion. Since

$$\frac{3k^2 + 3k}{2} + 3(k+1) = \frac{3k^2 + 9k + 6}{2},$$

the auxiliary proposition is proved.

By virtue of (A) and (B), the axiom of mathematical induction assures us that (1) is true for all positive integral values of n. That is, theorem 1 is proved.

The work may be arranged compactly as in the following example:

THEOREM 2. *If n is any positive integer, then*

$$(3) \qquad 2 + 5 + 8 + \ldots + (3n - 1) = \frac{n(3n+1)}{2}.$$

Proof. (A) We verify (3) for the case $n = 1$:

$$2 = \frac{1(3 \cdot 1 + 1)}{2}$$

$$2 = 2.$$

(B) *Hypothesis.* $2 + 5 + 8 + \ldots + (3k - 1) = \frac{k(3k+1)}{2}$

$$= \frac{3k^2 + k}{2}.$$

Conclusion.

$$2 + 5 + 8 + \ldots + (3[k + 1] - 1) = \frac{(k + 1)(3[k + 1] + 1)}{2},$$

or $2 + 5 + 8 + \ldots + (3k + 2) = \dfrac{3k^2 + 7k + 4}{2}$.

Proof. By hypothesis,

$$2 + 5 + 8 + \ldots + (3k - 1) = \frac{3k^2 + k}{2}.$$

Adding the $(k + 1)$th term, $(3k + 2)$, to both sides, we obtain

$$\begin{aligned} 2 + 5 + 8 + \ldots + (3k + 2) &= \frac{3k^2 + k}{2} + 3k + 2 \\ &= \frac{3k^2 + 7k + 4}{2}, \end{aligned}$$

which agrees with the right member of the conclusion.

This completes the proof of (B), and hence, by the axiom of mathematical induction, completes the proof of theorem 2.

Note 1. The analogy with the endless file of tin soldiers in section 100 makes clear that both part (A) and part (B) must be established. We give two algebraic examples to show that neither (A) nor (B) alone is sufficient. The student should fill in all details.

(*a*) Consider the false formula

$$3 + 6 + 9 + \ldots + 3n = \frac{3n(n + 1)}{2} + (n - 1).$$

Part (A) is correct, since $3 = \dfrac{3 \cdot 1(1 + 1)}{2} + 0$. But part (B) will not work. For, adding $3(k + 1)$ to $\dfrac{3k(k + 1)}{2} + (k - 1)$, we do not get $\dfrac{3(k + 1)(k + 2)}{2} + k$.

(*b*) Consider the false formula

$$3 + 6 + 9 + \ldots + 3n = \frac{3n(n + 1)}{2} + 1.$$

Here (B) is correct, since, if we add $3(k + 1)$ to $\dfrac{3k(k + 1)}{2} + 1$, we do get $\dfrac{3(k + 1)(k + 2)}{2} + 1$. But part (A) is false, since $3 \neq \dfrac{3 \cdot 1(1 + 1)}{2} + 1$.

Note 2. A proof by mathematical induction is deductive logic and must be distinguished from the induction or inductive logic used in experimental science. In experimental science we might test a formula like (1) for $n = 1, 2, 3, \ldots, 1000$. If it worked for 1000 trials we might conclude that it was (very) probably true. But we could not conclude more without further information. For the false formula

$$3 + 6 + 9 + \ldots + 3n$$
$$= \frac{3n(n+1)}{2} + (n-1)(n-2)(n-3)\ldots(n-1000)$$

would also work correctly for $n = 1, 2, 3, \ldots, 1000$, but would be false thereafter. Why?

Note 3. The axiom of mathematical induction is one of five axioms, stated by Peano in the 19th century, upon which all of elementary algebra can be based, but the idea of proving theorems by mathematical induction can be traced, in some form, to the 16th century.

Note 4. In the examples above and many of the exercises below, theorems giving the sum of the first n terms of a given sequence are verified by mathematical induction. But no hint is given as to how the right-hand member of the given formula is discovered. Many of these can be obtained by methods discussed in chapter XXI.

EXERCISES

(*a*) *In each of the following exercises, verify the correctness of the given formula for* $n = 1, 2, 3$. (*b*) *Does the work done in part* (*a*) *suffice to establish the correctness of the given formula for all positive integers n? Explain.* (*c*) *Prove by mathematical induction that the given formula is correct for all positive integers n.*

1. $1 + 3 + 5 + \ldots + (2n - 1) = n^2$.
2. $2 + 4 + 6 + \ldots + 2n = n(n + 1)$.
3. $4 + 8 + 12 + \ldots + 4n = 2n(n + 1)$.
4. $1 + 2 + 3 + \ldots + n = \dfrac{n(n+1)}{2}$.
5. $5 + 10 + 15 + \ldots + 5n = \dfrac{5n(n+1)}{2}$.
6. $5 + 7 + 9 + \ldots + (2n + 3) = n(n + 4)$.
7. $\dfrac{1}{1 \cdot 2} + \dfrac{1}{2 \cdot 3} + \dfrac{1}{3 \cdot 4} + \ldots + \dfrac{1}{n(n+1)} = \dfrac{n}{n+1}$.

8. $1 \cdot 2 + 2 \cdot 3 + 3 \cdot 4 + \ldots + n(n+1) = \dfrac{n(n+1)(n+2)}{3}$.

9. $1^2 + 2^2 + 3^2 + \ldots + n^2 = \dfrac{n(n+1)(2n+1)}{6}$.

10. $1^3 + 2^3 + 3^3 \ldots + n^3 = \dfrac{n^2(n+1)^2}{4}$.

11. $3 + 3^2 + 3^3 + \ldots + 3^n = \dfrac{3^{n+1} - 3}{2}$.

12. $1 + 5 + 9 + \ldots + (4n - 3) = n(2n - 1)$.

13. $5 + 8 + 11 + \ldots + (3n + 2) = \dfrac{n(3n+7)}{2}$.

14. $1 \cdot 3 + 2 \cdot 4 + 3 \cdot 5 + \ldots + n(n+2) = \dfrac{n}{6}(n+1)(2n+7)$.

15. $a + ar + ar^2 + \ldots + ar^{n-1} = \dfrac{a - ar^n}{1 - r}$ $(r \neq 1)$.

16. $a + (a + d) + (a + 2d) + \ldots + (a + [n - 1]d) = \dfrac{n(2a + [n - 1]d)}{2}$.

★**17.** $2^{n+3} < (n + 3)!$.

★**18.** Prove by mathematical induction that $x - y$ is a factor of $x^n - y^n$ for all positive integral values of n. (*Hint:* $x^{k+1} - y^{k+1} = (x^{k+1} - xy^k) + (xy^k - y^{k+1})$.)

★**19.** Prove by mathematical induction that

$$C(2,2) + C(3,2) + C(4,2) + \ldots + C(n+1, 2) = C(n+2, 3),$$

for all positive integral values of n.

★**20.** Prove by mathematical induction that the sum of the angles of an $(n + 2)$-sided simple polygon is $n \cdot 180°$.

★**21.** Prove by mathematical induction that

$$C(3,3) + C(4,3) + C(5,3) + \ldots + C(n+3,3) = C(n+4,4)$$

for all positive integral values of n.

★**22.** Prove by mathematical induction that

$$C(4,4) + C(5,4) + C(6,4) + \ldots + C(n+4,4) = C(n+5,5)$$

for all positive integral values of n.

23. Prove by mathematical induction that

$$\frac{1}{2} + \frac{1}{2^2} + \frac{1}{2^3} + \ldots + \frac{1}{2^n} = 1 - \frac{1}{2^n}$$

for all positive integral values of n.

★ May be omitted without disturbing continuity.

Chapter XVII

The binomial theorem

102. The Binomial Theorem for Positive Integral Exponents

By direct multiplication, the student may verify that

$$(a + b)^1 = a + b$$

$$(a + b)^2 = a^2 + 2ab + b^2$$

$$(a + b)^3 = a^3 + 3a^2b + 3ab^2 + b^3$$

$$(a + b)^4 = a^4 + 4a^3b + 6a^2b^2 + 4ab^3 + b^4$$

$$(a + b)^5 = a^5 + 5a^4b + 10a^3b^2 + 10a^2b^3 + 5ab^4 + b^5,$$

and so on. The right member is called the **expansion** of the left member. From these examples, we might guess that the expansion of $(a + b)^n$, where n is any positive integer, has the following properties:

(*A*) The expansion has $n + 1$ terms;

(*B*) The first term is a^n and the last is b^n;

(*C*) In each successive term after the first, the exponent of a decreases by 1 and the exponent of b increases by 1 so that the sum of the exponents in each term is n;

(*D*) The coefficients of terms equidistant from the ends of the expansion are equal;

(*E*) The coefficients may be determined from the triangular array

$$\begin{array}{ccccccccccc} & & & & 1 & & 1 & & & & \\ & & & 1 & & 2 & & 1 & & & \\ & & 1 & & 3 & & 3 & & 1 & & \\ & 1 & & 4 & & 6 & & 4 & & 1 & \\ 1 & & 5 & & 10 & & 10 & & 5 & & 1 \\ . & . & . & & & . & & & . & . & . \end{array}$$

in which each number, except those at the ends, is the sum of the two nearest numbers in the line above. This scheme is known as Pascal's triangle, although it was known, both in Europe and in China, considerably before Pascal's time (17th century). Pascal studied it in connection with permutations, combinations, and probability.

Example 1. From properties (A) to (E), we can write

$$(a+b)^6 = a^6 + 6a^5b + 15a^4b^2 + 20a^3b^3 + 15a^2b^4 + 6ab^5 + b^6.$$

To find the coefficients from Pascal's triangle is tedious if the exponent n is high. For example, to find the coefficients of $(a+b)^{150}$ by this scheme requires writing out the first 149 lines of Pascal's triangle before the 150th can be determined. A better way of writing the expansion is given by the following theorem, called the **binomial theorem.**

THEOREM. *If n is any positive integer,*

$$(1)\quad (a+b)^n = a^n + \frac{n}{1}a^{n-1}b + \frac{n(n-1)}{1\cdot 2}a^{n-2}b^2 + \frac{n(n-1)(n-2)}{1\cdot 2\cdot 3}a^{n-3}b^3 + \frac{n(n-1)(n-2)(n-3)}{1\cdot 2\cdot 3\cdot 4}a^{n-4}b^4 + \ldots + \frac{n(n-1)(n-2)\ldots(n-r+1)}{1\cdot 2\cdot 3\ldots r}a^{n-r}b^r + \ldots + b^n.$$

*Proof.** Recall that $(a+b)^n$, where n is a positive integer, stands for the product of n factors each equal to $(a+b)$. Hence each term in the result is a sum of terms each of which is a product of one letter from each factor $(a+b)$. Thus, the term involving b^r will be a sum of terms of the form $a^{n-r}b^r$, each of which is a product of r b's [one from each of r factors $(a+b)$] and $n-r$ a's [one from each of the remaining $n-r$ factors $(a+b)$]. There are as many such terms as there are choices of r factors from the n factors, from which to take a b. Hence there are $C(n,r)$ such terms and the total coefficient of $a^{n-r}b^r$ will be $C(n,r)$, or, by (3) of section 95, chapter XIV,

$$C(n,r) = \frac{n(n-1)(n-2)\ldots(n-r+1)}{1\cdot 2\cdot 3\ldots r}.$$

* Chapter XIV on Permutations and Combinations is needed for this proof. Another proof by mathematical induction will be given in the next section.

Note 1. The coefficient of the term involving b^r in the expansion of $(a + b)^n$ has the product of the first r positive integers (or $r!$) in the denominator, and r factors in the numerator beginning with n and successively decreasing by 1. That is, the **term involving b^r** is

$$(2) \quad C(n,r)a^{n-r}b^r = \frac{n(n-1)(n-2)\ldots(n-r+1)}{1 \cdot 2 \cdot 3 \ldots r} a^{n-r}b^r.$$

This is the **$(r + 1)$th term** of the expansion. The **rth term** would be, accordingly,

$$(3) \quad C(n, r-1)a^{n-r+1}b^{r-1} = \frac{n(n-1)(n-2)\ldots(n-r+2)}{1 \cdot 2 \cdot 3 \ldots (r-1)} a^{n-r+1}b^{r-1}.$$

The number of factors in either numerator or denominator of the coefficient is equal to the exponent of b.

Note 2. The **binomial formula** (1) may be written, for positive integral exponents, as

$$(4) \quad (a+b)^n = a^n + C(n,1)a^{n-1}b + C(n,2)a^{n-2}b^2 + \ldots + C(n,r)a^{n-r}b^r + \ldots + b^n.$$

In particular, setting $a = b = 1$ in (4), it follows that

$$C(n,0) + C(n,1) + C(n,2) + \ldots + C(n,n) = 2^n$$

where we have written $C(n,0) = C(n,n) = 1$. That is, *the total number of sets of any size that can be selected from a given set of n things is 2^n provided we include just 1 "empty" set $[C(n,0) = 1]$. If we do not wish to count this "empty" set, the number is $2^n - 1$.*

Note 3. If the product of the coefficient of any term and the exponent of a in that term is divided by the number of that term, we obtain the coefficient of the next term. For example, the coefficient of the 4th term in the expansion of $(a + b)^6$ (see example 1) may be obtained from the third term by calculating $15 \cdot 4/3 = 20$.

Note 4. If the entire expansion is desired, for a low value of n, the coefficients may be obtained by means of note 3, or (4), or (1), or Pascal's triangle. If a particular term only is desired, (2) or (3) will be useful.

Note 5. Since $a - b = a + [-b]$, *the expansion of $(a - b)^n$ will have alternating signs beginning with plus.* The minus signs will appear whenever the exponent of $[-b]$ is odd, namely in the even-numbered terms.

Example 2. Expand $\left(2x - \frac{b^2}{2}\right)^6$, and simplify each term.

Solution. By (1), or various other considerations above,

$$\left(2x - \frac{b^2}{2}\right)^6 = \left(2x + \left[-\frac{b^2}{2}\right]\right)^6$$
$$= (2x)^6 + \frac{6}{1}(2x)^5\left[-\frac{b^2}{2}\right] + \frac{6\cdot 5}{1\cdot 2}(2x)^4\left[-\frac{b^2}{2}\right]^2$$
$$+ \frac{6\cdot 5\cdot 4}{1\cdot 2\cdot 3}(2x)^3\left[-\frac{b^2}{2}\right]^3 + \frac{6\cdot 5\cdot 4\cdot 3}{1\cdot 2\cdot 3\cdot 4}(2x)^2\left[-\frac{b^2}{2}\right]^4$$
$$+ \frac{6\cdot 5\cdot 4\cdot 3\cdot 2}{1\cdot 2\cdot 3\cdot 4\cdot 5}(2x)\left[-\frac{b^2}{2}\right]^5 + \left[-\frac{b^2}{2}\right]^6$$
$$= 64x^6 - 96x^5b^2 + 60x^4b^4 - 20x^3b^6 + \frac{15}{4}x^2b^8 - \frac{3}{8}xb^{10} + \frac{b^{12}}{64}.$$

Example 3. Write and simplify the 8th term in the expansion of $\left(\frac{u}{2} - v^2\right)^{12}$.

Solution. The 8th term will involve b^7. Hence it is

$$\frac{\not{1}\not{2}\cdot 11\cdot \not{1}\not{0}\cdot 9\cdot 8\cdot \not{7}\cdot \not{6}}{1\cdot \not{2}\cdot \not{3}\cdot \not{4}\cdot \not{5}\cdot \not{6}\cdot \not{7}}\left(\frac{u}{2}\right)^5(-v^2)^7 = -99\cdot 8\,\frac{u^5}{2^5}\,v^{14} = -\frac{99}{4}u^5v^{14}.$$

Example 4. Write and simplify the term involving k^6 in the expansion of $(h^3 - k^2)^8$.

Solution. This is the fourth term since we must have b or k^2 raised to the third power. This term is

$$\frac{8\cdot 7\cdot 6}{1\cdot 2\cdot 3}(h^3)^5(-k^2)^3 = -56h^{15}k^6.$$

Example 5. By means of the binomial theorem, calculate $(1.02)^8$ to the nearest hundredth.

Solution. $(1.02)^8 = (1 + 0.02)^8 = 1^8 + \frac{8}{1}\cdot 1^7(0.02) + \frac{8\cdot 7}{1\cdot 2}\cdot 1^6(0.02)^2 + \frac{8\cdot 7\cdot 6}{1\cdot 2\cdot 3}\cdot 1^5(0.02)^3 + \ldots = 1 + 0.16 + 0.0112 + 0.000448 + \ldots = 1.171648 = 1.17$ to the nearest hundredth. It is intuitively clear that further terms will contribute negligible increases since higher powers of 0.02 are very small.

Example 6. Write and simplify the term involving x^2 in the expansion of $\left(2x - \frac{1}{x^2}\right)^8$.

Solution. The general term is $C(8,r)(2x)^{8-r}\left(-\frac{1}{x^2}\right)^r$. After simplification, the factor involving powers of x will be

$$\frac{x^{8-r}}{x^{2r}} = x^{8-r-2r} = x^{8-3r}.$$

Since we must have $8 - 3r = 2$, we must take $r = 2$. Hence the required term is

$$C(8,2)(2x)^6\left(-\frac{1}{x^2}\right)^2 = 1792x^2.$$

Note 6. Some elementary practical applications of the binomial theorem will be found in chapter XXIV. It will prove to be a useful theorem in more advanced courses in mathematics.

EXERCISES

Write the expansion of each of the following and simplify each term:

1. $(x + h)^7$. **2.** $(a + b)^8$. **3.** $(p + q)^5$. **4.** $(2a + 3b)^6$.

5. $(2a^3 + 3b^2)^5$. **6.** $\left(\frac{a}{2} + \frac{4b^2}{a}\right)^7$. **7.** $(a - b)^3$.

8. $(a - b)^4$. **9.** $(a - b)^5$. **10.** $\left(2a^3 - \frac{b^2}{2a}\right)^7$.

11. $\left(\frac{x}{y} - \frac{1}{x}\right)^4$. **12.** $\left(\frac{a^2}{2} - \frac{4}{a}\right)^6$. **13.** $\left(\frac{x^2}{2} - \frac{2}{x}\right)^4$.

14. $\left(\frac{ab^2}{2x} - \frac{4x}{a^2b}\right)^8$.

Write, in simplified form, only the specified term:

15. The term involving h^3 in the expansion of $(x + h)^{17}$.
16. The term involving b^6 in the expansion of $(a + b)^{24}$.
17. The sixth term of $(x + h)^{15}$.
18. The eighteenth term of $(x + h)^{20}$.

19. The term involving b^6 in the expansion of $\left(\frac{a^2}{2} + 2b^2\right)^{10}$.

20. The term involving q^8 in the expansion of $\left(\frac{p^2}{3} + 6q^2\right)^{12}$.

21. The middle term of $(a - b)^8$.

22. The middle terms of $\left(\frac{a}{2} - \frac{b}{3}\right)^{11}$.

23. The sixth term of $\left(\frac{2}{\sqrt{x}} - \frac{x^2}{2}\right)^9$.

24. The term involving x^8 in the expansion of $(2x^2 + \frac{1}{2})^{10}$.

25. The eighth term of $\left(2x^2 - \dfrac{1}{2x^2}\right)^{12}$.

26. The term involving y^2 in the expansion of $(\frac{1}{2}x^3 + 2\sqrt[3]{y})^8$.

27. The fourth term of $\left(\dfrac{x^2}{y} - \dfrac{2}{x}\right)^{10}$.

28. The sixth term of $\left(\dfrac{x^2}{3} - \dfrac{2}{x^{1/2}}\right)^9$.

29. The term involving x^6 in the expansion of $\left(x^3 - \dfrac{1}{2x}\right)^{10}$.

30. The term involving x^{14} in the expansion of $(2x^4 - \frac{1}{2}x^{-2})^8$.

31. The term involving x^7 in the expansion of $\left(\dfrac{x^2}{2} - \dfrac{4}{x}\right)^8$.

32. The term involving x^{-11} in the expansion of $(\sqrt{x} - 2x^{-2})^8$.

33. The term involving x^4 in the expansion of $\left(3x^2 - \dfrac{2}{x}\right)^8$.

34. The term involving a^{40} in the expansion of $\left(\dfrac{a^4}{3} - \dfrac{3}{2a}\right)^{15}$.

Compute by means of the binomial theorem, using only enough terms to get the result accurate to three decimal places:

35. $(1.01)^9 = (1 + 0.01)^9$. **36.** $(1.02)^7$.
37. $(0.99)^6 = (1 - 0.01)^6$. **38.** $(0.98)^6$.
39. $(1.02)^{11}$. **40.** $(0.99)^4$. **41.** $(0.98)^7$. **42.** $(1.03)^8$.

43. Explain the connection between Pascal's triangle and exercise 24 of section 95, chapter XIV.

44. Expand $(a + b + c)^4$ by the binomial theorem. (*Hint:* $a + b + c = [a + b] + c$.)

★103. Proof of the Binomial Theorem by Mathematical Induction

We give a second proof of theorem 1 or formula (4), section 102, for positive integral exponents, by mathematical induction.

Proof. (*A*) For $n = 1$ the formula (4), section 102, is true, since $(a + b)^1 = a + b$.

(*B*) If the formula (4) is true for $n = k$, then it must be true for $n = k + 1$.

Hypothesis. $(a + b)^k = a^k + C(k,1)a^{k-1}b + \ldots + C(k, r - 1)a^{k-r+1}b^{r-1} + C(k,r)a^{k-r}b^r + \ldots + b^k$.

★ May be omitted without disturbing continuity.

Conclusion. $(a + b)^{k+1} = a^{k+1} + C(k + 1, 1)a^k b + \ldots + C(k + 1, r)a^{k-r+1}b^r + \ldots + b^{k+1}$.

Proof. Multiply both sides of the hypothesis by $(a + b)$. The left member becomes $(a + b)^{k+1}$. In the right member let us consider a typical term in the product, say the term involving b^r. This will be the sum of two terms, the first being the product of a by the term involving b^r in the hypothesis, and the second being the product of b by the term involving b^{r-1} in the hypothesis. These two products are $C(k,r)a^{k-r+1}b^r$ and $C(k, r - 1)a^{k-r+1}b^r$ respectively. Hence their sum is

$$[C(k,r) + C(k, r - 1)]a^{k-r+1}b^r.$$

By direct calculation,* we find that $C(k,r) + C(k, r - 1) = C(k + 1, r)$. Hence the general term of the conclusion is $C(k + 1, r)a^{k-r+1}b^r$. This completes the proof.

104. The Binomial Series

Consider the binomial expansion

$$(1) \quad (1 + x)^n = 1 + \frac{n}{1}x + \frac{n(n-1)}{1 \cdot 2}x^2 + \frac{n(n-1)(n-2)}{1 \cdot 2 \cdot 3}x^3 + \ldots + \frac{n(n-1)\ldots(n-r+1)}{1 \cdot 2 \ldots r}x^r + \ldots.$$

If n is a positive integer, the term involving x^n is the last non-zero term, since if $r = n + 1, n + 2, \ldots$ the coefficient contains the factor zero in the numerator. But if n is not a positive integer, the expression (1) does not terminate. It becomes what is known as an **infinite series** which has significance under circumstances which will be explained in chapter XXVII.

This binomial series was used by Isaac Newton (English, 1642–1727) for negative and fractional values of n. The correctness of the expression (1) under suitable conditions was established for all complex values of n by N. H. Abel (Norwegian, 1802–1829).

Example 1. Write and simplify the first four terms of the expansion of $(1 + x)^{-1/3}$.

* See exercise 24, section 95, chapter XIV.

Solution.

$$(1 + x)^{-1/3} = 1^{-1/3} + \frac{(-\frac{1}{3})}{1}\,1^{-1/3-1}x + \frac{(-\frac{1}{3})(-\frac{1}{3} - 1)}{1 \cdot 2}\,1^{-1/3-2}x^2$$
$$+ \frac{(-\frac{1}{3})(-\frac{1}{3} - 1)(-\frac{1}{3} - 2)}{1 \cdot 2 \cdot 3}\,1^{-1/3-3}x^3 + \ldots$$
$$= 1 - \frac{1}{3}x + \frac{2}{9}x^2 - \frac{14}{81}x^3 + \ldots .$$

Example 2. Find to the nearest hundredth the positive cube root of 29.

Solution. We write $\sqrt[3]{29} = (27 + 2)^{1/3} = 3\left(1 + \frac{2}{27}\right)^{1/3}$. By (1), we have

$$\left(1 + \frac{2}{27}\right)^{1/3} = 1 + \frac{1}{3}\left(\frac{2}{27}\right) + \frac{\frac{1}{3}(\frac{1}{3} - 1)}{1 \cdot 2}\left(\frac{2}{27}\right)^2$$
$$+ \frac{\frac{1}{3}(\frac{1}{3} - 1)(\frac{1}{3} - 2)}{1 \cdot 2 \cdot 3}\left(\frac{2}{27}\right)^3 + \ldots$$
$$= 1 + 0.0247 - 0.0006 + 0.00002 - \ldots = 1.02412 \text{ approximately.}$$

Hence $\sqrt[3]{29} = 3(1.02412) = 3.07$ to the nearest hundredth.

EXERCISES

Write the first four terms in the expansion of:

1. $\sqrt{1 + x}$. (*Hint:* $\sqrt{1 + x} = (1 + x)^{1/2}$.)

2. $(1 + x)^{-3}$. **3.** $\sqrt[3]{1 + x}$. **4.** $\dfrac{1}{(1 + x)^2}$.

5. $\dfrac{1}{1 - x}$. **6.** $\dfrac{1}{\sqrt{1 + x}}$.

Using the binomial expansion, calculate to the nearest hundredth:

7. $\sqrt{17}$. **8.** $\sqrt[3]{10}$. **9.** $\sqrt[3]{65}$. **10.** $(1.01)^{-7}$
11. $\sqrt{28}$. **12.** $(1.02)^{-11}$.

By the binomial expansion, calculate to the nearest thousandth:

13. $\sqrt{50}$. **14.** $\sqrt{47}$. **15.** $\sqrt[3]{25}$. **16.** $\sqrt[3]{30}$.

Chapter XVIII

Complex numbers

105. The Standard Form of a Complex Number

It was proved, in section 16*, that no negative number has a real square root. The symbols $\pm\sqrt{N}$ where N is a negative number are called **pure imaginary numbers.** The symbol $\sqrt{-1}$ is denoted by the letter i and is called the **imaginary unit.** Thus

$$(1) \qquad i = \sqrt{-1} \text{ and } i^2 = -1.$$

We agree to express every pure imaginary number as a real multiple of i, as in the following examples.

Example 1. $\sqrt{-9} = \sqrt{9\cdot(-1)} = \sqrt{9}\sqrt{-1} = 3i.$

Example 2. $\sqrt{-2} = \sqrt{2(-1)} = \sqrt{2}\sqrt{-1} = \sqrt{2}\cdot i = i\sqrt{2}.$

In examples like the latter the i is usually written to the left of the radical sign to prevent it from slipping under the radical sign by mistake.

Thus every pure imaginary number can be written as bi, where b is real and different from 0, and $i = \sqrt{-1}$.

Any number which can be written in the form

$$(2) \qquad a + bi$$

where a and b are real and $i = \sqrt{-1}$, is called a **complex number.** When it is written in the form (2), it is said to be in **standard form.** This standard form of a complex number is often called the **rectangular form** for a reason which will appear below in section 107.

* The student should review section 16 at this time.

Example 3. $3 + \sqrt{-4}$ is a complex number. In standard form it is written as $3 + 2i$.

When written in standard form (2), the number a is called the **real part** and bi the **imaginary part** of the complex number. Thus 3 is the real part of $3 + 2i$ and $2i$ is its imaginary part. A complex number $a + bi$ for which $b \neq 0$ is called **imaginary.** A complex number for which $b = 0$ is called **real.*** Thus $3 + 2i$ is imaginary, $3 = 3 + 0i$ is real. The (real) number 0 may be written in standard form as $0 + 0i$.

Unless otherwise stated, complex numbers will be understood to be in standard form.

Two complex numbers $a + bi$ and $c + di$ are called **equal** if and only if $a = c$ and $b = d$; that is, their real parts must be equal and their imaginary parts must be equal. It follows that *$a + bi = 0$ if and only if $a = 0$ and $b = 0$.*

It will be understood in this chapter that all letters represent real numbers, except i, which always represents $\sqrt{-1}$, unless otherwise stated.

Example 4. Find x and y if $2x + 3yi = 4 + 15i$.
Solution. $2x = 4$ and $3y = 15$. Hence $x = 2$ and $y = 5$.

If i is raised to positive integral powers in succession, we obtain:

$i = \sqrt{-1}$	$i^7 = i^4i^3 = 1(-i) = -i$	
$i^2 = -1$	$i^8 = i^4i^4 = 1 \cdot 1 = 1$	
$i^3 = i^2i = -1i = -i$	$i^9 = i^8i = 1 \cdot i = i$	
$i^4 = i^2i^2 = (-1)(-1) = +1$	$i^{10} = i^8i^2 = 1(-1) = -1$	$\cdots$
$i^5 = i^4i = 1i = i$	$i^{11} = i^8i^3 = 1(-i) = -i$	
$i^6 = i^4i^2 = 1(-1) = -1$	$i^{12} = i^8i^4 = 1 \cdot 1 = 1$	

Hence the successive positive integral powers of i recur in cycles of four: i, -1, $-i$, $+1$.

The complex numbers $a + bi$ and $a - bi$ are called **conjugates** of each other. Thus, $2 + 3i$ and $2 - 3i$ are conjugates of each other. To form the conjugate of a complex number in standard form, reverse the sign preceding the imaginary part.

Note 1. In electrical theory, the letter i is usually reserved for the current, and j is used for $\sqrt{-1}$. In mathematical books the notation used here is standard.

* See the classification of numbers in section 16.

Note 2. The need for complex numbers is apparent from the fact that quadratic equations, even with real coefficients, often have complex imaginary roots.

EXERCISES

1–12. Work exercises 1–12 of section 16.

Simplify to either i, -1, $-i$, or 1:

13. i^7. **14.** i^9. **15.** i^{16}. **16.** i^{25}. **17.** i^{34}. **18.** i^{14}.

Write in standard form the conjugate of each number:

19. $3 + 4i$. **20.** $2 + \sqrt{-4}$. **21.** $5 - \sqrt{-4}$.
22. $3 - 7i$. **23.** $2i - 3$. **24.** $5i + 3$.
25. $\sqrt{-8} - 1$. **26.** $\sqrt{-24} + 5$. **27.** $3i + 7i^4$.
28. $3i^3 - 2i^2$. **29.** $2i^7 + 3i^6$. **30.** $4i^{14} + 5i^{15}$.
31–42. Do exercises 31–42 of section 16.

Find the real values of x and y satisfying the given equation:

43. $x + yi = 3 - 4i$. **44.** $x + yi = 5 - 3i$.
45. $2x - 4yi = 8$. **46.** $3x + 2yi = 6i$.
47. $(2x + 8) + (3y - 9)i = 0$.
48. $(x - 2) + (2y + 12)i = 0$.
49. $(x + y) + (x - y)i = 7 + i$.
50. $(2x + 3y) + (5x - 2y)i = 13 + 4i$.
51. $(2x - 3y) + (5x + 2y)i = 13 + 4i$.
52. $(2x - 3y) + (3x + y)i = 1 + 7i$.
53. Show that if a complex number is equal to its conjugate, then it is a real number. Is the converse true?

106. Elementary Operations with Complex Numbers

It is advisable to put complex numbers in the standard form $x + yi$ before performing any arithmetical operations with them. The **sum** of two complex numbers $a + bi$ and $c + di$ is defined as follows:

$$(1) \qquad (a + bi) + (c + di) = (a + c) + (b + d)i.$$

For example, $(2 + 3i) + (4 + 2i) = 6 + 5i$.

The **difference** $(a + bi) - (c + di)$ is, as always, the complex number $x + yi$ such that $(c + di) + (x + yi) = a + bi$. (Compare definition 1, section 5.) It follows that $c + x = a$ and

$d + y = b$, or

(2) $$(a + bi) - (c + di) = (a - c) + (b - d)i.$$

For example, $(5 + 6i) - (2 + 4i) = 3 + 2i$.

Thus addition and subtraction of complex numbers are performed without regard for the special meaning of the letter i.

Likewise, the **product** of two complex numbers is defined as the complex number obtained by multiplying them without regard for the special meaning of i, but to put the product into standard form it may be necessary to replace i^2 by -1. Thus

$$(a + bi)(c + di) = ac + bci + adi + bdi^2,$$

or, in standard form,

(3) $$(a + bi)(c + di) = (ac - bd) + (ad + bc)i.$$

For example, $(2 + 3i)(4 + 5i) = 8 + 12i + 10i + 15i^2 = 8 + 22i + 15(-1) = -7 + 22i$.

As before (see definition 3, section 5), the **quotient** of two complex numbers $(a + bi) \div (c + di)$ is defined as the complex number $x + yi$ such that $(c + di)(x + yi) = a + bi$. Thus

$$(cx - dy) + (cy + dx)i = a + bi$$

which implies that

$$\begin{cases} cx - dy = a, \\ dx + cy = b. \end{cases}$$

Solving this system of linear equations for x and y in terms of a, b, c, and d, we get

$$x = \frac{ac + bd}{c^2 + d^2} \text{ and } y = \frac{bc - ad}{c^2 + d^2}.$$

Hence we have the formula

(4) $$\frac{a + bi}{c + di} = \frac{ac + bd}{c^2 + d^2} + \frac{bc - ad}{c^2 + d^2}\, i.$$

In practice, this result can be obtained most conveniently by rationalizing the denominator of the left member of (4) just as was done with binomial quadratic surds in section 72. That is, *to express the quotient of two complex numbers in standard form, we multiply numerator and denominator by the conjugate of the denominator*. Thus

$$\frac{a+bi}{c+di} = \frac{a+bi}{c+di}\,\frac{c-di}{c-di} = \frac{(ac+bd)+(bc-ad)i}{c^2+d^2}$$

which is the same as the right member of (4).

For example,

$$\frac{2+3i}{4-5i} = \frac{2+3i}{4-5i}\,\frac{4+5i}{4+5i} = \frac{-7+22i}{41} = -\frac{7}{41} + \frac{22}{41}i.$$

It can be shown that complex numbers satisfy the usual laws of algebra. Compare sections 16 and 17.

EXERCISES

Perform the indicated operations and express the result in standard form:

1. $2i + 3i$. **2.** $5i + \sqrt{-4}$. **3.** $\sqrt{-9} + \sqrt{-4}$.

4. $\sqrt{-36} + \sqrt{-25}$. **5.** $(4i)(5i)$.

6. $3i\sqrt{-16}$. **7.** $\sqrt{-16}\sqrt{-4}$.

8. $\sqrt{-9}\sqrt{-25}$. **9.** $3i(4i^2)$.

10. $(3i)^3$. **11.** $(2i)^2(3i)^3$. **12.** $\sqrt{-2}\sqrt{-32}$.

13–30. Work exercises 13–30 of section 16.

31. $(3 + i\sqrt{2})(3 - i\sqrt{2})$. **32.** $(3 + \sqrt{-8})(3 - \sqrt{-8})$.

33. $(3 + i\sqrt{2})(1 + \sqrt{-8})$. **34.** $(3 + i\sqrt{2})(2 - \sqrt{-18})$.

35. $(4 - \sqrt{-18})(2 - \sqrt{-2})$. **36.** $(-1 + i\sqrt{3})^3$.

37. $\left(-\frac{1}{2} - i\frac{\sqrt{3}}{2}\right)^3$. **38.** $\left(\frac{1}{2} + i\frac{\sqrt{3}}{2}\right)^3$.

39. $\left(\frac{\sqrt{2}}{2} + i\frac{\sqrt{2}}{2}\right)^2$. **40.** $\left(-\frac{\sqrt{3}}{2} - \frac{1}{2}i\right)^3$.

41. $\left(\frac{\sqrt{3}}{2} - \frac{1}{2}i\right)^3$. **42.** $\left(-\frac{\sqrt{2}}{2} + i\frac{\sqrt{2}}{2}\right)^2$.

43. $\frac{2-3i}{3+2i}$. **44.** $\frac{3+2i}{2-i}$. **45.** $\frac{2+i}{2-3i}$. **46.** $\frac{1+i}{1-i}$.

47. $\frac{6i-2}{3-2i}$. **48.** $\frac{2-5i}{3i}$. **49.** $\frac{2i+3}{4i}$. **50.** $\frac{2i}{6+4i}$.

51. $\frac{1}{i}$. **52.** $\frac{3}{2i}$. **53.** $\frac{3}{2+i}$. **54.** $\frac{3+i}{2i}$. **55.** $\frac{3+\sqrt{-4}}{2-\sqrt{-9}}$.

56. $\frac{3-\sqrt{-16}}{1+\sqrt{-4}}$. **57.** $\frac{1}{2+\sqrt{-3}}$. **58.** $\frac{3+\sqrt{-3}}{2-\sqrt{-3}}$.

59. $\frac{3-\sqrt{-5}}{4+2\sqrt{-5}}$. **60.** $(2+3i)^{-2}$. **61.** $(2i-5)^{-1}$.

62. $(4 - \sqrt{-18})^{-2}$. **63.** $(3 - \sqrt{-2})^{-1}$.

64. $(2 - \sqrt{-3})^{-1}$.

Write the reciprocal of each number in standard form:

65. $1 + i$. **66.** $5 - 3i$. **67.** $2i - 3$. **68.** $3i$.

69. $-4i$. **70.** $2 - \sqrt{-5}$.

71. $\sqrt{-3} + 2$. **72.** $\sqrt{-3} - 2$.

73. Find the value of $2x^3 - 13x^2 + 32x - 13$ if $x = 3 - 2i$.

74. Find the value of $3x^3 - 4x^2 + 2x + 4$ if $x = 1 - i$.

Prove each of the following:

75. The conjugate of the sum of any two complex numbers equals the sum of their conjugates.

76. The conjugate of the difference of any two complex numbers equals the difference of their conjugates.

77. The conjugate of the product of any two complex numbers equals the product of their conjugates.

78. The reciprocal of the conjugate of a non-zero complex number equals the conjugate of its reciprocal.

79. The conjugate of the quotient of one complex number by a non-zero complex number equals the quotient of their conjugates.

80. A non-zero complex number is equal to the negative of its conjugate if and only if it is pure imaginary.

107. Graphical Representation of Complex Numbers

Let us introduce rectangular coordinates into the plane. The complex number whose standard form is $x + yi$ is usually represented graphically by the point whose rectangular coordinates are x and y. Thus (Fig. 42), $3 + 2i$ is represented by the point whose rectangular coordinates are (3,2). This is why the standard form is also called the rectangular form of the complex number. It follows that all real numbers lie on the x-axis and all pure imaginary numbers on the y-axis. The complex numbers thus correspond to all points of the plane.

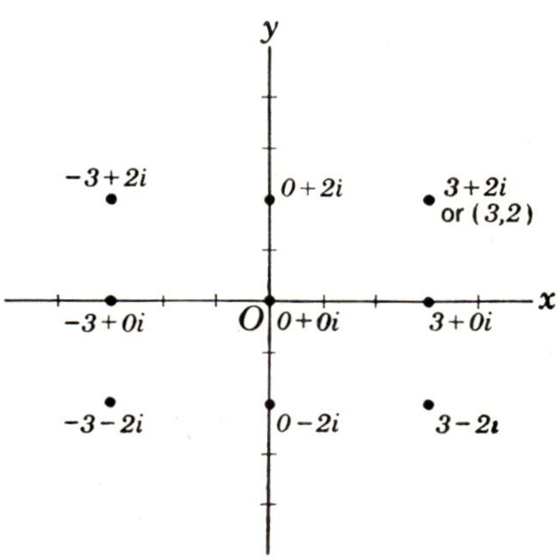

Fig. 42.

Another useful graphical representation of complex numbers is in terms of vectors, or directed line-segments. The number $x + iy$ is represented (Fig. 43) by the vector pointing from the origin to the point with rectangular coordinates (x,y). Thus $3 + 2i$ is pictured by a vector (arrow) pointing from the origin to the point whose rectangular coordinates are (3,2). Vectors are useful for the representation of quantities involving direction as well as magnitude, such as velocities and forces.

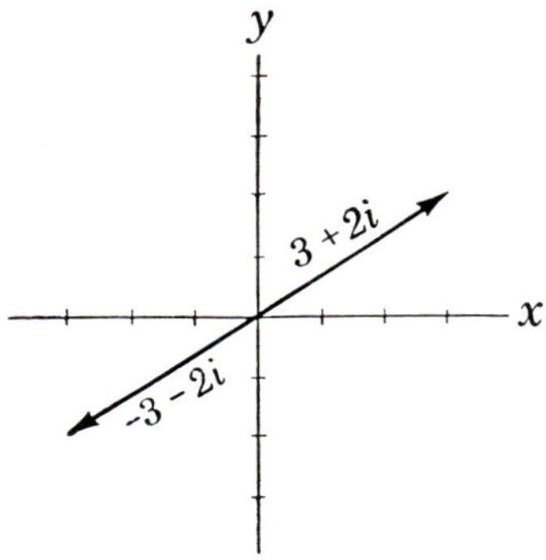

Fig. 43.

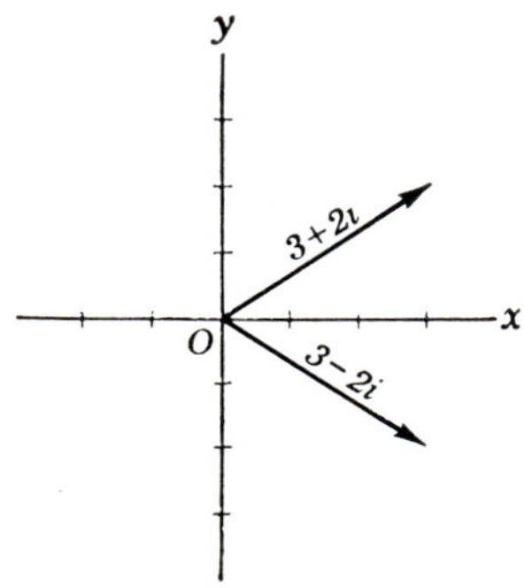

Fig. 44.

The negative of a complex number will be represented as a vector of the same length but opposite direction (Fig. 43). The conjugate of a complex number will be represented as a vector which is its reflection across the x-axis (Fig. 44).

EXERCISES

Represent each number graphically:

1. $4 - 3i$. **2.** $5 + i$. **3.** $-3 + 2i$.
4. $-4 - 3i$. **5.** 3. **6.** -3.
7. $3i$. **8.** $-3i$. **9.** 0.
10. $-3 + i\sqrt{2}$. **11.** $3i - 2$. **12.** $-\sqrt{2} - i\sqrt{3}$.

On the same coordinate system, plot the given number, the negative of the given number, and the conjugate of the given number:

13. $2 + 5i$. **14.** $4 - 2i$. **15.** $-2 + 5i$.
16. $-4 - i$. **17.** 5. **18.** $5i$.
19. -4. **20.** $-4i$. **21.** $2i - \sqrt{3}$.
22. $\sqrt{3} - 2i$. **23.** 0. **24.** $-\sqrt{5} - \frac{3}{2}i$.

★**25.** Prove that two non-zero complex numbers $a + bi$ and $c + di$ lie on a straight line through the origin if and only if their quotient is real.

108. Graphical Significance of Addition of Complex Numbers

The vector representing the sum of $a + bi$ and $c + di$ is the vector $\overrightarrow{OP}$ along the diagonal of the parallelogram whose adjacent sides are the vectors representing $a + bi$ and $c + di$, respectively (Fig. 45); for from Fig. 45 we see that the coordinates of P are

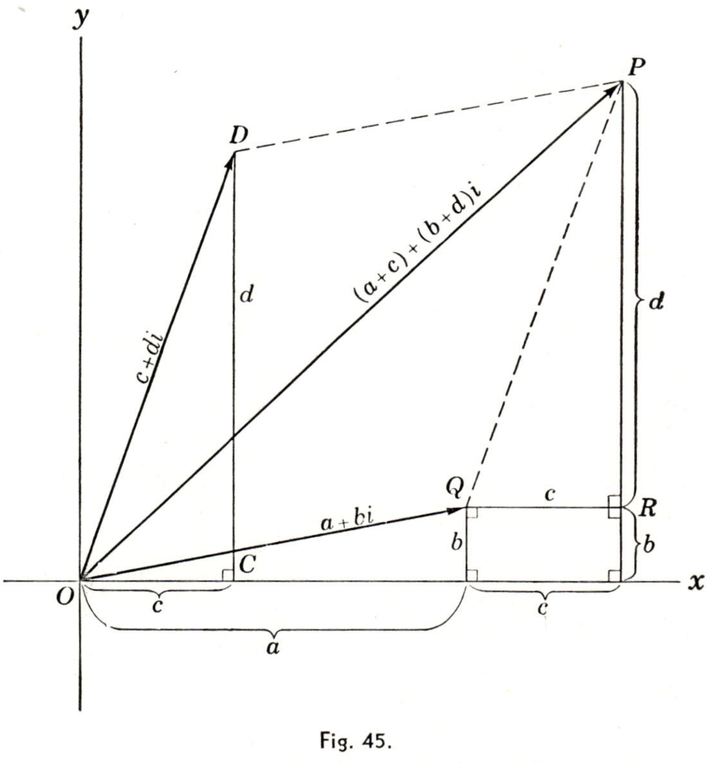

Fig. 45.

$(a + c,\ b + d)$, since right triangles OCD and QRP are congruent. Hence $\overrightarrow{OP}$ is the vector representing $(a + c) + (b + d)i$.

From the parallelogram law for the composition of vectors, we see that, if two vectors in a plane are represented by complex numbers, their resultant is represented by the sum of these complex numbers.

To subtract $c + di$ from $a + bi$ graphically, we add $-c - di$

to $a + bi$ graphically. Thus in Fig. 46, we subtract $-2 + 3i$ from $4 + 2i$ graphically by adding $2 - 3i$ to $4 + 2i$, obtaining $6 - i$.

EXERCISES

Construct each sum or difference graphically:

1. $(2 + 5i) + (4 + i)$.
2. $(5 + i) + (1 - 3i)$.
3. $(-3 + 2i) + (-1 - 4i)$.
4. $(-3 - 2i) + (2 - 3i)$.
5. $(2 + 3i) - (-3 + 4i)$.
6. $(5 + i) - (-1 + 4i)$.
7. $(-4 + i) - (5 + 2i)$.
8. $(-1 - 5i) - (4 - 2i)$.

9. Discuss the exceptional case where the given numbers $a + bi$ and $c + di$ are represented by vectors lying along the same straight line through the origin. Make up rules for constructing their sum and difference geometrically in this case.

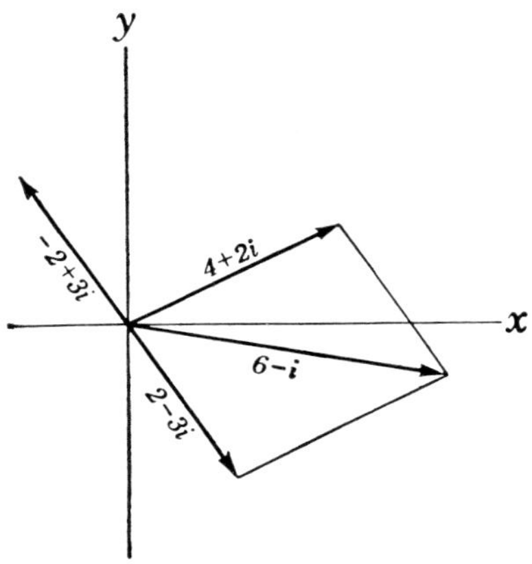

Fig. 46.

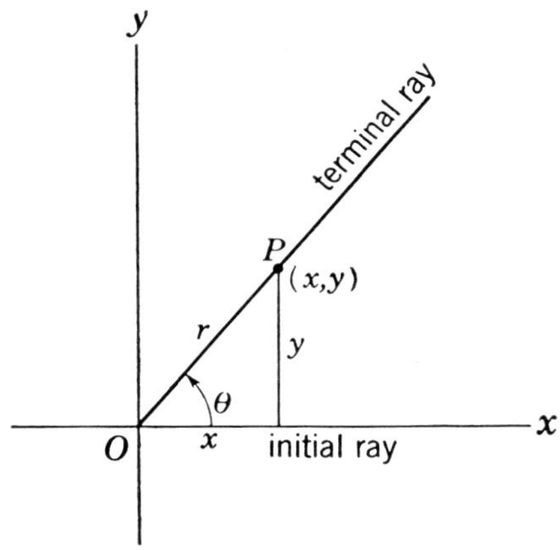

Fig. 47.

109. Review of Trigonometric Functions

We review briefly* some needed material concerning the trigonometric functions of a general angle. Recall that a general angle (of any number of degrees, positive, zero, or negative) represents a rotation, clockwise if negative, counterclockwise if positive. To define the trigonometric functions of a general angle θ, we introduce rectangular coordinates into the plane in such a way that the vertex of the angle is at the origin and the initial ray along the positive side of the x-axis. (See Fig. 47.)

* For a more extended treatment see M. Richardson, *Fundamentals of Mathematics,* Macmillan, 1941, chapter XII, or M. Richardson, *Plane and Spherical Trigonometry,* Macmillan, 1950.

We choose any point P we please on the terminal ray of the angle. If the coordinates of the point P are (x,y), then its distance r from the origin is given by $r = \sqrt{x^2 + y^2}$. We then define

$$\sin \theta = y/r, \cos \theta = x/r, \tan \theta = y/x.$$

Recall that the sides of a 30°-60°-90° triangle are in the ratio $1:\sqrt{3}:2$, and that the sides of a 45°-45°-90° triangle are in the ratio $1:1:\sqrt{2}$.

Example 1. To obtain the sine, cosine, and tangent of 150° pick a point P on the terminal ray 2 units from the origin (Fig. 48). Dropping a perpendicular from P to the x-axis, we form a so-called reference triangle, which is, in this example, a 30°-60°-90° triangle. Hence the coordinates of P are $(-\sqrt{3},1)$. Therefore $\sin 150° = 1/2$, $\cos 150° = -\sqrt{3}/2$, $\tan 150° = +1/-\sqrt{3} = -\sqrt{3}/3$.

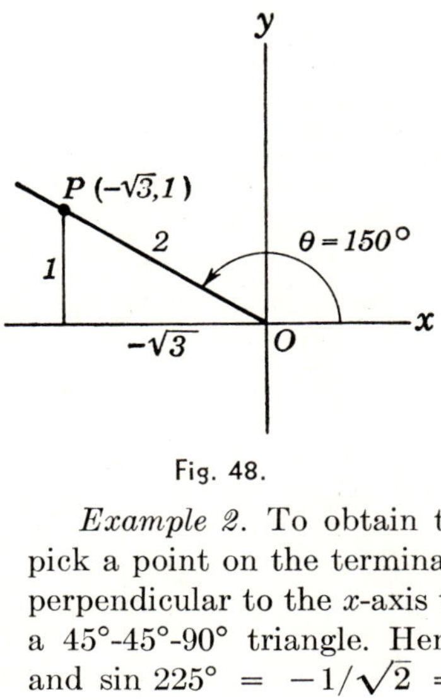

Fig. 48.

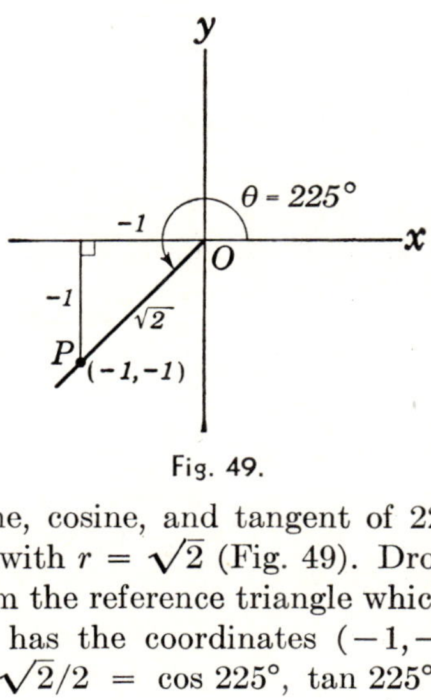

Fig. 49.

Example 2. To obtain the sine, cosine, and tangent of 225°, pick a point on the terminal ray with $r = \sqrt{2}$ (Fig. 49). Drop a perpendicular to the x-axis to form the reference triangle which is a 45°-45°-90° triangle. Hence P has the coordinates $(-1,-1)$, and $\sin 225° = -1/\sqrt{2} = -\sqrt{2}/2 = \cos 225°$, $\tan 225° = -1/-1 = +1$.

If the terminal ray of the angle θ lies along one of the coordinate axes, there is no reference triangle, but we do not need one.

Example 3. To find the sine, cosine, and tangent of 270°, we pick a point P with coordinates $(0,-1)$ on the terminal ray (Fig. 50). Then $r = 1$ and $\sin 270° = -1/1 = -1$, $\cos 270° = 0/1 = 0$, $\tan 270° = -1/0$ which does not exist.

The trigonometric functions of any angle which has a reference angle are the same as those of the reference angle except for sign. The sign may be determined by the quadrant in which the angle terminates (Fig. 51). The sine is positive in quadrants *I* and *II*, negative in *III* and *IV*. The cosine is positive in *I* and *IV*, negative in *II* and *III*. The tangent is positive in *I* and *III*, negative in *II* and *IV*.

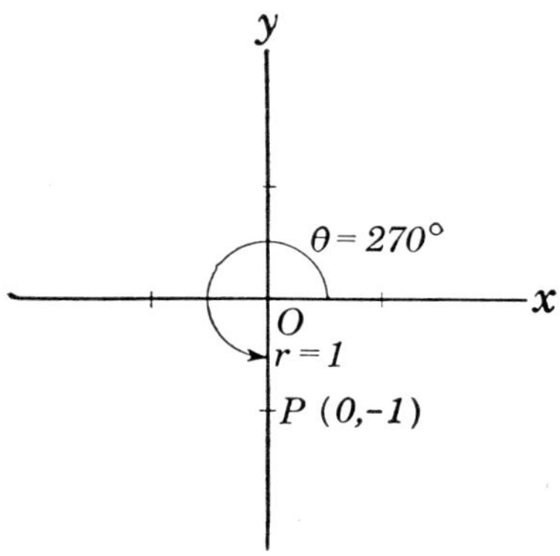

Fig. 50.

The trigonometric functions of certain angles are conveniently remembered as follows:

	0°	30°	45°	60°	90°
sin	$\sqrt{0}/2$	$\sqrt{1}/2$	$\sqrt{2}/2$	$\sqrt{3}/2$	$\sqrt{4}/2$
cos	$\sqrt{4}/2$	$\sqrt{3}/2$	$\sqrt{2}/2$	$\sqrt{1}/2$	$\sqrt{0}/2$

The sin and cos of angles which are equal to 30°, 45°, 60°, or have 30°, 45°, 60° as reference angles, or angles which terminate along

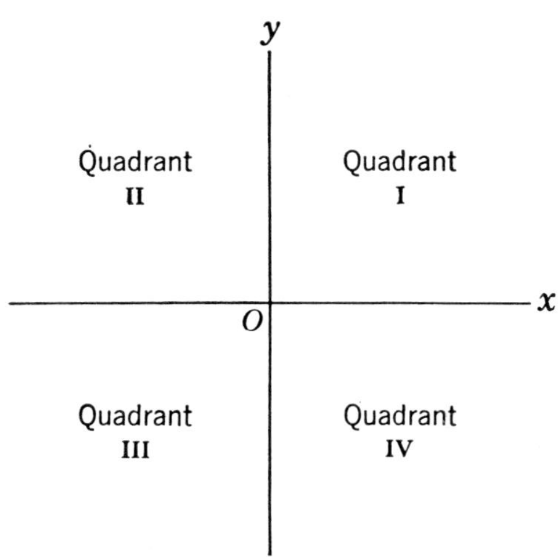

Fig. 51.

the axes, may be determined without tables as above. For other angles, we use table II, as follows.

Example 4. Find the sine and cosine of 160°. The reference angle is 20° and the angle 160° is in quadrant *II*. Hence sin 160° = sin 20° = 0.3420 and cos 160° = −cos 20° = −0.9397.

Angles which differ by integral multiples of 360° have the same trigonometric functions, since they have the same terminal ray. For example, the trigonometric functions of 750°, 390°, and 30° are the same.

EXERCISES

Find, without tables, the values of the sine, cosine, and tangent of the following angles:

1. 30°.	**2.** 45°.	**3.** 60°.	**4.** 120°.
5. 135°.	**6.** 150°.	**7.** 210°.	**8.** 225°.
9. 240°.	**10.** 300°.	**11.** 315°.	**12.** 330°.
13. 390°.	**14.** 405°.	**15.** −210°.	**16.** −750°.
17. 0°.	**18.** 90°.	**19.** 180°.	**20.** 270°.
21. 360°.		**22.** 450°.	

Find, using table II, the values of the sine and cosine of the following angles:

23. 20°.	**24.** 80°.	**25.** 110°.	**26.** 170°.
27. 200°.	**28.** 255°.	**29.** 290°.	**30.** 345°.
31. 380°.	**32.** −25°.	**33.** −130°.	**34.** 735°.
35. 516°.	**36.** 247°.	**37.** −235°.	**38.** 473°.
39. 605°.		**40.** 710°.	

110. Trigonometric or Polar Form of a Complex Number

A point may be specified by its x- and y-coordinates, which give directions for reaching it from the origin. But it may be specified equally well by stipulating its direction from the origin and the distance r one must go in that direction. The direction may be specified by the angle θ formed with the positive side of the x-axis (Fig. 52). The numbers (r,θ) are called **polar coordinates** of the point P, to distinguish them from the rectangular coordinates (x,y). The two systems of coordinates are related

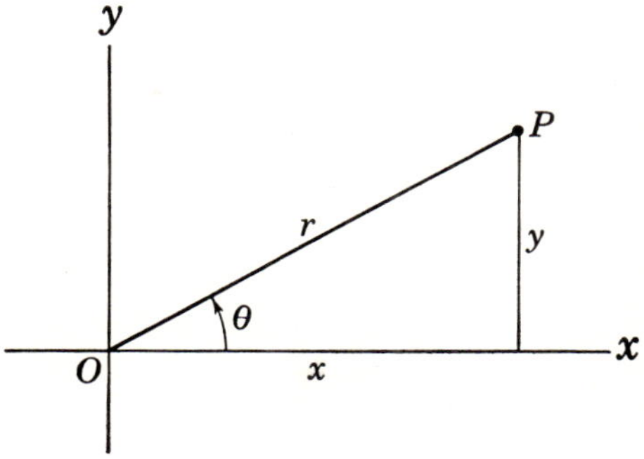

Fig. 52.

by the following equations, which are obvious from Fig. 52:

(1) $$r = \sqrt{x^2 + y^2},\ \sin\theta = y/r,\ \cos\theta = x/r.$$

(2) $$x = r\cos\theta,\ y = r\sin\theta.$$

Equations (2) enable us to determine x and y when r and θ are given. Equations (1) enable us to determine r and θ when x and y are given, for knowledge of $\sin\theta$ and $\cos\theta$ determine* the value of θ if $0° \leqq \theta < 360°$.

Using (2) we may write

$$x + yi = r\cos\theta + ir\sin\theta$$

or

(3) $$x + yi = r(\cos\theta + i\sin\theta).$$

The right member of (3) is called the **polar form** or **trigonometric form** of the complex number whose standard form is $x + yi$. The quantity r, which is always $\geqq 0$, is called the **modulus** or **length** or **absolute value** of the complex number. The angle θ is called the **amplitude** or **argument** or **angle** of the complex number. Note that the amplitude is not uniquely determined, for any integral multiple of 360° may be added without altering the sine and cosine. Hence the amplitude may be $\theta + k \cdot 360°$ where $k = 0, \pm 1, \pm 2, \pm 3, \ldots$.

The polar form exhibits clearly the polar coordinates of the point P, while the standard or rectangular form exhibits the rectangular coordinates of P. Hence equations (1) enable us to convert the standard form into the polar form, and equations (2) enable us to convert the polar form into the standard form.

Example 1. Write $-1 + i$ in polar form, using the smallest non-negative value of the amplitude.

Solution. $r = \sqrt{(-1)^2 + 1^2} = \sqrt{2}$, $\sin\theta = 1/\sqrt{2} = \sqrt{2}/2$, $\cos\theta = -1/\sqrt{2} = -\sqrt{2}/2$. Hence $\theta = 135°$. Therefore the polar form of $-1 + i = \sqrt{2}\,(\cos 135° + i\sin 135°)$.

Example 2. Write $2(\cos 300° + i\sin 300°)$ in standard form.

Solution. $\cos 300° = +1/2$, $\sin 300° = -\sqrt{3}/2$. Hence

$$2(\cos 300° + i\sin 300°) = 2\left(\frac{1}{2} + i\,\frac{-\sqrt{3}}{2}\right) = 1 - i\sqrt{3}.$$

* Many books suggest the use of $\tan\theta$ for this purpose. But $\tan\theta$, unassisted, does not determine the quadrant of θ, since $\tan(180° + \theta) = \tan\theta$; $\tan\theta$ may be used if the values of x and y are employed to determine the quadrant.

Example 3. Write the number -2 in polar form, using the smallest non-negative value of the amplitude.

Solution. In standard form, $-2 = -2 + 0i$. Hence $\theta = 180°$ and $r = 2$. Thus $-2 + 0i = 2(\cos 180° + i \sin 180°)$.

Example 4. Write $3(\cos 50° + i \sin 50°)$ in standard form.

Solution. From table II, we find $\cos 50° = 0.6428$ and $\sin 50° = 0.7660$, approximately. Hence $3(\cos 50° + i \sin 50°) = 3(0.6428 + 0.7660i) = 1.9284 + 2.2980i$, approximately.

If it is desired to exhibit all possible values of the amplitude, we may replace θ by $\theta + k \cdot 360°$ where $k = 0, \pm 1, \pm 2, \ldots$. Thus in example 1 above we would write

$$-1 + i = \sqrt{2}\,(\cos [135° + k \cdot 360°] + i \sin [135° + k \cdot 360°])$$

where k may be any integer whatever.

EXERCISES

Write each number in standard $(x + yi)$ form, and plot:

1. $2(\cos 30° + i \sin 30°)$.
2. $4(\cos 60° + i \sin 60°)$.
3. $8(\cos 405° + i \sin 405°)$.
4. $6(\cos 120° + i \sin 120°)$.
5. $1(\cos 135° + i \sin 135°)$.
6. $5(\cos 150° + i \sin 150°)$.
7. $3(\cos 90° + i \sin 90°)$.
8. $4(\cos 270° + i \sin 270°)$.
9. $4(\cos 180° + i \sin 180°)$.
10. $6(\cos 210° + i \sin 210°)$.
11. $3(\cos 0° + i \sin 0°)$.
12. $2(\cos 360° + i \sin 360°)$.
13. $4(\cos 225° + i \sin 225°)$.
14. $8(\cos 600° + i \sin 600°)$.
15. $5[\cos (-450°) + i \sin (-450°)]$.
16. $2(\cos 660° + i \sin 660°)$.
17. $2[\cos (-390°) + i \sin (-390°)]$.
18. $8[\cos (-45°) + i \sin (-45°)]$.

Write each of the following in polar form, using the value of θ lying in the interval $0° \leqq \theta < 360°$, and plot:

19. $1 + i\sqrt{3}$.
20. $2\sqrt{3} + 2i$.
21. $4 + 4i$.
22. $-2\sqrt{3} + 2i$.
23. $-2 + 2i\sqrt{3}$.
24. $3 + 3i$.
25. $-\sqrt{2} + i\sqrt{2}$.
26. $-3\sqrt{3} - 3i$.
27. $3i$.
28. $-5\sqrt{2} - 5i\sqrt{2}$.
29. 6.
30. $-5 - 5i\sqrt{3}$.
31. -3.
32. $4\sqrt{2} - 4i\sqrt{2}$.
33. $-5i$.
34. $3 - 3i\sqrt{3}$.
35. $2\sqrt{3} - 2i$.
36. $-1 - i$.

Write each of the following in polar form, exhibiting all possible values of θ, and plot:

37. 8. **38.** -16. **39.** $8i$. **40.** $-16i$.
41. $\sqrt{3} - i$. **42.** $-2\sqrt{2} - 2i\sqrt{2}$.

Write each of the following in standard form, approximately, using table II, and plot:

43. $2(\cos 20° + i \sin 20°)$. **44.** $3(\cos 170° + i \sin 170°)$.
45. $1(\cos 190° + i \sin 190°)$. **46.** $2(\cos 350° + i \sin 350°)$.

47. Find an amplitude of (*a*) any positive real number; (*b*) any negative real number; (*c*) any pure imaginary number yi where $y > 0$; (*d*) any pure imaginary number yi where $y < 0$.

48. Prove that the quotient of any non-zero complex number divided by its conjugate has modulus equal to 1.

49. Prove that any complex number $a + bi$ with modulus equal to 1 can be expressed as a quotient of a non-zero complex number divided by its conjugate. (*Hint:* if $a + bi = -1$, we have $-1 = \dfrac{0 + i}{0 - i}$; if $a + bi \neq -1$, we have $a + bi = \dfrac{1 + a + bi}{1 + a - bi}$.)

50. Prove that the product of any complex number with its conjugate is equal to the square of its modulus.

51. Prove that the modulus of the product of any two complex numbers is equal to the product of their moduli.

111. Graphical Meaning of Multiplication and Division of Complex Numbers

We have seen in section 108 that addition of complex numbers corresponds geometrically to the parallelogram law for the composition of vectors. To see what the geometric significance of multiplication of complex numbers is, we need the polar form. Let any two numbers be written in their polar forms as

$$r(\cos \alpha + i \sin \alpha), \tag{1}$$

$$R(\cos \beta + i \sin \beta). \tag{2}$$

Multiplying them we get

$$rR[(\cos \alpha \cos \beta - \sin \alpha \sin \beta) + i(\sin \alpha \cos \beta + \cos \alpha \sin \beta)]. \tag{3}$$

In trigonometry, it is proved that the first parenthesis is equal to $\cos(\alpha + \beta)$ while the second parenthesis is equal to $\sin(\alpha + \beta)$. Hence the product of (1) and (2) may be written as

$$rR[\cos(\alpha + \beta) + i \sin(\alpha + \beta)] \tag{4}$$

But this is in polar form. Hence we conclude that *the modulus of the product of two complex numbers is the product of their moduli, while the amplitude of the product is the sum of their amplitudes* (Fig. 53).

Example 1. The product of $2(\cos 20° + i \sin 20°)$ and $3(\cos 70° + i \sin 70°)$ is equal to $6(\cos 90° + i \sin 90°)$ or $6i$.

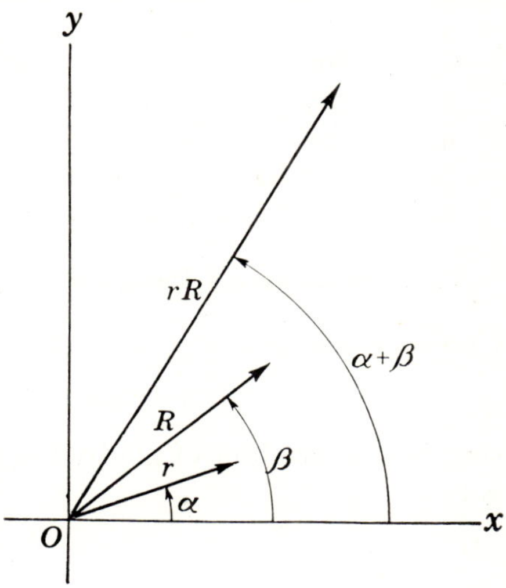

Fig. 53.

By definition, the quotient of one complex number $r_1(\cos \alpha + i \sin \alpha)$ by another $r_2(\cos \beta + i \sin \beta)$ is the complex number $R(\cos \theta + i \sin \theta)$ such that the second multiplied by the third equals the first. Hence

$$r_2R[\cos(\beta + \theta) + i \sin(\beta + \theta)] = r_1(\cos \alpha + i \sin \alpha)$$

which is true if $\quad r_2R = r_1$ and $\beta + \theta = \alpha$.

Hence we may write $\quad R = \dfrac{r_1}{r_2}$ and $\theta = \alpha - \beta$.

Therefore

$$(5) \quad \frac{r_1(\cos \alpha + i \sin \alpha)}{r_2(\cos \beta + i \sin \beta)} = \frac{r_1}{r_2}[\cos(\alpha - \beta) + i \sin(\alpha - \beta)].$$

That is, *the modulus of the quotient of two complex numbers is the quotient of their moduli, and the amplitude of the quotient is the difference of their amplitudes.*

Example 2. $\dfrac{6(\cos 50° + i \sin 50°)}{2(\cos 20° + i \sin 20°)} = 3(\cos 30° + i \sin 30°) = \dfrac{3\sqrt{3}}{2} + \dfrac{3}{2}i.$

EXERCISES

(*a*) *Perform the following multiplications and divisions by the above methods, leaving answers in polar form;* (*b*) *rewrite the answers in standard form:*

1. $3(\cos 110° + i \sin 110°) \cdot 2(\cos 70° + i \sin 70°)$.
2. $4(\cos 170° + i \sin 170°) \cdot 3(\cos 190° + i \sin 190°)$.
3. $3(\cos 10° + i \sin 10°) \cdot 4(\cos 125° + i \sin 125°)$.
4. $6(\cos 80° + i \sin 80°) \cdot 2(\cos 40° + i \sin 40°)$.
5. $6(\cos 110° + i \sin 110°) \div 3(\cos 20° + i \sin 20°)$.
6. $8(\cos 265° + i \sin 265°) \div 4(\cos 40° + i \sin 40°)$.
7. $10(\cos 155° + i \sin 155°) \div 2(\cos 35° + i \sin 35°)$.
8. $12(\cos 75° + i \sin 75°) \div 4(\cos 105° + i \sin 105°)$.

112. De Moivre's Theorem

Consider the square of a complex number

$$r(\cos \theta + i \sin \theta), \tag{1}$$

that is, the product of (1) by itself. By the preceding section we get

$$[r(\cos \theta + i \sin \theta)]^2 = r^2(\cos 2\theta + i \sin 2\theta). \tag{2}$$

Multiplying (2) by (1), we get

$$[r(\cos \theta + i \sin \theta)]^3 = r^3(\cos 3\theta + i \sin 3\theta), \tag{3}$$

and so on. This clearly suggests the following theorem.

De Moivre's Theorem: $[r(\cos \theta + i \sin \theta)]^n = r^n(\cos n\theta + i \sin n\theta)$.

For positive integral powers of n this can be proved by mathematical induction (see chapter XVI). It can also be proved for other kinds of exponents but we shall not give any proof here.

Example 1. $[2(\cos 30° + i \sin 30°)]^3 = 8(\cos 90° + i \sin 90°) = 8i$.

Example 2. $(1 + i)^4 = [\sqrt{2}\,(\cos 45° + i \sin 45°)]^4 = 4(\cos 180° + i \sin 180°) = -4$.

Abraham de Moivre (1667–1754) made important contributions to various branches of mathematics; he is especially famous for his work in the theory of probability.

EXERCISES

Find the indicated powers by de Moivre's theorem, leaving answers in polar form except where the standard form can be found without use of tables:

1. $[3(\cos 60° + i \sin 60°)]^3$. **2.** $[2(\cos 120° + i \sin 120°)]^3$.
3. $[2(\cos 50° + i \sin 50°)]^3$. **4.** $[3(\cos 45° + i \sin 45°)]^3$.
5. $[2(\cos 15° + i \sin 15°)]^5$. **6.** $[2(\cos 15° + i \sin 15°)]^4$.
7. $[2(\cos 45° + i \sin 45°)]^6$. **8.** $[\frac{1}{2}(\cos 15° + i \sin 15°)]^6$.
9. $[\frac{1}{2}(\cos 30° + i \sin 30°)]^4$. **10.** $[2(\cos 72° + i \sin 72°)]^5$.
11. $(1 + i)^8$. **12.** $(\sqrt{3} + i)^6$. **13.** $(1 - i)^3$.
14. $(2\sqrt{3} - 2i)^3$. **15.** $\left(\frac{\sqrt{3}}{2} + \frac{1}{2}i\right)^{12}$.
16. $(-3\sqrt{3} + 3i)^3$. **17.** $(\sqrt{2} + i\sqrt{2})^4$.
18. $(-\sqrt{3} - i)^6$. **19.** $\left(\frac{-\sqrt{3}}{2} + \frac{1}{2}i\right)^6$.
20. $\dfrac{(\sqrt{3} + i)^6(-1 + i)^3}{(1 - i)^4}$.

★21. Prove by mathematical induction that de Moivre's theorem is true for all positive integral values of n.

★22. By means of de Moivre's theorem and the binomial theorem derive the formulas

(*a*) $\cos 2\theta = \cos^2 \theta - \sin^2 \theta$, $\sin 2\theta = 2 \sin \theta \cos \theta$.

(*b*) $\cos 3\theta = 4 \cos^3 \theta - 3 \cos \theta$, $\sin 3\theta = 3 \sin \theta - 4 \sin^3 \theta$.

113. The *n*th Roots of a Complex Number

By means of de Moivre's theorem we can find the nth roots of any complex number.

Example 1. Find the cube roots of $8i$.

Solution. Expressing $8i$ in polar form, we have $8i = 8(\cos 90° + i \sin 90°)$. But $90°$ can be replaced by $90° + k \cdot 360°$ where k may be any integer, positive, negative, or zero. Hence we may write

$$8i = 8(\cos [90° + k \cdot 360°] + i \sin [90° + k \cdot 360°]). \tag{1}$$

Suppose $r(\cos \theta + i \sin \theta)$ is a cube root of $8i$. Then

$$[r(\cos \theta + i \sin \theta)]^3 = 8i \tag{2}$$

by definition. By de Moivre's theorem, the left member of (2) may be written as $r^3(\cos 3\theta + i \sin 3\theta)$. Hence, using (1) we get

$$r^3(\cos 3\theta + i \sin 3\theta) = 8(\cos [90° + k \cdot 360°] + i \sin [90° + k \cdot 360°]). \tag{3}$$

Therefore

$$r^3 = 8 \tag{4}$$

and

$$3\theta = 90° + k \cdot 360°. \tag{5}$$

From (4) we get $r = \sqrt[3]{8} = 2$, since r is necessarily a positive real number and therefore must be the principal cube root of 8. From (5), we get $\theta = 30° + k \cdot 120°$. Hence,

$$(6) \quad r(\cos\theta + i\sin\theta) = 2(\cos[30° + k \cdot 120°] + i\sin[30° + k \cdot 120°]).$$

If $k = 0$, (6) yields $2(\cos 30° + i\sin 30°) = 2\left(\frac{\sqrt{3}}{2} + i\frac{1}{2}\right) = \sqrt{3} + i$.

If $k = 1$, (6) yields $2(\cos 150° + i\sin 150°) = 2\left(-\frac{\sqrt{3}}{2} + i\frac{1}{2}\right) = -\sqrt{3} + i$.

If $k = 2$, (6) yields $2(\cos 270° + i\sin 270°) = 2(0 - 1i) = -2i$.

Taking $k = 3, 4, 5, \ldots$ will yield only repetitions of these answers, since adding 360° does not alter the values of the trigonometric functions. Hence, $8i$ has three distinct cube roots, namely $\sqrt{3} + i$, $-\sqrt{3} + i$, and $-2i$. The student should cube each of them, as a check, and should obtain $8i$ in each case.

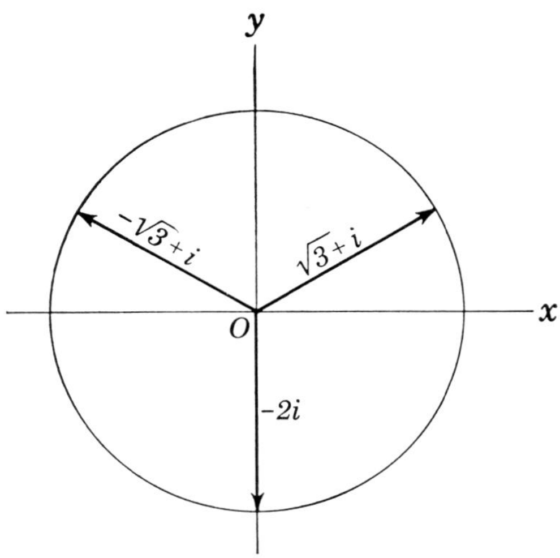

Fig. 54.

Note that if the three cube roots of $8i$ are plotted, the vectors which represent them graphically divide the circle of radius 2 with center at O into 3 equal sectors (Fig. 54).

In general, to obtain the nth roots of a given complex number, we first express the number in polar form

$$(7) \quad R(\cos \alpha + i \sin \alpha) = R(\cos [\alpha + k \cdot 360°] + i \sin [\alpha + k \cdot 360°]).$$

Let $r(\cos \theta + i \sin \theta)$ be an nth root of (7). Then, by definition of nth root,

$$[r(\cos \theta + i \sin \theta)]^n = R(\cos [\alpha + k \cdot 360°] + i \sin [\alpha + k \cdot 360°]).$$

By de Moivre's theorem, this becomes

$$r^n(\cos n\theta + i \sin n\theta) = R(\cos [\alpha + k \cdot 360°] + i \sin [\alpha + k \cdot 360°]).$$

Hence

$$(8) \qquad r^n = R, \text{ or } r = \sqrt[n]{R}$$

and

$$(9) \qquad n\theta = \alpha + k \cdot 360°, \text{ or } \theta = \frac{\alpha}{n} + k \cdot \frac{360°}{n}.$$

Therefore the nth roots of $R(\cos \alpha + i \sin \alpha)$ are given by the expression

$$(10) \qquad \sqrt[n]{R}\left(\cos\left[\frac{\alpha}{n} + k\frac{360°}{n}\right] + i \sin\left[\frac{\alpha}{n} + k\frac{360°}{n}\right]\right).$$

It is clear that any successive n integral values of k will yield n different nth roots, while further integral values of k will yield repetitions. Hence we can get the n distinct nth roots by substituting $k = 0, 1, 2, \ldots, n - 1$ successively in (10). The vectors representing the n distinct nth roots should divide the circle of radius $\sqrt[n]{R}$ with center at O into n equal sectors.

EXERCISES

Find the indicated roots, leaving answers in polar form except where the standard form can be found without tables:

1. Square roots of $4i$. **2.** Square roots of $-i$.
3. Cube roots of $64i$. **4.** Cube roots of $-27i$.
5. Cube roots of -8. **6.** Cube roots of 8.
7. Sixth roots of 1. **8.** Sixth roots of -64.
9. Fifth roots of 32. **10.** Fourth roots of -81.
11. Cube roots of $-4\sqrt{2} + 4i\sqrt{2}$.
12. Cube roots of $-4 - 4i\sqrt{3}$.
13. Cube roots of 1. **14.** Cube roots of -1.
15. Fourth roots of $-8 + i\,8\sqrt{3}$.

16. Fourth roots of 81. **17.** Fourth roots of 16.
18. Fourth roots of $-8 - 8i\sqrt{3}$.
19. Square roots of $8 + 8i\sqrt{3}$.
20. Square roots of $-2 + 2i\sqrt{3}$.
21. Sixth roots of $-64i$. **22.** Square roots of $-4i$.

23. (*a*) Show that if we let $\omega = -\frac{1}{2} + i\frac{\sqrt{3}}{2}$, one of the imaginary cube roots of 1, then the other imaginary cube root of 1 is ω^2.

(*b*) Show that if z is any complex number and $\sqrt[3]{z}$ denotes any one of its cube roots, then the other two cube roots of z are $\omega\sqrt[3]{z}$ and $\omega^2\sqrt[3]{z}$. (*Hint:* consider the polar forms of $\sqrt[3]{z}$ and ω and the geometric meaning of multiplication of complex numbers.)

114. Solution of Pure Equations of *n*th Degree

By a **pure equation** of nth degree is meant a polynomial equation of nth degree which has no terms of degree lower than n except the constant term, that is, an equation which can be written in the form

$$(1) \qquad x^n = A, \text{ or } x^n - A = 0,$$

where A is any (complex) constant. Clearly, any root of this equation is an nth root of A. Hence there are n different roots which may be found by the method of the preceding section.

Example. Solve the equation $x^3 - 8i = 0$.

Solution. Any root of this equation is, by definition, a cube root of $8i$, since $x^3 = 8i$. As in example 1 of the preceding section, we find that the roots are $\sqrt{3} + i$, $-\sqrt{3} + i$, and $-2i$.

Note. Hitherto we have found imaginary roots of polynomial equations only if they were quadratic. In this section, we have been able to find them no matter what the degree, provided the equation is of the special type (1). The further study of the solution of equations will be taken up in the next chapter.

EXERCISES

Find all the roots of each of the following equations, express them in standard form if possible without use of tables, and check:

1. $x^3 = 27$. **2.** $x^4 = 81$. **3.** $x^3 + 8i = 0$.
4. $x^6 + 64 = 0$. **5.** $x^2 = 4i$. **6.** $x^4 + 16 = 0$.
7. $x^3 - 27i = 0$. **8.** $x^6 = 64$.
9. $x^4 + 8 - 8i\sqrt{3} = 0$. **10.** $x^2 = -18 + 18i\sqrt{3}$.
11. $x^3 + 4\sqrt{2} - 4i\sqrt{2} = 0$. **12.** $x^3 + 4 + 4i\sqrt{3} = 0$.

Chapter XIX

Theory of equations

115. The Fundamental Theorem of Algebra

A **polynomial equation in x** or a **rational integral equation in x** of the nth degree is one that can be written in the form

$$(1)\quad a_0x^n + a_1x^{n-1} + a_2x^{n-2} + \dots + a_{n-1}x + a_n = 0 \quad (a_0 \neq 0)$$

where the coefficients $a_0, a_1, \dots, a_n$ are constants and n is a positive integer. A **root** or **solution** of such an equation is a value of x which satisfies the equation; that is, a value such that the equation becomes a true statement when this value is substituted for x. To **solve** an equation* means to find all its roots. Equations of degree 1, 2, 3, 4 are called **linear, quadratic, cubic,** and **quartic,** respectively. So far we have studied methods of solving equations of degree 1 and 2 only except for the pure equations of section 114. In this chapter, we shall attack the problem of solving equations of degree higher than 2. In (1) the coefficient of the highest power, a_0, is called the **leading coefficient,** while a_n is called the **constant term.**

To begin with, let us review the development of the complex number system, and see why we need complex numbers and no others.† We shall be guided by our desire that every polynomial equation in x possess a root. Suppose, however, that our number system contained only natural numbers or positive integers, no other numbers having yet been invented. Then,

* In this chapter the word *equation* will always mean *rational integral equation.*

† The student should review chapter I, especially sections 9–16, and familiarize himself thoroughly with the classification of numbers in section 16 without delay.

while the linear equation $x + 3 = 5$ would have the root 2, the linear equation

$$(1) \qquad x + 5 = 3$$

would have no root in our system of numbers. Therefore we would invent negative integers in order to provide a root for equations like (1). The equation

$$(2) \qquad x + 3 = 3$$

would still have no root among the positive or negative integers. Therefore we would invent the number zero. The linear equation

$$(3) \qquad 2x - 1 = 0$$

would still have no root, even though our number system were now enlarged to include all integers, or whole numbers, since the only solution of (3) is 1/2, which is not an integer. This would lead us to develop fractions or rational numbers, that is, numbers which can be expressed as quotients of integers. Even among the extensive system of all rational numbers, the simple quadratic equation

$$(4) \qquad x^2 - 2 = 0$$

would have no root, since the only roots are $\pm \sqrt{2}$, which are not rational as we proved in section 13. Hence we invent the irrational numbers, which together with the rational numbers compose the system of real numbers. The entire system of real numbers still contains no root of the quadratic equation

$$(5) \qquad x^2 + 4 = 0$$

as was proved in section 16. Hence we invent the pure imaginary numbers, such as $\pm 2i$. The quadratic equation

$$(6) \qquad x^2 + 2x + 4 = 0$$

has no root among either the real or the pure imaginary numbers, since its roots $-1 \pm i\sqrt{3}$ are neither real nor pure imaginary. Therefore we are led to invent the system of complex numbers.

So far, we have been impelled to enlarge our system of numbers again and again by considering only linear and quadratic equations. The student may well dread the possibility of having to invent still further kinds of numbers if we consider equations of degree higher than the second. That this will not be necessary was first proved by K. F. Gauss, a great German mathematician, in 1799, although it was suspected before. What Gauss proved is known as the fundamental theorem of algebra, and may be stated as follows.

FUNDAMENTAL THEOREM OF ALGEBRA. *Every polynomial equation of degree 1 or more has a root among the complex numbers (no matter how high its degree and no matter what complex numbers are its coefficients).*

For example, it is guaranteed that the equation $x^{1945} + 1918x^{1898} + 1865x^{1812} + 1776x^{1492} + \frac{7 + 3i}{2}x^7 - \frac{\sqrt{3}}{2}x + 1 = 0$ has a root among the complex numbers. A proof of the fundamental theorem is too difficult to be given here.* Unfortunately, it does not tell us how to find a root, but it does assure us of its existence. Therefore we do not find it necessary to invent more numbers in order to provide every polynomial equation with a root. The student should familiarize himself without delay with the facts about the various kinds of numbers discussed in sections 11–16 and with the classification of the complex numbers in section 16.

116. The Remainder Theorem and the Factor Theorem

If we divide the polynomial $F(x) = x^2 - 5x + 10$ by $x - 2$ we get the quotient $x - 3$ and the remainder 4:

$$
\begin{array}{r|l}
 & x \quad - 3 \\
x - 2 & x^2 - 5x + 10 \\
 & x^2 - 2x \\
 & \quad - 3x + 10 \\
 & \quad - 3x + \ \ 6 \\
 & \qquad\qquad 4.
\end{array}
$$

Also $F(2) = 2^2 - 5 \cdot 2 + 10 = 4$, the same as the remainder. That this is no mere coincidence is guaranteed by the following theorem:

* See, for example, L. E. Dickson, *Elementary Theory of Equations,* Wiley, 1914, or L. Weisner, *Introduction to the Theory of Equations,* Macmillan, 1938, or G. Birkhoff and S. MacLane, *A Survey of Modern Algebra,* Macmillan, 1941.

REMAINDER THEOREM. *If a polynomial $F(x)$ of degree 1 or more is divided by a divisor of the form $x - r$, then the remainder $R = F(r)$.*

Proof. By definition,* to divide $F(x)$ by any polynomial $D(x)$ means to find a quotient polynomial $Q(x)$ and a remainder polynomial $R(x)$ such that

$$(a)\ F(x) = D(x) \cdot Q(x) + R(x)$$

and

(b) the degree of $R(x)$ is less than the degree of $D(x)$.

Since our divisor $x - r$ is of degree 1, the remainder must be of degree zero, that is, a constant. Hence we may write

$$(1) \qquad F(x) = (x - r) \cdot Q(x) + R$$

where R is a constant. Since (1) is an identity, and therefore true for all values of x, it is true in particular when $x = r$. Substituting $x = r$ in (1) we obtain

$$F(r) = (r - r) \cdot Q(r) + R,$$

or

$$F(r) = 0 \cdot Q(r) + R,$$

or

$$F(r) = R,$$

which is what we had to prove.

If $R = 0$, then $x - r$ is a factor† of $F(x)$, (since $F(x) = (x - r) \cdot Q(x)$), and conversely. Hence we have the following as an immediate consequence of the remainder theorem.

THE FACTOR THEOREM. *If $x - r$ is a factor of $F(x)$, then r is a root of the equation $F(x) = 0$, and conversely.*

Proof. If $x - r$ is a factor of $F(x)$, then the remainder R obtained by dividing $F(x)$ by $x - r$ will be zero. But then $F(r) = 0$ which means that r is a root of the equation $F(x) = 0$. Conversely, if r is a root of $F(x) = 0$, then $F(r) = 0$. Hence the remainder R will be zero when $F(x)$ is divided by $x - r$, or $x - r$ is a factor of $F(x)$.

Example 1. $x - 2$ is a factor of $F(x) = x^2 - 5x + 6$, and 2 is a root of $x^2 - 5x + 6 = 0$.

* The student should review section 43 carefully at this point.

† No restriction on the kind of numbers occurring as coefficients in the factors will be made in this chapter. Unless otherwise specified, coefficients may be any complex numbers.

Example 2. $x + 2$ is a factor of $F(x) = x^2 + 5x + 6$ and -2 is a root of $x^2 + 5x + 6 = 0$. Note that $x + 2$ may be written as $x - (-2)$ to put it in the form $x - r$; here $r = -2$.

EXERCISES

Divide by long division (until a constant remainder is obtained) and then find the remainder again by means of the remainder theorem:

1. $(x^2 - 5x + 8) \div (x - 3)$. **2.** $(2x^2 - x - 5) \div (x - 4)$.
3. $(x^3 - 2x^2 + x - 3) \div (x - 2)$.
4. $(x^3 - 4) \div (x - 1)$. **5.** $(x^3 - 5x + 6) \div (x + 2)$.
6. $(x^3 + 2x^2 - 12) \div (x + 3)$.
7. $(x^3 - 2x^2 + x - 3) \div (x + 3)$.
8. $(x^3 + x - 5) \div (x + 1)$.

By means of the factor theorem, decide whether or not each of the following statements is true:

9. $x - 2$ is a factor of $x^3 - x^2 - 5x + 6$.
10. $x - 3$ is a factor of $2x^3 - 6x^2 - 5x + 15$.
11. $x + 2$ is a factor of $3x^3 + 10x^2 + 9x + 2$.
12. $x + 3$ is a factor of $3x^3 - x^2 - 22x + 24$.
13. $x + 2$ is a factor of $2x^3 - 5x^2 - x + 6$.
14. $x - 1$ is a factor of $8x^3 - 10x^2 - x + 3$.
15. $x - 2$ is a factor of $5x^3 + 12x^2 - 36x - 16$.
16. $x - 2$ is a factor of $3x^3 - 5x^2 - 16x + 12$.
17. $x - 3$ is a factor of $x^3 + 27$.
18. $x + 2$ is a factor of $x^4 + 16$.
19. $x + 2$ is a factor of $x^4 - 16$.
20. $x - 2$ is a factor of $x^4 - 16$.
21. $x - 3$ is a factor of $x^3 - 27$.
22. $x + 2$ is a factor of $x^3 + 8$.
23. $x + 2$ is a factor of $x^3 - 8$.
24. $x - 2$ is a factor of $x^4 + 16$.
25. (*a*) $x - y$ is a factor of $x^n - y^n$ if n is any natural number.
(*b*) $x - y$ is not a factor of $x^n + y^n$ if n is any natural number.
(*c*) $x + y$ is a factor of $x^n - y^n$ if n is even.
(*d*) $x + y$ is a factor of $x^n + y^n$ if n is odd.
(*e*) $x + y$ is not a factor of $x^n - y^n$ if n is odd.
(*f*) $x + y$ is not a factor of $x^n + y^n$ if n is even.

26. Verify that $x^2 + 1$ is a factor of $F(x) = x^3 - 2x^2 + x - 2$. May we conclude that therefore $F(-1) = 0$?

27. Verify that $2x - 1$ is a factor of $F(x) = 2x^3 - x^2 + 2x - 1$. May we conclude that therefore $F(1) = 0$?

28. Find the value of k for which $x - 3$ is a factor of $kx^3 - 6x^2 + 2kx - 12$.

29. Find the value of k for which $x + 2$ is a factor of $3x^3 + 2kx^2 - 4x - 8$.

30. Show that $x - 1$ is a factor of $1776x^{1956} - 1492x^{1865} - 284$.

117. Synthetic Division

Because of the connection between roots of equations and factors of the form $x - r$ of polynomials, we shall often have to divide polynomials by expressions of the form $x - r$. Whenever an operation has to be performed often, it is natural to try to arrange the work so as to involve the least possible waste motion. This will be done for division by $x - r$, as follows:

Example 1. Divide $3x^3 - x^2 - 3x + 1$ by $x - 2$.

Solution.

$$\begin{array}{rrrrrl}
 & 3x^2 & +\,5x & +\;\,7 & & = \text{quotient} \\
\text{divisor} = x - 2) & 3x^3 & -\;\,x^2 & -\;\,3x & +\;\,1 & = \text{dividend} \\
 & \mathbf{3}x^3 & -\,6x^2 & & & \\
\hline
 & & +\,5x^2 & -\;\,3x & +\;\,1 & \\
 & & \mathbf{5}x^2 & -\,10x & & \\
\hline
 & & & +\;\,7x & +\;\,1 & \\
 & & & \mathbf{7}x & -\,14 & \\
\hline
 & & & & \mathbf{15} & = \text{remainder.}
\end{array}$$

Examination of the steps in the division process show that (since the coefficient of x in the divisor is 1) the numbers in heavy type are the coefficients of the quotient and the remainder in succession. The powers of x and the entire quotient may be omitted, since we know, from the position of the other symbols, what they are. The first coefficient in the quotient must be the same as the first coefficient in the dividend, since the coefficient of x in the divisor is 1. The second coefficient is obtained by subtracting the product of -2 and 3 from -1; the next, by subtracting the product of -2 and 5 from -3; the next, by subtracting the product of -2 and 7 from 1. We shall multiply by 2 instead of -2 and add instead of subtracting. The work may be arranged as follows:

$$\begin{array}{r|rrrr}
 & \multicolumn{4}{c}{\overbrace{}^{\text{(coefficients of dividend)}}} \\
(r =)2 & 3 & -1 & -3 & 1 \\
 & & 6 & 10 & 14 \\
\hline
 & 3 & 5 & 7 & 15 \\
 & \multicolumn{3}{c}{\underbrace{}_{\text{(coefficients of quotient)}}} & \underbrace{}_{\text{(remainder)}}
\end{array}$$

Thus the quotient is $3x^2 + 5x + 7$ and the remainder is **15**.

When division by $x - r$ is done in the latter arrangement, it is called **synthetic division,** for which we have the following rule:

To divide a polynomial $f(x)$ by $x - r$ synthetically:

(*a*) Write the coefficients of $f(x)$ in order of descending powers, supplying zeros for missing powers, in the first line;

(*b*) Write the first coefficient in the third line, below its position in the first line;

(*c*) Write the product of r and this coefficient in the second line beneath the second coefficient in the first line, and add, putting the sum in the third line, and so on.

The number r may be called the **multiplier.**

The work may be arranged as in example 1.

Example 2. Divide $2x^3 + x^2 - 4$ by $x + 3$.

Solution.

$$\begin{array}{r|rrrr} -3 & 2 & 1 & 0 & -4 \\ & & -6 & 15 & -45 \\ \hline & 2 & -5 & 15 & -49 \end{array}$$

The quotient is $2x^2 - 5x + 15$ and the remainder is -49.

When dividing by $x - 2$, we multiply by 2. When dividing by $x + 3$, we multiply by -3. In general, when dividing by $x - r$, we multiply by r.

To evaluate $f(r)$, we may divide synthetically with r as multiplier, for the last number in the third line will be the remainder obtained when dividing $f(x)$ by $x - r$, or $f(r)$.

Example 3. If $f(x) = 2x^3 + x^2 - 4$, find $f(2)$.

Solution. $f(2)$ will be the remainder when $f(x)$ is divided by $x - 2$. Hence $f(2)$ will be the last number in the third line of synthetic division with 2 as multiplier:

$$\begin{array}{r|rrrr} 2 & 2 & 1 & 0 & -4 \\ & & 4 & 10 & 20 \\ \hline & 2 & 5 & 10 & 16. \end{array}$$

Thus, $f(2) = 16$.

Example 4. If $f(x) = 2x^3 + x^2 - 4$, find $f(-3)$.

Solution. The work done in example 2 shows that $f(-3) = -49$.

Example 5. Show that 2 is a root of the equation $f(x) = 2x^3 + x^2 - 8x - 4 = 0$.

Solution. By synthetic division

$$\begin{array}{r|rrrr} 2 & 2 & 1 & -8 & -4 \\ & & 4 & 10 & 4 \\ \hline & 2 & 5 & 2 & 0 \end{array}$$

we find that $f(2) = 0$, and hence 2 is a root of $f(x) = 0$.

Example 6. Show that $x - 2$ is a factor of $f(x) = 2x^3 + x^2 - 8x - 4$, and write another factor.

Solution. The work done in example 5 shows that $x - 2$ is a factor since $f(2) = 0$. Another factor is the quotient polynomial $2x^2 + 5x + 2$; that is, $f(x) = (x - 2)(2x^2 + 5x + 2)$.

EXERCISES

Find quotient and remainder by synthetic division, and check:

1. $(4x^3 - 2x + 5) \div (x - 2)$.
2. $(3x^3 - 4x^2 + 2) \div (x - 3)$.
3. $(x^3 + 4x^2 - 7) \div (x + 3)$.
4. $(2x^4 - 3x^2 + 1) \div (x + 2)$.
5. $(2x^3 + 3x^2 + 4x + 5) \div (x - \frac{1}{2})$.
6. $(x^4 - 10x^2 - 7) \div (x - \frac{2}{3})$.
7. $(6x^3 - x + 1) \div (x + \frac{1}{3})$.
8. $(2x^4 - 3x^2 + 4) \div (x + \frac{3}{2})$.

By means of synthetic division:

9. If $f(x) = x^3 - 2x^2 + 3x + 4$, find $f(3)$ and $f(-2)$.
10. If $f(x) = 4x^3 - 5x^2 - 3$, find $f(2)$ and $f(-3)$.
11. If $F(x) = 6x^4 - 10x^3 - 4x^2 + 3x + 1$, find $F(\frac{1}{2})$ and $F(-\frac{1}{3})$.
12. If $F(x) = 2x^3 - 13x^2 + 27x - 18$, find $F(\frac{3}{2})$ and $F(-\frac{1}{4})$.
13. If $g(x) = -2x^3 + x^2 - 7$, find $g(0.2)$ and $g(-1.3)$.
14. If $g(x) = -x^3 - 2x + 4$, find $g(1.3)$ and $g(-2.1)$.

By means of synthetic division decide whether each statement is true or false; if true, write another factor:

15. $x - 2$ is a factor of $3x^3 + 10x^2 + 9x + 2$.
16. $x + 3$ is a factor of $3x^3 - x^2 - 22x + 24$.
17. $x + 1$ is a factor of $2x^3 - 5x^2 - x + 6$.
18. $x + 2$ is a factor of $5x^3 - 12x^2 - 36x - 16$.
19. $x + 3$ is a factor of $x^4 + 81$.
20. $x - 3$ is a factor of $x^4 - 81$.
21. $x + 2$ is a factor of $x^5 + 32$.
22. $x + r$ is a factor of $x^5 + r^5$.
23. $x + r$ is a factor of $x^6 + r^6$.
24. $x - r$ is a factor of $x^6 - r^6$.

By synthetic division decide whether the statement is true or false:

25. 2 is a root of $2x^3 - 3x^2 - 4 = 0$.
26. -3 is a root of $2x^4 + 6x^3 + x^2 + 3x - 6 = 0$.
27. -1 is a root of $3x^3 - 4x^2 + 3x - 2 = 0$.
28. 4 is a root of $3x^3 + 10x^2 - 7x + 4 = 0$.
29. $\frac{3}{4}$ is a root of $8x^3 - 10x^2 - x + 3 = 0$.
30. $-\frac{2}{5}$ is a root of $5x^3 + 12x^2 - 36x - 16 = 0$.
31. $\sqrt{2}$ is a root of $x^4 + 2x^3 - 5x^2 - 4x + 6 = 0$.
32. $2i$ is a root of $2x^3 - 3x^2 + 4x - 6 = 0$.
33. $1 + \sqrt{2}$ is a root of $2x^4 - 4x^3 + x^2 - 6x - 3 = 0$.
34. $3 + 2i$ is a root of $2x^3 - 13x^2 + 32x - 13 = 0$.

Using synthetic division find a value of k such that:

35. -2 is a root of $3x^3 + 5x^2 + kx - 10 = 0$.
36. 3 is a root of $2x^3 + 3kx^2 - 5x + 15 = 0$.

118. The Depressed Equation

Suppose we have somehow found out that the equation $f(x) = a_0x^n + a_1x^{n-1} + \dots + a_n = 0$ of degree $n \geqq 1$ has the root r_1. Then $x - r_1$ is a factor of $f(x)$, and we may write the equation as

$$f(x) = (x - r_1)Q_1(x) = 0 \tag{1}$$

where the quotient $Q_1(x)$ is of degree $n - 1$ and begins with the same coefficient as $f(x)$, as can be seen from the process of synthetic division; that is,

$$Q_1(x) = a_0x^{n-1} + \dots .$$

Since the product of two factors can be zero when and only when one (or both) of the factors is itself zero, any root of $f(x) = 0$ must satisfy either

$$x - r_1 = 0 \tag{2}$$

or

$$Q_1(x) = 0. \tag{3}$$

The only root of $x - r_1 = 0$ is r_1, which we knew about at the start. Hence any further root of (1) must be a root of (3), which is called the **depressed equation** because its degree is lower than that of the original equation by one. The depressed equation is often easier to solve than the original equation for this reason.

If r_1 is known, the coefficients of the depressed equation may be found by synthetic division. If a depressed equation is quadratic, the equation may be solved completely.

Example 1. Given that 2 is a root of $f(x) = x^3 - 4x^2 + x + 6 = 0$, solve the equation.

Solution. Since 2 is a root of $f(x) = 0$, $x - 2$ is a factor of $f(x)$. By synthetic division

$$\begin{array}{r|rrrr} 2 & 1 & -4 & 1 & 6 \\ & & 2 & -4 & -6 \\ \hline & 1 & -2 & -3 & 0 \end{array}$$

we find that $f(x) = (x - 2)(x^2 - 2x - 3)$. Hence the depressed equation is $x^2 - 2x - 3 = 0$, which has the roots 3 and -1. Hence the roots of the given equation are 2, 3, -1. The student may check by substituting in the original equation.

If more than one root is known, the depressed equation may be depressed again, as in the following example.

Example 2. Given that 2 and 3 are roots of $f(x) = x^4 - 5x^3 + 5x^2 + 5x - 6 = 0$, solve the equation.

Solution. Since 2 is a root of $f(x) = 0$, $x - 2$ is a factor of $f(x)$. By synthetic division we find that $f(x) = (x - 2)(x^3 - 3x^2 - x + 3)$. But 3 is given to be a root of $f(x) = 0$ and is therefore a root of the depressed equation $Q_1(x) = x^3 - 3x^2 - x + 3 = 0$. By synthetic division we find that $Q_1(x) = (x - 3)(x^2 - 1) = 0$. The remaining roots of $f(x) = 0$ must be roots of $x^2 - 1 = 0$, or ± 1. Hence the roots of the given equation are 2, 3, and ± 1.

EXERCISES

Solve the equation:

1. $x^3 - 4x^2 + x + 6 = 0$, given that 3 is a root.

2. $x^3 - 7x + 6 = 0$, given that 2 is a root.

3. $x^3 + 4x^2 + 8x + 8 = 0$ given that -2 is a root.

4. $3x^3 - 4x^2 + 2x + 4 = 0$ given that $-\frac{2}{3}$ is a root.

5. $2x^4 - 3x^3 - 20x^2 + 27x + 18 = 0$ given that -3 and $-\frac{1}{2}$ are roots.

6. $8x^4 - 38x^3 + 20x^2 + 19x - 12 = 0$ given that 4 and $\frac{3}{4}$ are roots.

7. $3x^4 + 14x^3 - 4x^2 - 11x - 2 = 0$ given that 1 and $-\frac{2}{3}$ are roots.

8. $2x^4 - 9x^3 + 6x^2 + 51x - 26 = 0$ given that -2 and $\frac{1}{2}$ are roots.

9. $6x^4 - 11x^3 + 15x^2 - 22x + 6 = 0$ given that $\frac{3}{2}$ and $\frac{1}{3}$ are roots.

10. $x^4 + 2x^3 + 5x^2 + 2x + 4 = 0$ given that i and $-i$ are roots.

119. Factorization of Polynomials. The Number of Roots of an Equation

We shall prove several basic theorems.

THEOREM 1. *Every polynomial*

$$f(x) = a_0x^n + a_1x^{n-1} + \dots + a_n \qquad (a_0 \neq 0) \tag{1}$$

of degree 1 or more can be expressed as the product of n linear factors of the form

$$f(x) = a_0(x - r_1)(x - r_2) \dots (x - r_n). \tag{2}$$

Proof. By the fundamental theorem of algebra, the equation $f(x) = 0$ has a root r_1. Then $x - r_1$ is a factor of $f(x)$. Therefore

$$f(x) = (x - r_1) \cdot Q_1(x) \tag{3}$$

where

$$Q_1(x) = a_0x^{n-1} + \dots$$

may be found by division. If $n - 1 > 0$, the equation $Q_1(x) = 0$ must also have a root r_2 by the fundamental theorem of algebra. Therefore $x - r_2$ is a factor of $Q_1(x)$ and we may write

$$Q_1(x) = (x - r_2)Q_2(x) \tag{4}$$

where $Q_2(x) = a_0x^{n-2} + \dots$. Substituting (4) in (3) we have

$$f(x) = (x - r_1)(x - r_2) \cdot Q_2(x). \tag{5}$$

This process may be performed n times until the quotient $Q_n(x) = a_0$ of degree 0 is reached. Hence

$$f(x) = (x - r_1)(x - r_2) \dots (x - r_n) \cdot a_0$$

which was to be proved.

The numbers $r_1, r_2, \dots, r_n$ may or may not be all different from each other. By the factor theorem, each of them is a root of the equation $f(x) = 0$.

THEOREM 2. *No number r different from all the numbers $r_1, r_2, \dots, r_n$ found in theorem 1 can be a root of the equation $f(x) = 0$.*

Proof. By hypothesis $r - r_1$, $r - r_2$, $\ldots$, $r - r_n$ are all different from zero. Hence, substituting r for x in (2), we find

$$f(r) = a_0(r - r_1)(r - r_2) \ldots (r - r_n) \neq 0,$$

which proves the theorem.

The following is an immediate corollary.

COROLLARY 1. *Any equation of degree $n \geqq 1$ has at most n distinct roots.*

COROLLARY 2. *If two polynomials*

$$f(x) = a_0x^n + a_1x^{n-1} + \ldots + a_{n-1}x + a_n$$

and

$$g(x) = b_0x^n + b_1x^{n-1} + \ldots + b_{n-1}x + b_n,$$

of degrees not greater than n, are equal for more than n distinct values of x, then

$$a_0 = b_0,\ a_1 = b_1,\ \ldots,\ a_n = b_n, \tag{6}$$

and the two polynomials are identically equal.

Proof. Consider the equation

$$f(x) - g(x) = (a_0 - b_0)x^n + (a_1 - b_1)x^{n-1} + \ldots + (a_{n-1} - b_{n-1})x + (a_n - b_n) = 0. \tag{7}$$

If any of the coefficients $a_0 - b_0$, $a_1 - b_1$, $\ldots$, $a_{n-1} - b_{n-1}$ were different from zero, (7) would be an equation of degree d, where $1 \leqq d \leqq n$, with more than n distinct roots, contrary to corollary 1. Hence $a_0 = b_0$, $a_1 = b_1$, $\ldots$, $a_{n-1} = b_{n-1}$. But then the equation (7) reads $a_n - b_n = 0$, or $a_n = b_n$. This completes the proof.

THEOREM 3. *The factorization (2) is unique. That is, the polynomial (1) can be expressed as a product of factors of the form (2) in one and only one way, apart from the order in which the factors are written down.*

This theorem will be proved in the next section.

A number r is a root of $f(x) = 0$ of **multiplicity k** if the factor $(x - r)$ occurs exactly k times in the factorization (2). Roots of multiplicity 1, 2, and 3 are called **simple, double,** and **triple roots,** respectively.

Example 1. In the equation $f(x) = (x - 3)(x - 4)(x - 4)(x - 4)(x + 7)(x + 7) = 0$ of degree six, the number 3 is a simple root, 4 is a triple root, and -7 is a double root.

From theorems 1, 2, and 3 we have the following immediate consequence.

THEOREM 4. *Every equation of degree $n \geqq 1$ has exactly n roots provided a root of multiplicity k is counted as k roots.*

Thus the equation of example 1 has the six roots: 3, 4, 4, 4, -7, -7.

To form an equation with given roots, we have only to multiply together the appropriate factors, as in the following example.

Example 2. Form an equation with the roots $\frac{1}{2}$, $-\frac{3}{5}$, and 1, having integral coefficients.

Solution. An equation having the desired roots is

$$(x - \tfrac{1}{2})(x + \tfrac{3}{5})(x - 1) = 0.$$

The equation obtained by multiplying these factors together will have fractional coefficients, which may then be cleared away by multiplying both sides by the L.C.D. to obtain integral coefficients. However, it is less trouble to introduce the L.C.D., 10, as the value of a_0 and to write $10(x - \frac{1}{2})(x + \frac{3}{5})(x - 1) = 0$ or

$$2(x - \tfrac{1}{2}) \cdot 5(x + \tfrac{3}{5}) \cdot (x - 1) = 0,$$

or

$$(2x - 1)(5x + 3)(x - 1) = 0.$$

Multiplying the latter factors, we get the desired equation $10x^3 - 9x^2 - 4x + 3 = 0$.

Note. The idea that an equation of degree n has n roots is a fairly modern one. The great Greek algebraist Diophantus (3rd century A.D.) was able to find solutions of quadratic equations but was content with one root. He rejected negative and irrational roots as absurd; and, even when both roots are positive, he took only one of them. Even in the 16th century, great mathematicians such as Cardan and Vieta rejected negative and imaginary roots, although irrational roots were admitted, and Stifel asserted that except for the case of a quadratic with two positive roots, no equation has more than one root. In the 17th century, mathematicians such as Descartes began to have a better grasp of the subject.

EXERCISES

Solve each equation without multiplying the factors:

1. $(x - 3)(2x - 1)(x + 3) = 0.$
2. $(x - 2)(x + 5)(3x + 2) = 0.$
3. $(x - 1)(x - 1)(x^2 + 2x + 4) = 0.$
4. $(x^2 + 4x + 4)(x^2 + x + 1) = 0.$
5. $2(x + 3)(2x + 1)(3x - 2) = 0.$
6. $3(x - \frac{1}{3})(x + 3)(x^2 + 1) = 0.$
7. $x(x - 1)(x + 2)(x^2 - 3) = 0.$
8. $3x^2(2x + 3)(x^2 + 2) = 0.$

Form an equation with integral coefficients having the given numbers, and no others, as roots:

9. $1, 2, -3.$
10. $3, -4, 1.$
11. $\frac{1}{2}, 3, -2.$
12. $\frac{2}{3}, -2, -1.$
13. $2, 2, 0.$
14. $1, 1, -2, -2.$
15. $\frac{1}{2}, \frac{1}{2}, -3, 0.$
16. $\frac{3}{2}, -\frac{1}{3}, 0.$
17. $\pm\sqrt{3}, 0, 0.$
18. $\pm\sqrt{5}, \frac{4}{3}, 0.$
19. $1 \pm \sqrt{3}, \pm 2i.$
20. $2 \pm \sqrt{5}, \pm i\sqrt{3}.$
21. $1 \pm 2i, 2 \pm \sqrt{3}.$
22. $3 \pm i, 1 \pm i\sqrt{3}.$
23. $\frac{1}{2}, \frac{2}{3}, \pm 2i.$

24. Form a cubic equation with 3 as double root and -4 as a simple root.

25. Form a cubic equation with 2 as a triple root.

26. Form a quartic equation with -1 as a triple root and 4 as a simple root.

27. Form a quartic equation with 1 as a double root and 0 and -2 as simple roots.

Solve each of the following equations:

28. $18x^3 - 39x^2 + 8x + 16 = 0$ given that $\frac{4}{3}$ is a double root.

29. $4x^4 + x^2 - 3x + 1 = 0$ given that $\frac{1}{2}$ is a double root.

30. $9x^4 + 24x^3 + 49x^2 + 26x + 4 = 0$ given that $-\frac{1}{3}$ is a double root.

31. $9x^4 - 6x^3 - 2x^2 + 16x + 8 = 0$ given that $-\frac{2}{3}$ is a double root.

32. Form a cubic equation with ± 3 and i as simple roots.

33. Form a cubic equation with 3, -2, and $1 + 2i$ as simple roots.

34. Form a quartic equation one of the roots of which is 3 and the other three roots of which are the roots of $2x^3 - 7x + 5 = 0$.

★120. Proof of the Uniqueness of Factorization

We shall prove theorem 3 of the preceding section.

Proof. Suppose that there are d distinct numbers among the n numbers $r_1, r_2, \ldots, r_n$. Let these d distinct numbers be $\rho_1, \rho_2, \ldots, \rho_d$ where ρ_1 occurs α_1 times, ρ_2 occurs α_2 times, and so on. Then (2) of section 119 may be written as

$$(1) \qquad f(x) = a_0(x - \rho_1)^{\alpha_1}(x - \rho_2)^{\alpha_2} \ldots (x - \rho_d)^{\alpha_d},$$

where $\alpha_1 + \alpha_2 + \ldots + \alpha_d = n$. By theorem 2, section 119, no number r different from $\rho_1, \rho_2, \ldots, \rho_d$ can be a root of $f(x) = 0$, and hence no expression of the form $x - r$ other than those appearing in (1) can be a factor of $f(x)$. But it is still conceivable that the factors in (1) might also occur with different multiplicities; that is, that, besides (1), we might have

$$(2) \qquad f(x) = a_0(x - \rho_1)^{\beta_1}(x - \rho_2)^{\beta_2} \ldots (x - \rho_d)^{\beta_d},$$

where $\beta_1 + \beta_2 + \ldots + \beta_d = n$ but not all the equalities

$$(3) \qquad \alpha_1 = \beta_1, \alpha_2 = \beta_2, \ldots, \alpha_d = \beta_d$$

hold. Suppose that $\alpha_1 \neq \beta_1$ and in particular that $\alpha_1 > \beta_1$. Then, from (1) we have $f(x) = (x - \rho_1)^{\alpha_1}F(x)$, and from (2) we have $f(x) = (x - \rho_1)^{\beta_1}G(x)$ where $G(\rho_1) \neq 0$ since $x - \rho_1$ is not a factor of $G(x)$. Then

$$(x - \rho_1)^{\alpha_1}F(x) = (x - \rho_1)^{\beta_1}G(x)$$

is an identity. Dividing both sides by $(x - \rho_1)^{\beta_1}$ we have

$$(4) \qquad (x - \rho_1)^{\alpha_1 - \beta_1}F(x) = G(x),$$

for all values of x except $x = \rho_1$ since division by zero is excluded. But (4) is therefore an identity by corollary 2, section 119, because it is true for all values of x except $x = \rho_1$, hence for infinitely many values of x, and hence certainly for more than n values of x. But $G(x)$ is not divisible by $(x - \rho_1)$, and therefore the left member of (4) cannot have $(x - \rho_1)$ as a factor. There-

★ This section may be omitted upon first reading without disturbing the continuity of the chapter. It is included for the general reasons explained in the preface, and in particular because the definition of multiplicity of a root is no more than a pious wish without it.

fore $\alpha_1 - \beta_1 = 0$, or $\alpha_1 = \beta_1$. The argument can be repeated to show that all the equalities (3) must hold. This completes the proof.

121. Imaginary Roots of Equations with Real Coefficients*

The familiar formula for the roots of a quadratic equation shows clearly that if the quadratic has real coefficients and has an imaginary root, then the conjugate of this root is the other root. This result is generalized to apply to all equations with real coefficients no matter what the degree, by the following theorem.

THEOREM 1. *If all the coefficients of the equation $F(x) = 0$ are real and $a + bi$ $(b \neq 0)$ is an imaginary root, then its conjugate $a - bi$ is also a root.*

Proof. To show that $a - bi$ is a root of $F(x) = 0$ it will suffice to show that $x - (a - bi)$ is a factor of $F(x)$. This will be done by showing that

$$D(x) = [x - (a - bi)][x - (a + bi)] \tag{1}$$

is a factor of $F(x)$. Multiplying out the right member of (1), we find that

$$D(x) = x^2 - 2ax + a^2 + b^2, \tag{2}$$

which has real coefficients. Dividing† $F(x)$ by $D(x)$ we find a quotient $Q(x)$, and remainder $R(x)$ of degree not greater than 1. Writing $R(x) = cx + d$ we have

$$F(x) = D(x) \cdot Q(x) + cx + d. \tag{3}$$

Note that in the division process the coefficients of the dividend and divisor are combined only by multiplication, division, addition, and subtraction. Since the coefficients of $F(x)$ and $D(x)$ are all real, this process can yield nothing but real coefficients in the quotient and remainder. In particular, c and d are real. By hypothesis, $a + bi$ is a root of $F(x) = 0$. Hence substituting $a + bi$ in (3), we obtain

$$F(a + bi) = D(a + bi) \cdot Q(a + bi) + ca + cbi + d = 0. \tag{4}$$

From (1) we see that $D(a + bi) = 0$. Hence (4) implies that

$$ca + cbi + d = 0.$$

* The student should review carefully the classification of the number system given in section 16.

† See the definition of division in section 43.

This implies that

$$(5) \qquad ca + d = 0$$

and

$$(6) \qquad cb = 0.$$

Since $b \neq 0$ by hypothesis, (6) implies that $c = 0$. Substituting this in (5), we find $d = 0$. Hence $R(x) = 0$ and $D(x)$ is a factor of $F(x)$. This completes the proof.

COROLLARY 1. *Every polynomial $f(x)$ of degree $n \geqq 1$ with real coefficients can be expressed as a product of linear factors and quadratic factors having negative discriminants, all with real coefficients.*

Proof. We have $f(x) = a_0(x - r_1)(x - r_2) \ldots (x - r_n)$. Any root of $f(x) = 0$ that is real provides a linear factor with real coefficients. Any root that is imaginary can be paired with its conjugate root. The product of the factors arising from such a pair is of the form

$$[x - (a + bi)][x - (a - bi)] = x^2 - 2ax + a^2 + b^2,$$

which provides a quadratic factor with real coefficients and negative discriminant $-4b^2$. The constant a_0 can be absorbed into any of these factors. This completes the proof.

The quadratic formula shows similarly that if a quadratic with rational coefficients has a root $a + \sqrt{b}$ where a and b are rational but $\sqrt{b}$ is irrational, then $a - \sqrt{b}$ is also a root. This result can also be extended to equations of any degree by the following theorem:

THEOREM 2. *If an equation with rational coefficients has a binomial surd root of the form $a + \sqrt{b}$ (or $a - \sqrt{b}$) where a and b are rational but $\sqrt{b}$ is irrational, then the conjugate surd $a - \sqrt{b}$ (or $a + \sqrt{b}$) is also a root.*

The proof is similar to that of theorem 1 and is left to the reader.

EXERCISES

If all the coefficients of $f(x) = 0$ are real numbers and the given number is a root, write another number which must be a root of $f(x) = 0$:

1. $3 + 2i$. **2.** $-2 + 3i$. **3.** $-3 - 3i$. **4.** $3 - 4i$.
5. $3i$. **6.** $-2i$.

7. Form a cubic equation with real coefficients having $1 + 2i$ and 5 as roots.

8. Form a cubic equation with integral coefficients having $3 - 2i$ and $\frac{1}{2}$ as roots.

9. Form a quartic equation with integral coefficients having $1 + i$ as a simple root and $-\frac{1}{2}$ as a double root.

10. Form a quartic equation with real coefficients having 2 as a double root and $2 + i$ as a simple root.

11. Solve $x^4 - 2x^3 + 3x^2 - 2x + 2 = 0$, given that $1 + i$ is a root.

12. Solve $x^4 - 2x^3 - 6x^2 + 22x + 65 = 0$, given that $-2 + i$ is a root.

13. Solve $x^4 + 3x^3 + 2x^2 + 3x + 1 = 0$, given that i is a root.

14. Solve $x^4 + 3x^3 + 4x^2 + 27x - 45 = 0$, given that $-3i$ is a root.

15. Solve $x^6 + 3x^5 + 3x^4 + 6x^3 + 3x^2 + 3x + 1 = 0$, given that i is a double root.

16. Prove that a cubic equation with real coefficients has either three real roots, or one real and two conjugate imaginary roots.

17. Prove that a quartic equation with real coefficients has either 4 real roots, or 4 imaginary roots, or 2 real and 2 imaginary roots.

18. State and prove a theorem similar to those of exercises 16 and 17 for equations of degree: (*a*) five; (*b*) six.

19. Prove that an equation of odd degree with real coefficients has at least one real root.

20. Verify that the equation $x - (1 + i) = 0$ has the root $1 + i$, but the conjugate $1 - i$ is not a root. Does this contradict theorem 1? Explain.

Decide whether each of the following statements is true or false and explain:

21. If $1 + i$ is a root of $x^3 - (6 + i)x^2 + (11 + 5i)x - (6 + 6i) = 0$, then $1 - i$ is also a root.

22. If i is a root of $x^3 - \dfrac{\sqrt{3} + 1}{2} x^2 + x - \dfrac{\sqrt{3} + 1}{2} = 0$, then $-i$ is also a root.

23. Form a cubic equation with rational coefficients having $2 + \sqrt{3}$ and 1 as roots.

24. Form a cubic equation with integral coefficients having $1 - \sqrt{5}$ and $\frac{2}{3}$ as roots.

25. Form a quartic equation with rational coefficients having $1 + \sqrt{3}$, 2, and 0 as roots.

26. Form a quartic equation with integral coefficients having 0 as a double root and $\frac{3 - \sqrt{2}}{2}$ as a simple root.

27. Solve $x^4 - 4x^3 + 2x^2 - 4x + 1 = 0$, given that $2 + \sqrt{3}$ is a root.

28. Solve $x^4 - 5x^3 + 2x^2 + x + 7 = 0$, given that $3 - \sqrt{2}$ is a root.

★29. Prove theorem 1, above, by means of the results of exercise 53 of section 105, and exercises 75, 76, 77 of section 106.

122. Rational Roots of Equations with Integral Coefficients

In section 114, we learned how to find imaginary roots of a certain simple type of equation of the nth degree, namely the pure equation. Aside from this case, we shall find imaginary roots only when they arise from a quadratic depressed equation, because, except for certain advanced applications, the real* roots are what are generally needed for practical purposes. We are now ready to attack the problem of finding the real roots of an equation of any degree. Recall that real numbers are either rational or irrational, and that a rational number is one which can be expressed as a quotient or ratio of two integers. Hence all rational numbers are included in the following scheme:

$$
\begin{array}{l}
0, \pm \frac{1}{1}, \frac{2}{1}, \frac{3}{1}, \frac{4}{1}, \cdots \\
\quad \pm \frac{1}{2}, \frac{2}{2}, \frac{3}{2}, \frac{4}{2}, \cdots \\
\quad \pm \frac{1}{3}, \frac{2}{3}, \frac{3}{3}, \frac{4}{3}, \cdots \\
\quad \pm \frac{1}{4}, \frac{2}{4}, \frac{3}{4}, \frac{4}{4}, \cdots \\
\qquad \cdot \quad \cdot \quad \cdot \quad \cdot
\end{array}
$$

In this section, we give a method for finding the rational roots of equations with integral coefficients. Whether any particular rational number r is a root of a given equation $f(x) = 0$ or not can be decided by substituting it for x in the equation; this work is most conveniently done by synthetic division, the last number in the third line of synthetic division being $f(r)$. The number r

* The student should review carefully the classification of number system in section 16.

is a root if and only if $f(r) = 0$. Hence we might try to find all rational roots of the given equation simply by trying all rational numbers one at a time. (Compare section 13.) But since there are infinitely many rational numbers, this is a gloomy prospect. In fact, if we tried in this way to find the rational roots of an equation like $x^2 - 2 = 0$ which has no rational roots, we would waste eternity on this one problem without reaching any conclusion. This is considered to be too long an assignment. The following theorem has the virtue of cutting the number of trials from infinitely many to a limited number, thus making the method of trial possible in practice.

THEOREM. *If the coefficients of the equation*

$$(1) \quad a_0x^n + a_1x^{n-1} + a_2x^{n-2} + \dots + a_{n-1}x + a_n = 0 \quad (a_0 \neq 0)$$

are all integers, and if p/q is a rational root, reduced to lowest terms, then

(*a*) *the numerator p must be a factor of the constant term a_n, and*

(*b*) *the denominator q must be a factor of the leading coefficient a_0.*

Proof. Since p/q is a root, it must satisfy (1). Hence

$$a_0\frac{p^n}{q^n} + a_1\frac{p^{n-1}}{q^{n-1}} + a_2\frac{p^{n-2}}{q^{n-2}} + \dots + a_{n-1}\frac{p}{q} + a_n = 0.$$

Multiplying both sides by the L.C.D. q^n, we have

$$(2) \quad a_0p^n + a_1p^{n-1}q + a_2p^{n-2}q^2 + \dots + a_{n-1}pq^{n-1} + a_nq^n = 0.$$

Transposing the last term and taking the common factor p out of what remains in the left member, we have

$$(3) \quad p[a_0p^{n-1} + a_1p^{n-2}q + a_2p^{n-3}q^2 + \dots + a_{n-1}q^{n-1}] = -a_nq^n.$$

Since all the letters $a_0, a_1, \dots, a_n$, p and q represent integers, both members of (3) and the quantity in brackets are integers, because the sum, product, and difference of integers are always integers again. The left member of (3) has p as a factor. Therefore p is a factor of the right member $-a_nq^n$. But p has no factor in common with q, except ± 1, since p/q was reduced to lowest terms. Hence p has no factor in common with q^n (compare theorem A, section 13). By theorem 2, section 6, p must be a factor of a_n. This proves (*a*).

Transposing the first term of (2) and factoring q out of what remains in the left member, we get

$$(4)\quad q[a_1p^{n-1} + a_2p^{n-2}q + \ldots + a_{n-1}pq^{n-2} + a_nq^{n-1}] = -a_0p^n.$$

An argument, exactly like that above, now proves (*b*). The student should write out the details.

Example 1. Find the rational roots of $4x^4 - 4x^3 + 13x^2 - 12x + 3 = 0$, and if possible solve the equation (completely).

Solution. By the above theorem we find that the only:

possible numerators are ± 1, ± 3 (the factors of 3);
possible denominators are ± 1, ± 2, ± 4 (the factors of 4);
possible rational roots are ± 1, ± 3, $\pm \frac{1}{2}$, $\pm \frac{3}{2}$, $\pm \frac{1}{4}$, $\pm \frac{3}{4}$.

Trying the numbers in the preceding line in order, we find that $\frac{1}{2}$ is a root:

$$\begin{array}{r|rrrrr} \frac{1}{2} & 4 & -4 & 13 & -12 & 3 \\ & & 2 & -1 & 6 & -3 \\ \hline & 4 & -2 & 12 & -6 & 0. \end{array}$$

Any further root must be a root of the depressed equation $4x^3 - 2x^2 + 12x - 6 = 0$, or, dividing both members by 2, $2x^3 - x^2 + 6x - 3 = 0$. The only possible rational roots of the latter equation are ± 1, ± 3, $\pm \frac{1}{2}$, $\pm \frac{3}{2}$. Trying these in order, omitting those which failed before, we find $\frac{1}{2}$ is a root again; thus:

$$\begin{array}{r|rrrr} \frac{1}{2} & 2 & -1 & 6 & -3 \\ & & 1 & 0 & 3 \\ \hline & 2 & 0 & 6 & 0. \end{array}$$

Any further root is a root of the depressed equation $2x^2 + 6 = 0$, or $x^2 + 3 = 0$. The roots of this are $\pm i\sqrt{3}$. Hence the given equation has $\frac{1}{2}$ as a double root, and $\pm i\sqrt{3}$ as simple roots.

Always make a complete list of possible numerators, possible denominators, and possible rational roots, as in example 1. One cannot be sure of finding by elimination what is sought unless one is careful not to omit any possibilities.

Note 1. Our theorem cut the number of possible rational roots to be tried from infinitely many to twelve in this example.

Note 2. The work of example 1 shows also that the polynomial $4x^4 - 4x^3 + 13x^2 - 12x + 3$ can be expressed as $4(x - \frac{1}{2})(x - \frac{1}{2})(x - i\sqrt{3})(x + i\sqrt{3})$, by section 119, theorem 1. If we wish to express this polynomial as a product of factors with integral coefficients, we may write it as $2(x - \frac{1}{2}) \cdot 2(x - \frac{1}{2}) \cdot (x^2 + 3)$ or, finally, $(2x - 1)(2x - 1)(x^2 + 3)$.

COROLLARY 1. *If an equation has integral coefficients, and if the coefficient of the highest power of x is ± 1, then every rational root is an integer and a factor of the constant term.*

Proof. Immediate consequence of theorem. Left to reader.

Example 2. Solve $x^3 + 3x^2 + 4x + 12 = 0$, finding rational roots first.

Solution. The only possible rational roots are $\pm 1, 2, 3, 4, 6, 12$. Trying these in turn, we find that -3 is a root, thus:

$$\begin{array}{r|rrrr} -3 & 1 & 3 & 4 & 12 \\ & & -3 & 0 & -12 \\ \hline & 1 & 0 & 4 & 0. \end{array}$$

Further roots must be roots of the depressed equation $x^2 + 4 = 0$ or $x = \pm 2i$. Hence -3, $\pm 2i$ are the roots of the given equation.

Example 3. Factor the polynomial $x^3 - 2x^2 - x + 2$.

Solution. We find that the rational roots of the equation $x^3 - 2x^2 - x + 2 = 0$ are 2 and ± 1. Hence $x^3 - 2x^2 - x + 2 = (x - 2)(x - 1)(x + 1)$.

Note 3. The method of theorem 1 can be used for equations with rational coefficients also, as follows.

Example 4. Find the rational roots of $x^3 - 4x^2 - \frac{1}{2}x + 2 = 0$.

Solution. It is now incorrect to assert that a rational root must have a numerator which is a factor of 2 and a denominator which is a factor of 1, because not every coefficient of the given equation is an integer. *But every coefficient will become integral if we multiply both members by the L.C.D.* Thus our equation is equivalent to the equation $2x^3 - 8x^2 - x + 4 = 0$, and we can assert that a rational root must have a numerator which is a factor of 4 and a denominator which is a factor of 2. Hence the only possible rational roots are $\pm 1, 2, 4, \frac{1}{2}$. We find that 4 is a root, thus:

$$\begin{array}{r|rrrr} 4 & 2 & -8 & -1 & +4 \\ & & 8 & 0 & -4 \\ \hline & 2 & 0 & -1 & 0. \end{array}$$

The remaining roots are therefore $\pm \sqrt{2}/2$.

Note 4. Some theorems designed to shorten the labor by diminishing the number of necessary trials still further will be presented in sections 125 and 126. The instructor may delay the more tedious exercises until these are taken up if he wishes.

EXERCISES

Find all rational roots, and, if a quadratic depressed equation is obtained, solve the equation (completely):

1. $x^3 - 2x^2 - 5x + 6 = 0.$ **2.** $x^3 - 4x^2 + x + 6 = 0.$
3. $2x^3 - 13x^2 + 27x - 18 = 0.$
4. $4x^3 - 31x + 15 = 0.$ **5.** $3x^3 - x^2 - 22x + 24 = 0.$
6. $x^4 - x^3 - 19x^2 - 11x + 30 = 0.$
7. $8x^3 - 10x^2 - x + 3 = 0.$
8. $x^4 + 3x^3 - 12x^2 - 13x - 15 = 0.$
9. $9x^4 - 3x^3 + 7x^2 - 3x - 2 = 0.$
10. $3x^3 - 7x^2 - 3x + 2 = 0.$
11. $4x^4 - 7x^2 - 5x - 1 = 0.$
12. $12x^3 + 4x^2 - 13x + 4 = 0.$
13. $6x^4 + 2x^3 + 7x^2 + x + 2 = 0.$
14. $3x^4 + 2x^2 - 5 = 0.$ **15.** $x^4 - 2x^3 + x^2 - 2x = 0.$
16. $2x^4 - 5x^3 - 10x^2 + 3x = 0.$
17. $2x^4 + 5x^3 + 6x^2 + 2x = 0.$
18. $16x^4 + 8x^3 - 39x^2 + 18x = 0.$
19. $x^3 - 3x^2 - \frac{7}{2}x - 2 = 0.$ **20.** $2x^3 - \frac{25}{2}x^2 + \frac{7}{2}x - 3 = 0.$
21. $x^3 - \frac{17}{2}x^2 + \frac{9}{2}x - 4 = 0.$
22. $x^3 + \frac{17}{3}x^2 - \frac{5}{3}x + 2 = 0.$
23. $x^3 - \frac{9}{2}x^2 + \frac{7}{4}x + 1 = 0.$
24. $9x^4 - \frac{27}{2}x^3 + 11x^2 - \frac{27}{2}x + 2 = 0.$

Factor each of the following polynomials completely (that is, into linear factors):

25. $x^3 + 3x^2 - 10x + 24.$ **26.** $6x^3 - 13x^2 - 19x + 12.$
27. $18x^3 - 39x^2 + 8x + 16.$ **28.** $2x^3 + 5x^2 + 5x + 3.$
29. $2x^3 - 13x^2 + 32x - 13.$ **30.** $2x^3 - 5x^2 - 10x + 3.$
31. $4x^4 + 12x^3 + 17x^2 + 10x + 2.$
32. $x^4 - 4x^3 + x^2 - 6x + 36.$

Express each of the following as a product of linear and quadratic factors having real coefficients:

33. $2x^3 - x^2 + 8x - 4.$ **34.** $3x^3 + 10x^2 + 7x - 10.$
35. $2x^3 - x^2 + 4x + 15.$ **36.** $3x^3 + 2x^2 + 27x + 18.$
37. $6x^4 + 13x^3 + 24x^2 - 8.$
38. $4x^4 + 20x^3 + 97x^2 + 84x + 20.$

Decide whether each of the following statements is true or false and explain:

39. If p/q is a rational root, reduced to lowest terms, of $x^3 + \frac{7}{2}x^2 - \frac{5}{2}x + 2 = 0$, then p must be a factor of 2 and q must be a factor of 1.

40. Every rational root of $x^4 - \frac{37}{6}x^3 - 2x^2 + \frac{37}{3}x - 2 = 0$ must be an integer and a factor of 2.

123. Graphical Solution of Equations

The real roots, whether rational or irrational, of an equation $f(x) = 0$ with real coefficients, can be estimated approximately

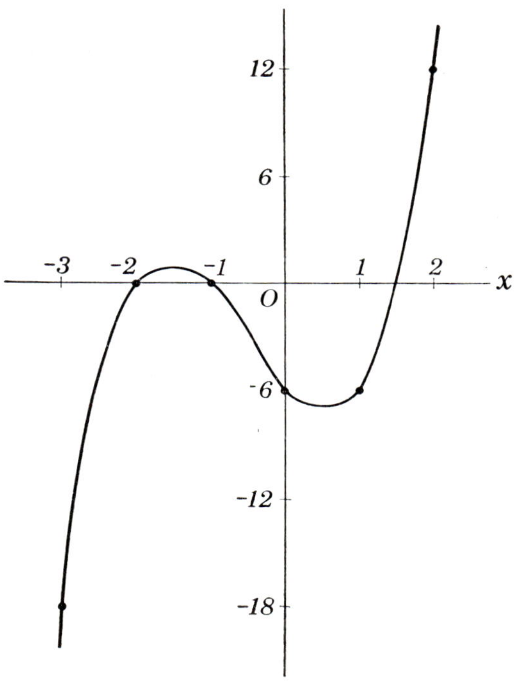

Fig. 55.

from the graph of the function $y = f(x)$, since they are the x-coordinates of the points where the graph meets the x-axis. (The student should review section 37.) The following fact is useful (see section 37):

If $f(x)$ is a polynomial with real coefficients, and $f(a)$ and $f(b)$ have opposite signs, a and b being two different real numbers, then the equation $f(x) = 0$ has at least one root between a and b.

Example. Solve graphically $2x^3 + 3x^2 - 5x - 6 = 0$.

Solution. From the function $y = 2x^3 + 3x^2 - 5x - 6$, we get the table

x	-3	-2	-1	0	1	2
y	-18	0	0	-6	-6	12

and the graph in Fig. 55. From the graph we estimate the roots to be -2, -1, and $3/2$.

It can be proved that at a multiple root, the graph is tangent to the x-axis.

EXERCISES

Plot the graphs of and estimate graphically the real roots of the equations in exercises **1–20**, *section 122, and of:*

21. $x^3 - 5x^2 - 8x + 12 = 0$. **22.** $x^3 + 3x^2 - 4x - 12 = 0$.

23. $x^4 + x^3 - 2x - 4 = 0$. **24.** $x^5 + x^3 - 2x^2 - 2 = 0$.

25. If $f(a)$ and $f(b)$ are both of the same sign, may we conclude that there exists no root between a and b? Explain.

26. If $f(a)$ and $f(b)$ are of opposite sign, may we conclude that there is only one root between a and b? Explain.

124. Transformation Reversing the Signs of the Roots

The roots $r_1, r_2 \ldots, r_n$ of the equation

$$f(x) = a_0x^n + a_1x^{n-1} + \ldots + a_n = 0 \tag{1}$$

or

$$f(x) = a_0(x - r_1)(x - r_2) \ldots (x - r_n) = 0 \tag{2}$$

may be obtained by setting each of the linear factors in (2) equal to zero. Clearly the equation $f(-x) = 0$, or

$$f(-x) = a_0(-x - r_1)(-x - r_2) \ldots (-x - r_n) = 0 \tag{3}$$

has the roots $-r_1, -r_2, \ldots, -r_n$, as may be seen by setting the linear factors of (3) equal to zero. Hence the equation $f(-x) = 0$ has as its roots the roots of $f(x) = 0$ each with its sign reversed. From (1) we see that

$$f(-x) = a_0(-x)^n + a_1(-x)^{n-1} + \ldots + a_n. \tag{4}$$

But $(-x)^k = x^k$ if k is even, and $(-x)^k = -x^k$ if k is odd. Hence to form the function $f(-x)$ we have only to reverse the signs of the odd-powered terms.

Example. As seen in section 123,

$$f(x) = 2x^3 + 3x^2 - 5x - 6 = 0$$

has the roots -2, -1, $3/2$. Hence

$$f(-x) = -2x^3 + 3x^2 + 5x - 6 = 0$$

has the roots $+2$, $+1$, $-3/2$.

Note. $f(-x) = 0$ must be distinguished from $-f(x) = 0$ which is obtained merely by multiplying both members of $f(x) = 0$ by -1 and consequently has exactly the same roots as $f(x) = 0$. See Fig. 56.

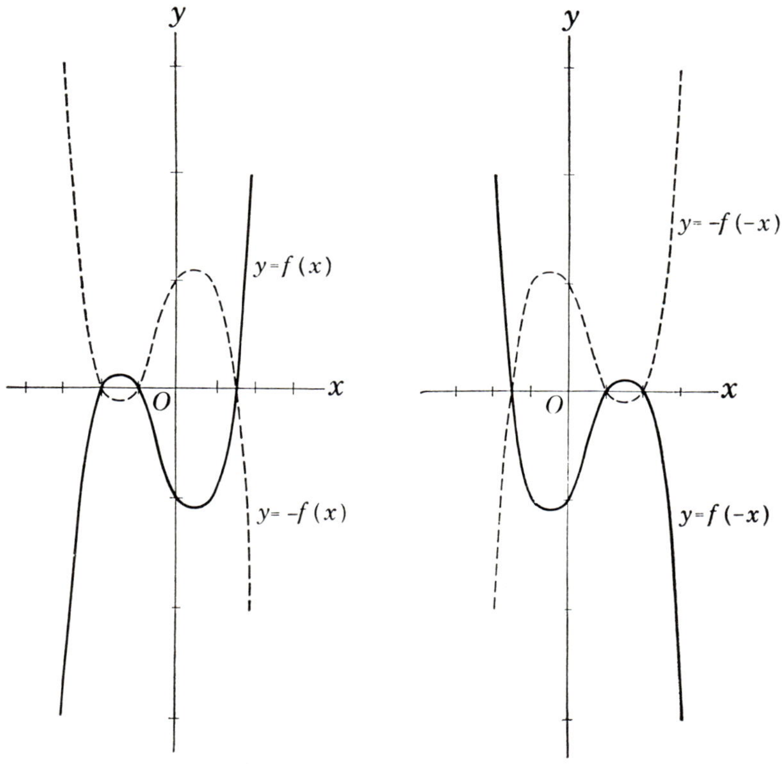

Fig. 56.

It follows that if we wish to avoid negative multipliers in synthetic division, *we may find the negative roots of* $f(x) = 0$ *by finding the positive roots of* $f(-x) = 0$ *and then reversing their signs.*

EXERCISES

Without solving, write an equation whose roots are those of the given equation, each with its sign reversed:

1. $x^3 - 2x^2 - 5x + 6 = 0.$ **2.** $x^3 - 5x^2 - 8x + 12 = 0.$
3. $x^3 - 4x^2 + x + 6 = 0.$ **4.** $x^3 + 3x^2 - 4x - 12 = 0.$
5. $9x^4 - 3x^3 + 7x^2 - 3x - 2 = 0.$
6. $4x^4 - 7x^2 - 5x - 1 = 0.$ **7.** $3x^4 + 2x^2 + 5 = 0.$
8. $3x^3 - 7x^2 - 3x + 2 = 0.$

Find the negative rational roots of the given equations by means of the positive roots of $f(-x) = 0$:

9. $5x^3 + 12x^2 - 36x - 16 = 0.$
10. $3x^3 + 4x^2 - 5x - 2 = 0.$
11. $6x^3 + 5x^2 - 7x - 4 = 0.$
12. $4x^3 - 17x^2 + 9x + 18 = 0.$

125. Upper and Lower Bounds for the Roots

The method of trial for finding rational roots (section 122) cuts the number of possible trials from infinitely many to a limited number. But if the coefficients are large, the number of trials may be unpleasantly large. For example, in the equation

$$(1) \qquad x^4 + x^3 + 70x^2 - 2x - 144 = 0$$

the possible rational roots are the factors of 144 or $\pm 1, 2, 3, 4, 6, 8, 9, 12, 16, 18, 24, 36, 48, 72, 144$, or thirty factors in all. If we tried them all, we would find that none of them satisfy the equation, which therefore has no rational roots. This work can be cut down by the following theorem:

THEOREM. *If $f(x) = a_0x^n + a_1x^{n-1} + \ldots + a_n = 0$ has real coefficients and has a positive leading coefficient a_0, and, if we divide synthetically, using a positive multiplier k, and obtain a third line containing no negative numbers, then no root of $f(x) = 0$ can be greater than k.*

Proof. Dividing $f(x)$ by $x - k$ we get

$$f(x) = (x - k)Q(x) + R.$$

By hypothesis the third line of synthetic division contains no negative numbers; that is, the coefficients of $Q(x)$ and R are not negative. Consider any value $x = m$ greater than k. Then

$m - k$ is positive; and since k is positive, so is m. Then substituting m in $Q(x)$, which has no negative coefficients, it is clear that $Q(m)$ is not negative. Also either $Q(m)$ or R must be positive, since they cannot both be zero. Hence $f(m)$ is positive and not zero. This completes the proof.

Note that any equation with real coefficients can be written with a_0 positive. (If it were negative, we could multiply both members by -1 without altering the roots.)

A number k such that no root is greater than k is called an **upper bound** for the roots. A **lower bound** for the negative roots can be found by finding an upper bound for the roots of $f(-x) = 0$ and reversing its sign.

Example. In the equation (1) above, we find that 2 is an upper bound for the roots, since no number in the third line of synthetic division is negative:

2 | 1	1	70	−2	−144
	2	6	152	300
1	3	76	150	156.

Therefore we need try no number greater than 2.

Here $f(-x) = x^4 - x^3 + 70x^2 + 2x - 144 = 0$ has 2 as an upper bound also, as can be seen from the synthetic division:

2 | 1	−1	70	2	−144
	2	2	144	292
1	1	72	146	148.

Hence -2 is a lower bound for the negative roots of $f(x) = 0$. That is, we need try no number less than -2.

If 2 is an upper bound for the roots, any greater number such as 3 or 100 is also an upper bound for the roots. It is advantageous to locate as low an upper bound as is conveniently possible in order to diminish the number of trials needed in the search for real roots.

EXERCISES

Find upper and lower bounds for the roots:

1. $x^4 + x^3 - 2x - 4 = 0$. **2.** $x^4 + 2x^3 - 6x - 9 = 0$.

3. $x^5 + x^3 - 2x^2 - 2 = 0$.

4. $x^4 + x^3 - 6x^2 - 8x - 16 = 0$.

Find all rational roots, using upper and lower bounds for the roots to cut down the number of trials:

5. $x^4 - 4x^3 + 20x^2 - 64x + 64 = 0.$
6. $x^4 - 2x^3 + 33x^2 - 72x - 108 = 0.$
7. $x^4 + 2x^3 + 22x^2 + 50x - 75 = 0.$
8. $2x^4 + 3x^3 + 70x^2 + 108x - 72 = 0.$
9. $x^4 - 2x^3 + 20x^2 - 32x + 64 = 0.$
10. $2x^4 + 2x^3 + 29x^2 + 9x + 90 = 0.$
11. $x^4 - 2x^3 + 29x^2 - 64x - 96 = 0.$
12. $3x^4 + 8x^3 + 105x^2 + 288x - 108 = 0.$
13. $x^4 + 2x^3 + 15x^2 - 6x - 54 = 0.$
14. $x^4 + x^3 + 10x^2 - 8x - 144 = 0.$
15. $x^6 + x^5 + 20x^4 + 4x^3 + 59x^2 - 5x - 80 = 0.$
16. $2x^6 + x^5 + 8x^4 - 32x^2 - 16x - 128 = 0.$

126. Descartes' Rule of Signs

Any information we can obtain concerning the roots of an equation before trying to solve it is likely to save us some labor. In this section we shall obtain some information concerning the number of positive roots and the number of negative roots we may expect to find.

When the terms of a polynomial with real coefficients are written in order of descending powers, we say that a **variation in sign** occurs if two successive terms have opposite signs, missing terms being ignored. For example, the polynomial $2x^3 - 4x^2 + x - 5$ has three variations in sign, since its coefficients taken in succession have the signs $+ - + -$. The polynomial $x^5 - 2x^3 - 3x^2 - 4x + 5$ has two variations in sign. The polynomial $x^3 + x + 5$ has no variations in sign. We shall prove the following theorem:

THEOREM 1. (*Descartes' rule of signs.*) *If the coefficients of $f(x) = 0$ are real, (a) the number of positive roots is not greater than the number of variations in sign; (b) the number of negative roots is not greater than the number of variations in sign of the polynomial $f(-x)$.*

Example 1. The number of positive roots of $f(x) = 2x^3 - 4x^2 + x - 5 = 0$ is three or less. There are no negative roots, since $f(-x) = -2x^3 - 4x^2 - x - 5$ has no variations in sign. That zero is not a root can be seen by inspection. Hence there are either three positive roots, or one positive root and two conjugate imaginary roots. (Why can there not be two positive and one imaginary?) As a result, we are saved the trouble of searching for negative roots.

Example 2. Show that $f(x) = x^4 - 10x + 1 = 0$ has exactly two imaginary roots.

Solution. By Descartes' rule, there are two or fewer positive roots, no negative roots, and zero is not a root. Hence there are either two imaginary and two positive roots, or four imaginary and no positive roots. The latter possibility is excluded by plotting the graph, for, since $f(0) = 1$ and $f(1) = -8$, there must be a real root between 0 and 1. Hence there must be two positive roots and two imaginary roots. This completes the proof.

Example 3. The number of positive roots of $x^4 - x^3 - x^2 - 7 = 0$ is one or none. The number of negative roots is one or none. Hence there are either two real roots and two imaginaries, or else four imaginaries.

Thus if one positive root in example 3 had already been located, we would not waste time looking for another. Similarly, if one negative root were already located, we would not look for more.

These remarks should make clear the practical value of Descartes' rule.

Before proving theorem 1, we need the following.

PRELIMINARY THEOREM. *If r is a positive root of the equation $f(x) = 0$ with real coefficients, then the quotient $Q(x)$ obtained by dividing $f(x)$ by $x - r$ has at least one less variation in sign than $f(x)$ has.*

Outline of proof. We write the equation $f(x) = 0$ so that the leading coefficient a_0 is positive; that is, if necessary, we multiply both members by -1, which has no effect on the roots or on the number of variations in sign. Now consider the process of synthetic division with r as multiplier. For example, consider the synthetic division of $x^8 + x^6 + 2x^5 - 4x^4 - 81x^2 + x + 2$ by $x - 2$:

2	+1	0	+1	+ 2	− 4	0	−81	+1	+2
		2	4	10	24	40	80	−2	−2
	+1	+2	+5	+12	+20	+40	− 1	−1	0.

It is clear from this example, and true in general, that the leading coefficient in $Q(x)$ has to be positive and that the coefficients of $Q(x)$ will remain positive at least until the first variation in sign of $f(x)$ and possibly longer. When a variation of $Q(x)$ takes place, from + to −, the succeeding coefficients of $Q(x)$ remain negative at least until the next variation in sign of $f(x)$ and possibly longer. But since the last number in the third line of synthetic division must be zero, the last coefficient of $Q(x)$ must have the opposite sign from that of the last coefficient in $f(x)$. Hence the largest number of variations $Q(x)$ can have is one less than the number of variations of $f(x)$, and it may have fewer than that.

Proof of theorem 1. (*a*) Let $r_1, r_2, \ldots, r_k$ be the positive roots of $f(x) = 0$. Dividing $f(x)$ by $x - r_1$ we get a quotient $Q_1(x)$ which has at least one less variation than $f(x)$. Dividing $Q_1(x)$ by $x - r_2$ we get a quotient $Q_2(x)$ which has at least two less variations than $f(x)$. And so on until we have $f(x) = (x - r_1)(x - r_2) \ldots (x - r_k)Q_k(x)$ where $Q_k(x)$ has at least k less variations than $f(x)$. But $Q_k(x)$ cannot have fewer than no variations at all. Hence $f(x)$ has at least as many variations as the number k of its positive roots.

(*b*) Part (*b*) follows at once from the fact that the positive roots of $f(-x) = 0$ correspond to the negative roots of $f(x) = 0$.

We get an even more useful result than theorem 1 by extending Descartes' rule as follows.

THEOREM 2. *If the number of positive roots of the equation $f(x) = 0$, with real coefficients, is fewer than the number of variations in sign, then it is fewer by an even number.*

> *Example 4.* From this theorem we see that the equation in example 3 above *must* have one positive and one negative root, since the number of positive roots cannot be one less than the number of variations in sign.

Outline of proof of theorem 2. By corollary 1, section 121, $f(x)$ can be expressed as a product of linear and quadratic factors, with real coefficients. The quadratic ones arise from pairs of conjugate imaginary roots. They are of the form $(x - [a + bi])(x - [a - bi]) = x^2 - 2ax + a^2 + b^2$ or $x^2 + px + q$ where $q = a^2 + b^2$ is necessarily positive. The linear ones arise from real roots, positive, negative, or zero. Let $r_1, r_2, \ldots, r_k$ be the positive roots, $-\rho_1, -\rho_2, \ldots, -\rho_h$ the negative ones. And let there be j pairs of conjugate imaginary roots, and g zero roots. Then

$$f(x) = a_0(x - r_1)(x - r_2) \ldots (x - r_k) \cdot (x + \rho_1)(x + \rho_2) \ldots (x + \rho_h) \cdot (x^2 + p_1x + q_1)(x^2 + p_2x + q_2) \ldots (x^2 + p_jx + q_j) \cdot x^g.$$

A little experimentation will indicate that the product of factors of the form $F(x) = (x - r_1) \ldots (x - r_k)$, arising from the positive roots, will have alternating signs. The product of all the other factors will be a polynomial $G(x)$ whose first and last coefficients are surely positive. Now $F(x)$ has k variations and its first and last coefficients will have the same sign if k is even and opposite signs if k is odd. Hence $f(x) = a_0 \cdot F(x) \cdot G(x)$ will be such that its first and last coefficients have the same sign if k is even and opposite signs if k is odd. But the number of variations in sign of a polynomial $f(x)$ is even if its first and last coefficients have the same sign and odd if the first and last coefficients have opposite signs. Hence the number V of variations in sign of $f(x)$ is even if the number k of positive roots is even and odd if the number k of positive roots is odd. Since the difference between two even numbers is even, and the difference between two odd numbers is even, the number $V - k$ is always even. This completes the proof.

Note. The method of theorem 1, section 122, can be used to prove that numbers such as $\sqrt{2}$, $\sqrt{3}$, $\sqrt{6}$, $\sqrt[3]{2}$, etc., are not rational, as in the following example:

Example 5. Prove that $\sqrt{2}$ is not rational.

Solution. Let $x = \sqrt{2}$. Then $x^2 = 2$, or $x^2 - 2 = 0$. The only possible rational roots of this equation are $\pm 1, 2$. Synthetic division will show that none of these are roots. Hence the equation $x^2 - 2 = 0$ has no rational roots. But by Descartes' rule, it has one positive and one negative root, namely $\pm \sqrt{2}$. Hence these real roots are not rational.

René Descartes (French, 1596–1650), to whom theorem 1 is credited, was a great mathematician and philosopher. Perhaps his most valuable contribution to human knowledge was his work in analytic geometry, of which he was the principal founder.

EXERCISES

Without solving, obtain whatever information you can concerning the number of positive, negative, and imaginary roots (a) by means of Descartes' rule (theorem 1) and previous theorems; (b) by means of the extended form of Descartes' rule (theorem 2):

1. $x^3 + 6x - 4 = 0$.
2. $3x^3 + 5x + 1 = 0$.
3. $x^4 + 2x^2 - 5 = 0$.
4. $x^4 + 3x^2 + 5 = 0$.

5. $x^4 + 3x^2 - 5x - 1 = 0.$ **6.** $x^3 - 2x^2 + 3x - 7 = 0.$
7. $x^4 - 3x^3 + 4x^2 + 5 = 0.$ **8.** $x^3 + 2x^2 + 3x = -4.$
9. $x^4 + 3x^2 + 4 = 0.$ **10.** $x^6 + 2x^4 + 5x^2 + 3 = 0.$
11. $x^4 - 3x^2 + 5 = 0.$ **12.** $x^6 + x^4 - 3x^2 - 4 = 0.$
13. $x^3 - 3x^2 + 4x - 5 = 0.$
14. $x^4 + 3x^3 - 2x^2 - 5x + 2 = 0.$
15. $x^5 + 32 = 0.$ **16.** $x^7 - 1 = 0.$
17. $x^6 - 64 = 0.$ **18.** $x^4 - 81 = 0.$
19. $x^3 + 4x = 0.$ **20.** $x^5 + 5x^3 + 2x = 0.$

Prove that each equation has exactly two real and two imaginary roots:

21. $x^4 - 10x^3 + 1 = 0.$ (*Hint:* consider the graph.)
22. $2x^4 - 15x + 3 = 0.$

Prove that each equation has exactly two real and four imaginary roots:

23. $x^6 - 14x^3 + x^2 + 3 = 0.$ **24.** $2x^6 + x^4 + 10x + 3 = 0.$

Prove that each equation has exactly two irrational roots:

25. $x^4 + x^2 - 6 = 0.$ **26.** $x^4 + x^3 - 2x^2 - 4 = 0.$

Find all rational roots, using Descartes' rule to cut down the number of trials:

27. $x^6 - x^5 - x^4 - x^3 + 30x^2 - 32x - 64 = 0.$
28. $x^6 + 2x^5 - 2x^4 + 2x^3 + 33x^2 + 72x - 108 = 0.$
29. Work the exercises of section 122, using the theorems of the present section to cut down the number of trials.

Show by the method of example 5 that each of the following numbers is not rational:

30. $\sqrt{3}.$ **31.** $\sqrt{6}.$ **32.** $\sqrt{8}.$ **33.** $\sqrt[3]{2}.$ **34.** $\sqrt[3]{3}.$
35. $\sqrt[3]{6}.$ **36.** $\sqrt[4]{6}.$ **37.** $\sqrt[5]{8}.$ **38.** $2 + \sqrt{3}.$
39. The lengths of the sides of a rectangular box, measured in inches, are three consecutive integers. If the volume is 210 cubic inches, find the sides.
40. After a slice 1 centimeter thick is cut off from one side of a cube, the volume which remains is 180 cubic centimeters. Find the side of the original cube.
41. An open box is to be made from a rectangular sheet of tin 10 inches by 14 inches by cutting equal squares from the corners and folding up the sides. Find one possible length of the side of the cutout square if the volume of the box is to be 120 cubic inches. How many solutions are there?

42. A rectangular box has the dimensions 3 inches, 4 inches, 5 inches. If the length, width, and height were increased by the same amount, the volume would be doubled. Find the amount by which the dimensions are to be increased.

43. An open box is to be made from a square sheet of tin 18 inches on each side by cutting out equal squares from the corners and folding up the sides. How long should the edge of the cutout square be if the volume of the box is to be 392 cubic inches?

127. Real Roots by Successive Approximations

In section 122 a method was given for finding exactly the rational roots of polynomial equations with rational coefficients. For practical purposes, an exact answer is seldom required. Rather, what is needed is a sufficiently good approximation. For example, in finding the distance of an enemy battleship from a shore battery we would not be interested in tenths of an inch. In drilling cylinders for an airplane engine we would, but there we would not care about millionths of an inch. Thus, for practical purposes it is sufficient if we provide *a method which will yield as many decimal places of the real roots of an equation as we happen to want,* no matter whether the roots are rational or irrational. Such a method for equations with real coefficients will now be explained by an example.

Example. Find the positive root of $f(x) = x^4 + x^3 - x^2 - 3x - 6 = 0$, correct to the nearest hundredth.

Solution. Substituting successive integers, we find that $f(0)$ and $f(1)$ are negative, while $f(2)$ is positive. (The substitution is most conveniently done by synthetic division.) That is,

$$\begin{array}{r|rrrrrl}
1 & 1 & 1 & -1 & -3 & -6 & \\
 & & 1 & 2 & 1 & -2 & \\
\hline
 & 1 & 2 & 1 & -2 & \mathbf{-8} & = f(1) \\
2 & 1 & 1 & -1 & -3 & -6 & \\
 & & 2 & 6 & 10 & 14 & \\
\hline
 & 1 & 3 & 5 & 7 & \mathbf{8} & = f(2).
\end{array}$$

By section 123, the change of sign indicates that the graph has crossed the x-axis; that is, the positive root r is between 1 and 2. Hence $r = 1. \ldots$.

To find the tenths place, we split the interval from 1 to 2 into tenths and try them in succession. We find that $f(1.1)$, $f(1.2)$, $f(1.3)$, $f(1.4)$, $f(1.5)$, $f(1.6)$, and $f(1.7)$ are negative, while $f(1.8)$ is positive. That is,

1.7	1	1	−1	−3	−6
		1.7	4.59	6.103	5.2751
	1	2.7	3.59	3.103	**−0.7249** $= f(1.7)$
1.8	1	1	−1	−3	−6
		1.8	5.04	7.272	7.6896
	1	2.8	4.04	4.272	**1.6896** $= f(1.8)$.

Therefore, the positive root is between 1.7 and 1.8, or $r = 1.7 \ldots$.

To find the hundredths place, we split the interval between 1.7 and 1.8 into hundredths and try them in succession. We find that $f(1.71)$, $f(1.72)$, and $f(1.73)$ are negative, while $f(1.74)$ is positive. That is,

1.73	1	1	−1	−3	−6
		1.73	4.7229	6.440617	5.95226741
	1	2.73	3.7229	3.440617	**−0.04773259** $= f(1.73)$
1.74	1	1	−1	−3	−6
		1.74	4.7676	6.555624	6.18678576
	1	2.74	3.7676	3.555624	**0.18678576** $= f(1.74)$.

Hence the root r lies between 1.73 and 1.74, or $r = 1.73 \ldots$.

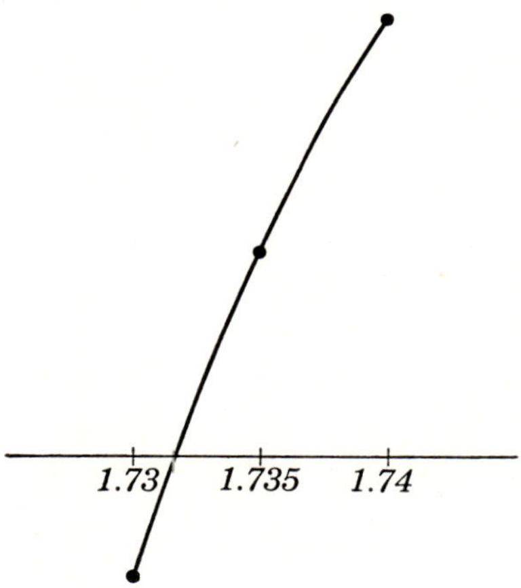

Fig. 57.

The nearest hundredth can be determined by finding the thousandths place; if it is 5 or more, we correct the hundredths place upwards by 1, and if it is less than 5 we leave the hundredths place as it is. But it is unnecessary to do this much work if one thinks of the graph. To correct to the nearest hundredth, we have only to try the 5 in the thousandths place; if $f(x)$ has the same sign as the lower hundredth, we must correct upward, if $f(x)$ has the same sign as the upper hundredth, we leave the hundredths place as it is (Fig. 57). Thus

1.735	1	1	−1	−3	−6
		1.735	4.745225	6.497965375	6.068969925625
	1	2.735	3.745225	3.497965375	**0.068969925625**
					$= f(1.735)$.

Since $f(1.735)$ has the same sign as $f(1.74)$, the crossing (Fig. 57) took place to the left of the middle point, and is nearer 1.73 than 1.74. Hence $r = 1.73$ to the nearest hundredth.

To find a negative root of $f(x) = 0$, we may use this process to find the corresponding positive root of $f(-x) = 0$ and then reverse the sign.

Note 1. The process of successive approximations is much like that of running down a base runner who tried to steal second base. We crowd the root r between narrowing bounds, first between successive integers, then between successive tenths, hundredths, thousandths, and so on. In the case of an irrational number or a rational number which happens to have an endless decimal expression, we never tag it exactly, but we approximate it more closely at each stage.

Note 2. This process may be used for any equation $f(x) = 0$ where $f(x)$ is a continuous function, and is not restricted to polynomial equations, except that for functions $f(x)$ which are not polynomials the substitutions have to be done directly instead of by synthetic division.

Note 3. Unless a calculating machine is available for the multiplications involved in the synthetic divisions, this process becomes tedious if it has to be carried beyond one or two decimal places. The next section will explain one way to avoid some of the labor. A further laborsaving device will be discussed in section 130.

EXERCISES

Approximate to the nearest hundredth:

1. The positive root of $x^4 + 4x^3 + 2x^2 - 12x - 15 = 0$.
2. The root of $x^4 - 11x^2 + 15x - 2 = 0$ lying between 0 and 1.
3. The positive root of $x^3 + 3x - 5 = 0$.
4. The positive root of $x^3 - 2x - 7 = 0$.
5. The smaller positive root of $x^4 - 5x^3 + 2x^2 + x + 7 = 0$.
6. The negative root of $x^4 - x^3 - 11x^2 + 12x - 12 = 0$.
7. The negative root of $x^4 - 2x^3 - 3x^2 - 2x - 4 = 0$.
8. $\sqrt[3]{10}$. (*Hint:* $\sqrt[3]{10}$ is the positive root of $x^3 - 10 = 0$.)
9. $\sqrt[3]{6}$. **10.** $\sqrt[3]{-11}$.

★128. Linear Interpolation

If a root r of an equation $f(x) = 0$ has been computed, by the method of the preceding section, say, up to a certain decimal place, an approximation of the next place may be obtained as follows.

Suppose that, in the example of the preceding section, we have found that $f(1) = -8$ and $f(2) = 8$. We want to find the value of r between 1 and 2 for which $f(r) = 0$. If we assume that the change in $f(x)$ is proportional to the change in x, we may say that the total change in $f(x)$ is $f(2) - f(1) = 16$, while the change in x is $2 - 1 = 1$; but the change in $f(x)$ from -8 to 0 is 8, while the change in x from 1 to r is called d. Then our assumption amounts to the proportion

$$\frac{d}{1} = \frac{8}{16} \text{ or } d = 0.5. \tag{1}$$

Thus we would conclude that $r = 1 + 0.5 = 1.5$ approximately.

The calculation may be arranged as follows:

$$\begin{array}{c|c} x & f(x) \\ \hline 1 & -8 \\ r = 1 + d & 0 \\ 2 & 8 \end{array}$$

(with d the change in x from 1 to r, 1 the change from 1 to 2; 8 the change in $f(x)$ from -8 to 0, 16 the change from -8 to 8)

Therefore $\dfrac{d}{1} = \dfrac{8}{16}$, or $d = 0.5$. Hence $r = \mathbf{1.5}$.

Note that this is not accurate, since the more careful calculation of the preceding section shows that, as far as the tenths place is concerned, $r = 1.7 \ldots$. But the present calculation would serve a purpose if only to save us the trouble of calculating $f(1.1), f(1.2), f(1.3), f(1.4)$ in the more accurate computation of the preceding section.

The proportion (1) rewritten in the form $8/d = 16/1$ amounts to the assumption that the graph between the points $(1,-8)$ and $(2,8)$ is a straight line of slope 16. That is, we are replacing the actual curve by the chord (straight line segment) joining these two points. For this reason, this process is called **linear interpolation** or **interpolation by proportional**

★ This section may be omitted if Horner's method is to be taken up.

parts. The accuracy of the result will thus depend on how closely the chord approximates the curve. Hence, the accuracy is likely to improve as the distance between the values of x diminishes. Thus, suppose we have found in the example of the preceding section that $f(1.7) = -0.7249$ and $f(1.8) = 1.6896$. Then

		x	$f(x)$		
0.1	d	1.7	-0.7249	0.7249	2.4145
		$r = 1.7 + d$	0		
		1.8	1.6896		

and
$$\frac{d}{0.1} = \frac{0.7249}{2.4145} = 0.30, \text{ approximately.}$$

Hence
$$d = 0.03, \text{ approximately,}$$

and
$$r = 1.73, \text{ approximately,}$$

which is accurate as far as the hundredths place is concerned.

The process of linear interpolation can also be done graphically if desired. Thus in the preceding calculation we could have plotted on a graph the points $(1.7, -0.7249)$ and $(1.8, 1.6896)$ as closely as possible, joined them by a straight line, and read from the graph the x-coordinate of the point where the straight line crosses the x-axis.

EXERCISES

1–10. Work the exercises 1–10 of the preceding section, using the method of linear interpolation to find the digit in the hundredths place.

Approximate to the nearest thousandth, using linear interpolation to find the digit in the thousandths place:

11. The positive root of $x^4 + 2x^3 - 3x^2 + 2x - 4 = 0$.
12. The negative root of $x^4 + 2x^3 - 3x^2 + 2x - 4 = 0$.
13. $\sqrt[3]{3}$. **14.** $\sqrt[3]{31}$. **15.** $\sqrt[3]{13}$. **16.** $\sqrt[3]{-15}$.

129. Transformations of Equations

It is often useful to transform an equation into a new equation whose roots have a certain relation to the roots of the given equation. One such transformation was taken up in section 124.

(*A*) *Transformation diminishing all the roots by a specified constant h.* From the form

(1) $$f(x) = a_0(x - r_1)(x - r_2) \dots (x - r_n) = 0$$

of the given equation

(2) $$f(x) = a_0x^n + a_1x^{n-1} + \dots + a_n = 0,$$

it is clear that substituting $x = X + h$ will yield the desired equation; for then the roots of

$$f(X + h) = a_0(X + h - r_1)(X + h - r_2) \dots (X + h - r_n) = 0$$

are obtained by setting each factor equal to zero, thus obtaining the required roots $X = r_1 - h$, $X = r_2 - h$, $\dots$, $X = r_n - h$. Hence we could obtain the desired equation from (2) by writing

(3) $$f(X + h) = a_0(X + h)^n + a_1(X + h)^{n-1} + \dots + a_n = 0,$$

expanding each power by multiplication, and collecting like terms, thus obtaining the desired equation

(4) $$F(X) = A_0X^n + A_1X^{n-1} + \dots + A_n = 0.$$

This process is tedious and we will calculate the coefficients of the desired equation (4) by a more efficient method, as follows.

Since $x = X + h$, and $f(x) = f(X + h) = F(X)$, we may write (4) as

$$f(x) = A_0(x - h)^n + A_1(x - h)^{n-1} + \dots + A_{n-1}(x - h) + A_n = 0.$$

Dividing the left member by $x - h$, we therefore must get the remainder $R_1 = A_n$. Dividing the resulting quotient

$$A_0(x - h)^{n-1} + A_1(x - h)^{n-2} + \dots + A_{n-1}$$

by $x - h$, we get the remainder $R_2 = A_{n-1}$, and so on until finally $A_0 = a_0$, as can be seen from (3). The successive remainders and a_0 are the coefficients we seek in reverse order. Writing x instead of X, we have the desired equation

(5) $$a_0x^n + R_nx^{n-1} + R_{n-1}x^{n-2} + \dots + R_2x + R_1 = 0.$$

The divisions may be arranged synthetically, as in the following example.

Example 1. Find an equation whose roots are the roots of $x^3 - 2x^2 - 5x + 6 = 0$, each diminished by 2.

Solution. Dividing synthetically by $x - 2$ repeatedly, we get

$$\begin{array}{r|rrrl} 2 & 1 & -2 & -5 & 6 \\ & & 2 & 0 & -10 \\ \hline & 1 & 0 & -5 & -4 = R_1 = A_3 \\ & & 2 & 4 & \\ \hline & 1 & 2 & -1 = R_2 = A_2 & \\ & & 2 & & \\ \hline a_0 = & \mathbf{1} & \mathbf{4} = R_3 = A_1. & & \end{array}$$

The desired equation is, according to (5), $x^3 + 4x^2 - x - 4 = 0$.

Solving both equations as a check, we find the original equation has the roots 1, 3, -2, while the new equation has the roots -1, 1, -4, as required.

Example 2. Find an equation whose roots are the roots of $x^3 - 2x^2 - 5x + 6 = 0$, each increased by 2.

Solution. Increasing by 2 is equivalent to diminishing by -2. Hence, from the calculation

$$\begin{array}{r|rrrr} -2 & 1 & -2 & -5 & 6 \\ & & -2 & 8 & -6 \\ \hline & 1 & -4 & 3 & \mathbf{0} \\ & & -2 & 12 & \\ \hline & 1 & -6 & \mathbf{15} & \\ & & -2 & & \\ \hline & \mathbf{1} & \mathbf{-8}, & & \end{array}$$

we find the equation $x^3 - 8x^2 + 15x = 0$. This equation has the roots 3, 5, 0, as required.

(*B*) *Transformation multiplying all the roots by a constant k.* Substituting $x = X/k$ in (1) will clearly produce an equation whose roots are $X = kr_1, X = kr_2, \ldots, X = kr_n$ as may be seen by setting each of the factors of

$$f\left(\frac{X}{k}\right) = a_0\left(\frac{X}{k} - r_1\right)\left(\frac{X}{k} - r_2\right) \cdots \left(\frac{X}{k} - r_n\right) = 0$$

equal to zero. From (2) we see that

$$f\left(\frac{X}{k}\right) = a_0\left(\frac{X}{k}\right)^n + a_1\left(\frac{X}{k}\right)^{n-1} + a_2\left(\frac{X}{k}\right)^{n-2} + \cdots + a_{n-1}\left(\frac{X}{k}\right) + a_n = 0.$$

Multiplying by the L.C.D. k^n, we find the desired equation

$$F(X) = a_0X^n + a_1kX^{n-1} + a_2k^2X^{n-2} + \cdots + a_{n-1}k^{n-1}X + a_nk^n = 0.$$

Hence the coefficients of the desired equation are obtained by multiplying each coefficient a_m $(m = 0, 1, 2, \ldots, n)$ of the given equation by k^m.

Example 3. Write an equation whose roots are those of $x^3 - 2x^2 - 5x + 6 = 0$, each multiplied by 2.

Solution. The desired equation is

$$X^3 - 2 \cdot 2X^2 - 5 \cdot 2^2X + 6 \cdot 2^3 = 0$$

or, writing x instead of X,

$$x^3 - 4x^2 - 20x + 48 = 0.$$

Since the original equation has the roots 1, 3, -2, the new equation should have the roots 2, 6, -4. The student should check.

Note. The transformation of section 124 is a special case of (B), above, taking $k = -1$.

EXERCISES

Write an equation whose roots are those of the given equation, each decreased by the given number h, and check:

1. $x^3 - 3x^2 - 4x + 12 = 0$; $h = 1$.
2. $x^3 - 3x + 2 = 0$; $h = 2$.
3. $x^3 - 4x^2 + x + 6 = 0$; $h = 2$.
4. $4x^4 + 12x^3 - 17x^2 - 3x + 4 = 0$; $h = -2$.
5. $2x^3 - 5x^2 - 4x + 3 = 0$; $h = -2$.
6. $x^3 - 6x^2 + 11x - 6 = 0$; $h = -1$.
7. $x^3 - 2x^2 - x + 2 = 0$; $h = 0.2$.
8. $x^3 - x^2 - 4x + 4 = 0$; $h = -0.3$.
9. $x^3 - 4x^2 + x + 6 = 0$; $h = 0.02$.
10. $x^4 + 7.8x^3 + 21.44x^2 + 21.922x - 0.7249 = 0$; $h = 0.02$.

Write an equation whose roots are those of the given equation each multiplied by the given number k:

11. $x^3 - 3x^2 - 4x + 12 = 0$; $k = 2$.
12. $x^3 - 3x + 2 = 0$; $k = 4$.
13. $x^3 - 4x^2 + x + 6 = 0$; $k = 10$.
14. $4x^4 + 12x^3 - 17x^2 - 3x + 4 = 0$; $k = 3$.
15. $x^4 + 7.8x^3 + 21.44x^2 + 21.922x - 0.7249 = 0$; $k = 10$.
16. $x^4 + 2.3x^3 + 1.43x^2 - 6.347x - 1.5782 = 0$; $k = 10$.

130. Horner's Method

The most tedious feature of the method of section 127 is the fact that the synthetic divisions with multipliers of two, three, or more digits require numerous multiplications which, in the absence of a calculating machine, demand much labor. This labor can be somewhat diminished by a method published by W. G. Horner in 1819,* the advantage of which is that all multipliers in synthetic divisions will consist of one digit only, so that all multiplications can be done mentally. We shall explain the method by reworking the example of section 127.

Example. Find the positive root of $f(x) = x^4 + x^3 - x^2 - 3x - 6 = 0$, correct to the nearest hundredth.

Solution. Substituting successive integers, we locate the positive root r between 1 and 2, since $f(1) = -8$ and $f(2) = 8$.

Now instead of substituting 1.1, 1.2, 1.3, . . ., which would involve 2-digit multipliers, we apply the transformation (A) of section 129, forming a new equation whose roots are equal to those of the original equation each diminished by 1, thus:

1	1	1	−1	−3	−6
		1	2	1	−2
	1	2	1	−2	**−8**
		1	3	4	
	1	3	4	**2**	
		1	4		
	1	4	**8**		
		1			
	1	**5.**			

The new equation $f_1(x) = x^4 + 5x^3 + 8x^2 + 2x - 8 = 0$ has a root between 0 and 1 which is exactly 1 less than the root r of the original equation. Hence in the new equation $f_1(x) = 0$ we may try successive tenths. We find that $f(0.7) = -0.7249$ and $f(0.8) = 1.6896$. Therefore $f_1(x)$ has a root 0.7 . . ., and the original equation has the root $r = 1.7$ . . . .

Instead of substituting 0.71, 0.72, 0.73, . . . in $f_1(x)$, we transform $f_1(x) = 0$ into a new equation whose roots are equal to those of $f_1(x) = 0$, each diminished by 0.7, thus:

* Horner's method, or something closely resembling it, was anticipated by Ruffini in 1804, and by the Chinese Ch'in Chiu-shao more than five centuries earlier.

0.7	1	5	8	2	−8
		0.7	3.99	8.393	7.2751
	1	5.7	11.99	10.393	**−0.7249**
		0.7	4.48	11.529	
	1	6.4	16.47	**21.922**	
		0.7	4.97		
	1	7.1	**21.44**		
		0.7			
	1	**7.8.**			

The new equation $f_2(x) = x^4 + 7.8x^3 + 21.44x^2 + 21.922x - 0.7249 = 0$ has a root between 0 and 0.1, exactly 0.7 less than the root of $f_1(x) = 0$, and exactly 1.7 less than the root r of $f(x) = 0$. Hence in $f_2(x)$ we substitute successive hundredths 0.01, 0.02, 0.03, and so on. We find that $f_2(0.03) = -0.04773259$ and $f_2(0.04) = 0.18678576$. Hence the positive root of $f_2(x) = 0$ is 0.03 . . ., the positive root of $f_1(x) = 0$ is 0.73 . . ., and the root r of $f(x) = 0$ is $r = 1.73$ . . . .

Instead of substituting 0.031, 0.032, 0.033, . . ., we may diminish the roots of $f_2(x) = 0$ by 0.03, thus:

0.03	1	7.8	21.44	21.922	−0.72490000
		0.03	0.2349	0.650247	0.67716741
	1	7.83	21.6749	22.572247	**−0.04773259**
		0.03	0.2358	0.657321	
	1	7.86	21.9107	**23.229568**	
		0.03	0.2367		
	1	7.89	**22.1474**		
		0.03			
	1	**7.92.**			

The equation $f_3(x) = x^4 + 7.92x^3 + 22.1474x^2 + 23.229568x - 0.04773259 = 0$ has a positive root between 0 and 0.01 exactly 0.03 less than that of $f_2(x)$ and 1.73 less than the root r of $f(x)$. If we wished to find the thousandths place of r, we would substitute 0.001, 0.002, and so on in $f_3(x)$. To find the value of r to the nearest hundredth, however, it is sufficient to substitute 0.005 in $f_3(x)$. By considering the graph (compare Fig. 57) it is obvious that if $f_3(0.005)$ has the same sign as at the left-hand end of the interval, we must correct the hundredths place upward; while if $f_3(0.005)$ has the same sign as at the right-hand end of the interval, we leave the hundredths place unaltered. In our example, we find that $f_3(0.005)$ is positive; hence $r = 1.73$ correct to the nearest hundredth.

Clearly this process will yield as many places as we wish in the decimal expression of a positive real root. *To find a negative root of $f(x) = 0$, we find the corresponding positive root of $f(-x) = 0$ and reverse the sign.*

Note 1. After the tenths place of the root has been determined, an estimate of the next decimal place can be made by neglecting the powers of x higher than the first, as follows. In the equation, $f_2(x) = 0$ above, we obtain the equation $21.922x - 0.7249 = 0$ by neglecting powers higher than the first. This yields

$$x = \frac{0.7249}{21.922} = 0.033 = 0.03 \text{ approximately,}$$

which is accurate for the next place. The student may verify that this method will give a very inaccurate prediction if applied to $f_1(x) = 0$, or $f(x) = 0$. The reason for this becomes apparent when we understand the justification of the method, which is as follows. If the root is very close to zero, then its square, cube, and so on are very much smaller. For example, if the root is 0.03, its square is 0.0009 and its cube is 0.000027, and so on. Hence higher powers of x will contribute a negligible amount to the value of the function unless their coefficients happen to be correspondingly large, which is not usually the case. While the estimate obtained in this way is not reliable, it may be used to save unnecessary trials. Thus if we get an estimate of 0.07 with this method, we would not bother to try 0.01, 0.02, . . ., 0.05 at first, but would start with 0.07. The accuracy of this estimate increases as we work with later decimal places.

Note 2. If it is desired to avoid the use of decimals, we may transform each equation after the original one into a new equation whose roots are equal to the roots of the preceding one, each multiplied by 10, by transformation (*B*) of section 129. Then instead of substituting 0.1, 0.2, and so on, we could substitute 1, 2, and so on.

Note 3. Horner's method incidentally provides a technique for calculating real nth roots. For $\sqrt[3]{2}$, for example, is nothing else than the positive root of $x^3 - 2 = 0$.

Note 4. Horner's process applies only to polynomial or integral rational equations, while the process of section 127 may be applied to a wider class of equations. A still more efficient method, devised by Isaac Newton (English, 1642–1727), applies to the wider class of equations, but requires some preliminary study of differential calculus and cannot be taken up here. For a discussion of Newton's method see textbooks on the theory of equations.

Note 5. In solving a polynomial equation with rational or integral coefficients, it is advisable to obtain information from Descartes' rule first, then search for rational roots, then approximate irrational roots.

EXERCISES

1–10. Using Horner's method, work exercises 1–10 of section 127.

11–16. Using Horner's method, work exercises 11–16 of section 128.

Using Horner's method, approximate to the nearest thousandth:

17. The larger real root of $x^4 - 4x^3 + 2x^2 - 4x + 1 = 0$.

18. The smaller positive root of $x^4 - 7x^3 + 4x^2 + 3x + 11 = 0$.

19. The two roots of $x^4 + 2x^3 - 6x^2 - 4x + 8 = 0$ lying between 1 and 2.

20. The negative root of $x^4 - x^3 - 5x^2 - 6x - 4 = 0$.

21. $\sqrt[3]{37}$. **22.** $\sqrt[3]{-18}$. **23.** $\sqrt[3]{-43}$. **24.** $\sqrt[5]{12}$.

25. All the real roots of $x^4 - 6x^3 + 8x^2 + 2x - 1 = 0$.

26. All the real roots of $x^4 - 10x^3 + 29x^2 - 18x - 6 = 0$.

27. In studying the performance of a certain airplane, it is necessary to solve the equation* $x - x^4 = Z$. If $Z = 0.2$, show that there are two real roots between 0 and 1 and find the larger one to the nearest hundredth.

28. An open box is to be made from a rectangular sheet of tin 8 inches by 10 inches by cutting equal squares from the corners and folding up the sides. Find the side of the cutout square to the nearest hundredth if the volume is to be 44 cubic inches.

29. In studying the lateral stability of a certain airplane, it is necessary to solve the equation* $x^4 + 45x^3 + 214x^2 + 1492x + 595 = 0$. Find the real roots to the nearest tenth.

30. Find the thickness of the wall of a hollow spherical shell of uniform thickness if its outer diameter is 10 inches and its capacity or volume is 300 cubic inches.

* The equations of exercises 27 and 29 are reprinted by permission from *Theory of Flight* by R. von Mises, copyrighted, 1945, by the McGraw-Hill Book Co., Inc.

131. Relations between Coefficients and Roots

Since the values of the coefficients of an equation determine the values of its roots, it is natural to expect the existence of some definite relations between the coefficients and the roots.

Consider the equation

$$(1) \quad a_0x^n + a_1x^{n-1} + a_2x^{n-2} + \ldots + a_{n-1}x + a_n = 0 \; (a_0 \neq 0)$$

of degree n, or, in factored form,

$$(2) \qquad a_0(x - r_1)(x - r_2) \ldots (x - r_n) = 0.$$

Dividing both members of (1) and of (2) by a_0, we get the equivalent equations

$$(3) \qquad x^n + \frac{a_1}{a_0}x^{n-1} + \frac{a_2}{a_0}x^{n-2} + \ldots + \frac{a_{n-1}}{a_0}x + \frac{a_n}{a_0} = 0,$$

and

$$(4) \qquad (x - r_1)(x - r_2) \ldots (x - r_n) = 0.$$

Since the left members of (3) and (4) are identically equal, it follows from corollary 2, section 119, that we may equate coefficients of like powers.

For simplicity, let us do this first for $n = 3$. Multiplying $(x - r_1)(x - r_2)(x - r_3)$, we obtain

$$x^3 + (-r_1 - r_2 - r_3)x^2 + (r_1r_2 + r_1r_3 + r_2r_3)x + (-r_1r_2r_3) = 0.$$

Comparing these coefficients with those of

$$x^3 + \frac{a_1}{a_0}x^2 + \frac{a_2}{a_0}x + \frac{a_3}{a_0} = 0,$$

we find the relations

$$(5) \qquad \begin{cases} -\dfrac{a_1}{a_0} = r_1 + r_2 + r_3 \\ \dfrac{a_2}{a_0} = r_1r_2 + r_1r_3 + r_2r_3 \\ -\dfrac{a_3}{a_0} = r_1r_2r_3. \end{cases}$$

Similarly, it can be shown in general that

$$
(6)\quad \begin{cases}
-\dfrac{a_1}{a_0} = r_1 + r_2 + \ldots + r_n & \text{(= the sum of all the roots taken one at a time)} \\[2ex]
\dfrac{a_2}{a_0} = r_1r_2 + r_1r_3 + \ldots + r_{n-1}r_n & \text{(= the sum of all possible products of the roots taken two at a time)} \\[2ex]
-\dfrac{a_3}{a_0} = r_1r_2r_3 + r_1r_2r_4 + \ldots + r_{n-2}r_{n-1}r_n & \text{(= the sum of all possible products of the roots taken three at a time)} \\
\vdots & \\
(-1)^n \dfrac{a_n}{a_0} = r_1r_2 \ldots r_n & \text{(= the product of the roots taken all at a time).}
\end{cases}
$$

Note 1. These relations (6) are a generalization of the relations of section 79 for quadratics.

Note 2. The first relation of (6) constitutes a convenient check on the correctness of the solution of an equation. If the roots have been found exactly, they should add up to $-a_1/a_0$. If they have been found approximately as decimals, they should add up to something very close to $-a_1/a_0$.

Example. Solve the equation $x^3 - 3x^2 + kx + 12 = 0$, given that one root is the negative of another.

First solution. Here $a_0 = 1$, $a_1 = -3$, $a_2 = k$, $a_3 = 12$, and $r_2 = -r_1$. Then by the first relation of (5) or (6), we have

$$-\frac{(-3)}{1} = r_1 + r_2 + r_3, \quad \text{or} \quad 3 = r_1 + (-r_1) + r_3, \quad \text{or} \quad r_3 = 3.$$

Hence 3 is a root. Therefore 3 must satisfy the equation, or $3^3 - 3 \cdot 3^2 + 3k + 12 = 0$. Hence $k = -4$. The remaining roots can be found by depressing the equation:

$$
\begin{array}{r|rrrr}
3 & 1 & -3 & -4 & +12 \\
 & & 3 & 0 & -12 \\
\hline
 & 1 & 0 & -4 & 0.
\end{array}
$$

Hence the remaining roots are found to be ± 2 from the depressed equation $x^2 - 4 = 0$.

Second solution. After finding $r_3 = 3$ as before, we may use the third relation of (5) thus:

$$-r_1(-r_1) \cdot 3 = 12, \text{ or } r_1^2 = 4.$$

Hence $r_1 = \pm 2$ and $r_2 = \mp 2$. As before, the roots are ± 2, 3.

EXERCISES

1. Without solving, find the sum and product of the roots of $2x^3 - 3x^2 + 5x + 4 = 0$.

2. By the method used in deriving equations (5), derive similar relations for the equation $a_0x^4 + a_1x^3 + a_2x^2 + a_3x + a_4 = 0$.

3. Solve $2x^3 - 13x^2 + kx - 18 = 0$ and find the value of k, given that the sum of two of the roots is 5.

4. Solve $12x^3 - 8x^2 + kx + 18 = 0$ and find the value of k, given that one root is the negative of another.

5. Solve $x^3 + kx^2 - 3x - 4 = 0$ and find the value of k, given that one root is 4.

6. Solve $x^3 + hx + k = 0$ and find the values of h and k, given that one root is 2 and the difference between the other roots is 4.

7. Solve $x^4 + 2x^3 - 3x^2 + kx + 4 = 0$ and find the value of k, given that there are two distinct double roots which are real.

8. Show that if one root of $x^3 + bx^2 + cx + d = 0$ is the negative of another, then $d = bc$.

9. Solve $2x^3 - 5x^2 + hx + k = 0$ and find the values of h and k, given that 2 and -1 are roots.

10. Solve $2x^3 + 3x^2 + hx + k = 0$ and find the values of h and k, given that -3 is the first root and the third root is twice the second.

132. Algebraic Solution of Equations

While the relations (6) of the preceding section between the coefficients and the roots are interesting and often useful, they do not suffice to determine the roots in terms of the coefficients. However, the formulas

$$r_1 = \frac{-b + \sqrt{b^2 - 4ac}}{2a} \quad \text{and} \quad r_2 = \frac{-b - \sqrt{b^2 - 4ac}}{2a}$$

give the roots of the general quadratic $ax^2 + bx + c = 0$ as algebraic expressions* in the coefficients. An equation with literal coefficients is said to be **solved algebraically** when the roots are written as algebraic expressions in the coefficients. The algebraic solution of the quadratic is attributed to the Hindus. It is natural to ask whether we can find formulas for the roots of the general equations of degree higher than 2 which similarly give the roots as algebraic expressions in the coefficients. Such formulas were obtained for the general cubic by Tartaglia, and

* The definition of *algebraic expression* is given in section 24.

possibly del Ferro, in the 16th century. They are sometimes called Cardano's formulas because they were first published in 1545 in the *Ars Magna* by Cardano, who obtained them from Tartaglia under a pledge of secrecy.* For the general equation of degree 4, similar formulas were obtained by Ferrari at about the same time. The ingenious solutions of the cubic and quartic will be sketched in the next sections. The next natural task is to find such formulas for equations of degree 5 or more. This question was attacked from the 16th century until 1824 when a brilliant young Norwegian mathematician, N. H. Abel, proved at the age of 22 that such formulas cannot exist for the general equation of degree higher than 4. Abel's proof is too advanced to be presented here. While the general equation of degree 5 or more cannot be solved by algebraic expressions, some equations of degree 5 or more can be solved by such formulas. For example, the pure equation of degree 5, $ax^5 + f = 0$, has a solution of the form $x = \sqrt[5]{-f/a}$, which is an algebraic expression in the coefficients. Hence a natural question is to find criteria for deciding which equations have such solutions and which have not. This task was completed by E. Galois, a young French mathematician, early in the 19th century. His work is likewise too advanced to be presented here. It is interesting to note that Abel died before he was 27, leaving behind him much original work which stimulated research for years afterward. Galois was killed in 1832 in a duel before he was 21, and before he received recognition for his remarkable work. It may comfort the student to know that Galois' ability was not appreciated by his teachers.

★133. Algebraic Solution of the General Cubic Equation

The general cubic equation $a_0x^3 + a_1x^2 + a_2x + a_3 = 0$ ($a_0 \neq 0$) can be written in the form

$$x^3 + bx^2 + cx + d = 0 \tag{1}$$

by dividing through by a_0. Setting

$$x = y - \frac{b}{3}, \tag{2}$$

* In those days it was customary to withhold a new discovery and challenge other mathematicians to solve the problem within a set time limit, say a month. This often gave the discoverer an opportunity to gloat, and at other times led to disputes over priority of discovery.

★ This section may be omitted on first reading.

equation (1) becomes

$$y^3 + py + q = 0 \tag{3}$$

where $p = c - \frac{b^2}{3}$ and $q = d - \frac{bc}{3} + \frac{2b^3}{27}\cdot$ Substituting

$$y = z - \frac{p}{3z} \tag{4}$$

in (3) and simplifying, we get $z^3 - \frac{p^3}{27z^3} + q = 0$. Multiplying by z^3, this becomes

$$z^6 + qz^3 - \frac{p^3}{27} = 0. \tag{5}$$

This is a quadratic in z^3. Solving it for z^3 by the quadratic formula, we get $\left(-q \pm \sqrt{q^2 + \frac{4p^3}{27}}\right)\Big/2$, or

$$z^3 = -\frac{q}{2} + \sqrt{\frac{q^2}{4} + \frac{p^3}{27}} \tag{6}$$

and

$$z^3 = -\frac{q}{2} - \sqrt{\frac{q^2}{4} + \frac{p^3}{27}}. \tag{7}$$

By exercise 23, section 113, any number N has three cube roots, such that if $\sqrt[3]{N}$ is one of them then the others are $\omega\sqrt[3]{N}$ and $\omega^2\sqrt[3]{N}$, where $\omega = -\frac{1}{2} + \frac{\sqrt{3}}{2}i$ is one of the imaginary cube roots of 1. Hence denoting one of the cube roots of the right member of (6) by α and one of the cube roots of the right member of (7) by β, we have, from (6) and (7), six values of z, namely α, $\omega\alpha$, $\omega^2\alpha$, β, $\omega\beta$, $\omega^2\beta$. It is easy to verify that α and β can be chosen so that $\alpha\beta = -p/3$, that is

$$\alpha = \frac{-p}{3\beta}, \ \omega\alpha = \frac{-p}{3\omega^2\beta}, \ \omega^2\alpha = \frac{-p}{3\omega\beta}, \text{ if both } \alpha, \beta \neq 0,$$

because the product of the right members of (6) and (7) is $(-p/3)^3$. By (4), the roots of (3) are

$$y_1 = \alpha + \beta, \ y_2 = \omega\alpha + \omega^2\beta, \ y_3 = \omega^2\alpha + \omega\beta.$$

It is easy to verify that the same results are obtained in the special case where either $\alpha = 0$ or $\beta = 0$ or both. Hence, by (2), the roots of (1) are

$$(8) \qquad r_1 = y_1 - \frac{b}{3}, \; r_2 = y_2 - \frac{b}{3}, \; r_3 = y_3 - \frac{b}{3}.$$

If one traces back the value of y_1, y_2, y_3, α, β, p, q, we get algebraic expressions for r_1, r_2, r_3 in terms of b, c, d.

Note 1. In practice, with given numerical coefficients, it is usually less tedious to solve a cubic by the methods explained in earlier sections of this chapter than by the method of this section. The principal interest of this section is that it proves that algebraic expressions for the roots of the general cubic in terms of the coefficients exist.*

EXERCISES

Following the steps in the text above, solve each of the following cubics:

1. $x^3 + 2x^2 + 2x + 1 = 0$. **2.** $x^3 - 2x^2 - x + 2 = 0$.

3. $y^3 - 3\omega y - 2 = 0$ where $\omega = -\frac{1}{2} + \frac{1}{2}i\sqrt{3}$ is an imaginary cube root of unity.

4. $y^3 - 6y - 2 = 0$. **5.** $x^3 - 3x^2 - 3x - 1 = 0$.

6. $x^3 + 6x^2 + 3x + 18 = 0$.

★134. Algebraic Solution of the General Quartic Equation

The general quartic† equation $a_0x^4 + a_1x^3 + a_2x^2 + a_3x + a_4 = 0$ $(a_0 \neq 0)$ can be written in the form

$$(1) \qquad x^4 + bx^3 + cx^2 + dx + e = 0$$

by dividing by a_0. Setting

$$(2) \qquad x = y - \frac{b}{4}$$

equation (1) becomes

$$(3) \qquad y^4 + gy^2 + hy + k = 0,$$

where g, h, and k are certain polynomials in b, c, d, e which should be calculated by the student. Adding and subtracting

* For a more detailed discussion of cubics, see textbooks on the theory of equations.

★ This section may be omitted on first reading.

† Or *biquadratic*, as it is often called.

$y^2z + \frac{z^2}{4}$, where z is a constant whose value is to be determined presently, we get

$$y^4 + y^2z + \frac{z^2}{4} - y^2z - \frac{z^2}{4} + gy^2 + hy + k = 0$$

or

$$(4) \qquad \left(y^2 + \frac{z}{2}\right)^2 - \left[(z - g)y^2 - hy + \left(\frac{z^2}{4} - k\right)\right] = 0.$$

The quantity in the brackets is a quadratic function of y which will be a perfect square if and only if its discriminant is zero, or

$$(5) \qquad h^2 - 4(z - g)\left(\frac{z^2}{4} - k\right) = 0.$$

Simplifying this, we get the so-called **resolvent cubic** equation

$$(6) \qquad z^3 - gz^2 - 4kz - (h^2 - 4gk) = 0.$$

Let z_1 be any root of (6), which may be found by the method of the preceding section or some other method. Then (4) becomes

$$(7) \qquad \left(y^2 + \frac{z_1}{2}\right)^2 - \left(y\sqrt{z_1 - g} - \frac{h}{2\sqrt{z_1 - g}}\right)^2 = 0$$

by virtue of (5). But an equation of the form $A^2 - B^2 = 0$ is equivalent to two equations $A + B = 0$ and $A - B = 0$. Hence (7) is equivalent to the two quadratics in y:

$$(8) \qquad y^2 + \frac{z_1}{2} + y\sqrt{z_1 - g} - \frac{h}{2\sqrt{z_1 - g}} = 0$$

and

$$(9) \qquad y^2 + \frac{z_1}{2} - y\sqrt{z_1 - g} + \frac{h}{2\sqrt{z_1 - g}} = 0.$$

Hence the four roots of (3) are the roots of (8) and (9). The four roots of (1) may be obtained by substituting these in (2).

Note. In practice, with given numerical coefficients, it is usually easier to solve quartic equations by the methods of earlier sections of this chapter than by the method of this section. The principal interest of this section is that it proves that there exist algebraic expressions for the roots of the general quartic in terms of the coefficients.*

EXERCISES

Following the steps in the text above, solve each of the following quartics:

1. $x^4 + 2x + \frac{1}{2} = 0$. (*Hint:* The resolvent cubic has a rational root.)

2. $x^4 + 4x^2 + 32x - 48 = 0$. **3.** $x^4 + 12x - 5 = 0$.

4. $x^4 + 4x^3 + 7x^2 + 10x + 3 = 0$.

5. The discussion in the text breaks down if $z_1 = g$. (Why?) Show that then g satisfies (6) so that $h = 0$, and (3) becomes a quadratic in y^2, which is thus easily solved.

* For a more detailed discussion of quartics, see textbooks on the Theory of Equations.

Chapter XX

Determinants and elimination theory

After studying the solution of polynomial equations in one unknown in the preceding chapter, it is natural to investigate systems of equations in more than one unknown. The technique of solving such systems involves eliminating all unknowns but one, and therefore the theory of such systems may be called **elimination theory.**

135. Solution of Linear Systems in Two Unknowns by Determinants

Let us solve the general system

$$a_1x + b_1y = k_1 \tag{1}$$

$$a_2x + b_2y = k_2 \tag{2}$$

by multiplication and addition (section 60). Multiplying (1) by b_2 and (2) by $-b_1$ and adding, we get

$$(a_1b_2 - a_2b_1)x = k_1b_2 - k_2b_1. \tag{3}$$

Multiplying (1) by $-a_2$ and (2) by a_1 and adding, we get

$$(a_1b_2 - a_2b_1)y = a_1k_2 - a_2k_1. \tag{4}$$

Hence if $a_1b_2 - a_2b_1 \neq 0$, we have, from (3) and (4),

$$x = \frac{k_1b_2 - k_2b_1}{a_1b_2 - a_2b_1} \tag{5}$$

$$y = \frac{a_1k_2 - a_2k_1}{a_1b_2 - a_2b_1}. \tag{6}$$

The student should check by substitution.

These formulas for the solutions can be written in a convenient pattern if we introduce a new symbol.

DEFINITION. *A symbol of the form* $\begin{vmatrix} a & b \\ c & d \end{vmatrix}$, *consisting of four numbers called* **elements** *arranged in two (horizontal) rows and two (vertical) columns flanked by vertical bars, will be called a* ***determinant of order two*** *and will denote the number ad − bc.* That is,

$$\begin{vmatrix} a & b \\ c & d \end{vmatrix} = ad - bc.$$

For example, $\begin{vmatrix} 2 & -3 \\ 4 & -2 \end{vmatrix} = 2(-2) - 4(-3) = 8.$

Using determinants of order 2, the solutions of the system of equations (1) and (2) may be written as

$$(7) \qquad x = \frac{\begin{vmatrix} k_1 & b_1 \\ k_2 & b_2 \end{vmatrix}}{\begin{vmatrix} a_1 & b_1 \\ a_2 & b_2 \end{vmatrix}}, \quad y = \frac{\begin{vmatrix} a_1 & k_1 \\ a_2 & k_2 \end{vmatrix}}{\begin{vmatrix} a_1 & b_1 \\ a_2 & b_2 \end{vmatrix}} \quad \text{if} \quad \begin{vmatrix} a_1 & b_1 \\ a_2 & b_2 \end{vmatrix} \neq 0.$$

The determinant in the denominators of (7) is denoted by Δ.

Formula (7) may be called **Cramer's rule** for the special case of a system of two equations in two unknowns. Note that the denominator is the same in the expressions for x and y. *If the given equations are written in the form of (1) and (2) with like terms under one another in the left member and the constant terms in the right member, then the denominator for both unknowns consists of the determinant* Δ *of the coefficients of the unknowns as they stand in the given equations. The numerator for either unknown is the same determinant except that the column of coefficients of the unknown being sought is replaced by the column of constant terms.* When determinants of orders higher than 2 have been defined, it will turn out that this rule applies to systems of n linear equations in n unknowns.

Example 1. Solve the system $\begin{cases} x = y + 1 \\ 2x + 3y - 12 = 0. \end{cases}$

Solution. Rewriting the equations as $\begin{cases} x - y = 1 \\ 2x + 3y = 12, \end{cases}$ we find

that
$$\Delta = \begin{vmatrix} 1 & -1 \\ 2 & 3 \end{vmatrix} = 3 - (-2) = 5$$

and
$$x = \frac{\begin{vmatrix} 1 & -1 \\ 12 & 3 \end{vmatrix}}{5} = \frac{15}{5} = 3, \quad y = \frac{\begin{vmatrix} 1 & 1 \\ 2 & 12 \end{vmatrix}}{5} = \frac{10}{5} = 2.$$

The student should check by substitution.

EXERCISES

1. Evaluate each determinant:

(*a*) $\begin{vmatrix} 2 & 4 \\ -1 & 3 \end{vmatrix}$; (*b*) $\begin{vmatrix} 3 & -2 \\ -1 & 4 \end{vmatrix}$; (*c*) $\begin{vmatrix} 3 & -2 \\ 4 & -3 \end{vmatrix}$; (*d*) $\begin{vmatrix} 1 & 2 \\ 3 & 4 \end{vmatrix}$;

(*e*) $\begin{vmatrix} 1 & 1 \\ 1 & -1 \end{vmatrix}$; (*f*) $\begin{vmatrix} a & b \\ b & a \end{vmatrix}$.

2–37. Work exercises 2–37 of section 60 by means of Cramer's rule, and check.

Solve each of the following equations and check:

38. $\begin{vmatrix} x & 13 \\ 2 & 3 \end{vmatrix} = 1.$

39. $\begin{vmatrix} 2 & 3 \\ 4 & x \end{vmatrix} = 2.$

40. $\begin{vmatrix} x - 10 & 2 \\ 2 + x & 2 + x \end{vmatrix} = 0.$

41. $\begin{vmatrix} 4 - x & 2 \\ -1 & 1 - x \end{vmatrix} = 0.$

42. Show that $\begin{vmatrix} a_1 & b_1 \\ a_2 & b_2 \end{vmatrix} = \begin{vmatrix} a_1 & a_2 \\ b_1 & b_2 \end{vmatrix}$.

43. Show that $\begin{vmatrix} ka_1 & b_1 \\ ka_2 & b_2 \end{vmatrix} = k \begin{vmatrix} a_1 & b_1 \\ a_2 & b_2 \end{vmatrix}$.

44. Show that $\begin{vmatrix} a_1 & b_1 \\ a_2 & b_2 \end{vmatrix} = - \begin{vmatrix} b_1 & a_1 \\ b_2 & a_2 \end{vmatrix}$.

45. Show that $\begin{vmatrix} a_1 + a_1' & b_1 \\ a_2 + a_2' & b_2 \end{vmatrix} = \begin{vmatrix} a_1 & b_1 \\ a_2 & b_2 \end{vmatrix} + \begin{vmatrix} a_1' & b_1 \\ a_2' & b_2 \end{vmatrix}$.

46. Show that $\begin{vmatrix} a_1 & ka_1 \\ a_2 & ka_2 \end{vmatrix} = 0.$

136. Determinants of Third Order

A **determinant of third order** is a symbol of the form

$$\text{(1)} \qquad \begin{vmatrix} a & b & c \\ d & e & f \\ g & h & k \end{vmatrix}$$

consisting of nine numbers called **elements** of the determinant, arranged in three (horizontal) rows and three (vertical) columns.

The symbol (1) stands for the number

$$aek + bfg + cdh - gec - hfa - dbk \tag{2}$$

which is called the **value of the determinant** (1). The expression (2) is called the **expansion of the determinant** (1).

By the **minor** of an element we mean the determinant of order 2 which remains when we cross out the row and column to which the element belongs. Thus in (1), the minor of a is $\begin{vmatrix} e & f \\ h & k \end{vmatrix}$; the minor of b is $\begin{vmatrix} d & f \\ g & k \end{vmatrix}$; the minor of c is $\begin{vmatrix} d & e \\ g & h \end{vmatrix}$; the minor of f is $\begin{vmatrix} a & b \\ g & h \end{vmatrix}$, and so on. Denote the minor of a by A, the minor of b by B and so on. A convenient way to obtain the value (2) is to take the algebraic sum

$$\begin{aligned} aA - bB + cC &= a\begin{vmatrix} e & f \\ h & k \end{vmatrix} - b\begin{vmatrix} d & f \\ g & k \end{vmatrix} + c\begin{vmatrix} d & e \\ g & h \end{vmatrix} \\ &= a(ek - hf) - b(dk - gf) + c(dh - ge) \\ &= aek - ahf - bdk + bgf + cdh - cge. \end{aligned} \tag{3}$$

Example. Evaluate the determinant $\begin{vmatrix} 2 & -1 & 3 \\ 1 & 2 & 3 \\ 3 & -2 & 1 \end{vmatrix}$.

Solution. By (3) we have

$$\begin{aligned} &2\begin{vmatrix} 2 & 3 \\ -2 & 1 \end{vmatrix} - (-1)\begin{vmatrix} 1 & 3 \\ 3 & 1 \end{vmatrix} + 3\begin{vmatrix} 1 & 2 \\ 3 & -2 \end{vmatrix} \\ &= 2(2 + 6) + 1(1 - 9) + 3(-2 - 6) = \mathbf{16 - 8 - 24 = -16.} \end{aligned}$$

EXERCISES

Evaluate each of the following determinants:

1. $\begin{vmatrix} 1 & 2 & 3 \\ 0 & 0 & -1 \\ 4 & -2 & 0 \end{vmatrix}$.

2. $\begin{vmatrix} 2 & -1 & 3 \\ 0 & 1 & 2 \\ 3 & -2 & 1 \end{vmatrix}$.

3. $\begin{vmatrix} -1 & -2 & -3 \\ 2 & 4 & 1 \\ -1 & 2 & 0 \end{vmatrix}$.

4. $\begin{vmatrix} -1 & 2 & -3 \\ 2 & 1 & 1 \\ 1 & 3 & 2 \end{vmatrix}$.

5. $\begin{vmatrix} 1 & 1 & 1 \\ 1 & 0 & -1 \\ 0 & 1 & -1 \end{vmatrix}$.

6. $\begin{vmatrix} a & b & b \\ b & a & b \\ b & b & a \end{vmatrix}$.

Solve each of the following equations and check:

7. $\begin{vmatrix} 1 & x & -2 \\ 2 & 3 & 2 \\ 2 & 6 & -4 \end{vmatrix} = 0.$ **8.** $\begin{vmatrix} 1 & 2x & -4 \\ 2 & x & 3 \\ -1 & -2 & 1 \end{vmatrix} = 0.$

9. $\begin{vmatrix} 3x & 3 & 2 \\ 1 & x & 2 \\ -1 & -1 & 1 \end{vmatrix} = 0.$ **10.** $\begin{vmatrix} -2 & -1 & -2 \\ 4 & x & 1 \\ 6 & 3 & x \end{vmatrix} = 0.$

In the determinant (1), above:

11. Write the minor G of g. **12.** Write the minor E of e.
13. Write the minor D of d. **14.** Write the minor H of h.
15. Write the minor K of k.

16. Verify that the value of the determinant (1) is obtained from the formulas:

(*a*) $aA - dD + gG.$

(*b*) $-dD + eE - fF.$

(*c*) $-bB + eE - hH.$

(*d*) $gG - hH + kK.$

(*e*) $cC - fF + kK.$

17. Using the notations of the preceding exercise, verify that

(*a*) $aD - bE + cF = 0.$

(*b*) $aG - bH + cK = 0.$

(*c*) $aB - dE + gH = 0.$

(*d*) $aC - dF + gK = 0.$

Evaluate each of the following determinants:

18. $\begin{vmatrix} a_1 & b_1 & c_1 \\ a_2 & b_2 & c_2 \\ a_3 & b_3 & c_3 \end{vmatrix}.$ **19.** $\begin{vmatrix} a_1 & k_1 & c_1 \\ a_2 & k_2 & c_2 \\ a_3 & k_3 & c_3 \end{vmatrix}.$

20. Show that $\begin{vmatrix} a_1 & b_1 & c_1 \\ a_2 & b_2 & c_2 \\ a_3 & b_3 & c_3 \end{vmatrix} = - \begin{vmatrix} c_1 & b_1 & a_1 \\ c_2 & b_2 & a_2 \\ c_3 & b_3 & a_3 \end{vmatrix}.$

21. Show that $\begin{vmatrix} a_1 & kb_1 & c_1 \\ a_2 & kb_2 & c_2 \\ a_3 & kb_3 & c_3 \end{vmatrix} = k \begin{vmatrix} a_1 & b_1 & c_1 \\ a_2 & b_2 & c_2 \\ a_3 & b_3 & c_3 \end{vmatrix}.$

22. Show that $\begin{vmatrix} a_1 & a_1 & c_1 \\ a_2 & a_2 & c_2 \\ a_3 & a_3 & c_3 \end{vmatrix} = 0.$

137. Solution of Linear Equations in Three Unknowns By Determinants

Let us solve the general system of three linear equations in three unknowns

$$(1)\qquad a_1x + b_1y + c_1z = k_1$$

$$(2)\qquad a_2x + b_2y + c_2z = k_2$$

$$(3)\qquad a_3x + b_3y + c_3z = k_3$$

by multiplication and addition. Multiplying (1) by c_2 and (2) by $-c_1$ and adding, we eliminate z, obtaining

$$(4)\qquad (a_1c_2 - a_2c_1)x + (b_1c_2 - b_2c_1)y = k_1c_2 - k_2c_1.$$

Multiplying (2) by c_3 and (3) by $-c_2$ and adding, we eliminate z, obtaining

$$(5)\qquad (a_2c_3 - a_3c_2)x + (b_2c_3 - b_3c_2)y = k_2c_3 - k_3c_2.$$

Solving the system of equations (4) and (5) for x by any previous method, we get

$$\begin{aligned}(a_1c_2b_2c_3 - a_2c_1b_2c_3 - a_1c_2b_3c_2 + a_2c_1b_3c_2 - a_2c_3b_1c_2 + a_2c_3b_2c_1 \\ + a_3c_2b_1c_2 - a_3c_2b_2c_1)x = b_2c_3k_1c_2 - b_2c_3k_2c_1 - b_3c_2k_1c_2 \\ + b_3c_2k_2c_1 - b_1c_2k_2c_3 + b_2c_1k_2c_3 + b_1c_2k_3c_2 - b_2c_1k_3c_2\end{aligned}$$

or, simplifying,

$$\begin{aligned}(6)\qquad c_2(a_1b_2c_3 - a_1b_3c_2 + a_2c_1b_3 - a_2b_1c_3 + a_3b_1c_2 - a_3b_2c_1)x \\ = c_2(k_1b_2c_3 - k_1b_3c_2 + k_2b_3c_1 - k_2b_1c_3 + k_3b_1c_2 - k_3b_2c_1)\end{aligned}$$

or*

$$(7)\qquad x = \frac{k_1b_2c_3 + k_2b_3c_1 + k_3b_1c_2 - k_3b_2c_1 - k_2b_1c_3 - k_1b_3c_2}{a_1b_2c_3 + a_2b_3c_1 + a_3b_1c_2 - a_3b_2c_1 - a_2b_1c_3 - a_1b_3c_2},$$

provided the denominator is not zero. Solving similarly for y and z, we get

$$(8)\qquad y = \frac{a_1k_2c_3 + a_2k_3c_1 + a_3k_1c_2 - a_3k_2c_1 - a_2k_1c_3 - a_1k_3c_2}{a_1b_2c_3 + a_2b_3c_1 + a_3b_1c_2 - a_3b_2c_1 - a_2b_1c_3 - a_1b_3c_2},$$

$$(9)\qquad z = \frac{a_1b_2k_3 + a_2b_3k_1 + a_3b_1k_2 - a_3b_2k_1 - a_2b_1k_3 - a_1b_3k_2}{a_1b_2c_3 + a_2b_3c_1 + a_3b_1c_2 - a_3b_2c_1 - a_2b_1c_3 - a_1b_3c_2}.$$

* This step is valid only if $c_2 \neq 0$. But the check by substitution reveals that the result is correct even if $c_2 = 0$. Another proof of these formulas will be given in section 142.

The student may check the solution (7), (8), and (9) by substitution in the original equations.

Comparing (7), (8), and (9) with the definition of determinant of order 3, we have the following result.

If $\Delta = \begin{vmatrix} a_1 & b_1 & c_1 \\ a_2 & b_2 & c_2 \\ a_3 & b_3 & c_3 \end{vmatrix} \neq 0$, *then the solution of the system of equations (1), (2), and (3) is given by*

$$(10) \qquad x = \frac{\begin{vmatrix} k_1 & b_1 & c_1 \\ k_2 & b_2 & c_2 \\ k_3 & b_3 & c_3 \end{vmatrix}}{\Delta}, \quad y = \frac{\begin{vmatrix} a_1 & k_1 & c_1 \\ a_2 & k_2 & c_2 \\ a_3 & k_3 & c_3 \end{vmatrix}}{\Delta}, \quad z = \frac{\begin{vmatrix} a_1 & b_1 & k_1 \\ a_2 & b_2 & k_2 \\ a_3 & b_3 & k_3 \end{vmatrix}}{\Delta}.$$

Formula (10) is known as **Cramer's rule** for the special case of three linear equations in three unknowns. Note that *the denominator for each unknown is the determinant* Δ *of the coefficients of the unknowns as they appear in the given equations when they are written with like terms under each other on the left and constant terms on the right. The numerator for any unknown is the same determinant* Δ *except that the column of coefficients of that unknown is replaced by the column of constant terms.* The same rule will apply verbatim to systems of n equations in n unknowns after we have defined determinants of order n.

Example. Solve the system

$$\begin{aligned} x + y + z &= 4 \\ 2x - y - 2z &= -1 \\ x - 2y - z &= 1. \end{aligned}$$

Solution. We have

$$\Delta = \begin{vmatrix} 1 & 1 & 1 \\ 2 & -1 & -2 \\ 1 & -2 & -1 \end{vmatrix} = -6,$$

$$x = \frac{\begin{vmatrix} 4 & 1 & 1 \\ -1 & -1 & -2 \\ 1 & -2 & -1 \end{vmatrix}}{-6} = \frac{-12}{-6} = 2,$$

$$y = \frac{\begin{vmatrix} 1 & 4 & 1 \\ 2 & -1 & -2 \\ 1 & 1 & -1 \end{vmatrix}}{-6} = \frac{6}{-6} = -1,$$

$$z = \frac{\begin{vmatrix} 1 & 1 & 4 \\ 2 & -1 & -1 \\ 1 & -2 & 1 \end{vmatrix}}{-6} = \frac{-18}{-6} = 3.$$

The student may check by substitution in the original equations.

EXERCISES

Solve by means of determinants, and check:

1–10. Exercises 1–10 of section 62.

11. In the electrical circuit pictured in the accompanying diagram, suppose $E_1 = 10$ volts, $E_2 = 20$ volts, $R_1 = 5$ ohms, $R_2 = 10$ ohms, $R_3 = 5$ ohms. We find by Kirchoff's laws, the system of equations

$$\begin{cases} i_1 - i_2 - i_3 = 0 \\ 5i_1 + 5i_3 = 10 \\ 10i_2 - 5i_3 = 20. \end{cases}$$

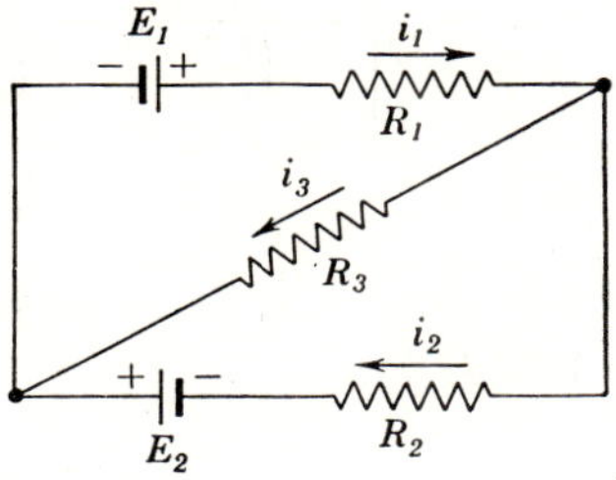

Find the currents i_1, i_2, i_3 in amperes.

★138. Area of a Triangle by Determinants

We digress to point out an interesting geometric application of determinants.

THEOREM. *If the vertices of a triangle have the rectangular coordinates* (x_1,y_1), (x_2,y_2), *and* (x_3,y_3), *the area of the triangle is given by the absolute value of*

$$\frac{1}{2}\begin{vmatrix} 1 & x_1 & y_1 \\ 1 & x_2 & y_2 \\ 1 & x_3 & y_3 \end{vmatrix}. \tag{1}$$

★ This section may be omitted on first reading.

Proof. Suppose that all three vertices are in the first quadrant, as in Fig. 58; it can be shown that the result is unaffected if they are anywhere else. Then the area of triangle ABC equals the area of trapezoid $ABED$ plus the area of trapezoid $BCFE$ minus the area of trapezoid $ACFD$. Now trapezoid $ABED$ has

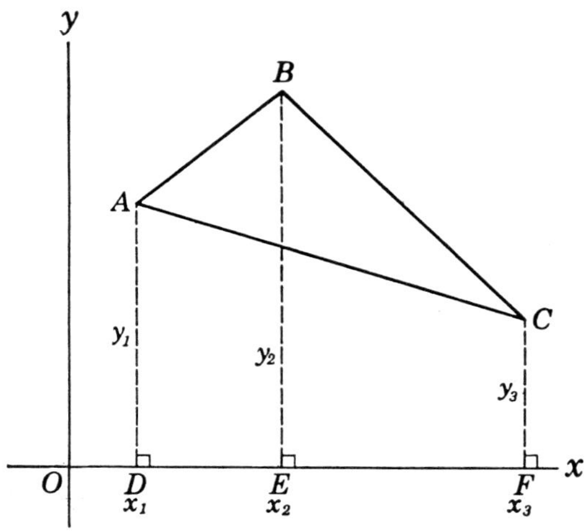

Fig. 58.

the altitude $DE = x_2 - x_1$ and the bases y_1 and y_2. Hence the area of $ABED$ is

$$\tfrac{1}{2}h(b + b') = \tfrac{1}{2}(x_2 - x_1)(y_1 + y_2) = \tfrac{1}{2}(x_2y_1 + x_2y_2 - x_1y_1 - x_1y_2).$$

Similarly the area of $BCFE$ is

$$\tfrac{1}{2}(x_3 - x_2)(y_2 + y_3) = \tfrac{1}{2}(x_3y_2 + x_3y_3 - x_2y_2 - x_2y_3),$$

and the area of $ACFD$ is

$$\tfrac{1}{2}(x_3 - x_1)(y_1 + y_3) = \tfrac{1}{2}(x_3y_1 + x_3y_3 - x_1y_1 - x_1y_3).$$

Hence the area of ABC is

$$\tfrac{1}{2}(x_2y_1 - x_1y_2 + x_3y_2 - x_2y_3 + x_3y_1 - x_1y_3).$$

But this is the expansion of (1), except for sign, which proves the theorem.

Corollary 1. *Three points* (x_1,y_1), (x_2,y_2), *and* (x_3,y_3) *are in the same straight line if and only if*

$$\begin{vmatrix} 1 & x_1 & y_1 \\ 1 & x_2 & y_2 \\ 1 & x_3 & y_3 \end{vmatrix} = 0.$$

Proof. Left to the student.

Corollary 2. *An equation of the straight line passing through two points* (x_1,y_1) *and* (x_2,y_2) *is*

$$\begin{vmatrix} 1 & x & y \\ 1 & x_1 & y_1 \\ 1 & x_2 & y_2 \end{vmatrix} = 0.$$

Proof. Left to the student.

EXERCISES

Find, by determinants, the area of the triangle having the following vertices:

1. $(-1,-2)$, $(3,-2)$, $(1,5)$.
2. $(0,0)$, $(0,-5)$, $(-14,2)$.
3. $(-2,3)$, $(-6,-3)$, $(1,1)$.
4. $(-4,-3)$, $(6,-5)$, $(-2,4)$.

Show by determinants that the following sets of points are collinear:

5. $(-3,-3)$, $(3,5)$, $(6,9)$.
6. $(1,2)$, $(2,6)$, $(4,14)$.

Write in determinant form an equation of the straight line joining the following pairs of points, expand and simplify:

7. $(2,3)$, $(5,6)$.
8. $(2,-3)$, $(-3,-4)$.
9. $(2,-3)$, $(2,5)$.
10. $(3,-2)$, $(-2,4)$.

139. Determinants of Order *n*

Consider any arrangement of the whole numbers 1, 2, 3, . . . , n in some order in a line. An **inversion** is said to occur whenever a larger number precedes a smaller one.

Example 1. The arrangement 14325 has three inversions because 4 precedes 3, 4 precedes 2, and 3 precedes 2.

In an arrangement of letters in a line, an **inversion** is said to occur whenever a letter precedes one which occurs earlier than it in alphabetical order.

Example 2. The arrangement $b\ e\ c\ d\ a$ has six inversions because b precedes a, e precedes c, e precedes d, e precedes a, c precedes a, and d precedes a.

DEFINITION. *A* ***determinant of order n*** *is a symbol consisting of n^2 numbers, called elements of the determinant, arranged in n (horizontal) rows and n (vertical) columns, written in the form*

$$(1)\qquad \begin{vmatrix} a_1 & b_1 & c_1 & \dots & h_1 \\ a_2 & b_2 & c_2 & \dots & h_2 \\ a_3 & b_3 & c_3 & \dots & h_3 \\ \cdot & \cdot & \cdot & & \cdot \\ \cdot & \cdot & \cdot & & \cdot \\ \cdot & \cdot & \cdot & & \cdot \\ a_n & b_n & c_n & \dots & h_n \end{vmatrix}.$$

The symbol (1) stands for the algebraic sum of all possible products of n factors each such that

(a) each product has as its factors one and only one element from each row and each column;

(b) each product is preceded by a plus or minus sign according as the number of inversions in the subscripts is even or odd, after the letters have been written in the order in which they appear in the first row of the determinant.

The algebraic sum of these products is called the **expansion of the determinant,** and each product is called a **term** of the expansion. The number resulting from calculating the expansion is called the **value of the determinant.** The elements lying in a straight line from the upper left-hand corner to the lower right-hand corner of the determinant constitute the **principal diagonal** of the determinant.

Example 3. Consider the determinant

$$D = \begin{vmatrix} a_1 & b_1 & c_1 \\ a_2 & b_2 & c_2 \\ a_3 & b_3 & c_3 \end{vmatrix}$$

of order 3. Every term in the expansion of D must be a product of an a, a b, and a c, and the subscripts must be the numbers 1, 2, 3 in some order. No matter which of the three numbers is chosen as the subscript for a, there are two choices for the subscript for b. Hence there are $3 \cdot 2 \cdot 1$ choices in all, since the subscript for c is determined by the choice of the other two. Therefore there are 6 terms in the expansion, namely

$$a_1b_2c_3,\ a_2b_3c_1,\ a_3b_1c_2,\ a_3b_2c_1,\ a_2b_1c_3, \text{ and } a_1b_3c_2.$$

The numbers of inversions in these terms are 0, 2, 2, 3, 1, 1, respectively. Hence

$$D = a_1b_2c_3 + a_2b_3c_1 + a_3b_1c_2 - a_3b_2c_1 - a_2b_1c_3 - a_1b_3c_2,$$

which is in agreement with the definition of section 136.

For $n > 3$, it is too tedious to expand a determinant directly from this definition. Better ways of evaluating determinants of order higher than 3 will emerge from the following sections.

A more systematic notation is obtained by using double subscripts, such as a_{32}, where the first subscript indicates the row and the second indicates the column. Thus a_{32} is the element in the third row and second column, while a_{23} is the element in the second row and third column. Then part (*b*) of the definition above should read: each term is preceded by a plus or minus sign according as the number of inversions in the row-subscripts is even or odd, after the factors have been written so that the column subscripts appear in order of magnitude. The determinant (1) is written in this notation as

$$\begin{vmatrix} a_{11} & a_{12} & a_{13} & \cdots & a_{1n} \\ a_{21} & a_{22} & a_{23} & \cdots & a_{2n} \\ a_{31} & a_{32} & a_{33} & \cdots & a_{3n} \\ \cdot & \cdot & \cdot & & \cdot \\ \cdot & \cdot & \cdot & & \cdot \\ \cdot & \cdot & \cdot & & \cdot \\ a_{n1} & a_{n2} & a_{n3} & \cdots & a_{nn} \end{vmatrix}.$$

Determinants were invented by G. W. Leibniz (German, 1646–1716) and also independently by Seki-Kowa (Japanese, 1642–1708). Because of the isolation of Japan, the latter's work had no influence outside of Japan.

EXERCISES

1. Write out the expansion of

$$\begin{vmatrix} a_1 & b_1 \\ a_2 & b_2 \end{vmatrix}$$

according to the definition of this section.

★**2.** Write out the expansion of

$$\begin{vmatrix} a_1 & b_1 & c_1 & d_1 \\ a_2 & b_2 & c_2 & d_2 \\ a_3 & b_3 & c_3 & d_3 \\ a_4 & b_4 & c_4 & d_4 \end{vmatrix}$$

according to the definition of this section.

140. Some Properties of Determinants

Some useful properties of determinants are given in the following theorems. While examples will be determinants of order 3, for convenience in writing, the theorems apply to determinants of any order.

THEOREM 1. *The expansion of a determinant of order n contains $n!$ terms.*

Proof. By (a) of the definition of section 139, in forming a term of the expansion, there are n choices for the subscript attached to the a-factor, $(n - 1)$ choices for the subscript attached to the b-factor, and so on. But there are n factors in each term. Hence there are $n(n - 1)(n - 2) \dots 1$ terms.

THEOREM 2. *The value of a determinant is unchanged if corresponding rows and columns are interchanged.*

Proof for order 3. The student should expand the determinants

$$D = \begin{vmatrix} a_1 & b_1 & c_1 \\ a_2 & b_2 & c_2 \\ a_3 & b_3 & c_3 \end{vmatrix} \text{ and } D' = \begin{vmatrix} a_1 & a_2 & a_3 \\ b_1 & b_2 & b_3 \\ c_1 & c_2 & c_3 \end{vmatrix}$$

by the scheme of section 136, and verify directly that the resulting expansions are equal.

Proof for order n. Suppose as above that D has the same letter for elements in the same column, while D' has the same letter for elements in the same row. By definition of determinant, D and D' have the same terms in their expansions except possibly for signs. (In reading the definition for D' we must of course reverse the role of letters and subscripts.) We shall show that corresponding terms of D and D' have the same sign by showing that they have the same number of inversions. Suppose

★ May be omitted without disturbing continuity.

a term of D is written with the letters in alphabetical order. The inversions in the subscripts can be removed by interchanging adjacent factors, repeatedly. But each such interchange introduces one inversion in the letters. Hence, when the subscripts have been restored to order of magnitude, we have introduced into the letters the same number of inversions that the subscripts had originally.

From theorem 2 it follows that *for every theorem concerning the columns of a determinant, there is a corresponding theorem concerning the rows, and conversely.* The proof of one of these theorems automatically suffices to prove the other.

THEOREM 3. *If all the elements of some column (or row) of a determinant are equal to zero, the value of the determinant is zero.*

Proof. By definition, each term of the expansion contains an element from the column of zeros as a factor. Hence each term of the expansion is equal to zero.

THEOREM 4. *If two columns (or rows) of a determinant are interchanged, the sign of the determinant is reversed.*

Proof. If we interchange two adjacent rows, we interchange two adjacent subscripts in each term of the expansion. This either increases or decreases the number of inversions in the subscripts of each term by one. Therefore the sign of each term and hence of the entire determinant is reversed.

Now suppose we interchange two rows separated by m intermediate rows. It will take m interchanges of adjacent rows to bring the lower row to just below the upper one, one to interchange them, and m more to bring the upper row to where the lower one was originally. Hence there have been $2m + 1$ interchanges of adjacent rows. Since $2m + 1$ is odd, this is equivalent to reversing the sign of the determinant an odd number of times, which leaves the sign reversed.

Example 1. If

$$D = \begin{vmatrix} a_1 & b_1 & c_1 \\ a_2 & b_2 & c_2 \\ a_3 & b_3 & c_3 \end{vmatrix} \text{ and } D' = \begin{vmatrix} c_1 & b_1 & a_1 \\ c_2 & b_2 & a_2 \\ c_3 & b_3 & a_3 \end{vmatrix}$$

then $D' = -D$. The student should verify this by expanding D and D'.

THEOREM 5. *If two columns (or rows) of a determinant are identical, the value of the determinant is zero.*

Proof. If we interchange the two identical columns of the determinant D, it is unaltered in value. But by theorem 4, its sign must be reversed. Hence $D = -D$, or $2D = 0$, or $D = 0$.

Example 2. By inspection,

$$\begin{vmatrix} 2 & 2 & 7 \\ 3 & 3 & -2 \\ 4 & 4 & 5 \end{vmatrix} = 0,$$

since the first two columns are identical. The student should verify this by expanding the determinant.

THEOREM 6. *If each of the elements of a column (or row) of a determinant is multiplied by the same quantity k, the value of the determinant is multiplied by k.*

Proof. Since each term in the expansion has as one factor an element from the column in question, each term is multiplied by k. The factor k is thus common to all the terms of the expansion and hence the entire determinant is multiplied by k.

Example 3.

$$\begin{vmatrix} ka_1 & b_1 & c_1 \\ ka_2 & b_2 & c_2 \\ ka_3 & b_3 & c_3 \end{vmatrix} = k \begin{vmatrix} a_1 & b_1 & c_1 \\ a_2 & b_2 & c_2 \\ a_3 & b_3 & c_3 \end{vmatrix}.$$

The student should verify this by expanding both determinants.

THEOREM 7. *If corresponding elements of two columns (or rows) of a determinant are proportional, the value of the determinant is zero.*

Proof. The hypothesis that corresponding elements of two columns are proportional means that there is a constant factor of proportionality k such that every element of one column is k times the corresponding element of the other column. Then, if this factor is taken out, by theorem 6, the two columns become identical, and the determinant is zero, by theorem 5. Illustrating this for a third order determinant, we have

$$\begin{vmatrix} kb_1 & b_1 & c_1 \\ kb_2 & b_2 & c_2 \\ kb_3 & b_3 & c_3 \end{vmatrix} = k \begin{vmatrix} b_1 & b_1 & c_1 \\ b_2 & b_2 & c_2 \\ b_3 & b_3 & c_3 \end{vmatrix} = k \cdot 0 = 0.$$

Example 4. By inspection,

$$\begin{vmatrix} 2 & 6 & 5 \\ 3 & 9 & 4 \\ -4 & -12 & 3 \end{vmatrix} = 0,$$

since the first two columns are proportional.

THEOREM 8. *If each element of a certain column* (*or row*) *of a determinant is expressed as a sum of two terms, the whole determinant may be expressed as a sum of two determinants, each of which has one of these terms in the corresponding position and both of which are identical with the given determinant in all other elements.*

Illustrating this for a determinant of third order, we may write

$$(1)\qquad \begin{vmatrix} (a_1 + a_1') & b_1 & c_1 \\ (a_2 + a_2') & b_2 & c_2 \\ (a_3 + a_3') & b_3 & c_3 \end{vmatrix} = \begin{vmatrix} a_1 & b_1 & c_1 \\ a_2 & b_2 & c_2 \\ a_3 & b_3 & c_3 \end{vmatrix} + \begin{vmatrix} a_1' & b_1 & c_1 \\ a_2' & b_2 & c_2 \\ a_3' & b_3 & c_3 \end{vmatrix}.$$

Proof. Consider a typical term of the left member of (1), such as $(a_1 + a_1')b_2c_3$. It is equal to $a_1b_2c_3 + a_1'b_2c_3$; but each of these two terms is a term of one of the determinants in the right member of (1). Hence the expansion of the left member of (1) is the sum of the expansions of the determinants in the right member of (1). While the proof has been given for the illustration, which is a third order determinant, it is clear that the argument applies to determinants of any order.

THEOREM 9. *If to every element of any column* (*or row*) *of a determinant is added the corresponding element of any other column* (*or row*), *each multiplied by the same quantity* k, *then the value of the determinant is unaltered.*

Illustrating this for a determinant of order 3, we may write

$$(2)\qquad \begin{vmatrix} a_1 + kc_1 & b_1 & c_1 \\ a_2 + kc_2 & b_2 & c_2 \\ a_3 + kc_3 & b_3 & c_3 \end{vmatrix} = \begin{vmatrix} a_1 & b_1 & c_1 \\ a_2 & b_2 & c_2 \\ a_3 & b_3 & c_3 \end{vmatrix}.$$

Proof. In the illustration,

$$\begin{vmatrix} a_1 + kc_1 & b_1 & c_1 \\ a_2 + kc_2 & b_2 & c_2 \\ a_3 + kc_3 & b_3 & c_3 \end{vmatrix} = \begin{vmatrix} a_1 & b_1 & c_1 \\ a_2 & b_2 & c_2 \\ a_3 & b_3 & c_3 \end{vmatrix} + \begin{vmatrix} kc_1 & b_1 & c_1 \\ kc_2 & b_2 & c_2 \\ kc_3 & b_3 & c_3 \end{vmatrix},$$

by theorem 8. By theorem 7 the last determinant is zero. This proves the theorem for the illustration. Clearly the argument applies in general.

Example 5. Write a determinant equal to

$$\begin{vmatrix} -2 & 3 & 1 \\ 1 & 2 & 3 \\ 2 & 3 & 3 \end{vmatrix}$$

but having zeros everywhere in the first row except the upper right-hand corner.

Solution. Multiplying the (elements of the) last column by $k = 2$ and adding to the (corresponding elements of the) first column, we get

$$\begin{vmatrix} 0 & 3 & 1 \\ 7 & 2 & 3 \\ 8 & 3 & 3 \end{vmatrix}.$$

Multiplying the last column by -3 and adding to the second column, we get

$$\begin{vmatrix} 0 & 0 & 1 \\ 7 & -7 & 3 \\ 8 & -6 & 3 \end{vmatrix}.$$

EXERCISES

1. Without expanding show that $\begin{vmatrix} 2 & 1 & 4 & 0 \\ 3 & 2 & 6 & 1 \\ 4 & 3 & 8 & -1 \\ -1 & 4 & -2 & 2 \end{vmatrix} = 0.$

2. Without expanding show that $\begin{vmatrix} 1 & 2 & -3 \\ 4 & 1 & 2 \\ 3 & 6 & -9 \end{vmatrix} = 0.$

3. Without expanding find the root of the equation

$$\begin{vmatrix} 1 & 2 & 4 \\ 3 & x & 12 \\ -1 & 2 & 1 \end{vmatrix} = 0.$$

4. Write a determinant equal to $\begin{vmatrix} 1 & 4 & -3 \\ 3 & 1 & -1 \\ 2 & -2 & 2 \end{vmatrix}$ but having zeros everywhere in the first row except the upper left-hand corner. (*Hint:* See example 5 above.)

5. Write a determinant equal to $\begin{vmatrix} 2 & 1 & -3 \\ -5 & 2 & 1 \\ 1 & 3 & 4 \end{vmatrix}$ but having zeros everywhere in the first column except the lower left-hand corner.

6. Write a determinant equal to $\begin{vmatrix} 2 & -1 & 0 & 3 \\ 1 & 2 & -1 & -3 \\ 3 & 0 & 2 & 0 \\ 4 & 3 & 1 & 2 \end{vmatrix}$ but having zeros everywhere in the first column except for the element in the second row.

7. Write a determinant equal to that given in exercise 6 but having zeros everywhere in the second row except for the element in the third column.

★8. Without expanding show that

$$\begin{vmatrix} 1 & x & x^2 \\ 1 & y & y^2 \\ 1 & z & z^2 \end{vmatrix} = (x - y)(y - z)(z - x).$$

141. Expansion by Minors

By the **minor of an element** in a determinant of order n is meant the determinant of order $n - 1$ obtained by removing both the row and the column to which the given element belongs. The minor of an element is denoted by the corresponding capital letter; thus the minor of b_1 is denoted by B_1, and so on. For example, in the determinant

$$D = \begin{vmatrix} a_1 & b_1 & c_1 & d_1 \\ a_2 & b_2 & c_2 & d_2 \\ a_3 & b_3 & c_3 & d_3 \\ a_4 & b_4 & c_4 & d_4 \end{vmatrix} \tag{1}$$

the minor of a_1 is

$$A_1 = \begin{vmatrix} b_2 & c_2 & d_2 \\ b_3 & c_3 & d_3 \\ b_4 & c_4 & d_4 \end{vmatrix},$$

the minor of b_3 is

$$B_3 = \begin{vmatrix} a_1 & c_1 & d_1 \\ a_2 & c_2 & d_2 \\ a_4 & c_4 & d_4 \end{vmatrix},$$

and so on.

★ May be omitted without disturbing continuity.

THEOREM 1. *The value of a determinant D may be obtained as follows:*

(a) Select any column (or row);

(b) Multiply each element in this column (or row) by its minor;

(c) Insert a plus or minus sign before each of these products obtained in (b) according to the following rule: if the element is in the ith row and jth column, we use a plus sign if $i + j$ is even and a minus sign if $i + j$ is odd;

(d) Then D is equal to the algebraic sum of the signed terms obtained in (c).

For example, the value of D in (1) may be obtained in any of the following ways:

$$D = a_1A_1 - a_2A_2 + a_3A_3 - a_4A_4,$$

or,
$$D = -b_1B_1 + b_2B_2 - b_3B_3 + b_4B_4,$$

or,
$$D = a_3A_3 - b_3B_3 + c_3C_3 - d_3D_3,$$

or,
$$D = -a_4A_4 + b_4B_4 - c_4C_4 + d_4D_4,$$

and so on.

The signs determined in (*c*) of theorem 1 may be found also by means of the so-called "checkerboard rule";

$$\begin{vmatrix} + & - & + & - & \cdots \\ - & + & - & + & \cdots \\ + & - & + & - & \cdots \\ - & + & - & + & \cdots \\ \cdot & \cdot & \cdot & \cdot & \end{vmatrix};$$

that is, the signs alternate, beginning with plus in the upper left-hand corner of the determinant.

Example 1. Evaluate by expanding by minors:

$$D = \begin{vmatrix} 2 & -1 & 4 \\ 5 & 1 & -2 \\ 3 & -3 & -4 \end{vmatrix}.$$

Solution. Selecting the third column to work with, we have

$$\begin{aligned} D &= 4\begin{vmatrix} 5 & 1 \\ 3 & -3 \end{vmatrix} - (-2)\begin{vmatrix} 2 & -1 \\ 3 & -3 \end{vmatrix} + (-4)\begin{vmatrix} 2 & -1 \\ 5 & 1 \end{vmatrix} \\ &= 4(-18) + 2(-3) - 4(7) = -72 - 6 - 28 = -106. \end{aligned}$$

Example 2. Evaluate by expanding by minors:

$$D = \begin{vmatrix} 1 & -1 & 3 & 2 \\ 2 & -2 & 2 & -1 \\ 3 & 0 & 0 & -2 \\ 4 & 3 & 1 & 1 \end{vmatrix}. \tag{2}$$

Solution. In order to simplify the calculation, we use the third row, since it contains two zeros. Thus

$$D = 3\begin{vmatrix} -1 & 3 & 2 \\ -2 & 2 & -1 \\ 3 & 1 & 1 \end{vmatrix} - 0\begin{vmatrix} 1 & 3 & 2 \\ 2 & 2 & -1 \\ 4 & 1 & 1 \end{vmatrix} + 0\begin{vmatrix} 1 & -1 & 2 \\ 2 & -2 & -1 \\ 4 & 3 & 1 \end{vmatrix} - (-2)\begin{vmatrix} 1 & -1 & 3 \\ 2 & -2 & 2 \\ 4 & 3 & 1 \end{vmatrix}.$$

The second and third minors need not be evaluated nor even written down, since they are to be multiplied by zero. Hence $D = 3(-22) + 2(28) = -66 + 56 = -10$.

Expansion by minors reduces the task of evaluating a determinant of order n to that of evaluating several determinants of order $n - 1$. *The technique of example 5, section 140, may be used to produce zeros by a judicious choice of the multiplier k, thus diminishing the number of minors to be worked out in an expansion by minors.*

Example 3. In (2) we may add -3 times the last row to the first row, and add -2 times the last row to the second row, obtaining

$$D = \begin{vmatrix} -11 & -10 & 0 & -1 \\ -6 & -8 & 0 & -3 \\ 3 & 0 & 0 & -2 \\ 4 & 3 & 1 & 1 \end{vmatrix}.$$

Using the third column, we have only one minor to evaluate, obtaining

$$D = -1\begin{vmatrix} -11 & -10 & -1 \\ -6 & -8 & -3 \\ 3 & 0 & -2 \end{vmatrix} = -10.$$

Proof of theorem 1. (*A*) The sum of all terms in the expansion of D which have a_1 as a factor is equal to a_1A_1. This is so because each such term is the product of a_1 by one element from each of the remaining rows and columns of D, that is, the product of a_1

by a term of the expansion of A_1; furthermore, the sign before each such term of D is the same as the corresponding term of A_1, for writing a_1 to the left of any term of A_1 cannot alter the number of inversions in either subscripts or letters.

(B) Let x be the element in the ith row and jth column of D, and let D' be the determinant formed from D by successive interchanges of rows with rows and columns with columns so that x is in the upper left-hand corner of D'. Since it takes $i - 1$ interchanges of adjacent rows to bring x to the top row, and $j - 1$ interchanges of adjacent columns to bring it to the first column, we have

$$D' = (-1)^{i+j-2}D = (-1)^{i+j}D,$$

since each such interchange multiplies the value of the determinant by -1. By (A), the sum of all terms in the expansion of D' which have x as a factor is equal to xX where X is the minor of x in D'. But the minor of x in D' is the same as its minor in D. Hence the sum of all terms having x as a factor in the expansion of D is equal to $(-1)^{i+j}xX$.

(C) Each term in the expansion of D contains as a factor one and only one element of a given column (or row). By (B), the sum of all the terms containing each element of the given column (or row) is equal to this element times its minor times $(-1)^{i+j}$ where the element is in the ith row and jth column. This completes the proof of theorem 1.

THEOREM 2. *If, in the expansion of a determinant by minors, using the elements of a certain column (or row), we replace the elements of this column (or row) by the corresponding elements of any other column (or row), the resulting expression has the value zero.*

Proof. Consider the expansion of (1):

$$D = -b_1B_1 + b_2B_2 - b_3B_3 + b_4B_4.$$

Replacing b_1, b_2, b_3, b_4 by a_1, a_2, a_3, a_4 respectively in the right member, we get

$$-a_1B_1 + a_2B_2 - a_3B_3 + a_4B_4. \tag{3}$$

But this is the expansion of the determinant

$$\begin{vmatrix} a_1 & a_1 & c_1 & d_1 \\ a_2 & a_2 & c_2 & d_2 \\ a_3 & a_3 & c_3 & d_3 \\ a_4 & a_4 & c_4 & d_4 \end{vmatrix}$$

according to minors of the second column, and is therefore equal to zero by theorem 5, section 140. Clearly, any expression of the type (3) may be treated similarly, and the argument obviously applies to determinants of any order.

Note. If x is the element in the ith row and jth column and X is its minor, then the quantity $(-1)^{i+j}X$ is often called the **cofactor** of x.

EXERCISES

1. Expand $\begin{vmatrix} a_1 & b_1 & c_1 \\ a_2 & b_2 & c_2 \\ a_3 & b_3 & c_3 \end{vmatrix}$ by minors of the second column and verify that the same result is obtained as when the scheme of section 136 is used.

2. Expand $\begin{vmatrix} a_1 & b_1 \\ a_2 & b_2 \end{vmatrix}$ by minors of the first column and verify that the same result is obtained as by the scheme of section 135. (We here regard an element as a determinant of order 1.)

Expand each of the following by minors:

3–6. The determinants of exercises 3–6 of section 136.

7. $\begin{vmatrix} 3 & -1 & 3 \\ 2 & 5 & -3 \\ 5 & 4 & -1 \end{vmatrix}.$ **8.** $\begin{vmatrix} 4 & -2 & 3 \\ 2 & -1 & 1 \\ -3 & 2 & 2 \end{vmatrix}.$

9. $\begin{vmatrix} 1 & 1 & 1 & 1 \\ 1 & 0 & -1 & 0 \\ 0 & 1 & 1 & -1 \\ 2 & 0 & -1 & -3 \end{vmatrix}.$ **10.** $\begin{vmatrix} 1 & 5 & 2 & -1 \\ 1 & 1 & 0 & 2 \\ 2 & -1 & 0 & 3 \\ -2 & -1 & -1 & 2 \end{vmatrix}.$

11. $\begin{vmatrix} 2 & -2 & 1 & 3 \\ 0 & 2 & -1 & -1 \\ 2 & -3 & 2 & 4 \\ 0 & -1 & 1 & 1 \end{vmatrix}.$ **12.** $\begin{vmatrix} 1 & -2 & -1 & 3 \\ 2 & 1 & 1 & -2 \\ 1 & 3 & 2 & -1 \\ -3 & -2 & 4 & 2 \end{vmatrix}.$

13. $\begin{vmatrix} 3 & -2 & 6 & 4 \\ 1 & 0 & 2 & -1 \\ 5 & 4 & 3 & 0 \\ 2 & 2 & -5 & 6 \end{vmatrix}.$ **14.** $\begin{vmatrix} 1 & 2 & -1 & 1 \\ -1 & -3 & 4 & 2 \\ 2 & 2 & 3 & -2 \\ 3 & 2 & 1 & 1 \end{vmatrix}.$

15. $D = \begin{vmatrix} -1 & 2 & -1 & 1 & 1 \\ -4 & 4 & -1 & 1 & 1 \\ -4 & 6 & -3 & 2 & 3 \\ 2 & -4 & 1 & 1 & 2 \\ -1 & 1 & -1 & 1 & 0 \end{vmatrix}.$ (*Hint:* First express D in terms of minors [of fourth order] of some row or column. Then evaluate each of these minors separately and insert their values.)

16. $\begin{vmatrix} 1 & 1 & 1 & 1 \\ 1 & 0 & 0 & -1 \\ 0 & 1 & 0 & -1 \\ 0 & 0 & 1 & -1 \end{vmatrix}.$ **17.** $\begin{vmatrix} c^2 & -bc & ac \\ -bc & b^2 & -ab \\ ac & -ab & a^2 \end{vmatrix}.$

18. $\begin{vmatrix} r & s & s & s \\ s & r & s & s \\ s & s & r & s \\ s & s & s & r \end{vmatrix}.$

★19. Evaluate the following determinants of order n:

$$(a)\ \begin{vmatrix} r & s & s & s & \cdot & \cdot & \cdot & \cdot & \cdot & s \\ s & r & s & s & \cdot & \cdot & \cdot & \cdot & \cdot & s \\ s & s & r & s & \cdot & \cdot & \cdot & \cdot & \cdot & s \\ s & s & s & r & \cdot & \cdot & \cdot & \cdot & \cdot & s \\ \cdot & \cdot & \cdot & \cdot & \cdot & & & & & \cdot \\ \cdot & \cdot & \cdot & \cdot & & \cdot & & & & \cdot \\ \cdot & \cdot & \cdot & \cdot & & & \cdot & & & \cdot \\ s & s & s & s & \cdot & \cdot & s & r & s \\ s & s & s & s & \cdot & \cdot & s & s & r \end{vmatrix};\ (b)\ \begin{vmatrix} 1 & 1 & 1 & \cdot & \cdot & \cdot & \cdot & \cdot & 1 & 1 \\ 1 & 0 & 0 & \cdot & \cdot & \cdot & \cdot & \cdot & 0 & -1 \\ 0 & 1 & 0 & \cdot & \cdot & \cdot & \cdot & \cdot & 0 & -1 \\ 0 & 0 & 1 & \cdot & \cdot & \cdot & \cdot & \cdot & 0 & -1 \\ \cdot & \cdot & \cdot & \cdot & & & & & \cdot & \cdot \\ \cdot & \cdot & \cdot & & \cdot & & & & \cdot & \cdot \\ \cdot & \cdot & \cdot & & & \cdot & & & \cdot & \cdot \\ 0 & 0 & 0 & \cdot & \cdot & \cdot & 1 & 0 & -1 \\ 0 & 0 & 0 & \cdot & \cdot & \cdot & 0 & 1 & -1 \end{vmatrix}$$

142. Solution of Linear Systems in *n* Unknowns by Determinants

Consider a system of n linear equations in n unknowns. For example, for $n = 4$,

$$\left.\begin{aligned} &(1) & a_1x + b_1y + c_1z + d_1t &= k_1 \\ &(2) & a_2x + b_2y + c_2z + d_2t &= k_2 \\ &(3) & a_3x + b_3y + c_3z + d_3t &= k_3 \\ &(4) & a_4x + b_4y + c_4z + d_4t &= k_4. \end{aligned}\right\}$$

Let Δ be the determinant of the coefficients as they appear in the equations when like terms are written below one another; that is,

$$\Delta = \begin{vmatrix} a_1 & b_1 & c_1 & d_1 \\ a_2 & b_2 & c_2 & d_2 \\ a_3 & b_3 & c_3 & d_3 \\ a_4 & b_4 & c_4 & d_4 \end{vmatrix}.$$

Then we have the following theorem, known as Cramer's rule, after G. Cramer (Swiss, 1704–1752).

★ May be omitted without disturbing continuity.

THEOREM 1. *If $\Delta \neq 0$, the solutions of the above system are given by*

$$x = \frac{N_1}{\Delta}, \; y = \frac{N_2}{\Delta}, \; z = \frac{N_3}{\Delta}, \; t = \frac{N_4}{\Delta} \tag{5}$$

where

$$N_1 = \begin{vmatrix} k_1 & b_1 & c_1 & d_1 \\ k_2 & b_2 & c_2 & d_2 \\ k_3 & b_3 & c_3 & d_3 \\ k_4 & b_4 & c_4 & d_4 \end{vmatrix}, \quad N_2 = \begin{vmatrix} a_1 & k_1 & c_1 & d_1 \\ a_2 & k_2 & c_2 & d_2 \\ a_3 & k_3 & c_3 & d_3 \\ a_4 & k_4 & c_4 & d_4 \end{vmatrix},$$

$$N_3 = \begin{vmatrix} a_1 & b_1 & k_1 & d_1 \\ a_2 & b_2 & k_2 & d_2 \\ a_3 & b_3 & k_3 & d_3 \\ a_4 & b_4 & k_4 & d_4 \end{vmatrix}, \quad \textit{and} \quad N_4 = \begin{vmatrix} a_1 & b_1 & c_1 & k_1 \\ a_2 & b_2 & c_2 & k_2 \\ a_3 & b_3 & c_3 & k_3 \\ a_4 & b_4 & c_4 & k_4 \end{vmatrix}.$$

Proof. (*A*) Multiplying equation (1) by A_1, (2) by $-A_2$, (3) by A_3, (4) by $-A_4$, and adding, we get

$$\begin{aligned}(a_1A_1 - a_2A_2 + a_3A_3 - a_4A_4)x + (b_1A_1 - b_2A_2 + b_3A_3 \\ - b_4A_4)y + (c_1A_1 - c_2A_2 + c_3A_3 - c_4A_4)z + (d_1A_1 - d_2A_2 \\ + d_3A_3 - d_4A_4)t = (k_1A_1 - k_2A_2 + k_3A_3 - k_4A_4).\end{aligned}$$

By theorems 1 and 2 of section 141, this reduces to $\Delta \cdot x = N_1$. Since $\Delta \neq 0$, this yields $x = N_1/\Delta$, which is the first equality in (5). Similarly, multiplying (1), (2), (3), (4) by $-B_1, B_2, -B_3, B_4$ respectively, and adding, we get $\Delta \cdot y = N_2$, and so on. This proves that if there exist values of x, y, z, t which satisfy the given equations, they can only be the values given by (5).

(*B*) To show, conversely, that these values really satisfy these equations, we may check by substitution, or proceed as follows.

The determinant of order 5

$$\begin{vmatrix} k_1 & a_1 & b_1 & c_1 & d_1 \\ k_1 & a_1 & b_1 & c_1 & d_1 \\ k_2 & a_2 & b_2 & c_2 & d_2 \\ k_3 & a_3 & b_3 & c_3 & d_3 \\ k_4 & a_4 & b_4 & c_4 & d_4 \end{vmatrix} = 0,$$

since the first two rows are identical. Expanding it by minors of the first row, we get

$$k_1\Delta - a_1N_1 - b_1N_2 - c_1N_3 - d_1N_4 = 0.$$

Transposing $k_1\Delta$ and dividing by $-\Delta(\neq 0)$, we get

$$a_1 \frac{N_1}{\Delta} + b_1 \frac{N_2}{\Delta} + c_1 \frac{N_3}{\Delta} + d_1 \frac{N_4}{\Delta} = k_1,$$

which proves that the values (5) satisfy equation (1). A similar device shows that they satisfy the other equations. This completes the proof for the case $n = 4$. It is clear that the argument applies as well for any value of n.

Cramer's rule can be stated in words as follows: *In a system of n linear equations in n unknowns, let Δ be the determinant of the coefficients as they appear in the equations after like terms have been written below each other in the left member, and constant terms in the right member. Then if $\Delta \neq 0$, the solution can be written down by expressing each unknown as a quotient of two determinants, the denominator being Δ and the numerator for a given unknown being the determinant obtained from Δ by replacing the column of coefficients of that unknown by the column of constant terms.*

Example. Solve the system $\begin{cases} x + 2y + z = 2 \\ 2x - 2z + t = 6 \\ 4y + 3z + 2t = -1 \\ -x + 6y - z - t = 2. \end{cases}$

Solution. The denominator $\Delta = \begin{vmatrix} 1 & 2 & 1 & 0 \\ 2 & 0 & -2 & 1 \\ 0 & 4 & 3 & 2 \\ -1 & 6 & -1 & -1 \end{vmatrix} = -92.$

The numerator for x is $N_1 = \begin{vmatrix} 2 & 2 & 1 & 0 \\ 6 & 0 & -2 & 1 \\ -1 & 4 & 3 & 2 \\ 2 & 6 & -1 & -1 \end{vmatrix} = -184.$

Hence $x = \dfrac{-184}{-92} = 2.$ Similarly $y = \dfrac{-46}{-92} = \dfrac{1}{2}$, $z = \dfrac{92}{-92} = -1$, and $t = \dfrac{0}{-92} = 0.$

The student may check by substituting in the given equations.

Note. While determinants are valuable for the theory of systems of linear equations, the methods explained here are too tedious for the numerical solution of a system of n equations in n unknowns if n is large. Some progress has been made recently in methods of solution suitable for use with calculating machines.

EXERCISES

Solve by determinants and check:

1. $\begin{cases} x + y + z + 3t = 3 \\ 3x + y - z = 0 \\ 2x - 2y - z + 6t = 4 \\ 4x - y - 2z - 3t = 0. \end{cases}$

2. $\begin{cases} x + y + z - w = 2 \\ 2x + 3y - 4z - w = 0 \\ x - y + w = 1 \\ 2x + 3y - z = 4. \end{cases}$

3. $\begin{cases} x + y + z + t = 3 \\ 2x - 2y - z + 2t = 0 \\ 3x - y + 2z + 2t = 2 \\ x - y - 2z + t = 0. \end{cases}$

4. $\begin{cases} x + y + z + t = 4 \\ 2x - y + t = 2 \\ 3x + y - z - t = 2 \\ x - 2y - 3z + t = -3. \end{cases}$

5. $\begin{cases} 2x + 3y + z + t = 1 \\ x - y - z + t = 1 \\ 3x + y + z + 2t = 0 \\ -x + z - t = -2. \end{cases}$

6. $\begin{cases} 4x + y + 2z + 4t = 14 \\ x - y - 8z = 7 \\ 3x - 2y - 4z + t = 12 \\ 3y + 4z + t = -3. \end{cases}$

7. $\begin{cases} x - 2y + z + t = 3 \\ x + z = t \\ 2y - z = t \\ x + 4y + 2z - t = 1. \end{cases}$

8. $\begin{cases} x + 2y - w = 0 \\ 3x + y + 4z + 2w = 3 \\ 2x - 3y - z + 5w = 1 \\ x + 2z + 2w = -1. \end{cases}$

143. Systems of Homogeneous Linear Equations

If in a linear equation in two or more unknowns the constant term is zero, the equation is called **homogeneous.** Clearly any homogeneous equation or system of several homogeneous equations is satisfied if we let every unknown equal zero; this solution is called the **trivial solution.** Consider a system of n linear homogeneous equations in n unknowns. The proof of the theorem of section 142 shows that if $\Delta \neq 0$, the trivial solution is the only solution. Hence we have the following theorem.

THEOREM 1. *A system of n homogeneous linear equations in n unknowns can have a non-trivial solution only if* $\Delta = 0$.

THEOREM 2. *Conversely, if, in a system of n homogeneous linear equations in n unknowns, we have* $\Delta = 0$*, then there exist infinitely many non-trivial solutions.*

The proof of theorem 2 is beyond the scope of this book.* We shall take up one particular case.

* See, for a more complete treatment of linear systems, Bôcher, *Introduction to Higher Algebra,* Macmillan, 1907, or Birkhoff-MacLane, *Survey of Modern Algebra,* Macmillan, 1941, or Schreier and Sperner, *Modern Algebra and Matrix Theory,* Chelsea, 1955, etc.

THEOREM 3. *If, in a system of n linear homogeneous equations with $\Delta = 0$, the minor of at least one element is not zero, then the system has infinitely many solutions.*

Proof. For simplicity, take $n = 3$. Suppose in the system

$$\begin{cases} a_1x + b_1y + c_1z = 0 \\ a_2x + b_2y + c_2z = 0 \\ a_3x + b_3y + c_3z = 0 \end{cases}$$

we have $\Delta = 0$ but $A_1 \neq 0$, say. Then we may solve the last two equations for y and z in terms of x, obtaining

$$y = \frac{\begin{vmatrix} -a_2x & c_2 \\ -a_3x & c_3 \end{vmatrix}}{A_1} = \frac{-xB_1}{A_1}, \quad z = \frac{\begin{vmatrix} b_2 & -a_2x \\ b_3 & -a_3x \end{vmatrix}}{A_1} = \frac{xC_1}{A_1}.$$

Substituting these values in the first equation, we have

$$a_1x - \frac{b_1B_1}{A_1}x + \frac{c_1C_1}{A_1}x = 0$$

or

$$x(a_1A_1 - b_1B_1 + c_1C_1) = 0,$$

which is true since the parenthesis has the value $\Delta = 0$. Hence for any value of x whatever, say $x = m$, the values $x = m$, $y = -mB_1/A_1$, and $z = mC_1/A_1$ satisfy the system.

The same proof may be applied for any value of n.

Example. Find infinitely many solutions of the system

$$\begin{cases} x + y + z = 0 \\ 4x + y + 2z = 0 \\ 3x - 3y - z = 0. \end{cases}$$

Solution. The student should verify that $\Delta = 0$ while the minor $A_1 \neq 0$. Hence we may solve the latter two equations for y and z in terms of x, obtaining $y = 2x$, $z = -3x$. Hence for any value $x = m$ whatever, the numbers $x = m$, $y = 2m$, $z = -3m$ will satisfy the given system. For example, $x = 1$, $y = 2$, $z = -3$, or $x = 4$, $y = 8$, $z = -12$, etc.

EXERCISES

Find an expression for infinitely many non-trivial solutions, and list two particular non-trivial solutions, or else show that no non-trivial solutions exist:

1. $\begin{cases} x + y + z = 0 \\ x + 2y + z = 0 \\ 2x + 3y + 2z = 0. \end{cases}$
2. $\begin{cases} 3x + 5y + 3z = 0 \\ 2x + 3y + 3z = 0 \\ x + 2y = 0. \end{cases}$

3. $\begin{cases} 2x + 2y + z = 0 \\ 2x - y - 2z = 0 \\ 4x + y - z = 0. \end{cases}$
4. $\begin{cases} x + 5y + 2z = 0 \\ 2x + 3y - z = 0 \\ -x + 2y + 3z = 0. \end{cases}$

5. $\begin{cases} 2x + y + z = 0 \\ 3x + 2y + 4z = 0 \\ x - 2y - 3z = 0. \end{cases}$
6. $\begin{cases} -x + 3y - 3z = 0 \\ 3x + y - 2z = 0 \\ 4x - 2y + z = 0. \end{cases}$

Find the value which k must have in order that non-trivial solutions may exist:

7. $\begin{cases} 3kx + 4y + 3z = 0 \\ kx + 2y + z = 0 \\ 2x + 3y + 2z = 0. \end{cases}$
8. $\begin{cases} 5x + ky + z = 0 \\ x - 4ky + 3z = 0 \\ 3x + 3ky - z = 0. \end{cases}$

144. Systems of *m* Linear Equations in *n* Unknowns

A system of m linear equations in n unknowns is called **consistent** if all the equations of the system possess at least one solution in common; otherwise **inconsistent.** By section 142, if $m = n$ and $\Delta \neq 0$, the system is consistent.

THEOREM 1. *If $m = n$, and $\Delta = 0$, and at least one of the numerator determinants N_1, N_2, N_3, ... is not zero, then the system is inconsistent.*

Proof. By the proof of theorem 1, section 142, if there were a common solution, x, y, z, ..., we would have

$$\Delta \cdot x = N_1,\ \Delta \cdot y = N_2,\ \Delta \cdot z = N_3,\ \ldots . \tag{1}$$

But the left members are all zero, while at least one of the right members is not, by hypothesis. Hence this is impossible.

If $m > n$, the system is usually inconsistent, although under special circumstances it may be consistent. If it is possible to solve n of the equations for the n unknowns, then the values thus obtained can be substituted into the remaining $m - n$ equations; if these are satisfied, the system is consistent. A special case is given by the following theorem.

THEOREM 2. *If a system of m linear equations in $m - 1$ unknowns is written with the constant terms in the left member, it is inconsistent unless the resulting determinant is zero.*

Proof. For simplicity, we take $m = 3$. The given system

$$(2)\qquad \begin{cases} a_1x + b_1y + c_1 = 0 \\ a_2x + b_2y + c_2 = 0 \\ a_3x + b_3y + c_3 = 0 \end{cases}$$

can have a solution only if the system

$$(3)\qquad \begin{cases} a_1x + b_1y + c_1z = 0 \\ a_2x + b_2y + c_2z = 0 \\ a_3x + b_3y + c_3z = 0 \end{cases}$$

of homogeneous equations has a non-trivial solution with $z = 1$. By section 143, this implies that

$$(4)\qquad \begin{vmatrix} a_1 & b_1 & c_1 \\ a_2 & b_2 & c_2 \\ a_3 & b_3 & c_3 \end{vmatrix} = 0.$$

This completes the proof.

Note 1. This does not imply that the vanishing of the determinant (4) guarantees the consistency of the system (2).

Note 2. If the system (2) is consistent, the equations represent straight lines having a point in common.

Example 1. If possible, solve the system

$$\begin{cases} x + y - 7 = 0 \\ x - y - 1 = 0 \\ 2x - 3y + 1 = 0. \end{cases}$$

Solution. Since the determinant $\begin{vmatrix} 1 & 1 & -7 \\ 1 & -1 & -1 \\ 2 & -3 & +1 \end{vmatrix} = 0$, it is possible that the system may be consistent. Solving the first two equations, we find $x = 4$, $y = 3$. Substituting these values in all three equations, we find that they are satisfied.

If $m < n$, then usually the system has infinitely many solutions, although, under special circumstances, it may be inconsistent. To find the solutions, when they exist, we may solve for m of the unknowns in terms of the remaining $n - m$ unknowns, which may then have arbitrary values,* as in the following example.

* For a more complete discussion of the material in this section, see the works of Bôcher, Birkhoff-MacLane, and Schreier-Sperner, cited in section 143. Also, Dresden, *Solid Analytical Geometry and Determinants*, Wiley, 1930.

Example 2. Solve the system

$$\begin{cases} x + y - 2z = 3 \\ x - 2y + z = 1. \end{cases}$$

Solution. Solving for x and y in terms of z we obtain $x = z + \frac{7}{3}$, $y = z + \frac{2}{3}$. Hence, z may be given an arbitrary value a, and $x = a + \frac{7}{3}$, $y = a + \frac{2}{3}$, $z = a$ will satisfy the system.

EXERCISES

Show, by means of theorem 1, that the following systems are inconsistent:

1. $\begin{cases} x + 2y - z = 3 \\ 2x + 4y + 3z = 1 \\ 3x + 6y - 2z = 2. \end{cases}$ **2.** $\begin{cases} 2x - y + 3z = 1 \\ 4x - 2y + 6z = 3 \\ 3x + 4z = 2. \end{cases}$

Solve each system, or else show that it is inconsistent:

3. $\begin{cases} x + y - 7 = 0 \\ x - y - 3 = 0 \\ x - 2y - 1 = 0. \end{cases}$ **4.** $\begin{cases} x - 3y = 0 \\ x - 4y - 1 = 0 \\ 2x - 5y + 1 = 0. \end{cases}$

5. $\begin{cases} x + y = 1 \\ 2x - y = 11 \\ x - y = 3. \end{cases}$ **6.** $\begin{cases} 2x + y = 0 \\ x - 2y + 10 = 0 \\ x + y - 2 = 0. \end{cases}$

7. $\begin{cases} 2x - y + 3 = 0 \\ 4x - 2y + 5 = 0 \\ 6x - 3y + 1 = 0. \end{cases}$ **8.** $\begin{cases} 8x + 5y = 19 \\ x - 2y = 5 \\ 2x + 5y = 1. \end{cases}$

9. $\begin{cases} x + y - 5z = 0 \\ x - y + 3z = 4. \end{cases}$ **10.** $\begin{cases} 3x + y - z = 1 \\ x - 2y - 2z = -2. \end{cases}$

11. $\begin{cases} x + 2y + 3z = 4 \\ 2x + 4y + 6z = 5. \end{cases}$ **12.** $\begin{cases} 2x - y + 3z = 1 \\ 4x - 2y + 6z = 3. \end{cases}$

13. Prove that if x, y, and z satisfy the system

$$\begin{cases} a_1x + b_1y + c_1z = 0 \\ a_2x + b_2y + c_2z = 0 \end{cases}$$

and if at least one of the three determinants below is different from zero, then

$$x:y:z = \begin{vmatrix} b_1 & c_1 \\ b_2 & c_2 \end{vmatrix} : \begin{vmatrix} c_1 & a_1 \\ c_2 & a_2 \end{vmatrix} : \begin{vmatrix} a_1 & b_1 \\ a_2 & b_2 \end{vmatrix}.$$

14. Prove that if x_1, y_1, z_1 and x_2, y_2, z_2 are two solutions of the system

$$\begin{cases} a_1x + b_1y + c_1z = d_1 \\ a_2x + b_2y + c_2z = d_2 \end{cases}$$

and if none of the determinants below are zero, then

$$\frac{x_1 - x_2}{\begin{vmatrix} b_1 & c_1 \\ b_2 & c_2 \end{vmatrix}} = \frac{y_1 - y_2}{\begin{vmatrix} c_1 & a_1 \\ c_2 & a_2 \end{vmatrix}} = \frac{z_1 - z_2}{\begin{vmatrix} a_1 & b_1 \\ a_2 & b_2 \end{vmatrix}}.$$

★145. The Resultant of Two Polynomials

Having studied the simultaneous solution of linear equations, it is natural to investigate when equations of degree higher than the first have a solution in common.

THEOREM 1. *If* $f(x) = a_0x^m + a_1x^{m-1} + \ldots + a_m$ *and* $g(x) = b_0x^n + b_1x^{n-1} + \ldots + b_n$, *and if not both* a_0 *and* b_0 *are equal to zero, then* $f(x)$ *and* $g(x)$ *have a common factor involving* x *if and only if there exist polynomials* $h(x)$ *and* $k(x)$ *of degrees not greater than* $n - 1$ *and* $m - 1$ *respectively, not both identically zero, such that*

$$f(x) \cdot h(x) - g(x) \cdot k(x) = 0 \tag{1}$$

identically.

Proof. (*A*) If (1) holds, and say $a_0 \neq 0$, then $f(x)$ has m linear factors of which at most $m - 1$ can be factors of $k(x)$. Therefore at least one of these factors is a factor of $g(x)$. A similar argument can be given if $b_0 \neq 0$.

(*B*) Conversely, if $f(x)$ and $g(x)$ have a common factor $F(x)$, then writing $h(x) = g(x)/F(x)$ and $k(x) = f(x)/F(x)$, we see by substitution that condition (1) is satisfied. This completes the proof.

Now let the polynomials $h(x)$ and $k(x)$ be given by $h(x) = \alpha_0x^{n-1} + \alpha_1x^{n-2} + \ldots + \alpha_{n-1}$ and $k(x) = \beta_0x^{m-1} + \beta_1x^{m-2} + \ldots + \beta_{m-1}$, respectively. Then condition (1) says that

$$\begin{aligned}(a_0x^m + a_1x^{m-1} + \ldots + a_m)(\alpha_0x^{n-1} + \alpha_1x^{n-2} + \ldots \\ + \alpha_{n-1}) - (b_0x^n + b_1x^{n-1} + \ldots + b_n)(\beta_0x^{m-1} + \beta_1x^{m-2} \\ + \ldots + \beta_{m-1}) = 0\end{aligned} \tag{2}$$

identically. But a polynomial can be identically zero only if all its coefficients are zero. Hence multiplying (2) out and equating the total coefficient of each power of x to zero, we get

★ May be omitted without disturbing continuity.

$$
(3)\qquad \begin{cases}
a_0\alpha_0 - b_0\beta_0 = 0 \\
a_1\alpha_0 + a_0\alpha_1 - b_1\beta_0 - b_0\beta_1 = 0 \\
\cdot \\
\cdot \\
\cdot \\
a_m\alpha_{n-2} + a_{m-1}\alpha_{n-1} - b_n\beta_{m-2} - b_{n-1}\beta_{m-1} = 0 \\
a_m\alpha_{n-1} - b_n\beta_{m-1} = 0.
\end{cases}
$$

This is a system of $m + n$ linear homogeneous equations in the $m + n$ unknowns $\alpha_0, \alpha_1, \ldots, \alpha_{n-1}, -\beta_0, -\beta_1, \ldots, -\beta_{m-1}$ with the determinant

$$
\left|\begin{array}{cccccccccc}
a_0 & & & & \mathbf{0} & b_0 & & & & \mathbf{0} \\
a_1 & a_0 & & & & b_1 & b_0 & & & \\
\cdot & a_1 & \cdot & & & \cdot & b_1 & \cdot & & \\
\cdot & \cdot & & \cdot & & \cdot & \cdot & & \cdot & \\
\cdot & \cdot & & & \cdot & \cdot & \cdot & & & \cdot \\
a_m & \cdot & & & a_0 & & \cdot & & & b_0 \\
 & a_m & & & a_1 & b_n & & & & b_1 \\
 & & \cdot & & \cdot & & b_n & & & \\
 & & & \cdot & \cdot & & & \cdot & & \cdot \\
 & & & & \cdot & & & & \cdot & \cdot \\
 & & & & & & & & \cdot & \cdot \\
\mathbf{0} & & & & a_m & \mathbf{0} & & & & b_n
\end{array}\right|,
$$

where the large zeros indicate that all other elements are zero. It is customary to interchange the rows and columns, and to denote the determinant thus arrived at by R. Thus,

$$
(4)\quad R = \left|\begin{array}{ccccccccccc}
a_0 & a_1 & \cdot & \cdot & \cdot & a_m & & & & & \mathbf{0} \\
 & a_0 & a_1 & \cdot & \cdot & \cdot & a_m & & & & \\
 & & \cdot & & & & & \cdot & & & \\
 & & & \cdot & & & & & \cdot & & \\
\mathbf{0} & & & & \cdot & & & & & \cdot & \\
 & & & & & a_0 & a_1 & \cdot & \cdot & \cdot & a_m \\
b_0 & b_1 & \cdot & \cdot & \cdot & & b_n & & & & \\
 & b_0 & b_1 & \cdot & \cdot & \cdot & & b_n & & & \mathbf{0} \\
 & & \cdot & & & & & & \cdot & & \\
 & & & \cdot & & & & & & \cdot & \\
 & & & & \cdot & & & & & & \cdot \\
\mathbf{0} & & & & & b_0 & b_1 & \cdot & \cdot & \cdot & b_n
\end{array}\right|
\begin{array}{l}
\left.\vphantom{\begin{array}{c}1\\1\\1\\1\\1\\1\end{array}}\right\} n \text{ rows} \\
\left.\vphantom{\begin{array}{c}1\\1\\1\\1\\1\\1\end{array}}\right\} m \text{ rows.}
\end{array}
$$

By theorem 1 above, and theorems 1 and 2 of section 143, we have the following theorem.

THEOREM 2. *Either $f(x)$ and $g(x)$ have a common factor involving x, or $a_0 = b_0 = 0$, if and only if $R = 0$.*

The determinant R given by (4) is called the **resultant** of the polynomials $f(x)$ and $g(x)$.

Example. Show, by means of the resultant, that the equations $f(x) = x^2 - 5x + 6 = 0$ and $g(x) = 2x - 6 = 0$ have a common solution.

Solution. The resultant is

$$R = \begin{vmatrix} 1 & -5 & 6 \\ 2 & -6 & 0 \\ 0 & 2 & -6 \end{vmatrix} = 0.$$

Since $a_0 = 1 \neq 0$ and $b_0 = 2 \neq 0$, $f(x)$ and $g(x)$ have a common factor involving x. Therefore $f(x) = 0$ and $g(x) = 0$ have a common solution.

EXERCISES

1. Find the resultant of $a_0x + a_1$ and $b_0x^2 + b_1x + b_2$.

Show by means of the resultant that the following pairs of equations have a common root:

2. $\begin{cases} 2x - 4 = 0 \\ x^2 - 5x + 6 = 0. \end{cases}$

3. $\begin{cases} 3x - 2 = 0 \\ 3x^2 + x - 2 = 0. \end{cases}$

4. $\begin{cases} 3x^2 - 5x + 2 = 0 \\ 6x^2 - x - 2 = 0. \end{cases}$

5. $\begin{cases} 2x^2 - 3x - 2 = 0 \\ 6x^2 + x - 1 = 0. \end{cases}$

6. $\begin{cases} x^3 - 8 = 0 \\ x^2 + 2x + 4 = 0. \end{cases}$

7. $\begin{cases} x^3 - 8 = 0 \\ x^2 - 5x + 6 = 0. \end{cases}$

★146. Systems of Higher Degree in Two Unknowns

Nothing in the preceding section requires that the coefficients a_0, $a_1, \ldots, a_m, b_0, b_1, \ldots, b_n$ be constants. Hence we may use theorem 2 of section 145 to solve systems of two equations of degree higher than the first in two unknowns x and y, by writing the equations as polynomial equations in x with coefficients which are polynomials in y, as in the following example.

★ May be omitted without disturbing continuity.

Example. Solve the system

$$\begin{cases} x^2 - 2y^2 + x - 8y - 8 = 0 \\ 2x^2 - 5y^2 + 4x - 17y - 12 = 0. \end{cases}$$

Solution. Writing the given equations as polynomial equations in x with coefficients which are polynomials in y, we have

$$\begin{cases} x^2 + x + (-2y^2 - 8y - 8) = 0 \\ 2x^2 + 4x + (-5y^2 - 17y - 12) = 0. \end{cases}$$

Hence

$$R = \begin{vmatrix} 1 & 1 & (-2y^2 - 8y - 8) & 0 \\ 0 & 1 & 1 & (-2y^2 - 8y - 8) \\ 2 & 4 & (-5y^2 - 17y - 12) & 0 \\ 0 & 2 & 4 & (-5y^2 - 17y - 12) \end{vmatrix}.$$

Evaluating R and setting it equal to zero, we have

$$R = y^4 + 2y^3 - 13y^2 - 38y - 24 = 0,$$

which has the roots $y = -1, -2, -3, +4$.

For $y = -1$, the given equations reduce to $\begin{cases} x^2 + x - 2 = 0 \\ 2x^2 + 4x = 0 \end{cases}$ which have the root $x = -2$ in common.

For $y = -2$, the given equations reduce to $\begin{cases} x^2 + x = 0 \\ 2x^2 + 4x + 2 = 0 \end{cases}$ which have the root $x = -1$ in common.

For $y = -3$, the given equations reduce to $\begin{cases} x^2 + x - 2 = 0 \\ 2x^2 + 4x - 6 = 0 \end{cases}$ which have the root $x = 1$ in common.

For $y = 4$, the given equations reduce to $\begin{cases} x^2 + x - 72 = 0 \\ 2x^2 + 4x - 160 = 0 \end{cases}$ which have the root $x = 8$ in common.

Hence the common solutions of the given equations are

x	-2	-1	1	8
y	-1	-2	-3	4

The real solutions are the coordinates of the points of intersection of the graphs of the given equations.

The form of the resultant used in the method of elimination of one unknown from a pair of equations explained here is attributed to J. J. Sylvester (English, 1814–1897) who was one of the first great mathematicians to teach in an American university. Sylvester made many important contributions to mathematics.

For a further study of resultants and systems of equations see Bôcher, *Introduction to Higher Algebra,* Macmillan, 1907.

EXERCISES

Find the common solutions of each of the following systems of equations:

1. $\begin{cases} x^2 - 2y^2 - x = 0 \\ 2x^2 - 5y^2 + 3y = 0. \end{cases}$ **2.** $\begin{cases} x^2 - 3x - 2y^2 - 4y = 0 \\ 2x^2 - 4x - 5y^2 - 7y = 0. \end{cases}$

3. $\begin{cases} x^2 - 3xy + 2y^2 - 16x - 28y = 0 \\ x^2 - xy - 2y^2 - 5x - 5y = 0. \end{cases}$

4. $\begin{cases} x^2 - 4x + y^2 = 1 \\ x^2 + y^2 - 2y = 9. \end{cases}$

5. Find the (real) coordinates of the points of intersection of the graphs of the equations

$$\begin{cases} y^2 - x^3 = 0 \\ (y - 1)^2 + x^2 - 1 = 0. \end{cases}$$

Solve the system of equations given in:

6. Exercise 1, section 87. **7.** Exercise 3, section 87.
8. Exercise 7, section 87. **9.** Exercise 9, section 87.
10. Exercise 15, section 87. **11.** Exercise 17, section 87.
12. Exercise 21, section 87. **13.** Exercise 23, section 87.
14. Exercise 25, section 87. **15.** Exercise 27, section 87.

Chapter XXI

Progressions

147. Arithmetic Progressions

A sequence* of numbers in which each term is obtained from the preceding one by addition of a constant number d is called an **arithmetic progression.** The number d is called the **common difference.** If a is the first term of an arithmetic progression with difference d, the first n terms are

$$(1) \qquad a,\ a + d,\ a + 2d,\ a + 3d,\ \ldots,\ a + (n - 1)d.$$

The ***n*th term** or **last term** l is

$$(2) \qquad l = a + (n - 1)d.$$

Example 1. The first five terms of the arithmetic progression with first term equal to 3 and common difference equal to 2 are 3, 5, 7, 9, 11. The nth term is $3 + (n - 1)2 = 2n + 1$.

Example 2. Given that 4, 7, 10 are the first three terms of an arithmetic progression, find the 11th term.

Solution. Here $d = 7 - 4 = 3$. Hence the eleventh term is $a + (11 - 1)d = 4 + 30 = 34$.

Let s_n stand for the sum of the first n terms of the arithmetic progression (1). That is,

$$(3) \qquad s_n = a + (a + d) + (a + 2d) + (a + 3d) + \cdots + (l - d) + l.$$

Reversing the order of the terms, we have also

$$(4) \qquad s_n = l + (l - d) + (l - 2d) + (l - 3d) + \cdots + (a + d) + a.$$

* See section 99, chapter XVI, for the general definition of a sequence.

Adding terms written under each other in (3) and (4), we obtain

$$(5) \quad 2s_n = (a + l) + (a + l) + (a + l) + (a + l) + \dots + (a + l) + (a + l)$$

where there are n parentheses in the right member. Hence, $2s_n = n(a + l)$, or

$$(6) \qquad s_n = \frac{n}{2}(a + l).$$

Using (2), we can rewrite (6) as

$$(7) \qquad s_n = \frac{n}{2}[2a + (n - 1)d].$$

Example 3. Find the sum of the first 20 terms of the arithmetic progression in example 1.

First solution. Here $a = 3$, $d = 2$, $n = 20$. Hence, from (7),

$$s_{20} = \frac{20}{2}(2 \cdot 3 + [20 - 1]2) = 440.$$

Second solution. The twentieth term is $3 + (20 - 1)2 = 41$. From, (6),

$$s_{20} = \frac{20}{2}(3 + 41) = 440.$$

Example 4. Find the sum of the first 50 positive integral multiples of 3.

Solution. The positive integral multiples of 3 form an arithmetic progression with $a = 3$, and $d = 3$. Hence $s_{50} = \frac{50}{2}[2 \cdot 3 + (50 - 1)3] = 3825$.

In an arithmetic progression of n terms, the first and last (nth) terms are called the **extremes,** while those in between are called **arithmetic means.**

Example 5. Insert five arithmetic means between the numbers 4 and 22.

Solution. After the five arithmetic means are inserted, the number of terms in the arithmetic progression is $n = 7$. We have $a = 4$, $l = 22$. From (2) we have $22 = 4 + (7 - 1)d$; hence $d = 3$. Therefore the arithmetic means are 7, 10, 13, 16, 19.

Note 1. If we insert one arithmetic mean between a and b, we find $d = \frac{b - a}{2}$. Hence the arithmetic mean is $a + \frac{b - a}{2} =$

$\frac{a+b}{2}$. The number $\frac{a+b}{2}$ is often called **the arithmetic mean** (or **average**) of a and b. Generalizing this definition, the number $\frac{a_1 + a_2 + a_3 + \ldots + a_n}{n}$ is called **the arithmetic mean** (or **average**) of the numbers $a_1, a_2, \ldots, a_n$.

Note 2. Further applications of arithmetic progressions will be found in chapter XXIV.

EXERCISES

Find the common difference, the last term, and the sum of each of the following arithmetic progressions:

1. 2, 5, 8, . . . to fifteen terms.
2. 3, 7, 11, . . . to sixteen terms.
3. 10, 7, 4, . . . to eight terms.
4. $-11, -8, -5, \ldots$ to fourteen terms.
5. 2, $3\frac{2}{3}$, $5\frac{1}{3}$, . . . to twelve terms.
6. 9, 6.6, 4.2, . . . to thirteen terms.
7. Find the sum of the first 80 positive integers divisible by 4.
8. Find the sum of the first 70 positive integers divisible by 6.
9. (a) How many even numbers lie between 11 and 65? (b) Find their sum.
10. (a) How many numbers divisible by 4 lie between 65 and 193? (b) Find their sum.
11. (a) How many numbers divisible by 6 lie between 75 and 190? (b) Find their sum.
12. (a) How many numbers divisible by 7 lie between 100 and 400? (b) Find their sum.
13. Insert five arithmetic means between 2.9 and 15.5.
14. Insert six arithmetic means between 6 and 34.
15. Insert seven arithmetic means between -5 and 15.
16. Insert eight arithmetic means between 18 and -12.
17. Find the sum of the first n even positive integers.
18. Find the sum of the first n odd positive integers.
19. Find the sum of the first n positive integral multiples of 4.
20. Find the sum of the first n positive integral multiples of 5.

In each of the following problems, some of the quantities a, d, l, n, s_n, are given; find the remaining ones:

21. $d = 3, n = 13, s_n = 286$.
22. $d = 4, n = 12, s_n = 300$.
23. $a = 7, l = 75, n = 18$.
24. $a = 8, l = 38, n = 16$.
25. $a = 5, l = 25, d = 2$.
26. $a = 5, l = 56, d = 3$.

27. $d = 5$, $n = 13$, $l = 63$.
28. $d = -3$, $n = 13$, $l = -16$.
29. $a = 6$, $s_n = 720$, $l = 74$.
30. $a = 2$, $S_n = 572$, $l = 86$.

Find a formula for:

★**31.** d in terms of a, s_n, and l.
★**32.** s_n in terms of n, l, and d.
★**33.** d in terms of s_n, l, and n.
★**34.** l in terms of d, n, and s_n.
★**35.** a in terms of d, n, and s_n.

Find the arithmetic mean of:

36. 14 and 20.
37. 60, 72, 87.
38. 50, 75, 82, 95.

39. The first year a man is employed he saves \$100. In each succeeding year he saves \$50 more than the year before. How much has he accumulated at the end of 15 years?

40. A man pays a \$1000 debt by paying \$100 at the end of each year plus 5% interest on the amount unpaid during that year. Find his total payment.

41. A man receives a salary of \$1600 per annum with the understanding that his salary will increase by \$160 per year up to and including the tenth year. (*a*) What will his salary be for the tenth year? (*b*) What are his total earnings during the 10 years?

42. A body falls 16 feet during the first second, 48 feet during the second second, 80 feet during the third second, and so on in arithmetic progression. (*a*) How far does it fall during the tenth second? (*b*) How far does it fall during the first 10 seconds?

43. Show that if the nth term of a sequence is a linear function of n, say $pn + q$, where p and q are constants, then the sequence is an arithmetic progression. Find expressions for a, d, l, and s_n in terms of p, q, and n.

148. Geometric Progressions

A sequence of numbers in which each term is obtained from the preceding term by multiplication by a constant factor r is called a **geometric progression.** The constant factor r is called the **common ratio.** If a is the first term of a geometric progression, with common ratio r, the first n terms are

$$a, ar, ar^2, ar^3, \ldots, ar^{n-1}. \tag{1}$$

The **nth term** or **last term** l is

$$l = ar^{n-1}. \tag{2}$$

★May be omitted without disturbing continuity.

Example 1. The first five terms of the geometric progression with first term equal to 3 and common ratio equal to 2 are 3, 6, 12, 24, 48. The nth term is $3 \cdot 2^{n-1}$.

Example 2. Given that 16, 8, 4 are the first three terms of a geometric progression, find the fifth term.

Solution. Here $r = 8/16 = 1/2$. Hence the fifth term is $16 \cdot (\frac{1}{2})^4 = 1$.

Let s_n stand for the sum of the first n terms of the geometric progression (1). That is,

$$s_n = a + ar + ar^2 + ar^3 + \dots + ar^{n-1}. \tag{3}$$

Multiplying both members of (3) by r, we obtain

$$rs_n = ar + ar^2 + ar^3 + ar^4 + \dots + ar^n. \tag{4}$$

Subtracting (4) from (3), we get

$$(1 - r)s_n = a - ar^n. \tag{5}$$

Dividing (5) by* $1 - r$, we have the formula

$$s_n = \frac{a - ar^n}{1 - r}. \tag{6}$$

Using (2) we may rewrite (6) as

$$s_n = \frac{a - rl}{1 - r}. \tag{7}$$

Example 3. Find the sum of the first seven terms of the geometric progression whose first three terms are 3, 6, 12.

First solution. Here $r = 6/3 = 2$, $a = 3$. Hence by (6),

$$s_7 = \frac{3 - 3 \cdot 2^7}{1 - 2} = \frac{3 - 3 \cdot 128}{-1} = 381.$$

Second solution. The last term is $l = 3 \cdot 2^6 = 3 \cdot 64 = 192$.

Hence by (7), $s_7 = \dfrac{3 - 2 \cdot 192}{1 - 2} = 381.$

In a geometric progression of n terms, the first and last (nth) terms are called the **extremes,** while those in between are called **geometric means.** *We shall assume that the common ratio r is real.*

* It is assumed that $r \neq 1$. If $r = 1$, $s_n = na$.

Example 4. Insert five geometric means between the numbers 3 and 192.

Solution. After the five geometric means are inserted, the number of terms in the geometric progression is $n = 7$. Hence, from (2), $192 = 3r^6$, or $r^6 = 64$. Hence $r = \pm 2$. Therefore the five geometric means are 6, 12, 24, 48, 96, or -6, 12, -24, 48, -96.

Note 1. Inserting one geometric mean between the numbers a and b, we find that if x is a geometric mean then $\frac{x}{a} = \frac{b}{x}$. Thus $x = \pm \sqrt{ab}$. That is, x is a mean proportional between a and b. Either quantity $\pm \sqrt{ab}$ is called **a geometric mean** of a and b. Generalizing this definition, $\sqrt[n]{a_1 a_2 a_3 \ \ldots \ a_n}$ is called **a geometric mean** of $a_1, a_2, \ldots, a_n$.

Note 2. Further applications of geometric progressions will be found in chapter XXIV.

EXERCISES

Find the common ratio, the last term, and the sum of each of the following geometric progressions:

1. 2, 6, 18, . . ., to five terms.
2. 192, 96, 48, . . ., to six terms.
3. 3, -6, 12, . . ., to six terms.
4. 8, 6, 9/2, . . ., to six terms.
5. 3, $-3/2$, 3/4, . . ., to five terms.
6. 12, -18, 27, . . ., to six terms.
7. Insert four geometric means between 5 and 160.
8. Insert four geometric means between 729 and 3.
9. Insert four geometric means between 162 and $-2/3$.
10. Insert three geometric means between 3 and 768. How many sets of answers are there?

In each of the following problems, some of the quantities a, r, n, l, s_n are given; find the remaining ones:

11. $a = 512$, $r = 1/2$, $l = 16$. **12.** $s_n = 248$, $l = 128$, $r = 2$.
13. $a = 36$, $r = 1/3$, $l = 4/9$.
14. $s_n = 3333.33$, $r = 10$, $a = 0.03$.
15. Find the seventh term of a geometric progression whose fourth term is 4 and common ratio is 2.
16. Find the sixth term of a geometric progression whose third term is 1 and common ratio -3.

Find a geometric mean of each set of numbers:

17. 2 and 32. **18.** 4 and 144. **19.** 1/5, 4, 10.

20. $\frac{1}{2}$, 6, 9.

21. If each bacterium in a culture divides into 2 bacteria every hour, how many bacteria will be present at the end of 6 hours if there are 4 bacteria at the start?

22. A man deposits 15 cents in a toy bank at the beginning of the month and doubles the size of his deposit every month thereafter. How much has he in the bank at the end of a year?

23. Suppose a dropped ball always rebounds 1/2 the height it falls. If it is dropped from a height of 128 feet, how far has it travelled when it reaches the top of the 5th bounce?

24. At the end of each year the value of a certain machine depreciates by 20% of the value it had at the beginning of the year. If it was worth $1000 new, what is its value at the end of 5 years?

25. A man receives a salary of $1600 per annum for the first year and a 10% raise every year for 10 years. What is his salary during the 5th year?

26. If $1000 is deposited in a bank at 3% interest compounded annually, what is the amount in the account at the end of 5 years?

27. Suppose an investment depreciates in value to 20% of the original value during the first year, and then depreciates during the second year to 80% of the value it had at the start of the second year. What steady annual rate of decrease would yield the same resulting value at the end of 2 years?

28. Suppose a house costing $12,000 depreciates 10% in value each year. What is it worth at the end of 6 years?

29. A tank full of alcohol is emptied of 1/3 of its contents and then filled up with water and mixed. If this is done 6 times, what fraction of the original volume of alcohol remains?

30. Show that the reciprocals of the terms of a geometric progression also form a geometric progression.

149. Harmonic Progressions

A sequence of numbers is said to form a **harmonic progression** provided their reciprocals form an arithmetic progression.

Example 1. The numbers $\frac{1}{a}, \frac{1}{a+d}, \frac{1}{a+2d}, \ldots, \frac{1}{a+(n-1)d}$ form a harmonic progression, provided none of the denominators is zero.

The first and last terms of a harmonic progression of n terms are called the **extremes,** while the terms in between are called

harmonic means. To insert k harmonic means between a and b, it is sufficient to find k arithmetic means between $1/a$ and $1/b$; the reciprocals of these arithmetic means will be the required harmonic means.

Example 2. Insert five harmonic means between 1/4 and 1/22.

Solution. Five arithmetic means between 4 and 22 are 7, 10, 13, 16, 19. Hence 1/7, 1/10, 1/13, 1/16, 1/19 are the required harmonic means.

Note. If one harmonic mean H is inserted between a and b, H is called **the harmonic mean** of a and b. Then

$$\frac{1}{H} = \frac{\frac{1}{a} + \frac{1}{b}}{2} \quad \text{or} \quad H = \frac{2ab}{a + b}.$$

Generalizing this definition **the harmonic mean** of $a_1, a_2, \ldots, a_n$ is the number H given by

$$\frac{1}{H} = \frac{\frac{1}{a_1} + \frac{1}{a_2} + \ldots + \frac{1}{a_n}}{n}.$$

EXERCISES

1. Find the eighth term of the harmonic progression $\frac{1}{2}, \frac{1}{5}, \frac{1}{8}, \ldots$.
2. Find the twelfth term of the harmonic progression $\frac{1}{3}, \frac{1}{8}, \frac{1}{13}, \ldots$.
3. Insert three harmonic means between 1/4 and 1/24.
4. Insert four harmonic means between 5/2 and 10/19.
5. Insert three harmonic means between 1/14 and $-1/2$.
6. Insert four harmonic means between $-2/5$ and 1/5.

Find the harmonic mean of each set of numbers:

7. 5, 2. **8.** 10, 20. **9.** 8, 3, 1/2. **10.** 1/3, 6, 9, 12.

11. Show that if A, G, and H are respectively the arithmetic, geometric, and harmonic means of two numbers a and b, then $G^2 = AH$.

12. Show that if a car travels a certain distance at the rate of x miles per hour and makes the return trip at the rate of y miles per hour, then the average rate for the entire journey is the harmonic mean of x and y.

150. Infinite Geometric Series. Periodic Decimals

The sum of the terms of an unending geometric progression

(1) $$S = a + ar + ar^2 + \ldots + ar^{n-1} + \ldots,$$

where the number of terms in the right member is infinite, is an example of a so-called **infinite series.** It is called a **geometric series.** If S_n is the sum of the first n terms, the sum of an infinite series is defined as $S = \lim_{n\to\infty} S_n$, provided this limit exists. See chapter XXVII for a general discussion of infinite series.

If r is between -1 and $+1$, the sum of the infinite geometric series (1) is given by the formula

$$(2) \qquad S = \frac{a}{1-r}.$$

For,

$$(3) \qquad S_n = \frac{a - ar^n}{1-r} = \frac{a}{1-r} - \frac{a}{1-r}\cdot r^n.$$

But it can be proved that if $-1 < r < 1$, then $\lim_{n\to\infty} r^n = 0$; for example, $\lim_{n\to\infty} (\frac{1}{2})^n = 0$, since successive powers of $\frac{1}{2}$ clearly approach zero as n increases, or $\frac{1}{2}, \frac{1}{4}, \frac{1}{8}, \frac{1}{16}, \frac{1}{32}, \ldots, \frac{1}{2^n}, \ldots \to 0.$ Hence the last term of (3) has zero as a limit as n increases indefinitely. Hence $S = \frac{a}{1-r}.$

Example 1. Find the sum S of the infinite geometric series

$$1 + \frac{1}{2} + \frac{1}{4} + \frac{1}{8} + \ldots + \frac{1}{2^{n-1}} + \ldots .$$

Solution. Here $a = 1$, $r = \frac{1}{2}$. Hence $S = \frac{1}{1 - \frac{1}{2}} = 2.$

An unending decimal is called **periodic** if the same block of digits repeats itself endlessly. See section 14.

Example 2. Write in lowest terms the rational number which has the periodic decimal expression 0.1818181818 . . . .

Solution. The given decimal may be considered as an infinite geometric series

$$\frac{18}{100} + \frac{18}{100}\left(\frac{1}{100}\right) + \frac{18}{100}\left(\frac{1}{100}\right)^2 + \ldots + \frac{18}{100}\left(\frac{1}{100}\right)^{n-1} + \ldots$$

with $a = 18/100$ and $r = 1/100$. By (2), its sum is

$$\frac{18/100}{1 - 1/100} = \frac{18/100}{99/100} = \frac{18}{99} = \frac{2}{11}.$$

It is clear, from example 2, that one could prove that every periodic decimal represents a rational number, which may be found by the method of example 2.

EXERCISES

Find the sum of each of the following infinite geometric series:

1. $1 + \frac{1}{3} + \frac{1}{9} + \dots$. **2.** $16 + 12 + 9 + \dots$.
3. $1 - \frac{1}{2} + \frac{1}{4} - \dots$. **4.** $4 - 0.4 + 0.04 - \dots$.
5. $3 + 2 + \frac{4}{3} + \dots$. **6.** $5 - 3 + 9/5 - \dots$.

Find, in simplest form, the rational number which has the given periodic decimal expression, the repeating block of digits being written below only three times:

7. 0.666 . . . **8.** 0.777 . . . . **9.** 0.181818 . . . .
10. 0.353535 . . . . **11.** 0.1666 . . . . **12.** 3.262626 . . . .
13. 0.230769230769230769 . . . .
14. 0.142857142857142857 . . . .
15. 2.090909 . . . . **16.** 3.373737 . . . .

17. Suppose a ball rebounds one-half the distance it falls. If it is dropped from a height of 40 feet, how far does it travel before* coming to rest?

18. Suppose that each swing of a pendulum bob is 80% as long as the preceding swing. If the first swing is 20 inches long, how far does the bob travel before coming to rest?

19. Suppose a ball rebounds 3/4 the distance it falls. If it is dropped from a height of 40 feet, how far does it travel before coming to rest?

20. Suppose that each swing of a pendulum bob is 90% as long as the preceding swing. If the first swing is 20 inches long, how far does the bob travel before coming to rest?

★151. Differences of rth Order

Consider a sequence of numbers

$$(1) \qquad a_0, a_1, a_2, a_3, \dots, a_n, a_{n+1}, \dots$$

where a_n is the $(n + 1)$th term. The **differences of first order**

* Assume infinitely many bounces. This is a convenient idealization of actual reality which gives a very good approximation, such as is always made in applying mathematics to concrete situations.

★ May be omitted without disturbing continuity.

are defined as the terms of the sequence

$$(2)\quad \Delta a_0 = a_1 - a_0,\ \Delta a_1 = a_2 - a_1,\ \Delta a_2 = a_3 - a_2,\ \ldots,\ \Delta a_n = a_{n+1} - a_n,\ \ldots.$$

The **differences of second order** are defined as the differences of first order of the sequence (2) of differences of the first order. In symbols, the differences of the second order are

$$(3)\quad \Delta^2 a_0 = \Delta a_1 - \Delta a_0,\ \Delta^2 a_1 = \Delta a_2 - \Delta a_1,\ \Delta^2 a_2 = \Delta a_3 - \Delta a_2,\ \ldots.\ \Delta^2 a_n = \Delta a_{n+1} - \Delta a_n,\ \ldots.$$

In general, the **differences of rth order** are defined as the differences of first order of the sequence of differences of the $(r-1)$th order. Thus,

$$(4)\quad \Delta^r a_0 = \Delta^{r-1} a_1 - \Delta^{r-1} a_0,\ \Delta^r a_1 = \Delta^{r-1} a_2 - \Delta^{r-1} a_1,\ \ldots,\ \Delta^r a_n = \Delta^{r-1} a_{n+1} - \Delta^{r-1} a_n,\ \ldots,$$

for $r = 2, 3, 4, \ldots$, and $n = 0, 1, 2, \ldots$.

Example 1. From the sequence

$$1, 8, 27, 64, 125, \ldots, n^3, \ldots$$

we obtain the differences of

1st order: $7, 19, 37, 61, \ldots, (n+1)^3 - n^3 = 3n^2 + 3n + 1, \ldots;$
2nd order: $12, 18, 24, \ldots, 3(n+1)^2 + 3(n+1) + 1 - (3n^2 + 3n + 1) = 6n + 6, \ldots;$
3rd order: $6, 6, \ldots, 6(n+1) + 6 - (6n+6) = 6, \ldots;$
all higher orders: $0, 0, \ldots, 0, \ldots$.

By direct calculation, we obtain from the definitions (2), (3), and (4) above, the relations

$$(5)\quad \begin{cases} a_1 = a_0 + \Delta a_0 \\ a_2 = a_1 + \Delta a_1 = (a_0 + \Delta a_0) + (\Delta a_0 + \Delta^2 a_0) \\ \qquad = a_0 + 2\Delta a_0 + \Delta^2 a_0 \\ a_3 = a_0 + 3\Delta a_0 + 3\Delta^2 a_0 + \Delta^3 a_0 \\ \cdot \\ \cdot \\ \cdot \\ a_n = a_0 + C(n,1)\Delta a_0 + C(n,2)\Delta^2 a_0 + \ldots + C(n,n)\Delta^n a_0 \\ \cdot \\ \cdot \\ \cdot \end{cases}$$

which express the successive terms of the given sequence in terms of its first term and the first terms in the successive rows of differences. The relations (5) can be summed up as follows. For $n = 1, 2, 3, \ldots$, the $(n + 1)$th term of the sequence (1) is given by

$$(6)\quad a_n = a_0 + n\Delta a_0 + \frac{n(n-1)}{2!}\Delta^2 a_0 + \frac{n(n-1)(n-2)}{3!}\Delta^3 a_0 + \ldots + \Delta^n a_0,$$

where the coefficients on the right are the coefficients in the binomial expansion of $(x + y)^n$. Formula (6) may be proved for all positive integral values of n by mathematical induction; this will not be done here.

Consider the sequence (1) and let s_k be the sum of its first k terms; that is

$$(7)\qquad s_k = a_0 + a_1 + \ldots + a_{k-1}.$$

Then the sequence

$$(8)\qquad s_0 = 0, s_1, s_2, s_3, \ldots, s_k, \ldots$$

is a sequence whose differences of first order are exactly the terms of the sequence (1), for

$$\Delta s_0 = s_1 - s_0 = a_0,\ \Delta s_1 = s_2 - s_1 = a_1,\ \ldots,\ \Delta s_k = s_{k+1} - s_k = a_k,\ \ldots .$$

Consequently the rth differences of the sequence (8) are the $(r - 1)$th differences of the sequence (1). By (5), we may write

$$(9)\quad s_k = s_0 + C(k,1)\Delta s_0 + C(k,2)\Delta^2 s_0 + C(k,3)\Delta^3 s_0 + \ldots + C(k,k)\Delta^k s_0.$$

Hence,

$$(10)\quad s_k = C(k,1)a_0 + C(k,2)\Delta a_0 + C(k,3)\Delta^2 a_0 + \ldots + C(k,k)\Delta^{k-1} a_0.$$

Note that in general the expression (10) for s_k has k terms.

★152. Arithmetic Progressions of *r*th Order

A sequence for which all the differences of rth order are equal, and this is true for no lower order, is called an **arithmetic progression of *r*th order.**

★ May be omitted without disturbing continuity.

For example, an arithmetic progression of first order is what we have formerly called an arithmetic progression. The sequence of example 1, section 151, is an arithmetic progression of third order.

For an arithmetic progression of rth order, all differences of $(r+1)$th and higher orders are zero. Hence (10), section 151, yields a terminating expression for the sum of the first n terms of such a progression. That is

$$(1) \quad s_n = C(n,1)a_0 + C(n,2)\Delta a_0 + C(n,3)\Delta^2 a_0 + \dots + C(n, r+1)\Delta^r a_0$$

or,

$$(2) \quad s_n = na_0 + \frac{n(n-1)}{2!}\Delta a_0 + \frac{n(n-1)(n-2)}{3!}\Delta^2 a_0 + \dots + \frac{n(n-1)\dots(n-r)}{(r+1)!}\Delta^r a_0.$$

Example 1. In the sequence of example 1, section 151, we have $a_0 = 1$, $\Delta a_0 = 7$, $\Delta^2 a_0 = 12$, $\Delta^3 a_0 = 6$, $\Delta^4 a_0 = 0 = \Delta^5 a_0 = \dots$. Hence the sum s_n of the first n terms of the given sequence is

$$s_n = n + \frac{n(n-1)}{2}\cdot 7 + \frac{n(n-1)(n-2)}{6}\cdot 12 + \frac{n(n-1)(n-2)(n-3)}{24}\cdot 6.$$

Or,

$$s_n = \frac{n^4 + 2n^3 + n^2}{4} = \frac{n^2(n+1)^2}{4}.$$

Example 2. Find the sum of the first n terms of the sequence $1, 4, 9, 16, \dots, n^2, \dots$.

Solution. The given sequence and its successive differences may be arranged as follows:

sequence: $1, 4, 9, 16, \dots, n^2, (n+1)^2, (n+2)^2, \dots$
1st differences: $3, 5, 7, \dots, 2n+1, 2n+3, \dots$
2nd differences: $2, 2, \dots, 2, \dots$

Hence we have an arithmetic progression of 2nd order. The sum of the first n terms is

$$s_n = n + \frac{n(n-1)}{2}\cdot 3 + \frac{n(n-1)(n-2)}{6}\cdot 2 = \frac{2n^3 + 3n^2 + n}{6} = \frac{n(n+1)(2n+1)}{6}$$

The illustrative examples suggest that the general or nth term of an arithmetic progression of rth order is a polynomial in n of degree r. We shall prove this and the converse.

THEOREM 1. *The general term of an arithmetic progression of rth order is a polynomial in n of degree r.*

Proof. By definition, all differences of order $r + 1$ or higher are zero; that is, $\Delta^{r+1}a_0 = \Delta^{r+2}a_0 = \ldots = 0$. Hence (6) of section 151 becomes

$$a_n = a_0 + n\Delta a_0 + \frac{n(n-1)}{2!}\Delta^2 a_0 + \ldots + \frac{n(n-1)\ldots(n-r+1)}{r!}\Delta^r a_0.$$

The numerator of the last coefficient has r factors and hence a_n is a polynomial of degree r in the variable n.

THEOREM 2. *If $a_n = f(n)$ where $f(n) = c_0n^r + c_1n^{r-1} + \ldots + c_r$ is a polynomial of degree r in the variable n, then the sequence $a_0, a_1, a_2, \ldots, a_n, \ldots$ is an arithmetic progression of rth order.*

Proof. By definition,

$$\begin{aligned}\Delta a_n &= a_{n+1} - a_n = f(n+1) - f(n)\\ &= c_0(n+1)^r + c_1(n+1)^{r-1} + \ldots + c_r - c_0n^r - c_1n^{r-1} - \ldots - c_r\\ &= c_0n^r + rc_0n^{r-1} + \ldots + c_1n^{r-1} + \ldots + c_r - c_0n^r - c_1n^{r-1} - \ldots - c_r\\ &= rc_0n^{r-1} + \ldots.\end{aligned}$$

Hence Δa_n is a polynomial in n of degree $r - 1$. It follows that $\Delta^r a_n$ is a constant (a polynomial of degree zero) and hence $\Delta^{r+1}a_n = \Delta^{r+2}a_n = \ldots = 0$.

EXERCISES

(a) Determine the order of each arithmetical progression; (b) find a formula in terms of n for the sum of the first n terms of each progression:

1. $1 \cdot 2 + 2 \cdot 3 + 3 \cdot 4 + \ldots + n(n+1)$.

2. $1 \cdot 4 + 2 \cdot 5 + 3 \cdot 6 + \ldots + n(n+3)$.

3. $1 + 3 + 6 + \ldots + \dfrac{n(n+1)}{2}$.

4. $4 + 14 + 36 + \ldots + n(n^2+3)$.

5. $1 + 16 + 81 + \ldots + n^4$.

6. $4 + 13 + 34 + \ldots + (n^3 + 2n + 1)$.

7. Find the total number of spherical shot piled in a pyramid with an equilateral triangle as base, having just one shot in the top layer (*a*) if there are n layers; (*b*) if there are 18 layers.

8. A grocer displays cans of soup in a pyramid, having a rectangular base, whose top layer consists of one row of 5 cans. Find the total number of cans (*a*) if there are n layers; (*b*) if there are 12 layers.

9. Find the total number of spherical shot piled in a pyramid with a square base, having just one shot in the top layer, (*a*) if there are n layers; (*b*) if there are 18 layers.

10. Find the total number of spherical shot piled in a pyramid with a square base, having 16 shot in the top layer if there are 15 layers.

(a) By formula (6) of section 151, find a possible nth term of the sequence whose first five terms are given. (b) Find another possible nth term yielding the same given first five terms. (c) Is there another nth term different from that formed in (a) whose degree is not greater than that of the answer to (a)? Explain. (Hint: consider corollary 2 of section 119.)

11. 1, 3, 7, 13, 21, ... 　　**12.** 7, 16, 29, 46, 67, ...

13. 1, 5, 19, 49, 101, ... 　　**14.** 1, 1, 7, 27, 69, ...

★15. Prove, by mathematical induction, that formula (6) of section 151 is true for all positive integral values of n.

★ May be omitted without disturbing continuity.

Chapter XXII

Inequalities

153. Inequalities

If a and b are any real numbers, then one and only one of the three relations $a > b$, $a = b$, $a < b$ holds. *In this chapter we consider only real numbers.* By definition $a > b$, read "a is greater than b," means that $a - b$ is positive, or that $a = b + k$ where k is positive. Graphically, $a > b$ means that a lies to the right of b if they are plotted on the x-axis, or that a is above b if they are plotted on the y-axis. For example, $-2 > -4$, $0 > -2$, $2 > -4$. The symbol $a \geqq b$ means that a is greater than or equal to b, or a is not less than b. Similarly $a \leqq b$ means that a is less than or equal to b, or a is not greater than b. Of course, $b < a$ means the same thing as $a > b$.

A statement involving one of the symbols $>$, $<$, $\geqq$, $\leqq$ is called an **inequality.** An inequality which involves only constants, or one which is true for all permissible values of the variables involved, is called an **absolute inequality.** If an inequality fails to be true for some permissible value of the variables involved, it is called a **conditional inequality.** To **solve** a conditional inequality means to find those values of the variables for which it is true.

Examples. $a^2 \geqq 0$, $2 > -4$, $\pi < 22/7$ are absolute inequalities. But $x + 2 < 7$ is a conditional inequality which is true only if $x < 5$. Similarly $x^2 < 9$ is conditional, being true only if $-3 < x < 3$.

The **absolute value** of the real number x is defined as x itself if $x \geqq 0$, and as the positive number $-x$ if $x < 0$. The absolute value of x is denoted by $|x|$.

Examples. $|3| = 3$, $|-3| = 3$, $|0| = 0$.

Two inequalities are said to have the **same sense** if their symbols for inequality point in the same direction; they are said to have **opposite senses** if their symbols for inequality point in opposite directions.

Examples. $x > y$ and $u > v$ have the same sense, as do $y < x$ and $v < u$. But $x < y$ and $v > u$ have opposite senses.

154. Properties of Inequalities

The following theorems concerning inequalities are intuitively plausible, and can be proved from the axioms of section 17, chapter I.

THEOREM 1. *The sense of an inequality is not changed if the same real number is added to (or subtracted from) both sides. That is, if* $a > b$, *then* $a + c > b + c$ (*and* $a - c > b - c$).

Proof. By hypothesis, $a > b$, or $a - b = k$ where k is positive. That is, $a = b + k$. Adding c to both sides, we have $(a + c) = (b + c) + k$. Therefore $(a + c) - (b + c) = k$ where k is positive. Hence $a + c > b + c$.

Since subtracting c is equivalent to adding $-c$, we have similarly $a - c > b - c$.

THEOREM 2. *The sense of an inequality is not changed if both sides are multiplied or divided by the same positive number. That is, if* $a > b$ *and* $c > 0$, *then* $ac > bc$ *and* $a/c > b/c$.

Proof. By hypothesis, $a = b + k$ where k is positive. Multiplying both sides by c, we have $ac = bc + kc$. But kc is positive. Hence $ac > bc$.

Since dividing by c is equivalent to multiplying by the positive number $1/c$, we have similarly $a/c > b/c$.

THEOREM 3. *The sense of an inequality is reversed if both sides are multiplied or divided by the same negative number* c. *That is, if* $a > b$ *and* $c < 0$, *then* $ac < bc$ *and* $a/c < b/c$.

Proof. By hypothesis, $a = b + k$, where k is positive. Then $ac = bc + kc$ where kc is negative. Or, transposing, $-kc = bc - ac$ where $-kc$ is positive. Hence, $bc > ac$, or $ac < bc$. The proof of $a/c < b/c$ is left to the student.

THEOREM 4. *If* $a > b$ *and if* $ab > 0$, *then* $\frac{1}{a} < \frac{1}{b}$.

Proof. Suppose a and b are positive. Let

$$A = \frac{1}{a}, \qquad B = \frac{1}{b}.$$

Then A and B are positive. By hypothesis, $a = b + k$ where k is positive. Multiplying both sides by AB, we get $aAB = bBA + kAB$, or $B = A + kAB$. But kAB is positive. Hence

$$B > A, \quad \text{or} \quad \frac{1}{b} > \frac{1}{a}, \quad \text{or} \quad \frac{1}{a} < \frac{1}{b}.$$

The case where a and b are both negative is left to the student.

The proofs of the remaining theorems are left as exercises.

THEOREM 5. *If unequals are added to unequals in the same sense, the results are unequal in the same sense. That is, if $a > b$ and $c > d$, then $a + c > b + d$.*

THEOREM 6. *If a, b, c, d are positive and if $a > b$ and $c > d$, then $ac > bd$.*

THEOREM 7. *If a and b are positive and $a > b$, then $a^n > b^n$ and $\sqrt[n]{a} > \sqrt[n]{b}$ where n is any positive integer.*

155. Solution of Inequalities

By the use of the properties of the preceding section we may verify absolute inequalities and solve conditional inequalities, much as we treat equalities.

Example 1. Find those values of x for which $3x + 4 < x + 10$.

Solution. If x is such that $3x + 4 < x + 10$, then subtracting x and 4 from both sides we have $2x < 6$. Dividing both sides by 2 we have $x < 3$. We have proved that if x satisfies $3x + 4 < x + 10$ then $x < 3$. The converse "if $x < 3$ then $3x + 4 < x + 10$" can now be proved by reversing each step in the above discussion, since each step is reversible. Hence $3x + 4 < x + 10$ for all values of $x < 3$.

Example 2. Show that if $x \neq y$ then $x^2 + y^2 > 2xy$.

Solution. If x and y satisfy $x^2 + y^2 > 2xy$, then they also satisfy $x^2 - 2xy + y^2 > 0$ or $(x - y)^2 > 0$. But the latter inequality is true if $x - y \neq 0$ or $x \neq y$. We have proved that if $x^2 + y^2 > 2xy$ then $x \neq y$. Again each step is reversible and hence the converse proposition, which we had to prove, is true.

EXERCISES

Find the values of x for which the following inequalities are true:

1. $2x - 8 < 0.$ **2.** $3x - 12 > 0.$

3. $4x - 1 < x + 8.$ **4.** $3x - 2 < \frac{10 - x}{2}.$

5. $\frac{4x - 3}{2} \geqq \frac{2 - x}{3}.$ **6.** $\frac{x + 3}{2} \geqq \frac{2 - x}{-2}.$

7. $x^2 < 4.$ **8.** $16x^2 < 25.$ **9.** $x^2 \geqq 16.$ **10.** $9x^2 \geqq 16.$

11. $(x - 3)(x - 2) < 0.$ (*Hint:* the factors must have opposite signs.)

12. $(x - 4)(x - 1) < 0.$ **13.** $(x - 2)(x - 4) > 0.$

14. $(x - 1)(x - 5) > 0.$ **15.** $(x - 1)(x + 2) < 0.$

16. $(x - 1)(x + 3) > 0.$ **17.** $x^2 - 5x + 6 \leqq 0.$

18. $x^2 - x - 12 < 0.$ **19.** $(x + 1)(x - 3)(x - 4) < 0.$

20. $(x + 2)(x - 1)(x - 3) > 0.$

21. $|x - 3| \leqq 1.$ **22.** $|x - 1| < 3.$

Assuming that all letters represent positive numbers, prove the following inequalities:

23. $a + b > \frac{4ab}{a + b}$ if $a \neq b.$

24. $a + \frac{1}{a} > 2$ if $a \neq 1.$ **25.** $\frac{a}{b} + \frac{b}{a} > 2$ if $a \neq b.$

26. $a^3 + 3ab^2 > b^3 + 3a^2b$ if $a > b.$

27. $a^3 + b^3 > a^2b + ab^2$ if $a \neq b.$

28. If $a \neq b$ and if A, G, and H are respectively the arithmetic, geometric, and harmonic means of a and b, then $A > G > H$. (*Hint:* See exercise 11, section 149.)

If all letters represent any real numbers, prove that:

★29. $|x| + |y| \geqq |x + y|.$ **★30.** $|x| - |y| \leqq |x - y|.$

★31. $|ab| = |a||b|.$

★32. $(a_1b_1 + a_2b_2)^2 \leqq (a_1^2 + a_2^2)(b_1^2 + b_2^2)$. (*Hint:* the quadratic equation $(b_1x - a_1)^2 + (b_2x - a_2)^2 = (b_1^2 + b_2^2)x^2 - 2(a_1b_1 + a_2b_2)x + (a_1^2 + a_2^2) = 0$ cannot have distinct real roots; therefore its discriminant must be $\leqq 0$.)

★33. $\sqrt{(x_1 - x_2)^2 + (y_1 - y_2)^2} \leqq \sqrt{x_1^2 + y_1^2} + \sqrt{x_2^2 + y_2^2}$. (*Hint:* use exercise 32.)

★34. $\sqrt{(x_1 - x_2)^2 + (y_1 - y_2)^2} + \sqrt{(x_2 - x_3)^2 + (y_2 - y_3)^2} \geqq \sqrt{(x_1 - x_3)^2 + (y_1 - y_3)^2}.$

★ May be omitted without disturbing continuity.

★**35.** $\left|\sqrt{x_1^2 + y_1^2} - \sqrt{x_2^2 + y_2^2}\right| \leqq \sqrt{(x_1 - x_2)^2 + (y_1 - y_2)^2}$.

★**36.** $\left|\sqrt{(x_1 - x_3)^2 + (y_1 - y_3)^2} - \sqrt{(x_2 - x_3)^2 + (y_2 - y_3)^2}\right| \leqq \sqrt{(x_1 - x_2)^2 + (y_1 - y_2)^2}$.

37–40. Work exercises 37–40 of section 77.

156. Graphical Significance of Inequalities

Any conditional inequality involving one variable x can be written in the form $f(x) > 0$ or $f(x) \geqq 0$. Then the values of x

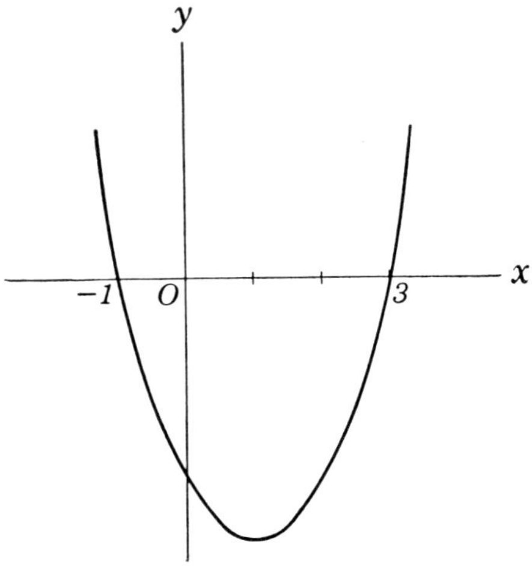

Fig. 59.

for which $f(x) > 0$ are those values of x for which the graph of $y = f(x)$ lies above the x-axis.

Example. Find graphically the values of x for which $x^2 > 2x + 3$.

Solution. We have $f(x) = x^2 - 2x - 3 > 0$. The graph of $y = x^2 - 2x - 3$ crosses the x-axis at $x = -1$ and at $x = 3$. From the graph (Fig. 59) we see that $y = f(x) > 0$ for $x > 3$ and for $x < -1$.

An inequality involving two variables, x and y, describes a region of the plane. Thus the straight line $x + y = 3$ divides the rest of the plane into two regions, namely the region consisting of all points (x,y) such that $x + y > 3$, and the region consisting of all points (x,y) such that $x + y < 3$ (Fig. 60). Similarly, the circle $x^2 + y^2 = 4$ divides the rest of the plane

★ May be omitted without disturbing continuity.

into two regions, one for which $x^2 + y^2 < 4$, and the other for which $x^2 + y^2 > 4$ (Fig. 61).

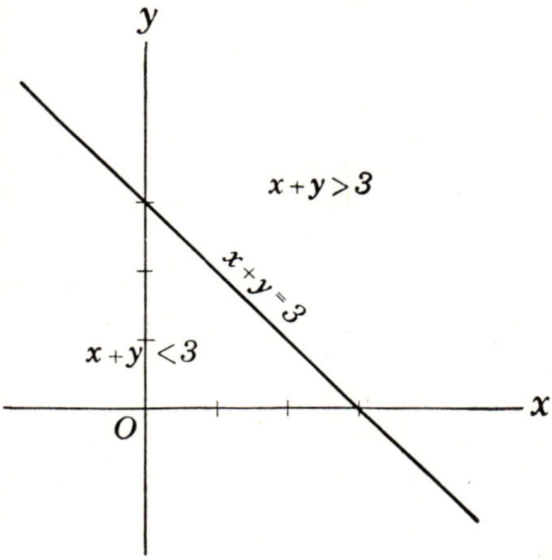

Fig. 60.

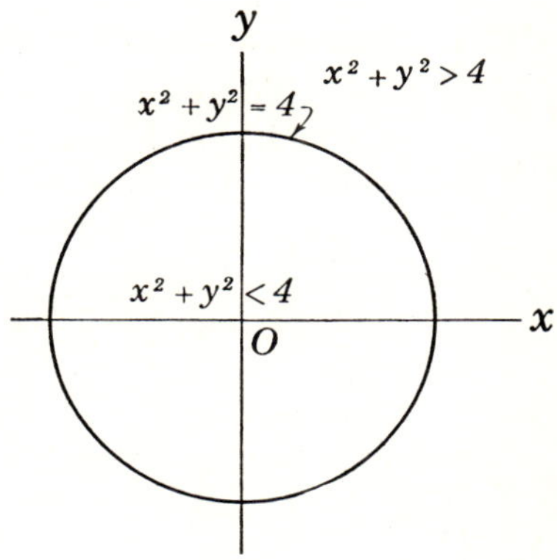

Fig. 61.

EXERCISES

Find graphically the values of x for which the following inequalities are true:

1–20. Exercises 1–20, inclusive, of section 155.

21. $x^2 + 3x - 4 > 0$. **22.** $x^2 < 7x - 10$.

23. $x^2 < 3 - 2x$. **24.** $x^2 > 10 - 3x$.

Show on a graph the region described by the following inequalities or sets of inequalities:

25. $x - y < 1$. **26.** $2x + y < 3$. **27.** $4y < x^2$.

28. $4y > x^2$. **29.** $x^2 + y^2 < 1$. **30.** $x^2 + y^2 > 1$.

31. $x^2 + y^2 < 25$, $x + y < 1$.

32. $y > x^2$, $x + y < 1$. **33.** $x > y^2$, $x < y$.

34. $x \geqq 0$, $y \geqq 0$, $x + y \leqq 4$.

35. $x \geqq 0$, $y \geqq 0$, $2x + 3y \leqq 12$, $2x + y \leqq 8$.

36. $x \geqq 0$, $y \geqq 0$, $x \leqq 4$, $y \leqq 3$, $x + y \leqq 5$.

37. $x \geqq 0$, $y \geqq 0$, $3x + y \geqq 3$, $x + 3y \geqq 3$, $2x + 2y \geqq 5$.

38. $x + y \leqq 2$, $y - x \leqq 1$, $2y - x + 2 \geqq 0$.

39. $x^2 + y^2 \leqq 1$, $y \geqq x$, $x + y + 1 \geqq 0$.

157. Linear Programming

The preceding graphical ideas can be applied to solve some elementary economic problems in a modern subject called **linear programming.**

Example 1. A manufacturer makes two different models of can opener, model C, the Colossal, and a better model D, the Deluxe. Each model must be processed by two machines, A and B. To make one unit of model C, machine A must work 2 hours and machine B must work 4 hours. To make one unit of model D, machine A must work 4 hours and machine B 2 hours. No machine may work more than 24 hours per day. Due to the decline of the art of cooking in Western civilization, the manufacturer can sell all the units of both models that he chooses to make. He makes a profit of \$3 on each unit of model C and a profit of \$5 on each unit of model D. How should he allocate his production? That is, how many of each model should be produce daily in order to maximize his profit?

Solution. Let x be the number of units of model C and y the number of units of model D which he makes. His profit P is then

$$P = 3x + 5y. \tag{1}$$

We wish to find the values of x and y which will maximize P subject to the constraints

$$x \geqq 0, \tag{2}$$

$$y \geqq 0, \tag{3}$$

$$2x + 4y \leqq 24, \tag{4}$$

$$4x + 2y \leqq 24. \tag{5}$$

The constraints (inequalities) (2) and (3) assert merely that it is impossible to make a negative number of either model. Constraint (4) indicates that machine A may not work more than 24 hours in a day; constraint (5) indicates the same for machine B. Any point (x,y) in the plane, satisfying the constraints (2), (3), (4), (5), is called a **feasible point.** The set of feasible points fills the shaded region in Fig. 62, including its boundary. For a given value of P, say $P = 15$, the points (x,y) yielding $P = 15$ lie on a straight line (Fig. 62). For a larger value of P, say $P = 30$, they lie on a line with a higher y-intercept, since the y-intercept of (1) is $P/5$. All the lines (1) with various values of P are parallel. It is intuitively clear, and can be proved, that the parallel line with highest y-intercept (highest P) that contains a feasible point must intersect the feasible (shaded) region at a corner point. In this example it can be seen visually that (4,4) is the point required, but at worst one might wish to test the other corners (0,6) and (6,0) to see which yields the highest profit. In this case $x = 4$, $y = 4$ yields the profit $P = 12 + 20 = 32$, while the others, (0,6) and (6,0), yield profits of 30 and 18, respectively. Hence this manufacturer should make equal numbers of both models, i.e., he should make them in the ratio 4:4. Note that a hasty decision to make more of model D, because of the larger profit from model D, would fail to yield the maximum profit. The point (4,4) which maximizes (1), subject to the given constraints, is called **optimal.**

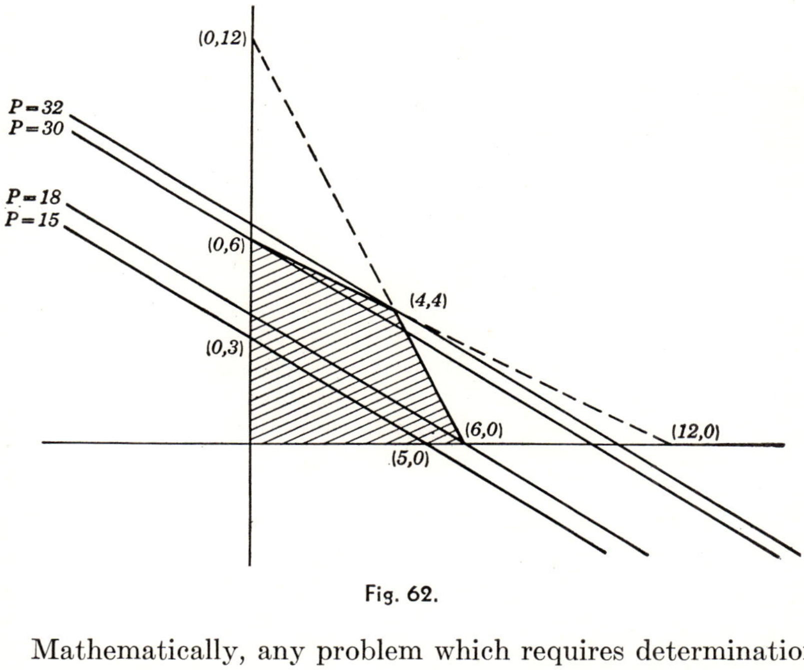

Fig. 62.

Mathematically, any problem which requires determination of several unknowns x, y, z, ... so as to maximize or minimize a linear expression

$$ax + by + cz + \ldots, \tag{6}$$

subject to linear inequalities like those above, is called a **linear programming problem.** The points satisfying the inequalities are called **feasible.** A feasible point which maximizes or minimizes the expression (6) is called **optimal.**

When the number of unknowns is more than 2, the simple graphical method used above is inadequate. The general theory is beyond the scope of this book.* The subject has practical applications in large scale business, supply, and other problems.

EXERCISES

1. Find x and y so as to maximize $Q = x + y$, subject to the constraints $x \geqq 0$, $y \geqq 0$, $2x + 3y \leqq 12$, $2x + y \leqq 8$.

2. Find x and y so as to maximize $C = x + 2y$, subject to the constraints $x \geqq 0$, $y \geqq 0$, $x \leqq 4$, $y \leqq 3$, $x + y \leqq 5$.

3. Find x and y so as to maximize $K = 2x + y$, subject to the constraints in exercise 2.

4. Find x and y so as to minimize $Q = x + y$, subject to the constraints $x \geqq 0$, $y \geqq 0$, $2x + 3y \geqq 12$, $2x + y \geqq 8$.

5. Find x and y so as to maximize $C = x + 3y$, subject to the constraints $x \geqq 0$, $y \geqq 0$, $x \leqq 4$, $y \leqq 3$, $2x + y \leqq 10$.

6. Find x and y so as to maximize $C = 2x + 5y$, subject to the constraints $x \geqq 0$, $y \geqq 0$, $x \leqq 4$, $y \leqq 3$, $x + 2y \leqq 8$.

7. A feed for animals is to be a mixture of two cereals A and B. Each ounce of cereal A contains 12 grams of protein, 0.1 grams of fat, and 14 grams of carbohydrates. Each ounce of cereal B contains 4 grams of protein, 0.8 grams of fat, and 21 grams of carbohydrates. The costs per ounce of A and B are 4 cents and 2 cents, respectively. Each bag of the resulting feed is to contain at least 48 grams of protein, 1.6 grams of fat, and 140 grams of carbohydrates. Find the amounts in ounces, x and y, of cereals A and B, respectively, that will produce a feed satisfying these minimum requirements (constraints) so as to minimize the cost.

8. A manufacturer makes two models S and T of a product. Each model must be worked on by 3 machines A, B, and C. The number of hours each machine works on each model is given by the table:

Machine	*Model* S	*Model* T
A	1	10
B	5	6
C	8	1

No machine may work more than 40 hours per week. The manufacturer's profit on each unit of model S is \$20; on each unit of model T, \$30. How many of each model should he make per week in order to maximize his profit?

* See S. Vajda, *Theory of Games and Linear Programming,* Methuen, 1956; J. McKinsey, *Introduction to Theory of Games,* McGraw-Hill, 1952; T. J. Koopmans, *Activity Analysis of Production and Allocation,* Wiley, 1951.

Chapter XXIII

Logarithms

158. Definition of Logarithm

We have noted (see chapter V) the simplifications of arithmetic due to the use of Hindu-Arabic notation and decimal fractions. In the 17th century, the computations required in astronomy and other sciences became extremely tedious, and further simplifications were sought. In answer to this need, the mathematicians of the 17th century produced logarithms, mechanical calculating machines, and slide rules. Slide rules are convenient but severely limited in accuracy. While calculating machines are much used today in office work, they are not available to everyone. Since they are expensive and not portable, they still do not replace completely the logarithms which will be explained here.* Quite apart from numerical calculations, logarithms are essential in higher mathematics. *All numbers in this chapter are understood to be real.*

DEFINITION. *If b is any positive number, different from 1, and if $b^y = x$, then the exponent y is called the* **logarithm of x to the base b.** *In symbols* $\mathbf{y = \log_b x}$. *That is, the logarithm of x to the base b is that exponent y to which b must be raised in order to get x.*

Example. $3 = \log_2 8$ since $2^3 = 8$. Similarly $2 = \log_3 9$, since $3^2 = 9$.

Note 1. It can be proved that every positive real number x can be expressed as b^y, where y is a real number. *We shall study here only logarithms of positive numbers.*

Exercise. Why do we make the restriction that $b \neq 1$?

* For a brief explanation of calculating machines and slide rules, see M. Richardson, *Fundamentals of Mathematics,* Macmillan, 1941.

Note 2. If we did not restrict b to be positive, we would introduce technical difficulties which would serve no useful purpose here. For example, if b were negative, b^2 would be positive, b^3 negative, $b^{1/2}$ imaginary, etc.

EXERCISES

Express in logarithmic notation:

1. $5^2 = 25$. **2.** $4^3 = 64$. **3.** $10^2 = 100$.
4. $16^{1/2} = 4$. **5.** $8^{1/3} = 2$. **6.** $8^{-1/3} = 1/2$.
7. $5^{-2} = 1/25$. **8.** $2^{-3} = 1/8$. **9.** $5^0 = 1$.
10. $10^{-2} = 1/100$.

Express in exponential notation:

11. $\log_2 16 = 4$. **12.** $\log_{10} 10{,}000 = 4$.
13. $\log_{10} 0.01 = -2$. **14.** $\log_{10} 10 = 1$.
15. $\log_{10} 1 = 0$. **16.** $\log_2 32 = 5$.
17. $\log_5 125 = 3$. **18.** $\log_8 4 = 2/3$.
19. $\log_{10} 0.1 = -1$. **20.** $\log_4 8 = 3/2$.

Find the value of:

21. $\log_2 64$. **22.** $\log_{10} 0.001$. **23.** $\log_{27} 9$.
24. $\log_5 1$. **25.** $\log_5 5$. **26.** $\log_b b$. **27.** $\log_b 1$.
28. $\log_b (b^4)$. **29.** $\log_{25} 125$. **30.** $\log_9 27$.
31. $\log_{10} 1000$. **32.** $\log_{10} (10^3)$. **33.** $\log_{10} (10^5)$.
34. $\log_b (b^5)$. **35.** $\log_8 16$. **36.** $\log_{16} 8$.
37. $\log_8 \frac{1}{2}$. **38.** $\log_{16} \frac{1}{4}$. **39.** $\log_{27} \frac{1}{3}$. **40.** $\log_9 \frac{1}{3}$.

Find the value of the unknown letter:

41. $\log_b 16 = 2$. **42.** $\log_6 x = -2$. **43.** $\log_4 x = -3$.
44. $\log_a 64 = 3$. **45.** $\log_{16} N = 1/4$.
46. $\log_{16} N = -1/4$. **47.** $x = \log_9 81$.
48. $\log_x \frac{1}{16} = -2$. **49.** $\log_x 32 = 5/2$.
50. $\log_{16} N = 3/2$.

159. Properties of Logarithms

The simplifications of arithmetic accomplished by logarithms are due to the three properties given in the following theorems.

THEOREM 1. $\log_b (xw) = \log_b x + \log_b w$. *That is, the logarithm of a product is the sum of the logarithms of the factors.*

THEOREM 2. $\log_b (x/w) = \log_b x - \log_b w$. *That is, the logarithm of a quotient is the logarithm of the numerator minus the logarithm of the denominator.*

THEOREM 3. $\log_b (x^r) = r \log_b x$. *That is, the logarithm of a quantity raised to a power equals the exponent times the logarithm of the quantity.*

Note. We write $r \log_b x$ rather than $log_b\, x \cdot r$ because the latter expression would give rise to confusion as to whether it meant $\log_b (x \cdot r)$ or $(\log_b x) \cdot r$.

Examples. Suppose $\log_{10} 2 = 0.3010$ and $\log_{10} 3 = 0.4771$ approximately. Then, by theorem 1, $\log_{10} 6 = \log_{10} (2 \cdot 3) = \log_{10} 2 + \log_{10} 3 = 0.3010 + 0.4771 = 0.7781$, approximately. By theorem 2, $\log_{10} (3/2) = \log_{10} 3 - \log_{10} 2 = 0.1761$, approximately. By theorem 3, $\log_{10} \sqrt[3]{2} = \log_{10} (2^{1/3}) = \frac{1}{3} \log_{10} 2 = \frac{1}{3}(0.3010) = 0.1003$, approximately.

Since logarithms are exponents, the theorems above may be proved by translating from exponential language into logarithmic language the corresponding three laws of exponents, namely: (*a*) $b^y b^u = b^{y+u}$; (*b*) $b^y/b^u = b^{y-u}$; (*c*) $(b^y)^r = b^{yr}$. Let

(1) $$y = \log_b x \text{ and } u = \log_b w \quad \text{or}$$

(2) $$b^y = x \text{ and } b^u = w.$$

Proof of theorem 1. By (2), $xw = b^y b^u$. Hence, by (*a*), $xw = b^{y+u}$. By definition, this says that $\log_b (xw) = y + u$. Substituting from (1), we get the theorem.

Proof of theorem 2. By (2), $x/w = b^y/b^u$. Hence, by (*b*), $x/w = b^{y-u}$. By definition, $\log_b (x/w) = y - u$. Substituting from (1), we have the theorem.

Proof of theorem 3. By (2), $x^r = (b^y)^r$. Hence, by (*c*), $x^r = b^{yr}$. By definition, $\log_b (x^r) = yr$. Substituting from (1), we get the theorem.

EXERCISES

Given $\log_{10} 2 = 0.3010$, $\log_{10} 3 = 0.4771$, *and* $\log_{10} 7 = 0.8451$, *find, without using tables, the value of:*

1. $\log_{10} 14$. **2.** $\log_{10} 49$. **3.** $\log_{10} (7/2)$. **4.** $\log_{10} 21$.
5. $\log_{10} 12$. **6.** $\log_{10} 30$. **7.** $\log_{10} 0.2$. **8.** $\log_{10} 28$.
9. $\log_{10} \sqrt{2}$. **10.** $\log_{10} \sqrt[3]{3}$. **11.** $\log_{10} (\sqrt[3]{6}/2)$.

12. $\log_{10} (7/\sqrt{6})$. **13.** $\log_{10} \sqrt[3]{42}$. **14.** $\log_{10} \sqrt{\frac{7}{2}}$.

15. $\log_{10} 5$. **16.** $\log_{10} \left(\frac{21}{3\sqrt{2}}\right)$. **17.** $\log_{10} \left(\frac{49}{3\sqrt[3]{2}}\right)$.

18. $\log_{10} \frac{12\sqrt[3]{7}}{\sqrt{14/3}}$. **19.** $\log_{10} \frac{108}{7\sqrt[3]{7}}$. **20.** $\log_{10} \frac{\sqrt[3]{24/7}}{98}$.

160. Common Logarithms

While any positive number $b \neq 1$ can be used as a base for a system of logarithms, the base $b = 10$ is most convenient for computation. Logarithms with the base 10 are called **common logarithms** or **Briggsian logarithms** (after H. Briggs, who first used them). *Hereafter we write log x, without any base indicated, to mean* $\log_{10} x$. Thus we write $\log 100 = 2$ because $10^2 = 100$, $\log 0.1 = -1$ because $10^{-1} = 1/10 = 0.1$, etc. The student should learn to construct the following rudimentary table of logarithms:

Table A

x	...	0.0001	0.001	0.01	0.1	1	10	100	1000	10000	...
$\log x$	...	-4	-3	-2	-1	0	1	2	3	4	...

This table suggests that *if $m < n$ then $\log m < \log n$.* This can be proved, but we shall assume it here. Hence we expect log 532 to be between 2 and 3 since 532 is between 10^2 and 10^3. It can also be proved, and we shall assume here, that every positive number can be expressed as the product of a number between 1 and 10 and an integral power of 10. For example,

$$53.2 = 5.32(10^1),$$
$$532 = 5.32(10^2),$$
$$5.32 = 5.32(10^0),$$
$$0.00532 = 5.32(10^{-3}).$$

By theorem 1, section 159, we have

$$\log 532 = \log 5.32 + \log (10^2).$$

By definition of logarithm, or by table A, above,

$$\log 532 = \log 5.32 + 2. \tag{1}$$

Similarly,

$$\text{(2)} \qquad \log 53.2 = \log 5.32 + \log (10^1) = \log 5.32 + 1,$$

$$\text{(3)} \qquad \log 5.32 = \log 5.32 + \log (10^0) = \log 5.32 + 0,$$

$$\text{(4)} \quad \log 0.00532 = \log 5.32 + \log (10^{-3}) = \log 5.32 + (-3).$$

Clearly we could write down the logarithm of any number if we only knew *the logarithms of numbers between 1 and 10;* the latter are called **mantissas.** A table of mantissas approximated to four decimal places (table III) is found at the back of the book. Mantissas are decimal fractions, since the logarithm of a number between 1 and 10 must be between 0 and 1.

From (1), (2), (3), and (4), we observe that *the mantissa depends only on the sequence of digits beginning with the first non-zero digit;* all numbers which have the same sequence of digits have the same mantissas. From the table we find that

$$\begin{aligned} \log 532 &= 2 + 0.7259 = 2.7259 \\ \log 53.2 &= 1 + 0.7259 = 1.7259 \\ \log 5.32 &= 0 + 0.7259 = 0.7259 \\ \log 0.00532 &= (-3) + 0.7259. \end{aligned}$$

When a logarithm is expressed so that its decimal part is written positively, the integral part of the logarithm is called its **characteristic.**

For example, the characteristic of log 532 is 2, and the characteristic of log 0.00532 is -3, not -2 (although $\log 0.00532 = -2.2741 = -2 - 0.2741$). *When a characteristic is negative, it is convenient to add and subtract a suitable multiple of 10;* for example, we could write

$$\log 0.00532 = 7.7259 - 10 \text{ (adding and subtracting 10)},$$

or,

$$\log 0.00532 = 17.7259 - 20 \text{ (adding and subtracting 20)}.$$

To find the logarithm of a number N, we determine the characteristic by inspection, and look up the mantissa in table III.

The following rule may be used to determine the characteristic.

Rule 1. Refer to table A above. If N lies between two entries in this table, the characteristic of $\log N$ is the logarithm of the lower of the two entries. In other words, if N lies between 10^k and 10^{k+1}, $(10^k < N < 10^{k+1})$, then the characteristic of $\log N$ is k.

Note 1. The characteristic may also be determined by means of any one of the following rules:

Rule 2. Let N be written in decimal form. If $N \geqq 1$, then the characteristic of $\log N$ is one less than the number of digits in N to the left of the decimal point; if $N < 1$, and if the first non-zero digit of N is in the kth decimal place, then $-k$ is the characteristic of $\log N$.

Rule 3. Express the number N in the standard form of section 66, chapter IX, that is, as a product of a number between 1 and 10 and an integral power of 10. Then the exponent of 10 is the characteristic of $\log N$.

Rule 4. Write N in decimal form. If moving the decimal point k places to the left will make the resulting number between 1 and 10, then the characteristic of $\log N$ is k; if moving the decimal point k places to the right will make the resulting number between 1 and 10, then the characteristic of $\log N$ is $-k$.

The justification of these rules is left as an exercise.

Example 1. Find log 630.

Solution. By rule 1, the characteristic is 2. In table III we find the first two digits 63 in the N column and the third digit 0 at the top of the page. In this column and row, we find the mantissa 0.7993. Hence $\log 630 = 2.7993$.

Example 2. Find log 0.0234.

Solution. By rule 1, the characteristic is -2, or $8 - 10$. From table III, the mantissa corresponding to the digits 234 is 0.3692. Hence $\log 0.0234 = 8.3692 - 10$.

To find a number N whose logarithm is known, we look up the mantissa in table III, write the digits corresponding to it, and place the decimal point according to the characteristic.

Example 3. Find N if $\log N = 1.5752$.

Solution. From table III, the digits corresponding to the mantissa 0.5752 are 376. Since the characteristic is 1, N must be between 10^1 and 10^2. Hence $N = 37.6$.

Example 4. Find N if $\log N = 8.5752 - 10$.

Solution. From table III, the digits are 376. Since the characteristic is $8 - 10 = -2$, N must be between $10^{-2} = 0.01$ and $10^{-1} = 0.1$. Hence $N = 0.0376$.

Note 2. Table III gives for three-digit numbers the corresponding mantissas correct to four decimal places. If we want the mantissa for a four-digit number or the digits corresponding to a mantissa not found exactly in the table, we may proceed in several ways: (*a*) we may take the nearest thing in the table as a sufficiently good approximation; (*b*) we may use a larger table giving mantissas correct to more than four decimal places for numbers of more than three digits; (*c*) we may use the process of interpolation by proportional parts described in the next section.

Note 3. If $\log N = L$, then N may be called the **antilogarithm** of L; in symbols, $N = \text{antilog } L$.

Note 4. The statement $\log 37.6 = 1.5752$ means that $10^{1.5752} = 37.6$ approximately, or $\sqrt[10000]{10^{15752}} = 37.6$ approximately. However, the tables are not constructed by means of this relation directly. How they are calculated will be indicated in section 165.

EXERCISES

Find log N if N is equal to:

1. 124. **2.** 12.4. **3.** 0.0124. **4.** 356. **5.** 39.6.
6. 6.72. **7.** 0.0435. **8.** 0.367. **9.** 0.000346.
10. 208. **11.** 360. **12.** 800. **13.** 0.005.

Find N if log N is equal to:

14. 2.5211. **15.** 1.5211. **16.** $8.5211 - 10$.
17. $9.5211 - 10$. **18.** 0.5211. **19.** 1.6972.
20. 0.8915 **21.** $9.7364 - 10$. **22.** $8.8000 - 10$.
23. 3.9717. **24.** 2.3766.

25. Express the statement $\log 862 = 2.9355$ in (*a*) exponential notation; (*b*) radical notation.

26. Express the statement $\log 631 = 2.8000$ in (*a*) exponential notation; (*b*) radical notation.

161. Interpolation by Proportional Parts

Instead of taking the nearest entry in the table, we may obtain greater accuracy from a given table by the process of **interpolation by proportional parts,** as in the following examples.

Example 1. Find log 36.14.

Solution. The characteristic is 1. From table III, we find that the mantissa for the digits 361 is 0.5575 and the mantissa for 362 is 0.5587. Since 36.14 is 0.4 of the way from 36.1 to 36.2, we take a mantissa 0.4 of the way from 0.5575 to 0.5587. The difference between 0.5575 and 0.5587 is 0.0012; this is called the **tabular difference.** We add 0.4 of the tabular difference, that is $0.4(0.0012) = 0.00048 = 0.0005$ approximately, to 0.5575, obtaining 0.5580. Hence $\log 36.14 = 1.5580$.

The work may be arranged briefly as follows.

$$
10\left[\begin{array}{c|c} \textit{digits} & \textit{mantissa} \\ \hline 4\left[\begin{array}{c} 3610 \\ \\ 3614 \end{array}\right. & \left.\begin{array}{c} 0.5575 \\ \\ \end{array}\right] x \\ \\ 3620 & 0.5587 \end{array}\right] \begin{array}{l} \\ \\ 0.0012 = \textit{tabular difference} \\ \underline{\quad 0.4 \quad} \\ 0.00048 = 0.0005 \end{array}
$$

Then $\dfrac{x}{0.0012} = \dfrac{4}{10}$, or $x = 0.00048 = 0.0005$ to the nearest ten-thousandth. Hence the mantissa to be put in the blank space is 0.5580. Thus, $\log 36.14 = 1.5580$.

Example 2. Find N if $\log N = 9.5579 - 10$.

Solution. The digits corresponding to the mantissas 0.5575 and 0.5587 are 3610 and 3620 respectively. We find that the mantissa 0.5579 is 4/12 of the way from 0.5575 to 0.5587. Hence we take the fourth digit to be 4/12 of the way between 10 and 20. Since 4/12 of 10 is $3\frac{1}{3}$ or 3 to the nearest integer, the digits desired are 3613. Since the characteristic is $9 - 10 = -1$, $N = 0.3613$.

The work may be arranged briefly as follows.

$$
10\left[\begin{array}{c|c} \textit{digits} & \textit{mantissa} \\ \hline x\left[\begin{array}{c} 3610 \\ \\ \end{array}\right. & \left.\begin{array}{c} 0.5575 \\ \\ 0.5579 \end{array}\right] 4 \\ 3620 & 0.5587 \end{array}\right] 12
$$

Then $\dfrac{x}{10} = \dfrac{4}{12}$, or $x = 3.3\ldots = 3$ to the nearest integer. Hence $N = 0.3613$.

Note. It can be proved that if $p < N < q$ then $\log p < \log N < \log q$. But the assumption on which the above procedure is based states further that changes in logarithms are proportional to changes in the corresponding numbers. This is not true, but results obtained by using this assumption are sufficiently accurate for many practical purposes, especially when the difference between the successive values between which we are interpolating is sufficiently small. The significance of this assumption will be discussed in chapter XXVIII.

EXERCISES

Using interpolation by proportional parts, find log N if N is equal to:

1. 241.2. **2.** 38.65. **3.** 2.957. **4.** 0.3874.
5. 0.08513. **6.** 0.02566. **7.** 0.4267. **8.** 7.832.
9. 4.015. **10.** 506.8.

Using interpolation by proportional parts, find N if log N is equal to:

11. 2.1917. **12.** 1.3457. **13.** 0.5374. **14.** 0.9508.
15. 9.7995 − 10. **16.** 9.7206 − 10.
17. 8.9035 − 10. **18.** 8.9891 − 10.
19. 8.31048 − 10. **20.** 3.46345.

162. Significant Figures

When a number is written in the usual decimal form, the **significant figures** or **significant digits** are the digits beginning with the first non-zero digit and ending with the last digit written.

> *Examples.* The numbers 341, 34.1, and 0.00341 all have three significant figures, namely 3, 4, and 1. But 3410, 341.0, 34.10, and 0.003410 all have four significant figures, namely 3, 4, 1, and 0.

In dealing with measurements or other approximate numbers, 34.10 means something different from 34.1. For 34.1 signifies a number known to be between 34.05 and 34.15; that is, $34.05 \leqq 34.1 < 34.15$. But 34.10 signifies a number known to be between 34.095 and 34.105; that is, $34.095 \leqq 34.10 < 34.105$. In other words, 34.1 is correct to the nearest tenth, while 34.10 is correct to the nearest hundredth. Or 34.1 signifies a number which is nearer to 34.1 than it is to 34.0 or to 34.2; while 34.10 is a number nearer to 34.10 than to 34.09 or to 34.11.

Note. If a number has zeros just preceding the decimal point and no digits after the decimal point, such as 341,000., the number of significant figures is six according to the definition above. If the number is not known to six-digit accuracy, the proper number of significant digits must be specifically stated. One way to do that is to use powers of ten as in section 66, chapter IX. Thus if 341,000 is known to only three significant digits, it may be written as 341×10^3; if it is known to four significant digits, it may be written as $3{,}410 \times 10^2$.

A number is said to be **rounded off** to four significant figures if it is replaced by the number of four significant digits to which it is nearest. If the fifth digit is a 5 we make the convention that the fourth digit is to be increased by one. A similar definition of rounding off applies in general.

Example. The number 3.14159265 is written as 3.142 rounded off to four significant figures, as 3.14 rounded off to three significant figures, and as 3.1416 rounded off to five significant figures.

In calculating with approximate numbers, the answer should be rounded off to the number of decimal places justified by the data. In general, the answer cannot be expected to be more accurate than the least accurate number in the data. However, it is advisable to carry one more place throughout the computation and do the rounding off at the end.

Example. The sum of the approximate numbers 3.43 and 1.124 should be written as 4.55 rather than as 4.554. Actually, all that is known, since $3.425 \leqq 3.43 < 3.435$ and $1.1235 \leqq 1.124 < 1.1245$, is that the sum s is such that $4.5485 \leqq s < 4.5595$. Hence, rounding off to three significant figures, $4.55 \leqq s < 4.56$. Therefore $s = 4.55$ rounded off to three significant figures gives somewhat more accuracy than is justified by the data. The answer 4.554 would give a false illusion of greater accuracy.

If the numbers entering a computation are considered to be exact, then, of course, all places may be retained.

EXERCISES

Round off the number:

1. 3.206 to three significant figures.
2. 16.146 to four significant figures.
3. 14.195 to four significant figures.
4. 3.2104 to four significant figures.

5. 4.9997 to four significant figures.

6. 2.71828 to five significant figures.

How many significant figures are there in:

7. 0.0003. **8.** 0.00230. **9.** 32×10^6.

10. 3.20×10^{-5}.

11. Explain the difference in meaning between the approximate numbers 3.2 and 3.20.

12. Explain the difference in meaning between the approximate numbers 4.70 and 4.700.

163. Computation with Logarithms

The simplifications in computation obtained by use of logarithms are due to theorems 1, 2, and 3 of section 159. The following examples will suffice to make the method clear.

Note. No more than four-figure accuracy is usually justified when using a four-place table of logarithms, even if the data of the problem are exact.

Example 1. Calculate $\dfrac{(8.34)(65.2)}{376}$.

Solution. Let N be the result. By theorems 1 and 2, section 159, $\log N = \log \dfrac{(8.34)(65.2)}{376} = \log 8.34 + \log 65.2 - \log 376$.

Now,

$$\begin{aligned} \log 8.34 &= 0.9212 \\ \log 65.2 &= 1.8142 \\ \log 8.34 + \log 65.2 &= 2.7354 \\ \log 376 &= 2.5752 \\ \log N &= 0.1602 \end{aligned}$$

Hence $N = 1.45$ approximately. Since the data are given to three significant figures, so is the answer.

Example 2. Find $\sqrt[3]{473.0}$.

Solution. Let $x = \sqrt[3]{473.0} = (473.0)^{1/3}$. By theorem 3, section 159, $\log x = \frac{1}{3}\log 473.0 = \frac{1}{3}(2.6749) = 0.8916$. Interpolating for the fourth significant figure, we get $x = 7.792$ approximately.

Example 3. If P dollars is invested at an interest rate of r (expressed as a decimal) compounded n times, the amount is given by the formula*

$$A_n = P(1 + r)^n. \tag{1}$$

* This formula will be discussed in detail in chapter XXIV.

If \$100 is invested as a trust fund for a child at 4% per annum interest compounded annually, how much will it amount to in 21 years?

Solution. From (1), we have $A_{21} = 100(1.04)^{21}$. Hence

$$\log A_{21} = \log 100 + 21 \log 1.04$$
$$= 2 + 21(0.0170) = 2.3570.$$

Hence $A_{21} = \$227.50$ approximately.

Example 4. If the interest in example 3 is compounded semi-annually, what will the amount be?

Solution. Interest at 4% per annum compounded semi-annually means that 2% interest is compounded twice a year. Hence in 21 years, interest is compounded forty-two times. Hence $A_{42} = 100(1.02)^{42}$. Thus

$$\log A_{42} = \log 100 + 42 \log 1.02$$
$$= 2 + 42(0.0086) = 2.3612.$$

Hence $A_{42} = \$229.70$ approximately.

Note 1. The time-saving property of logarithms becomes obvious if you consider how long it would take to do example 4 by ordinary arithmetic, multiplying 100 by 1.02 forty-two times in succession. Other simple applications will be found in the exercises.

Note 2. The incorrect formula $\log (x + y) = \log x + \log y$ appears plausible at first glance. But if true, we would need no table of logarithms, for all logarithms would be equal to 0. For $\log 1 = 0$. Hence $\log 2 = \log (1 + 1)$ would be, according to this wrong formula, $\log 1 + \log 1$ or $0 + 0 = 0$; $\log 3 = \log (2 + 1) = \log 2 + \log 1 = 0 + 0 = 0$, and so on. The resemblance of this incorrect formula to the correct distributive law is no more than superficial, for we are here not *multiplying* by the *word* log. Actually, of course, $\log x + \log y = \log xy$.

Note 3. To compute logarithmically with negative numbers, compute the result obtained by ignoring all minus signs, and then determine the sign of the result by the usual rule of signs.

EXERCISES

Compute by means of logarithms:

1. $\dfrac{(34.2)(1.57)}{31.3}$. **2.** $\sqrt[3]{35.2}$. **3.** $\sqrt[5]{117}$.

4. $212(1.04)^{16}$. **5.** $21.3/27.2$. **6.** $\dfrac{(827)(52.4)}{(2.26)(85.8)}$.

7. $\sqrt[3]{673 \times 10^2}$. **8.** $\sqrt[5]{863 \times 10^2}$. **9.** $376(1.06)^{21}$.

10. $\dfrac{(67.6)(57.4)}{33.7}$. **11.** $\sqrt[3]{0.00857}$. **12.** $\sqrt[5]{0.0868}$.

13. $\dfrac{86.42}{2.560 \times 3.120}$. **14.** $\dfrac{\sqrt{535.7}}{6.420 \times (0.1380)^{-3}}$.

15. $\dfrac{(23.54)(0.8765)}{1.760 \sqrt[3]{6.380}}$. **16.** $\dfrac{(3.363)(5.466) \sqrt[3]{12.30}}{0.06620 \sqrt{82.30}}$.

17. $(-43.27)(732.2)$. **18.** $\dfrac{(-4.345)(-8.921)}{-5.427}$.

19. $\dfrac{47.20(1.025)^{10}}{(6.520)(3.480)}$. **20.** $\sqrt{\dfrac{8.27 \times 0.528}{0.360 \times 4.01}}$.

21. $\sqrt[3]{\dfrac{4.23}{0.0842 \times 6.94}}$. **22.** $\sqrt[3]{\dfrac{-625.3}{(3.692)(-22.68)}}$.

★**23.** $\dfrac{\sqrt{45.2} - (1.37)^5}{(1.63)^7 + 4.25}$. ★**24.** $\dfrac{\sqrt{(22.3)(36.1)} - \sqrt{3.24}}{(1.03)^6 + 1}$.

Use formula (1), above, in each of the following problems:*

25. If \$100 is invested at 4% interest compounded quarterly, how much will it amount to in 21 years?

26. If \$550 is deposited in a bank at 2% interest compounded semi-annually, what will the amount be in 6 years?

27. It is desired to have a sum of \$1000 in the bank 10 years from now. If the bank pays 4% interest compounded annually, how much should we deposit now?

28. If the \$24 which the Indians received in 1626 for Manhattan had been deposited in a bank paying 4% interest compounded annually, what would it have amounted to in 1946?

29. How long will it take a sum of money to become doubled if it is invested at 4% compounded annually?

30. If an investment of \$80 will amount to \$100 in 10 years, what rate of interest, compounded annually, is being paid?

31. The area of the surface of a sphere is $4\pi r^2$ where r is the radius. Taking $\pi = 3.14$ and assuming that the earth is a sphere of radius 3960 miles, find the area of the earth's surface.

32. The volume of a sphere is $4\pi r^3/3$ where r is the radius. Using the data of exercise 31, find the volume of the earth.

★ May be omitted without disturbing continuity.

* Further applications to financial problems will be found in the next chapter.

33. If a person bets 2 cents on April 1st and doubles his bet each day thereafter, how much will his bet be on April 30th?

34. The volume of a right circular cylinder is $\pi r^2 h$ where r is the radius and h is the height. Taking $\pi = 3.14$, find the volume of a right circular cylinder with $r = 8.32$ inches and $h = 15.7$ inches.

35. The period T, measured in seconds, of a simple pendulum (that is, the time required for a complete oscillation) of length k is given by the formula $T = 2\pi \sqrt{k/g}$. If $k = 3.26$ feet, $g = 32.2$, $\pi = 3.14$, find T.

36. Using the formula and data of exercise 35, find the length of a simple pendulum whose period is 1 second.

37. Suppose that an automobile costing \$1000 depreciates at the rate of 20% per year; that is, its value at the end of each year is 80% of its value at the beginning of that year. Find its value at the end of (*a*) 5 years; (*b*) 10 years.

38. If a, b, c are the lengths of the sides of a triangle, the area K of the triangle is given by the formula $K = \sqrt{s(s-a)(s-b)(s-c)}$ where $s = \frac{1}{2}(a + b + c)$. Find the area of a triangle with sides 372.3 feet, 417.8 feet, and 292.4 feet, respectively.

39. Using the formula of exercise 38, find the area of a triangle whose sides are 515.6 inches, 742.2 inches, and 469.6 inches, respectively.

40. If a, b, c are the lengths of the sides of a triangle, the radius r of the inscribed circle is given by the formula

$$r = \sqrt{\frac{(s-a)(s-b)(s-c)}{s}},$$

where $s = \frac{1}{2}(a + b + c)$. Find the radius of the inscribed circle for a triangle whose sides are 212.5 feet, 334.7 feet, and 422.1 feet, respectively.

41. Find the geometric mean of the numbers 469, 72.5, 3.42, 0.658, and 0.0566. (*Hint:* See note 1, section 148, for the definition of geometric mean.)

42. In a polytropic atmosphere, temperature T and pressure p are related by the following equation, where T_0 is the temperature at pressure p_0:

$$\frac{T}{T_0} = \left(\frac{p}{p_0}\right)^{\frac{k-1}{k}}.$$

Show that $$k = \frac{\log p - \log p_0}{\log p - \log p_0 - \log T + \log T_0}.$$

43. From the equation

$$\log p - \log p_0 = \frac{1}{lR}[\log (T_0 - lz) - \log T_0],$$

which occurs in the study of the atmosphere, show that

$$p = p_0\left(1 - \frac{lz}{T_0}\right)^{1/lR}.$$

★164. Cologarithms

By the **cologarithm** of N is meant the logarithm of the reciprocal of N; in symbols,

$$\text{colog } N = \log \frac{1}{N} = -\log N.$$

Hence colog N is found by subtracting log N from 0 which may be written as $10 - 10$ for convenience.

Example 1. Since log 376 = 2.5752, colog 376 = 7.4248 − 10.

Since division by N is equivalent to multiplication by $1/N$, we may add colog N instead of subtracting log N, whenever the latter operation is indicated.

Example 2. Do example 1, section 163, by means of cologarithms.

Solution. As before, log N = log 8.34 + log 65.2 − log 376. Hence log N = log 8.34 + log 65.2 + colog 376. Now,

$$\begin{aligned} \log 8.34 &= 0.9212 \\ \log 65.2 &= 1.8142 \\ \text{colog } 376 &= \underline{7.4248 - 10} \\ \log N &= 10.1602 - 10 = 0.1602. \end{aligned}$$

Therefore, $N = 1.45$ approximately.

EXERCISES

Find the cologarithm of:

1. 2.03. **2.** 67.4. **3.** 0.891. **4.** 0.0546.
5. 558.3. **6.** 0.3843.

Work the following exercises, using cologs:

7. Exercise 1, section 163. **8.** Exercise 6, section 163.
9. Exercise 13, section 163. **10.** Exercise 14, section 163.
11. Show that $\log(x^{-r}) = r$ colog x.

★ This section may be omitted without disturbing the continuity of the chapter.

165. Natural Logarithms. Computation of Tables

Theoretically any positive number except 1 can be used as a base for a system of logarithms, although we have so far used only the base 10. Logarithms with any base can be converted into logarithms with any other base merely by multiplying by a suitable constant factor. For, suppose $y = \log_b x$ and $z = \log_a x$. Then

$$(1) \qquad b^y = a^z,$$

since both are equal to x. Taking the logarithm with the base a of both members of (1), we get $y \log_a b = z$ or,

$$(2) \qquad \log_b x \cdot \log_a b = \log_a x.$$

Hence the logarithm of any number x with the base a can be obtained from the logarithm of x with the base b by multiplying by the constant $\log_a b$. The number $\log_a b$ is called the **modulus** of the system of logarithms to the base a with respect to the system of logarithms to the base b.

Aside from common logarithms (with the base 10), the only system in actual use is the system of **natural logarithms** whose base is the irrational number $e = 2.71828 \ldots$. Natural logarithms are natural for reasons which become clear only when one studies the calculus. For example, it can be shown by calculus that

$$(3) \quad \log_e \frac{1+x}{1-x} = 2\left[x + \frac{x^3}{3} + \frac{x^5}{5} + \ldots\right], \text{ for } -1 < x < 1.$$

The right member of (3) is an infinite series (see chapter XXVII). For example, for $x = 1/3$, we obtain from (3)

$$(4) \qquad \log_e 2 = 2\left[\frac{1}{3} + \frac{1}{81} + \frac{1}{1215} + \ldots\right].$$

Clearly the successive terms of the series get small rapidly, and it is plausible and true that a good approximation is obtained by neglecting all but the first few terms. Thus from (4) we get

$$(5) \quad \log_e 2 = 2[0.333333 + 0.012345 + 0.000823 + \ldots] = 0.693002$$

which is correct to the nearest thousandth. By (2),

$$(6) \qquad \log_{10} x = \log_e x \cdot \log_{10} e.$$

The number

$$(7) \qquad M = \log_{10} e = 0.434294 \ldots .$$

Hence, using (5), (6), and (7), we have

$$(8) \qquad \log_{10} 2 = 0.301 \ldots$$

to the nearest thousandth. This, with technical modifications, is the way tables of logarithms are computed.

Note 1. Logarithms were invented in the 17th century by J. Napier (Scottish, 1550–1617) and independently by J. Bürgi (Swiss, 1552–1632). Their systems were closely related to the system of natural logarithms. Strangely enough, the invention of logarithms preceded that of exponents. Common logarithms are credited to Napier and H. Briggs (English, 1556–1631). The first table of common logarithms of numbers was laboriously calculated by Briggs and A. Vlacq (Dutch, c. 1600–1667).

Note 2. The natural logarithm $\log_e x$ is often written as $\ln x$.

EXERCISES

Using (3) and (7), find (a) the natural logarithm and (b) the common logarithm of each of the following numbers to the nearest thousandth:

1. 3. (*Hint:* take $x = \frac{1}{2}$.) **2.** 4/3. **3.** 3/2. **4.** 4.

Using (7), convert each of the following common logarithms into the corresponding natural logarithms, or vice versa:

5. $\log_{10} x = 0.6021$. **6.** $\log_{10} x = 0.8031$.
7. $\log_e x = 1.936$. **8.** $\log_e x = 2.493$.

166. Exponential and Logarithmic Functions and Equations

The exponential function $y = b^x$ and the logarithmic function $y = \log_b x$, where b is positive and different from 1, occur frequently in applied mathematics. Their graphs are given in Figs. 63 and 64 for $b > 1$.

Equations involving exponential functions (that is, in which the unknown occurs as an exponent) can sometimes be solved by taking logarithms, as in the following examples.

Example 1. Solve the equation $3^{x+1} = 5(2^x)$.

Solution. Taking the logarithm of both sides of the given equation, we obtain $(x + 1) \log 3 = \log 5 + x \log 2$. Hence $x =$

$\dfrac{\log 5 - \log 3}{\log 3 - \log 2}$, or $x = 0.2219/0.1761$. Therefore $\log x = \log 0.2219 - \log 0.1761 = 0.1004$. Hence $x = 1.26$ approximately.

Example 2. The number n of bacteria in a certain culture at the end of t hours is given by $n = 100 \times 10^{0.243t}$. Find (*a*) the number of bacteria at $t = 0$; (*b*) at $t = 3$; (*c*) the number of hours required for the number of bacteria to double itself.

Solution. (*a*) At $t = 0$, $n = 100 \times 10^0 = 100$. (*b*) At $t = 3$, $n = 100 \times 10^{0.729}$. Now, $\log n = \log 100 + 0.729 \log 10 = 2.729$. Hence, $n = 536$ approximately. (*c*) t must satisfy the equation $200 = 100 \times 10^{0.243t}$. Hence $\log 200 = \log 100 + 0.243t \log 10$. Hence $2.301 = 2 + 0.243t$, or $t = \dfrac{0.301}{0.243} = 1.24$ hours, approximately.

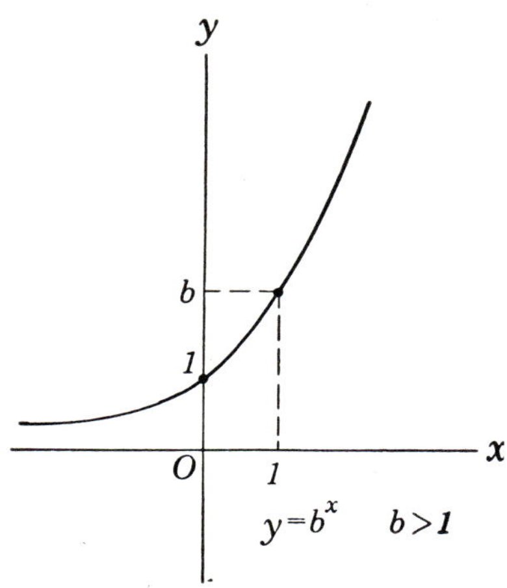

Fig. 63.

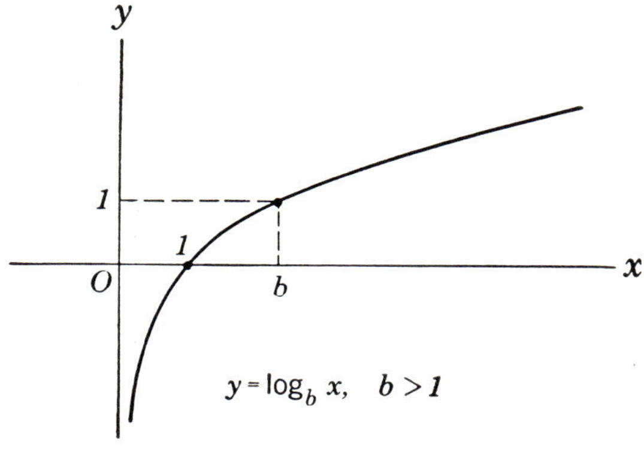

Fig. 64.

EXERCISES

1. Plot the graph of $y = \log_{10} x$ from $x = 0.01$ to $x = 20$.
2. Plot the graph of $y = 10^x$ from $x = -2$ to $x = 1.5$.

Solve each of the following equations for x approximately:

3. $5^x = 10$. **4.** $6^x = 24$.
5. $3^{2x+1} = 2(5^x)$. **6.** $3^{3x-1} = 6(2^{x+1})$.
7. $\log x^2 - \log \dfrac{2x}{5} = 2.641$. **8.** $\log 3x^2 - \log \dfrac{6x}{5} = 1.463$.

Solve each of the following formulas for n:

9. $A = P(1 + r)^n$. **10.** $S = \dfrac{a - ar^n}{1 - r}$.
11. $s = \dfrac{(1 + i)^n - 1}{i}$. **12.** $a = \dfrac{1 - (1 + i)^{-n}}{i}$. **13.** $pv^n = c$.

14. Show that if x takes a succession of values in geometric progression, then $y = \log_b x$ $(b \neq 1)$ takes correspondingly a succession of values in arithmetic progression.

15. Show that if x takes a succession of values in arithmetic progression, then $y = b^x$ $(b \neq 1)$ takes correspondingly a succession of values in geometric progression.

16. The number n of bacteria in a certain culture at the end of t hours is given by $n = 1000 \times 10^{0.164t}$. Find (*a*) the number present at $t = 2$; (*b*) the time required for the number to be doubled.

17. Suppose radium decomposes in such a way that, of x milligrams of it, y milligrams will remain at the end of t centuries where $y = x(0.97)^t$. How much will remain of 2 milligrams after 1000 years?

18. If $pv^n = 4.354$ and $p = 10$ when $v = 3$, find n.

19. Suppose a radioactive substance decomposes so that the amount y remaining of an original amount x after t years is given by $y = xe^{-rt}$ where $e = 2.71828 \ldots$. If $x = 2$ grams, $r = 0.035$, and $t = 100$, find y.

20. Suppose the number y of bacteria present in a certain culture after t hours is given by $y = xe^{rt}$ where $x = 200$, $r = 0.35$, $e = 2.71828 \ldots$. (*a*) If $t = 24$, find y. (*b*) Find the value of t at which there will be 300 bacteria present.

Solve each of the following equations for x:

21. $\log_2 (x - 2) + \log_2 x = 3$.
22. $\log_6 (x + 9) + \log_6 x = 2$.
23. $\log_2 (6x + 5) + \log_2 x = 2$.
24. $\log_5 (6x + 7) + \log_5 x = 1$.

Chapter XXIV

Mathematics of investment

Since few people are able to follow Polonius' advice* to Laertes, problems concerning loans, investments, annuities, etc., arise in everyday life. Such problems may be solved by applying certain principles of elementary algebra already taken up.

167. Simple Interest and Simple Discount

Money paid by one party for the use of another's money is called **interest.** The parties concerned may be individuals, companies, banks, governments, etc. Rates of interest are usually stated in percent per annum. For example, if Mr. Borrower borrows \$100 from Mr. Lender for one year with the understanding that at the end of that period he will repay \$106, we say that the rate of interest is 6% per annum. We shall write rates of interest as decimals for the purpose of actual computation. Thus 6% will be written as 0.06.

When interest is to be proportional to the length of time, it is called **simple interest.** The sum loaned is called the **principal.** The sum to be repaid at the end of the period is called the **amount.**

Example 1. Mr. B borrows \$100 from Mr. L for 2 years at simple interest of 6% per annum. Then the interest will be \$12.

If the principal P dollars is invested at simple interest at the rate r per annum, the interest at the end of n years will be

$$I = Prn, \tag{1}$$

and the amount will be

$$A = P + Prn = P(1 + rn). \tag{2}$$

* "Neither a borrower nor a lender be." (W. Shakespeare, *Hamlet.*)

Example 2. Mr. B borrows \$200 from Mr. L for $2\frac{1}{2}$ years at simple interest of 6% per annum. Then the interest is \$200(0.06)(2.5) = \$30. The amount is \$230.

Example 3. Mr. B borrows \$1000 from Mr. L with the understanding that he will pay at the end of each month \$100 on the principal and simple interest at 6% per annum on the principal outstanding during each month. Find the total amount to be paid.

Solution. There will be 10 payments. The interest due with the first payment is \$1000(0.06)(1/12) = \$5; with the second payment, the interest due is \$900(0.06)(1/12) = \$4.50; with the third payment, the interest due is \$800(0.06)(1/12) = \$4.00; . . . ; with the tenth payment, the interest due is \$100(0.06)(1/12) = \$0.50. The total interest is the sum of the arithmetic progression \$0.50 + \$1.00 + . . . + \$5.00 = $\frac{10}{2}$(\$0.50 + \$5.00) = \$27.50. The total amount paid is therefore \$1027.50.

Often confused with simple interest is the concept of simple discount. For example, if Mr. B borrows \$100 from a banker for one year at 6%, he may be given only \$94 and is expected to repay \$100 at the end of the year. Here the 6% is called a **simple discount rate.** Actually, the interest received by the banker on his investment is more than 6%, for he receives \$6 interest plus the principal at the end of a year, but the principal invested is only \$94. Hence the rate of interest is 6/94 = 0.0638 approximately.

EXERCISES

Find (a) the simple interest and (b) the amount on a loan of:

1. \$500 at 6% per annum for 2 years.

2. \$800 at 5% per annum for 6 months.

3. \$750 at $2\frac{1}{2}$% per annum for 15 months.

4. \$700 at $4\frac{1}{2}$% per annum for 10 months.

5. A man borrows \$250 for 1 year at a simple discount rate of 5% per annum. (*a*) How much does he actually receive? (*b*) What rate of interest is he actually paying?

6. A man borrows \$500 for 6 months at a simple discount rate of 6% per annum. (*a*) How much does he actually receive? (*b*) What rate of interest is he actually paying?

7. A man borrows \$1000 for 18 months at a simple discount rate of $4\frac{1}{2}$% per annum. (*a*) How much does he actually receive? (*b*) What rate of simple interest is he actually paying?

8. A man borrows some money at a simple discount rate at 6% per annum for 1 year. If he wishes to receive \$350 in cash, how much must he borrow?

9. A man borrows some money at a simple discount rate of 6% per annum for 8 months. If he wishes to receive \$200 in cash, how much must he borrow?

10. Show that if i is the actual simple interest rate paid when money is borrowed for 1 year at a simple discount rate of d, then

$$i = \frac{d}{1 - d} \quad \text{and} \quad d = \frac{i}{1 + i}.$$

where both i and d are rates expressed as decimals.

11. A loan of \$1200 is to be repaid in monthly installments of \$100 plus simple interest at the rate of 6% per annum on the principal outstanding during each month. Find the total amount paid.

12. A man borrows \$600 for 10 months and agrees to pay back \$60 per month plus simple interest at the rate of 5% per annum on the principal outstanding during each month. Find the total amount paid.

13. Solve the formula $A = P(1 + rn)$ for (*a*) P; (*b*) r; (*c*) n.

14. A mortgage of \$4500 is to be repaid in monthly installments of \$50 plus simple interest at the rate of 5% per annum on the principal outstanding during each month. Find the total amount paid.

168. Compound Interest and Compound Discount

If Mr. B borrows \$100 from Mr. L for two years at simple interest 6%, it is usually understood that at the end of the first year he is to pay \$6 interest and at the end of the second year \$6 more plus the \$100 principal. If at the end of the first year B fails to pay the \$6 interest, L would be justified in saying that for the second year he has loaned \$6 more to B. Thus he would demand, at the end of the two years

$$\$100(1.06) + \$100(1.06)(0.06) = \$100(1.06)^2 = \$112.36.$$

In general, if the interest due is added to the principal, or converted into principal, at stated intervals of time, and thereafter itself earns interest as new principal, then the sum by which the original principal has been increased at the end of any time is called **compound interest.** A common illustration is a savings account where the depositor seldom withdraws the interest he is entitled to but prefers to let it be converted into principal, which in turn will draw interest thenceforth. The time between successive additions of interest to principal (or conversions of interest into principal) is called the **conversion period** or **interest period.** At the end of each conversion period the new principal, consisting of the original principal plus the compound interest, is called the **compound amount.** The interest rate is usually stated in percent per annum, but the conversion period may be shorter than a year, usually 6 or 3 months. In the latter cases, interest is said to be compounded semi-annually or quarterly, as the case may be. Interest at 6% per annum compounded semi-annually means that 3% interest is added every 6 months.

Suppose P dollars is invested at compound interest at a rate r per conversion period; as usual, r will be expressed as a decimal. Then at the end of the first period the compound amount will be

$$A_1 = P + Pr = P(1 + r).$$

At the end of the second period the amount will be

$$A_2 = A_1 + A_1 r = A_1(1 + r) = P(1 + r)^2.$$

At the end of the third period the amount will be

$$A_3 = A_2 + A_2 r = A_2(1 + r) = P(1 + r)^3.$$

Clearly *at the end of the nth conversion period the compound amount will be*

$$A = A_n = P(1 + r)^n. \tag{1}$$

The basic formula (1) connects four quantities, namely A, P, r, and n. If any three of them are known, the fourth may be found.

Example 1. If \$100 is invested at 6% per annum compounded semi-annually, what will it amount to in 3 years?

Solution. The rate per conversion period is 3%. Hence, by (1), the amount at the end of 3 years or 6 conversion periods will be

$$A_6 = 100(1.03)^6 \text{ dollars.} \tag{2}$$

To evaluate this quantity by multiplying 100 six times by 1.06 is extremely tedious. This calculation may be avoided by the following three methods.

First method. From table IV, we find that $(1.03)^6 = 1.1941$ approximately. Hence $A_6 = \$119.41$ approximately.

Second method. If table IV is not available, we may use logarithms as follows. From (2), and table III, we find

$$\log A_6 = \log 100 + 6 \log 1.03 = 2 + 6(0.0128) = 2.0768.$$

Hence, A_6 = \$119.40 approximately.

Third method. If no tables are available, we may calculate $(1.03)^6$ by the binomial theorem, as follows.

$$\begin{aligned}(1 + 0.03)^6 &= 1^6 + 6 \cdot 1^5 \cdot (0.03) + 15 \cdot 1^4(0.03)^2 + 20 \cdot 1^3(0.03)^3 \\ &\quad + 15 \cdot 1^2(0.03)^4 + 6 \cdot 1 \cdot (0.03)^5 + (0.03)^6 \\ &= 1 + 0.18 + 0.0135 + 0.00054 + 0.00001215 + \ldots \\ &= 1.1941 \text{ approximately.}\end{aligned}$$

Hence A_6 = \$119.41 approximately.

Note. For financial purposes, approximations to the nearest hundredth of a dollar are all that are required. But the four- or five-place tables included in this book may be insufficient to yield such accuracy. More extensive tables may be obtained. The binomial theorem method, of course, may be carried out to any desired accuracy. Calculating machines are useful, if available.

The quantity P is called the **present value** of A_n dollars due at the end of n conversion periods at the interest rate r per conversion period. Clearly

$$P = A_n \cdot \frac{1}{(1 + r)^n} = A_n(1 + r)^{-n}. \tag{3}$$

Example 2. How much must be deposited in a trust fund for a child at 4% compounded annually, if it is desired to have \$1000 in the fund at the end of 10 years?

Solution. By (3), the present value is

$$P = 1000(1.04)^{-10}.$$

From table V, $(1.04)^{-10} = 0.67556$. Hence P = \$675.56 approximately.

Note. Logarithms or the binomial theorem may also be used as in the second and third methods in example 1 above.

To discount a sum A for n conversion periods, at the rate r per conversion period, means to find the present value P of A at a time n periods before the payment A is due. The difference $A - P$ is called the **discount.** The present value of \$1 at the rate r per period for n periods is called **compound discount;** its value is $(1 + r)^{-n}$.

Example 3. Mr. X holds a note from Mr. Y promising to pay \$1000 at a date 10 years from now. Mr. X sells this note to Mr. Z. If discount is reckoned at 4% compounded annually, Mr. Z will pay Mr. X \$675.56 approximately for the note. (See example 2.)

EXERCISES

1. Find the amount if \$500 is invested for 7 years at 4% per annum compounded (*a*) annually; (*b*) semi-annually.

2. Find the amount if \$400 is invested for 10 years at 4% per annum compounded (*a*) annually; (*b*) semi-annually.

3. Find the amount at the end of 21 years if \$100 is invested at 6% per annum compounded (*a*) annually; (*b*) semi-annually.

4. Find the amount at the end of 12 years if \$250 is invested at 6% per annum compounded (*a*) annually; (*b*) semi-annually.

5. How much must be deposited in a bank now in order to have \$500 at the end of 10 years if interest is paid at the rate of 4% per annum compounded (*a*) annually; (*b*) semi-annually.

6. How much must be invested now in order to have \$750 at the end of 5 years if interest is paid at the rate of 4% per annum compounded (*a*) annually; (*b*) semi-annually.

7. A note promising to pay \$600 in 3 years is sold. How much should be paid for it if discount is reckoned at 7% per annum compounded (*a*) annually; (*b*) semi-annually?

8. A note promising to pay \$800 in 5 years is sold. How much should be paid for it if discount is reckoned at 6% per annum compounded (*a*) annually; (*b*) semi-annually.

9. How many years will it take for a deposit of \$100 to amount to \$253.16 if interest is paid at the rate of $3\frac{1}{2}$% per annum compounded annually?

10. How many years will it take for a deposit of \$250 to amount to \$475 if interest is paid at the rate of $3\frac{1}{2}$% per annum compounded annually?

11. How many years will it take for a deposit of \$100 to amount to \$196 if interest is paid at the rate of 4% per annum compounded semi-annually?

12. How many years will it take for an investment to double itself if interest is paid at the rate of 4% per annum compounded annually?

13. How many years will it take for an investment to triple itself if interest is paid at the rate of $3\frac{1}{2}$% per annum compounded annually?

14. A bond costing \$75 will be redeemed in 10 years for \$100. If interest is compounded annually, what rate of interest is being paid?

15. A deposit of \$100 will amount to \$266.58 in 25 years. If interest is compounded annually, what rate of interest is being paid?

16. A deposit of \$100 will amount to \$163.86 in 10 years. If interest is compounded semi-annually, what rate of interest is being paid?

17. A deposit of \$500 will amount to \$1048.80 in 15 years. Find the interest rate if interest is compounded semi-annually.

18. A deposit of \$500 will amount to \$671.95 in 5 years. If interest is compounded semi-annually, find the interest rate.

169. Effective and Nominal Rates

The interest rate i per annum compounded m times per year stated in describing an arrangement of compound interest is called the **nominal rate.** The **effective rate** is the rate j which, if compounded annually, would yield the same interest. For interest compounded once a year the nominal rate and the effective rate are the same. Equating the amounts at the end of one year under both arrangements, we have

$$P(1+j) = P\left(1+\frac{i}{m}\right)^m.$$

Hence

$$1+j = \left(1+\frac{i}{m}\right)^m \tag{1}$$

or

$$j = \left(1+\frac{i}{m}\right)^m - 1. \tag{2}$$

Example. Find the effective rate if the nominal rate is 8% compounded quarterly.

Solution. We have $1 + j = (1.02)^4 = 1.0824$ from table IV. Hence the effective rate is 8.24% approximately.

EXERCISES

Find the effective rate if interest is compounded semi-annually at the nominal rate of:

1. 4% per annum.
2. 5% per annum.
3. 6% per annum.
4. 3% per annum.
5. 8% per annum.
6. 2% per annum.

Find the effective rate if interest is compounded quarterly at the nominal rate of:

7. 4% per annum.
8. 3% per annum.
9. 6% per annum.
10. 2% per annum.

170. Annuities

Many business transactions are arranged so that equal payments are made at regular intervals of time. Such a series of payments is called an **annuity.** The interval of time at the end* of which payments are to be made is called the **payment period.** Although the word *annuity* suggests annual payments, the payment period of an annuity may be of any length of time. Familiar examples of annuities are installment buying, premiums for insurance, amortizing a mortgage, etc. The **term** of the annuity is the time between the beginning of the first payment period and the end of the last. Each payment is called **rent** and is denoted by R. The sum of the payments made in one year is called the **annual rent.**

The **amount** of an annuity is the total amount which would be accumulated at the end of the term if each payment were invested at a given rate of compound interest at the time of payment. *We treat only the case where the conversion period of the compound interest coincides with the payment period of the annuity.* The amount of an annuity which pays \$1 at the end of each payment period for n periods is denoted by $s_{\overline{n|}}$, read "s angle n." The **present value** of an annuity is the sum of the present values of all the payments at the beginning of the term. The present value of an annuity which pays \$1 at the end of each payment period for n periods is denoted by $a_{\overline{n|}}$, read "a angle n."

The amount or the present value of an annuity can be determined by working out a schedule of the compound amount or present value of all payments, and adding, but if the number of payments is large this is very tedious. Instead we shall calculate formulas for $s_{\overline{n|}}$ and $a_{\overline{n|}}$ when compound interest is computed at the rate i per period.

At the end of the term, the last payment of \$1 will have just been made and will amount to \$1. The payment before the last will have drawn interest for one period and will amount to $(1 + i)$. The payment before that, the $(n - 2)$th payment, will amount to $(1 + i)^2$. And so on. The first payment will have

* If payments are made at the beginnings of the payment periods, slight adjustments must be made in the subsequent formulas. The same *principles*, however, may be applied.

drawn interest for $n - 1$ periods and will amount to $(1 + i)^{n-1}$. Hence

$$s_{\overline{n|}} = 1 + (1 + i) + (1 + i)^2 + \cdots + (1 + i)^{n-1}.$$

This is a geometric progression with first term 1 and common ratio $1 + i$. By section 148, chapter XXI, the sum is therefore

$$s_{\overline{n|}} = \frac{(1 + i)^n - 1}{i}. \tag{1}$$

Values of this quantity for various values of n and i are given in table VI.

If the rent, or periodic payment, is $\$R$ (instead of $1), it is easily seen that *the amount of the annuity is*

$$R \cdot s_{\overline{n|}} \text{ dollars.} \tag{2}$$

To find $a_{\overline{n|}}$, we compute the present values of the payments at the beginning of the term and add. The present value of the first payment of $1, which is due at the end of the first period, is $(1 + i)^{-1}$. The present value of the second payment is $(1 + i)^{-2}$, and so on. The present value of the last (nth) payment is $(1 + i)^{-n}$. Hence

$$a_{\overline{n|}} = (1 + i)^{-1} + (1 + i)^{-2} + \cdots + (1 + i)^{-n}.$$

This is a geometric progression with first term $(1 + i)^{-1}$ and common ratio $(1 + i)^{-1}$. By section 148, chapter XXI, we find

$$a_{\overline{n|}} = \frac{1 - (1 + i)^{-n}}{i}. \tag{3}$$

Values of this quantity are given in table VII.

If the rent, or periodic payment, is $\$R$ (instead of $1), it is easily seen that *the present value of the annuity is*

$$R \cdot a_{\overline{n|}} \textit{ dollars.} \tag{4}$$

Example. Find the amount and present value of an annuity paying $100 every 6 months for 10 years, interest being compounded semi-annually at 6% per annum.

First solution. The rate i is 0.03, and the number of periods is $n = 20$. From the tables VI and VII, $s_{\overline{20|}} = 26.8704$ and $a_{\overline{20|}} = 14.8775$. Hence the amount is $2687.04 approximately and the present value is $1487.75 approximately.

Second solution. If tables VI and VII are not available, logarithms may be used as follows. We have

$$(5) \qquad s_{\overline{20|}} = \frac{(1.03)^{20} - 1}{0.03}.$$

By logarithms we find $(1.03)^{20} = 1.8061$ approximately. Hence $s_{\overline{20|}} = \dfrac{0.8061}{0.03} = 26.87$ approximately. Hence the amount is \$2687 approximately. A similar method can be used for the present value.

Third solution. In (5), the quantity $(1.03)^{20} = (1 + 0.03)^{20}$ can be evaluated by means of the binomial theorem, as in section 168, example 1, third method.

As before, answers obtained by use of tables will be approximate because of the limitations of the tables. More extensive tables are available.

EXERCISES

Solve the following problems using the methods above rather than formulas.

1. Find (*a*) the amount, and (*b*) the present value of an annuity of \$300 per year for 10 years at $3\frac{1}{2}\%$ per annum.

2. Find (*a*) the amount and (*b*) the present value of an annuity of \$250 per year for 25 years at 4% per annum.

3. A man deposits \$200 in a bank on the first day of each year for 20 years. If the bank pays interest at $2\frac{1}{2}\%$ per annum, compounded annually, what is the amount at the end of the 20th year? (*Hint:* note that payments are made at the *beginning*, not at the end, of periods. See footnote on page 417.)

4. A man deposits \$300 in a bank on the first day of each year for 25 years. If the bank pays interest at $3\frac{1}{2}\%$ per annum, compounded annually, what is the amount at the end of the 25th year?

5. A man deposits \$200 every 6 months in a bank paying interest at 4% per annum compounded semi-annually. How much is in his account just after he makes his 30th deposit?

6. A man deposits \$150 every 6 months in a bank paying interest at 3% per annum compounded semi-annually. How much is in his account just after he makes his 40th deposit?

7. A man wishes to make equal annual deposits at the beginning of each year for 30 years so as to accumulate \$10,000 at the end of the 30th year. If the bank pays interest at 2% per annum, compounded annually, what must his annual deposit be?

8. A machine costing \$5000 will wear out in 20 years. What equal annual deposits must be made at the end of each year to pay for its replacement if interest is paid at $2\frac{1}{2}\%$ per annum compounded annually?

9. Show that if a debt (such as a mortgage) of D dollars, at an interest rate i per annum compounded annually, is paid off (amortized) in n years by equal annual payments of E dollars, then $E = D/a_{\overline{n}|}$ or $E \cdot a_{\overline{n}|} = D$.

10. A man pays off a \$10,000 mortgage, interest and principal, by equal payments, made at the end of each year for 15 years. Find his annual payment if interest is paid at 5% per annum compounded annually. Make a schedule showing the part of each payment which goes for interest and principal respectively. (*Hint:* Use the result of exercise 9.)

11. A man pays off a \$5000 mortgage, interest and principal, by equal payments, made at the end of each year for 10 years. Find his annual payment if interest is paid at 6% per annum compounded annually.

12. A mortgage of \$8000 is paid off, interest and principal, by equal payments, made at the end of each year, for 20 years. Find the annual payment if interest is paid at 5% per annum compounded annually.

13. A man pays premiums of \$23.24 for an insurance policy at the beginning of each year for 20 years. Just after paying the 20th premium he surrenders his policy, receiving \$327.58 from the company. What was the actual total cost of being insured if money is worth 3%?* (*Hint:* The answer will be the amount of the annuity of which the premium is the periodic payment minus the cash received at surrender of the policy.)

14. A man pays premiums of \$164.50 for an insurance policy at the beginning of each year for 20 years. Just after paying the 20th premium, he surrenders the policy, receiving \$562.15 from the company. What was the actual total cost of being insured if money is worth 4%?

* This phrase means that interest is to be computed at the rate of 3% per annum compounded annually.

Chapter XXV

The Euclidean algorithm

171. The Euclidean Algorithm for Positive Integers

Recall* that **to divide** a positive integer a by a positive integer b, where $b \leqq a$, means to find a positive integer q, the quotient, and a non-negative integer r such that

$$(1) \qquad a = bq + r \text{ where } 0 \leqq r < b.$$

If $r = 0$, we say b is a **factor** of a, or a is **exactly divisible** by b, or a is a **multiple** of b.†

It can be proved that for a given a and b, the q and r are uniquely determined.

THEOREM 1. *Any common factor* c *of* a *and* b *is a factor of* r. *Also any common factor* d *of* b *and* r *is a factor of* a.

Proof. By hypothesis, $a = cx$ and $b = cy$ where x and y are positive integers. Then by (1),

$$cx = cyq + r$$

or

$$r = c(x - yq).$$

Since $x - yq$ is a non-negative integer, c is a factor of r. This completes the proof of the first statement in the theorem.

If $b = du$ and $r = dv$ then $a = bq + r = duq + dv = d(uq + v)$. This proves the second statement in the theorem.

It follows from theorem 1 that *the highest common factor (H.C.F.) of* a *and* b *is also the highest common factor of* b *and* r.

* See section 43, chapter V.

† The definitions and procedures of this section are easily extended to all integers, positive, negative, or zero.

Given two positive integers a and b $(a > b)$, their H.C.F. can be found by the following process, known as the **Euclidean algorithm.** Divide a by b, obtaining

$$a = bq_1 + r_1 \qquad (0 \leqq r_1 < b). \tag{2}$$

If $r_1 > 0$, divide b by r_1, obtaining

$$b = r_1 q_2 + r_2 \qquad (0 \leqq r_2 < r_1). \tag{3}$$

If $r_2 > 0$, divide r_1 by r_2, obtaining

$$r_1 = r_2 q_3 + r_3 \qquad (0 \leqq r_3 < r_2). \tag{4}$$

If $r_3 > 0$, divide r_2 by r_3, obtaining

$$r_2 = r_3 q_4 + r_4 \qquad (0 \leqq r_4 < r_3), \tag{5}$$

and so on until a remainder r_{n+1} equal to 0 is reached:

$$r_{n-1} = r_n q_{n+1}. \tag{6}$$

THEOREM 2. *The last non-zero remainder r_n is the H.C.F. of a and b.*

Proof. By theorem 1, (2) implies that the H.C.F. of a and b is equal to the H.C.F. of b and r_1. Similarly (3) implies that the H.C.F. of b and r_1 is the H.C.F. of r_1 and r_2. And so on until we conclude that the H.C.F. of a and b is equal to the H.C.F. of r_{n-1} and r_n. But since r_n is a factor of r_{n-1} by (6), this H.C.F. is r_n itself.

Example 1. Find the H.C.F. of 66 and 180.

Solution. The successive divisions may be arranged compactly, as follows.

```
       2
  66)180
     132  1
     48)66
        48  2
        18)48
           36  1
           12)18
              12  2
               6)12
                 12
                  0.
```

Therefore 6 is the H.C.F. of 66 and 180.

Several consequences of the Euclidean algorithm are worth noting.

THEOREM 3. *If h is the H.C.F. of a and b, there exist integers s and t such that $h = sa + tb$.*

Proof. To simplify the discussion, suppose $n = 4$ in (6), so that $h = r_4$. Then from (2) we have

$$r_1 = a - bq_1.$$

From (3),

$$r_2 = b - r_1q_2 = b - (a - bq_1)q_2 = (-q_2)a + (1 + q_1q_2)b.$$

From (4),

$$\begin{aligned} r_3 = r_1 - r_2q_3 &= (a - bq_1) - [(-q_2)a + (1 + q_1q_2)b]q_3 \\ &= (1 + q_2q_3)a + (-q_1 - q_3 - q_1q_2q_3)b. \end{aligned}$$

From (5),

$$\begin{aligned} r_4 = r_2 - r_3q_4 = (-q_2 - q_4 - q_2q_3q_4)a \\ + (1 + q_1q_2 + q_1q_4 + q_3q_4 + q_1q_2q_3q_4)b. \end{aligned}$$

Since the parentheses on the right represent integers, we have $r_4 = sa + tb$, which was to have been proved. The proof for any value of n is similar.

DEFINITION 1. *If a positive integer greater than 1 has no factors except itself and 1, it is called a* ***prime*** *or a* ***prime number.***

Examples. The first few primes are 2, 3, 5, 7, 11, ...

DEFINITION 2. *Two different positive integers are* ***relatively prime*** *if they have no common factor except 1.*

Examples. 4 and 9 are relatively prime. So are 10 and 21; and 6 and 35. But 6 and 10 are not relatively prime because they have the common factor 2.

Clearly, *if two positive integers are relatively prime, their H.C.F. is 1, and conversely.*

THEOREM 4. *If a and b are relatively prime but a is a factor of bc, then a is a factor of c.*

Proof. The H.C.F. of a and b is 1. Hence, by theorem 3, there exist integers s and t such that

$$1 = sa + tb.$$

Multiplying by c, we have

$$c = sac + tbc. \tag{7}$$

By hypothesis, there exists an integer x such that $bc = ax$. Substituting in (7), we have

$$c = sac + tax = a(sc + tx).$$

Hence a is a factor of c.

Theorem 4 is useful in the theory of equations; see section 122, chapter XIX. Compare also theorem 2, section 6, chapter I.

EXERCISES

By means of the Euclidean algorithm, find the H.C.F. of each of the following pairs of numbers:

1. 60, 80. **2.** 60, 75. **3.** 555, 925. **4.** 204, 85.
5. 86, 75. **6.** 165, 286. **7.** 156, 306. **8.** 851, 656.

9. Prove that if p is prime and p is a factor of ab, where a and b are positive integers, then either p is a factor of a or p is a factor of b.

10. Prove that if a, b, c are positive integers, a and b being relatively prime, and if a and b are both factors of c, then ab is a factor of c.

★11. Prove that the q and r in (1) above are uniquely determined. That is, if $a = bq + r$, $0 \leqq r < b$, and $a = bq' + r'$, $0 \leqq r' < b$, then $q = q'$ and $r = r'$.

★12. Prove that any positive integer can be expressed as a product of prime factors in one and only one way apart from the order of the factors. (*Hint:* use the theorem of exercise 9.)

172. Euclidean Algorithm for Polynomials

We shall confine ourselves to polynomials with real coefficients, and factors with real coefficients, although this restriction is not essential. Recall that if a and b are polynomials, to divide a by b means to find polynomials q and r such that $a = bq + r$ and such that r has lower degree than b. If $r = 0$, b and q are factors of a. We shall consider as **essentially the same** polynomials which differ only by a constant factor; such polynomials are also termed **associates** of each other. For example, we shall regard $x + 2$, $2x + 4$, $3x + 6$, $\frac{1}{2}x + 1$, etc., as essentially the same. Thus we shall say that $x^2 - 1 = (x - 1)(x + 1) = (2x - 2)(\frac{1}{2}x + \frac{1}{2}) = (3x - 3)(\frac{1}{3}x + \frac{1}{3})$ are essentially the same fac-

★ May be omitted without disturbing continuity.

torization. If we merely interpret the letters a, b, q_1, r_1, q_2, r_2, . . . as polynomials instead of integers, and $<$ as referring to degree, it can be seen that the Euclidean algorithm will yield an H.C.F.* of the polynomials a and b. That is, theorems 1 and 2 of the preceding section and their proofs apply to polynomials as well as to integers.

Example 1. Find an H.C.F. of $x^4 - 3x^2 + 2$ and $x^4 + x^3 - x - 1$.

Solution. By the Euclidean algorithm,

$$\begin{array}{llll}
 & 1 & & \\
x^4 - 3x^2 + 2\,) & x^4 + x^3 \qquad - x - 1 & & \\
 & x^4 \qquad - 3x^2 \qquad + 2 & x - 3 & \\
 & \overline{x^3 + 3x^2 - x - 3} &)\,x^4 \qquad - 3x^2 \qquad + 2 & \\
 & & x^4 + 3x^3 - x^2 - 3x & \\
 & & \overline{-3x^3 - 2x^2 + 3x + 2} & \\
 & & -3x^3 - 9x^2 + 3x + 9 & \frac{1}{7}x + \frac{3}{7} \\
 & & \overline{7x^2 \qquad\qquad - 7} &)\,x^3 + 3x^2 - x - 3 \\
 & & & x^3 \qquad - x \\
 & & & \overline{3x^2 \qquad - 3} \\
 & & & 3x^2 \qquad - 3 \\
 & & & \overline{\qquad\qquad 0}
\end{array}$$

Hence $x^2 - 1$ (which is essentially the same as $7x^2 - 7$) is an H.C.F. of the given polynomials.

Similarly, theorem 3 of the preceding section and its proof may be extended to polynomials as follows.

THEOREM. *If $h(x)$ is an H.C.F. of two polynomials $a(x)$ and $b(x)$, then there exist polynomials $s(x)$ and $t(x)$ such that $h(x) = s(x)a(x) + t(x)b(x)$.*

A polynomial may be called **prime** or **irreducible**† if it has no factors except constants (which are all essentially the same as 1) and polynomials essentially the same as itself. Two polynomials are called **relatively prime** if they have no common factors, except constants (which are all essentially the same as 1). Clearly, if two polynomials are relatively prime, their H.C.F. is essentially the same as 1. Hence we have the following immediate corollary of the above theorem.

* An H.C.F. is a common factor of a and b which is exactly divisible by any common factor of a and b. All H.C.F.'s are essentially the same.

† More precisely, prime or irreducible over the domain of real coefficients.

COROLLARY. *If polynomials $a(x)$ and $b(x)$ are relatively prime, then there exist polynomials $s(x)$ and $t(x)$ such that* $1 = s(x)a(x) + t(x)b(x)$.

Note. By corollary 1, section 121, chapter XIX, every polynomial (with real coefficients) can be expressed as a product of prime factors (with real coefficients) of the first or second degree.

EXERCISES

By means of the Euclidean algorithm, find a H.C.F. of each of the following pairs of polynomials:

1. $x^2 - 9,\ x^2 - 6x + 9$.
2. $5x^3 + 12x^2 - 36x - 16,\ 25x^2 - 4$.
3. $x^2 - x - 2,\ x^2 + x - 6$.
4. $6x^3 - 13x^2 - 19x + 12,\ 9x^2 - 16$.
5. $2x^3 - x^2 + 6x - 3,\ 8x^3 - 2x^2 + 17x - 9$.
6. $2x^3 - 13x^2 + 27x - 18,\ 2x^4 - 3x^3 - 6x^2 + 5x + 6$.
7. $8x^3 - 10x^2 - x + 3,\ 2x^3 - 5x^2 - x + 6$.
8. $3x^3 + 10x^2 + 9x + 2,\ 6x^3 - 13x^2 - 19x + 12$.

Chapter XXVI

Partial fractions

173. The Theorem and Its Use

To add two or more rational expressions, as

$$\frac{3}{x-2}+\frac{2}{x-3}=\frac{5x-13}{x^2-5x+6},$$

is a matter of routine. For certain purposes, particularly in connection with a more advanced subject called *integral calculus*, it is advantageous to be able to start with the sum and find out the so-called **partial fractions** of which it is the sum. This inverse process of "un-adding" the sum is more difficult than the direct process of addition, just as factoring is harder than multiplying and chemical analysis is harder than making compounds.

Every rational function or fractional expression can be expressed as a quotient of two polynomials. We confine ourselves to expressions of the form $N(x)/D(x)$ where $N(x)$ and $D(x)$ are polynomials with real coefficients. If the degree of the numerator $N(x)$ is less than the degree of the denominator $D(x)$, the expression is called a **proper** fraction; otherwise, **improper.** Every improper fraction can be reduced to the sum of a polynomial and a proper fraction, for if the degree of the numerator is equal to or greater than the degree of the denominator, we can divide $N(x)$ by $D(x)$ obtaining a quotient polynomial $Q(x)$ and a remainder polynomial $R(x)$ whose degree is less than that of $D(x)$. Thus

$$\frac{N(x)}{D(x)}=Q(x)+\frac{R(x)}{D(x)}$$

where $R(x)/D(x)$ is a proper fraction. For example,

$$\frac{x^3 - 4x^2 + 6x - 7}{x^2 - 5x + 6} = x + 1 + \frac{5x - 13}{x^2 - 5x + 6}.$$

Hence *we confine ourselves in what follows to proper fractions.*

By corollary 1, section 121, chapter XIX, the denominator $D(x)$ can be expressed as a product of linear and quadratic factors, with real coefficients, which cannot be factored further* into linear factors with real coefficients. We suppose this done; these are the factors referred to throughout, below. Then every proper fraction can be resolved into partial fractions in accordance with the following theorem, whose proof will be given in the following section.

THEOREM. *Every proper fraction $N(x)/D(x)$ can be resolved into an algebraic sum of partial fractions, as follows:*

I. If a linear factor $(ax + b)$ occurs once as a factor of $D(x)$, there is a partial fraction of the form $\frac{A}{ax + b}$, where A is a constant;

II. If a linear factor $(ax + b)$ occurs k times as a factor of $D(x)$, there are k partial fractions of the form

$$\frac{A_1}{ax + b} + \frac{A_2}{(ax + b)^2} + \cdots + \frac{A_k}{(ax + b)^k},$$

where $A_1, A_2, \ldots, A_k$ are constants;

III. If a quadratic factor† $(ax^2 + bx + c)$ occurs once as a factor of $D(x)$, there is a partial fraction of the form $\frac{Cx + K}{ax^2 + bx + c}$ where C and K are constants;

IV. If a quadratic factor $(ax^2 + bx + c)$ occurs k times as a factor of $D(x)$, there are k partial fractions of the form

$$\frac{C_1x + K_1}{ax^2 + bx + c} + \frac{C_2x + K_2}{(ax^2 + bx + c)^2} + \cdots + \frac{C_kx + K_k}{(ax^2 + bx + c)^k},$$

where $C_1, K_1, C_2, K_2, \ldots, C_k, K_k$ are constants.

* That is, they are prime. Factors differing only by a constant factor are regarded as essentially the same. See section 172, chapter XXV.

† More precisely, a prime or irreducible quadratic factor; that is, with $b^2 - 4ac < 0$. In fact the theorem as stated does not require that the factors be irreducible. But it is often advantageous to use irreducible factors.

The process of determining the unknown constants will be illustrated in the following examples. *In every case the result may be checked by adding the partial fractions to get the original expression.*

*Case I. When the factors of the denominator $D(x)$ are all linear and distinct.**

Example 1. Resolve $\dfrac{9x^2 - 9x + 6}{(x-1)(2x-1)(x+2)}$ into partial fractions.

Solution. By I of the theorem, we may write

$$(1)\quad \frac{9x^2 - 9x + 6}{(x-1)(2x-1)(x+2)} = \frac{A}{x-1} + \frac{B}{2x-1} + \frac{C}{x+2}.$$

This is an identity; that is, it is true for all values of x except possibly for the values $x = 1, \frac{1}{2}$, and -2, which make the denominator vanish. Hence, multiplying both members of (1) by the least common denominator $(x-1)(2x-1)(x+2)$, we get

$$(2)\quad 9x^2 - 9x + 6 = A(2x-1)(x+2) + B(x-1)(x+2) + C(x-1)(2x-1).$$

We may determine the values of the constants A, B, C by either of two methods.

First method. Rewriting (2) we get

$$(3)\quad 9x^2 - 9x + 6 = (2A + B + 2C)x^2 + (3A + B - 3C)x + (-2A - 2B + C).$$

Since (2) is true for all values of x except possibly $x = 1, \frac{1}{2}$, and -2, it follows from corollary 2, section 119, chapter XIX, that (2) is an identity. Hence we may equate coefficients of like powers of x in (3) obtaining

$$(4)\quad \begin{cases} 2A + B + 2C = 9 \\ 3A + B - 3C = -9 \\ -2A - 2B + C = 6. \end{cases}$$

Solving the system (4), we get $A = 2$, $B = -3$, $C = 4$. Hence, from (1), we have

$$(5)\quad \frac{9x^2 - 9x + 6}{(x-1)(2x-1)(x+2)} = \frac{2}{x-1} - \frac{3}{2x-1} + \frac{4}{x+2}.$$

Second method. Since (2) is true for all values of x except possibly $x = 1, \frac{1}{2}, -2$, it follows from corollary 2, section 119,

* That is, no two are essentially the same.

chapter XIX, that (2) is an identity and hence true for all values of x including $x = 1, \frac{1}{2}$, and -2. Substituting $x = 1$ in (2), we find that $6 = 3A$, or $A = 2$. Substituting $x = \frac{1}{2}$ in (2), we find that $15/4 = -5B/4$, or $B = -3$. Substituting $x = -2$ in (2), we find that $60 = 15C$, or $C = 4$. Hence we have (5) as before.

EXERCISES

Resolve into partial fractions:

1. $\dfrac{x + 12}{(2x - 1)(x + 2)}$. **2.** $\dfrac{10x + 2}{(3x + 2)(x - 4)}$.

3. $\dfrac{-x - 12}{2x^2 - x - 6}$. **4.** $\dfrac{9x - 7}{6x^2 + 14x + 4}$.

5. $\dfrac{2x^2 + 10x + 2}{(x + 1)(x - 2)(x + 3)}$. **6.** $\dfrac{7x^2 - 22}{(2x - 3)(x - 2)(x + 1)}$.

7. $\dfrac{x^4 + 5x^3 + 7x^2 + 7x + 13}{x^2 + 5x + 6}$.

8. $\dfrac{x^4 + 3x^3 - 5x^2 - 4x - 2}{x^2 + 2x - 8}$.

9. $\dfrac{5x^3 + 36x^2 + 77x + 48}{(x + 1)(x^2 + 5x + 6)}$. **10.** $\dfrac{2x^4 - 3x^3 - 19x^2 + 9x + 2}{(x^2 + x - 2)(x - 4)}$.

Case II. When all the factors of $D(x)$ are linear but some are repeated.

Example 2. Resolve $\dfrac{x^2 - 3x + 1}{(x - 1)^2(x - 2)}$ into partial fractions.

Solution. By *I* and *II* of the theorem, we may write

$$\frac{x^2 - 3x + 1}{(x - 1)^2(x - 2)} = \frac{A}{x - 1} + \frac{B}{(x - 1)^2} + \frac{C}{x - 2}. \tag{6}$$

Multiplying both members by the L.C.D., we obtain

$$x^2 - 3x + 1 = A(x - 1)(x - 2) + B(x - 2) + C(x - 1)^2, \tag{7}$$

or,

$$x^2 - 3x + 1 = (A + C)x^2 + (-3A + B - 2C)x + (2A - 2B + C). \tag{8}$$

First method. Since (8) is an identity, we may equate coefficients of like powers, obtaining

$$\begin{cases} A + C = 1 \\ -3A + B - 2C = -3 \\ 2A - 2B + C = 1. \end{cases} \tag{9}$$

Solving (9), we get $A = 2$, $B = 1$, $C = -1$. Hence

$$(10)\qquad \frac{x^2 - 3x + 1}{(x-1)^2(x-2)} = \frac{2}{x-1} + \frac{1}{(x-1)^2} - \frac{1}{x-2}.$$

Second method. Since (7) is an identity, we may substitute $x = 1$ in (7), obtaining $-1 = -B$, or $B = 1$. Similarly substituting $x = 2$ in (7), we get $-1 = C$. Substituting these values for B and C and substituting for x any value different from 1 or 2, say $x = 3$, in (7), we get $1 = 2A + 1 - 4$, or $A = 2$. Hence we have (10) as before.

EXERCISES

Resolve into partial fractions:

11. $\dfrac{3x+5}{(x+1)^2}$. **12.** $\dfrac{2x-5}{x^2-6x+9}$.

13. $\dfrac{2x^2-9x+11}{x^3-6x^2+12x-8}$. **14.** $\dfrac{2x^2+7x+9}{(x+1)^3}$.

15. $\dfrac{5x^2+18x+14}{(2x+1)(x^2+6x+9)}$. **16.** $\dfrac{5x^2-25x+8}{(3x+2)(x-3)^2}$.

17. $\dfrac{5x^3-6x^2-3x-1}{(x+1)^2(x-2)^2}$. **18.** $\dfrac{8x^3-37x^2+39x-18}{x^4-6x^3+9x^2}$.

19. $\dfrac{6x^3-29x^2-51x+278}{(2x+4)(x-4)^2}$. **20.** $\dfrac{8x^3-8x^2+9x-9}{2x^4-3x^3}$.

Case III. When $D(x)$ has (irreducible) quadratic factors, none repeated.

Example 3. Resolve $\dfrac{4x^2+x+1}{(x^2+1)(x-1)}$ into partial fractions.

Solution. By *I* and *III* of the theorem, we may write

$$(11)\qquad \frac{4x^2+x+1}{(x^2+1)(x-1)} = \frac{Ax+B}{x^2+1} + \frac{C}{x-1}.$$

Hence,

$$(12)\qquad 4x^2 + x + 1 = (Ax+B)(x-1) + C(x^2+1)$$

or,

$$(13)\qquad 4x^2 + x + 1 = (A+C)x^2 + (-A+B)x + (-B+C).$$

First method. Equating coefficients in (13), we get

$$(14)\qquad \begin{cases} A + C = 4 \\ -A + B = 1 \\ -B + C = 1. \end{cases}$$

Solving (14), we get $A = 1$, $B = 2$, $C = 3$. Hence

$$(15) \qquad \frac{4x^2 + x + 1}{(x^2 + 1)(x - 1)} = \frac{x + 2}{x^2 + 1} + \frac{3}{x - 1}.$$

Second method. Since (12) is an identity, we may set x equal to any three values in (12). Substituting $x = 1$, we get at once $6 = 2C$, or $C = 3$. Substituting $C = 3$, and any value of x, say $x = 0$, in (12) we get $1 = -B + 3$, or $B = 2$. Substituting $C = 3$, $B = 2$, and any other value for x, say $x = 2$, in (12), we get $19 = (2A + 2) + 15$, or $A = 1$. Hence, we have (15) as before.

EXERCISES

Resolve into partial fractions:

21. $\dfrac{x^2 - 6x + 2}{(x - 1)(x^2 + x + 1)}$. **22.** $\dfrac{8x^2 - 5x + 6}{(3x - 2)(x^2 + x + 2)}$.

23. $\dfrac{5x^2 + 1}{x^3 + 1}$. **24.** $\dfrac{5x^2 - 6x}{x^3 - 27}$. **25.** $\dfrac{5x^2 + 2x + 9}{x^3 + x^3 + 3x}$.

26. $\dfrac{x^2 + 14x - 15}{2x^3 + 6x}$. **27.** $\dfrac{3x^2 - 2x + 5}{(x^2 + 1)(2x^2 - x + 3)}$.

28. $\dfrac{-2x^3 + 4x + 15}{(x^2 + 3x + 4)(x^2 + x + 1)}$.

29. $\dfrac{x^5 + x^4 + 4x^3 + 7x^2 - 2x + 19}{(x^2 + 3)(x^2 - x + 2)}$.

30. $\dfrac{x^5 + 7x^3 + 6x - 2}{x^4 + 2x^2}$. **31.** $\dfrac{4x^4 + 7x^3 + 5x^2 + x - 6}{x^4 + x^3 + 2x^2}$.

32. $\dfrac{3x^4 - 5x^3 + 16x^2 - 28x + 4}{(x - 1)^2(x^2 + 4)}$.

Case IV. When $D(x)$ has repeated (irreducible) quadratic factors.

Example 4. Resolve $\dfrac{4x^4 + 3x^3 + 6x^2 + 5x}{(x - 1)(x^2 + x + 1)^2}$ into partial fractions.

Solution. By *I* and *IV* of the theorem, we may write

$$(16) \qquad \frac{4x^4 + 3x^3 + 6x^2 + 5x}{(x - 1)(x^2 + x + 1)^2} = \frac{A}{x - 1} + \frac{Bx + C}{x^2 + x + 1} + \frac{Dx + E}{(x^2 + x + 1)^2}.$$

Multiplying both members of (16) by the L.C.D., we obtain

$$(17) \qquad 4x^4 + 3x^3 + 6x^2 + 5x = A(x^2 + x + 1)^2 + (Bx + C)(x - 1)(x^2 + x + 1) + (Dx + E)(x - 1),$$

or,

$$(18)\quad 4x^4 + 3x^3 + 6x^2 + 5x = (A + B)x^4 + (2A + C)x^3 + (3A + D)x^2 + (2A - B - D + E)x + (A - C - E).$$

First method. Equating coefficients in (18), we obtain

$$(19)\quad \begin{cases} A + B = 4 \\ 2A + C = 3 \\ 3A + D = 6 \\ 2A - B - D + E = 5 \\ A - C - E = 0. \end{cases}$$

Solving (19), we obtain $A = 2$, $B = 2$, $C = -1$, $D = 0$, $E = 3$. Hence,

$$(20)\quad \frac{4x^4 + 3x^3 + 6x^2 + 5x}{(x - 1)(x^2 + x + 1)^2} = \frac{2}{x - 1} + \frac{2x - 1}{x^2 + x + 1} + \frac{3}{(x^2 + x + 1)^2}.$$

Second method. Since (17) is an identity, we may substitute any five values for x. Setting $x = 1$ in (17) we get $18 = 9A$, or $A = 2$. Using $A = 2$, and setting $x = 0, -1, 2, -2$ successively in (17), we get

$$(21)\quad \begin{cases} 0 = 2 - C - E & (x = 0) \\ 2 = 2 + 2B - 2C + 2D - 2E & (x = -1) \\ 122 = 98 + 14B + 7C + 2D + E & (x = 2) \\ 54 = 18 + 18B - 9C + 6D - 3E & (x = -2). \end{cases}$$

Solving (21), we get $B = 2$, $C = -1$, $D = 0$, $E = 3$, and hence (20) as before.

EXERCISES

Resolve into partial fractions:

33. $\dfrac{x^2 + 2x - 2}{(x^2 + 1)^2}$.

34. $\dfrac{2x^3 - x^2 + 3x + 1}{(x^2 - x + 2)^2}$.

35. $\dfrac{3x^4 + 15x^2 + 3x + 18}{x(x^2 + 3)^2}$.

36. $\dfrac{x^4 + 2x^3 - 5x^2 + 8x - 3}{(x - 2)(x^2 + 1)^2}$.

37. $\dfrac{3x^5 + 2x^4 + 6x^3 + 6x^2 + 3x + 5}{(x^2 + 1)^2(x^2 + 2)^2}$.

38. $\dfrac{2x^4 + 14x^2 + 25}{(x^2 + 3)^2(x^2 + 4)^2}$.

39. $\dfrac{3x^4 - 2x^3 + 4x^2 - 3x + 2}{(x - 1)(x^2 + 1)^2}.$

40. $\dfrac{2x^4 - x^3 + 15x^2 - 2x + 28}{(x + 1)(x^2 + 3)^2}.$

*174. Proof of the Theorem

We shall indicate briefly a proof of the theorem that every rational expression $N(x)/D(x)$, where $N(x)$ and $D(x)$ are polynomials with real coefficients, can be expressed as the sum of a polynomial and partial fractions of the forms described in the theorem of the preceding section. The proof will be broken up into a number of auxiliary theorems.

THEOREM 1. *A rational expression (fraction) $N(x)/D(x)$, whose denominator $D(x) = a(x)b(x)$ where $a(x)$ and $b(x)$ are relatively prime polynomials, can be written as the sum of two rational expressions (fractions) with denominators $a(x)$ and $b(x)$ respectively.*

Proof. By the corollary of section 172, chapter XXV, there exist polynomials $s(x)$ and $t(x)$ such that

$$1 = s(x)a(x) + t(x)b(x). \tag{1}$$

Multiplying both members of (1) by $\dfrac{N(x)}{a(x)b(x)}$, we get†

$$\frac{N}{ab} = \frac{Ns}{b} + \frac{Nt}{a}. \tag{2}$$

This proves theorem 1.

THEOREM 2. *If $D(x) = [c(x)]^p$ where $c(x)$ is a polynomial and p is a positive integer, then the rational expression $N(x)/D(x)$ can be written as a sum of a polynomial and several rational expressions whose denominators are powers of $c(x)$ with exponents $\leqq p$ and whose numerators have lower degree than $c(x)$.*

Proof. Dividing $N(x)$ by $c(x)$, we obtain a quotient $Q_0(x)$ and a remainder $R_0(x)$ whose degree is less than that of $c(x)$; thus we write

$$N = cQ_0 + R_0. \tag{3}$$

* Chapter XXV is required for this section.

† For brevity, we write N, a, b, ... instead of $N(x)$, $a(x)$, $b(x)$, ... respectively.

Dividing Q_0 by c, we get similarly

$$Q_0 = cQ_1 + R_1 \tag{4}$$

where the degree of R_1 is less than that of c. Substituting (4) into (3), we obtain

$$N = c^2Q_1 + cR_1 + R_0. \tag{5}$$

Dividing Q_1 by c, we get $Q_1 = cQ_2 + R_2$, and substituting in (5), we get

$$N = c^3Q_2 + c^2R_2 + cR_1 + R_0. \tag{6}$$

After p such steps, we have finally

$$N = c^pQ_{p-1} + c^{p-1}R_{p-1} + c^{p-2}R_{p-2} + \ldots + cR_1 + R_0. \tag{7}$$

Dividing both sides by $D = c^p$, we have

$$\frac{N}{D} = Q_{p-1} + \frac{R_{p-1}}{c} + \frac{R_{p-2}}{c^2} + \ldots + \frac{R_1}{c^{p-1}} + \frac{R_0}{c^p}. \tag{8}$$

This completes the proof of theorem 2.

Proof of the theorem of section 173. From corollary 1, section 121, chapter XIX, we see that the denominator $D(x)$ of any rational expression can be expressed as a product of the form

$$\begin{aligned} D(x) = a_0(x - r_1)^{m_1}(x - r_2)^{m_2} \ldots \\ (x - r_h)^{m_h}(x^2 + u_1x + v_1)^{n_1} \ldots (x^2 + u_kx + v_k)^{n_k} \end{aligned} \tag{9}$$

where the discriminants of the quadratic factors are negative. The distinct factors in (9) are all prime or irreducible and hence certainly relatively prime. We may apply theorem 1 with $a(x) = (x - r_1)^{m_1}$ and $b(x)$ equal to the product of the remaining factors in (9), obtaining

$$\frac{N}{D} = \frac{A_1(x)}{(x - r_1)^{m_1}} + \frac{B(x)}{b(x)}.$$

Applying theorem 1 similarly to $B(x)/b(x)$, we get

$$\frac{N}{D} = \frac{A_1(x)}{(x - r_1)^{m_1}} + \frac{A_2(x)}{(x - r_2)^{m_2}} + \frac{C(x)}{c(x)}.$$

Continuing in this way, we express N/D as a sum of rational expressions whose denominators are $(x - r_1)^{m_1}$, $(x - r_2)^{m_2}$, $\ldots$, $(x^2 + u_kx + v_k)^{n_k}$, respectively. To each of these rational expressions we may apply theorem 2. This completes the proof.

Chapter XXVII

Infinite series

175. Introduction

Dividing 1 by $1 - x$ by the usual process,* we obtain the expression

$$(1) \qquad \frac{1}{1-x} = 1 + x + x^2 + x^3 + \ldots + x^{n-1} + \ldots$$

where the dots indicate that the process does not terminate. What can (1) mean? If $x = \frac{1}{2}$ is substituted in (1), we get

$$\frac{1}{1-\frac{1}{2}} = 1 + \frac{1}{2} + \left(\frac{1}{2}\right)^2 + \left(\frac{1}{2}\right)^3 + \ldots + \left(\frac{1}{2}\right)^{n-1} + \ldots$$

or,

$$(2) \qquad 2 = 1 + \frac{1}{2} + \frac{1}{4} + \frac{1}{8} + \ldots + \frac{1}{2^{n-1}} + \ldots .$$

Now the right member of (2) does seem to add up to 2 in the sense that (Fig. 65) if you add up enough of its terms you can

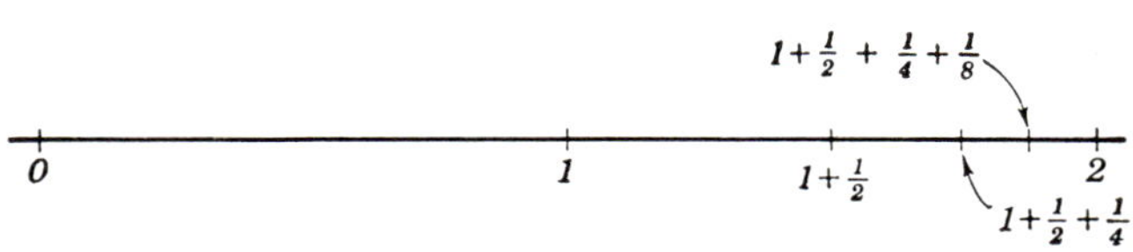

Fig. 65.

get an answer as close as you please to 2; for example, the sum of the first four terms is $1 + \frac{7}{8}$ while the sum of the first five terms is $1 + \frac{15}{16}$. However, substituting $x = 1$ in (1), we get the senseless result

* The expression (1) can also be obtained by expanding $(1 - x)^{-1}$ by the binomial theorem.

(3) $$\frac{1}{0} = 1 + 1 + 1 + 1 + \dots + 1 + \dots;$$

and substituting $x = 2$ in (1), we get the false result

(4) $$-1 = 1 + 2 + 4 + 8 + \dots + 2^{n-1} + \dots.$$

Hence it is clear that the right member of (1) does not always have a sense and that the equation (1) is not always correct.

Expressions like the right members of (1), (2), (3), or (4) are called infinite series. In general, an **infinite series** is an expression of the form

$$a_1 + a_2 + a_3 + \dots + a_n + \dots$$

where the number of terms is infinite. Only series whose terms are real numbers will be considered here. We shall now examine more precisely under what circumstances such an expression has a meaning, and shall indicate some uses for them. Infinite series are among the most powerful tools in advanced mathematics, both pure and applied.

176. Limit of an Infinite Sequence

By an **infinite sequence** is meant an unending succession of terms $s_1, s_2, s_3, \dots, s_n, \dots$. The subscript n indicates the position of the term s_n in the sequence. The term s_n is called the ***n*th term** or **general term.** An infinite sequence may be defined by stating its general term.

Example 1. The first four terms of the sequence whose nth term is $1/n$ are $1/1, 1/2, 1/3, 1/4, \dots$.

Example 2. The first four terms of the sequence whose nth term is $2 - \dfrac{1}{10^n}$ are 1.9, 1.99, 1.999, 1.9999, ...

Example 3. The first four terms of the sequence whose nth term is $2n$ are 2, 4, 6, 8, ...

Definition. *The number S will be called the* ***limit of the sequence*** *$s_1, s_2, s_3, \dots, s_n, \dots$ provided that, given any positive number h, no matter how small, there exists a term s_N of the sequence such that every succeeding term s_m, $m > N$, lies between $S - h$ and $S + h$ (that is $|s_m - S| < h$ for all $m > N$). If there is such a limit S, we call the sequence* ***convergent,*** *and we write* $\lim_{n\to\infty} s_n = S$.

In this case, we say that as n increases indefinitely, s_n approaches S as a limit. A sequence which has no limit is called ***divergent.***

It can be shown that if a sequence has a limit, the value of the limit is uniquely determined.

Example 4. The limit of the sequence in example 1 is 0. For, given any positive h, we have only to take $N > \frac{1}{h}$, and clearly $0 - h < \frac{1}{m} < 0 + h$ for all $m > N$. We write $\lim_{n \to \infty} \frac{1}{n} = 0$ (see Fig. 66).

It is intuitively clear that the limit of the sequence in example 2 is 2 (Fig. 66), and that the sequence of example 3 is divergent.

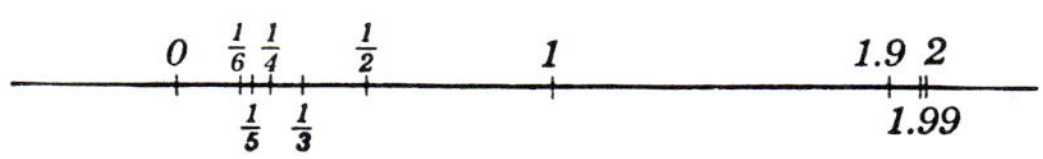

Fig. 66.

The following theorems are proved in more advanced books. We assume them here.

Theorem 1. *If* $\lim_{n \to \infty} s_n = S$, *and* $\lim_{n \to \infty} t_n = T$, *then*

$$(a)\ \lim_{n \to \infty} (s_n + t_n) = S + T$$

$$(b)\ \lim_{n \to \infty} (s_n - t_n) = S - T$$

$$(c)\ \lim_{n \to \infty} (s_n \cdot t_n) = ST$$

$$(d)\ \lim_{n \to \infty} \frac{s_n}{t_n} = \frac{S}{T} \text{ provided } T \neq 0.$$

Theorem 2.* *If* $s_n \leqq s_{n+1}$ *for all* $n = 1, 2, 3, \ldots,$ *and if there exists a number B such that* $s_n \leqq B$ *for all* $n = 1, 2, 3, \ldots,$ *then the sequence* $s_1, s_2, s_3, \ldots, s_n, \ldots$ *has a limit* $S \leqq B$.

Theorem 3. *If* $|r| < 1$, *then* $\lim_{n \to \infty} r^n = 0$.

Theorem 4. *If* $s_n = c$, *where c is any constant, for every* $n = 1, 2, 3, \ldots,$ *then* $\lim_{n \to \infty} s_n = c$.

* This theorem may be taken as one of the fundamental axioms of the system of real numbers.

THEOREM 5. *If s_n increases beyond all bounds as $n \to \infty$, then* $\lim_{n\to\infty} \frac{c}{s_n} = 0$, *where c is any constant.*

Example 5. Show that $\lim_{n\to\infty} \frac{n^2 + 3n + 4}{2n^2 - n - 5} = \frac{1}{2}$.

Solution. Using theorems **1**, 4, and 5,

$$\lim_{n\to\infty} \frac{n^2 + 3n + 4}{2n^2 - n - 5} = \lim_{n\to\infty} \frac{1 + \frac{3}{n} + \frac{4}{n^2}}{2 - \frac{1}{n} - \frac{5}{n^2}} = \frac{1}{2}.$$

THEOREM 6. *If $|s_n| \geqq nc$, where c is any positive constant, then the sequence $s_1, s_2, s_3, \ldots, s_n, \ldots$ diverges.*

EXERCISES

(*a*) *Write the first four terms of the sequence whose nth term is given;* (*b*) *using the theorems above, find its limit:*

1. $3 + \left(\frac{1}{10}\right)^n$. **2.** $3 - \left(\frac{1}{2}\right)^n$. **3.** $2 + \left(\frac{-1}{2}\right)^n$.

4. $\frac{2n + 1}{3n}$. (*Hint:* divide numerator and denominator by n.)

5. $\frac{n^2 + 1}{3n^2 - n + 2}$. **6.** $\frac{3n}{n + 1}$. **7.** $\frac{n!n}{(n + 1)!}$.

8. $\frac{3n^2 - 3n + 2}{2n^2 + n + 1}$. **9.** $\frac{2^n - 1}{2^n + 1}$. **10.** $\frac{5n^2 - n + 2}{2n^3 + 1}$.

11. $\frac{n^2 + 2n + 1}{2n + 3}$. **12.** $\frac{2 \cdot 3^n - 1}{3^n + 1}$.

177. Convergence of Infinite Series

Consider the infinite series

$$(1) \qquad u_1 + u_2 + u_3 + \ldots + u_n + \ldots.$$

Let $s_1 = u_1$, $s_2 = u_1 + u_2$, $s_3 = u_1 + u_2 + u_3$, $\ldots$, $s_n = u_1 + u_2 + \ldots + u_n$, $\ldots$. The numbers $s_1, s_2, \ldots, s_n, \ldots$ are called the **partial sums** of the series (1).

DEFINITION. *If the sequence of partial sums has the limit S, that is, if $\lim_{n\to\infty} s_n = S$, then the series (1) is said to have the sum S and is said to be* ***convergent,*** *or to* ***converge to*** *S. A series which is not convergent is called* ***divergent.***

Note. The "sum" of an infinite series is not a sum in the ordinary sense, but is the limit of such sums.

Example 1. Consider the series $\frac{1}{2} + \frac{1}{4} + \frac{1}{8} + \ldots + \frac{1}{2^n} + \ldots$. Here $s_1 = \frac{1}{2}$, $s_2 = \frac{3}{4}$, $s_3 = \frac{7}{8}$, $\ldots$, $s_n = \frac{2^n - 1}{2^n} = 1 - \frac{1}{2^n}$. Hence $\lim_{n \to \infty} s_n = 1$. The series converges to the "sum" 1.

THEOREM 1. *If the series (1) converges, then* $\lim_{n \to \infty} u_n = 0$.

Proof. Suppose the series (1) converges to S. Then $\lim_{n \to \infty} s_n = S$ and $\lim_{n \to \infty} s_{n-1} = S$. But $s_n = s_{n-1} + u_n$ or $u_n = s_n - s_{n-1}$. Hence $\lim_{n \to \infty} u_n = \lim_{n \to \infty} (s_n - s_{n-1}) = \lim_{n \to \infty} s_n - \lim_{n \to \infty} s_{n-1} = S - S = 0$.

Note. The converse of theorem 1 is false, as will be seen in the next section.

COROLLARY 1. *If* $\lim_{n \to \infty} u_n \neq 0$, *then the series (1) does not converge.*

Proof. An immediate consequence of theorem 1.

THEOREM 2. *The* ***geometric series*** $a + ar + ar^2 + \ldots + ar^{n-1} + \ldots$ *converges to* $\frac{a}{1 - r}$ *if* $|r| < 1$ *and diverges if* $|r| \geqq 1$.

Proof. By the formula for geometric progressions, we have $s_n = \frac{a - ar^n}{1 - r} = \frac{a}{1 - r} - \frac{a}{1 - r} r^n$. By the theorems of section 176, if $|r| < 1$, we have $\lim_{n \to \infty} s_n = \frac{a}{1 - r}$. By corollary 1, above, the series diverges if $|r| \geqq 1$, for then $\lim_{n \to \infty} ar^{n-1} \neq 0$.

EXERCISES

1. Show that the series $\frac{1}{3} + \frac{1}{9} + \frac{1}{27} + \ldots + \frac{1}{3^n} + \ldots$ converges to $\frac{1}{2}$.

2. Show that the series $1 + 0.1 + 0.01 + \ldots + \left(\frac{1}{10}\right)^{n-1} + \ldots$ converges to 10/9.

3. Show that the series $1 + 2 + 3 + \ldots + n + \ldots$ diverges.

4. Show that the series $1 - 1 + 1 - \ldots + (-1)^{n-1} + \ldots$ diverges.

5. Show that the series $1 + \frac{3}{2} + \frac{9}{4} + \ldots + \left(\frac{3}{2}\right)^{n-1} + \ldots$ diverges.

6. Show that the series $25[1 + 0.01 + (0.01)^2 + \ldots + (0.01)^{n-1} + \ldots]$ converges to 2500/99.

178. The Comparison Tests

Consider two series

$$a_1 + a_2 + a_3 + \ldots + a_n + \ldots, \tag{1}$$

$$b_1 + b_2 + b_3 + \ldots + b_n + \ldots, \tag{2}$$

with non-negative terms.

THEOREM 1. *If $b_n \leqq a_n$ for all $n = 1, 2, 3, \ldots$, and if the series (1) converges, then the series (2) converges.*

Proof. Let $s_n = a_1 + a_2 + \ldots + a_n$ and $t_n = b_1 + b_2 + \ldots + b_n$. Then, by hypothesis, $\lim_{n\to\infty} s_n = S$ exists. But since the terms of (1) and (2) are non-negative, we have $t_n \leqq s_n \leqq S$ and $t_n \leqq t_{n+1}$ for all $n = 1, 2, 3, \ldots$. Hence by theorem 2, section 176, $\lim_{n\to\infty} t_n = T \leqq S$ exists.

THEOREM 2. *If $b_n \geqq a_n$ for all $n = 1, 2, 3, \ldots$, and if the series (1) diverges, then the series (2) diverges.*

Proof. If (2) converged, then theorem 1 would imply that (1) converged, contrary to hypothesis.

Note. In applying these theorems to test convergence of series, any finite number of terms at the beginning of the series may be omitted, for if the series $u_{k+1} + u_{k+2} + u_{k+3} + \ldots$ converges to U, then the series $u_1 + u_2 + u_3 + \ldots + u_k + u_{k+1} + \ldots$ converges to $u_1 + u_2 + \ldots + u_k + U$.

Any series known to be convergent or divergent may be used, according to the above theorems, to test the convergence or divergence of certain other series.

Example 1. The series

$$1 + \frac{1}{2^2} + \frac{1}{3^3} + \ldots + \frac{1}{n^n} + \ldots \tag{3}$$

is convergent, for, comparing it with the convergent geometric series $1 + \frac{1}{2} + \frac{1}{2^2} + \ldots + \frac{1}{2^{n-1}} + \ldots$ with $r = \frac{1}{2}$, we find that $\frac{1}{n^n} \leqq \frac{1}{2^{n-1}}$ since $n^n \geqq 2^{n-1}$ for all $n = 1, 2, 3, \ldots$. Hence the series (3) is convergent by theorem 1.

In addition to the geometric series, the series

$$(4) \qquad 1 + \frac{1}{2^p} + \frac{1}{3^p} + \ldots + \frac{1}{n^p} + \ldots$$

known as the ***p*-series** will be found often useful for comparison. The special case of the p-series with $p = 1$, that is, the series

$$(5) \qquad 1 + \frac{1}{2} + \frac{1}{3} + \ldots + \frac{1}{n} + \ldots$$

is known as the **harmonic series.**

THEOREM 3. *The p-series (4) is convergent for $p > 1$ and divergent for $p \leqq 1$.*

Proof. Case 1. $p > 1$. We write (4) as

$$(6) \quad 1 + \left(\frac{1}{2^p} + \frac{1}{3^p}\right) + \left(\frac{1}{4^p} + \frac{1}{5^p} + \frac{1}{6^p} + \frac{1}{7^p}\right) + \left(\frac{1}{8^p} + \ldots + \frac{1}{15^p}\right) + \ldots$$

where there are 2 terms in the first parentheses, 4 terms in the next, 8 terms in the next, ..., 2^k terms in the kth parentheses. Now

$$1 \leqq 1,$$

$$\frac{1}{2^p} + \frac{1}{3^p} \leqq \frac{1}{2^p} + \frac{1}{2^p} = \frac{2}{2^p} = \frac{1}{2^{p-1}},$$

$$\frac{1}{4^p} + \frac{1}{5^p} + \frac{1}{6^p} + \frac{1}{7^p} \leqq \frac{1}{4^p} + \frac{1}{4^p} + \frac{1}{4^p} + \frac{1}{4^p} = \frac{4}{4^p} = \frac{1}{4^{p-1}} = \left(\frac{1}{2^{p-1}}\right)^2,$$

$$\frac{1}{8^p} + \ldots + \frac{1}{15^p} \leqq \frac{1}{8^p} + \ldots + \frac{1}{8^p} = \frac{8}{8^p} = \frac{1}{8^{p-1}} = \left(\frac{1}{2^{p-1}}\right)^3,$$

and so on. The terms on the right form a geometric series

$$(7) \quad 1 + \frac{1}{2^{p-1}} + \left(\frac{1}{2^{p-1}}\right)^2 + \left(\frac{1}{2^{p-1}}\right)^3 + \ldots + \left(\frac{1}{2^{p-1}}\right)^{n-1} + \ldots$$

with ratio $r = \frac{1}{2^{p-1}}$. Since $p > 1$, $r < 1$ and (7) converges to a sum S. Because of the above inequalities, the sum s_n of any number of terms of (4) is $\leqq S$. Hence $\lim_{n \to \infty} s_n$ exists, and (4) converges.

Case 2. $p = 1$. We write (5) as

$$(8) \quad 1 + \frac{1}{2} + \left(\frac{1}{3} + \frac{1}{4}\right) + \left(\frac{1}{5} + \frac{1}{6} + \frac{1}{7} + \frac{1}{8}\right) + \left(\frac{1}{9} + \dots + \frac{1}{16}\right) + \dots .$$

Now,

$$1 + \frac{1}{2} > 1$$

$$\frac{1}{3} + \frac{1}{4} > \frac{1}{4} + \frac{1}{4} = \frac{1}{2}$$

$$\frac{1}{5} + \frac{1}{6} + \frac{1}{7} + \frac{1}{8} > \frac{1}{8} + \frac{1}{8} + \frac{1}{8} + \frac{1}{8} = \frac{1}{2}$$

$$\frac{1}{9} + \dots + \frac{1}{16} > \frac{1}{16} + \dots + \frac{1}{16} = \frac{1}{2},$$

and so on. Hence for any n,

$$1 + \frac{1}{2} + \frac{1}{3} + \dots + \frac{1}{n} > 1 + \frac{1}{2} + \frac{1}{2} + \frac{1}{2} + \dots + \frac{1}{2}.$$

Since the right member increases beyond any bound as n increases, (5) diverges.

Case 3. $p < 1$. Since $p < 1$, $\frac{1}{n^p} \geqq \frac{1}{n}$ because $n^p \leqq n$. Hence by comparison with the harmonic series (5), (4) diverges in this case.

EXERCISES

Prove that the following series converge:

1. $\frac{1}{1 \cdot 3} + \frac{1}{2 \cdot 3^2} + \frac{1}{3 \cdot 3^3} + \dots + \frac{1}{n \cdot 3^n} + \dots .$

2. $\frac{1}{1 + 2} + \frac{1}{1 + 2^2} + \frac{1}{1 + 2^3} + \dots + \frac{1}{1 + 2^n} + \dots .$

3. $\frac{1}{3 \cdot 1^2} + \frac{1}{3 \cdot 2^2} + \frac{1}{3 \cdot 3^2} + \dots + \frac{1}{3 \cdot n^2} + \dots .$

4. $\frac{1}{1+1^3} + \frac{1}{2+2^3} + \frac{1}{3+3^3} + \ldots + \frac{1}{n+n^3} + \cdots.$

5. $\frac{1}{1 \cdot 2} + \frac{1}{2 \cdot 3} + \frac{1}{3 \cdot 4} + \ldots + \frac{1}{n(n+1)} + \cdots.$

Prove that the following series diverge:

6. $\frac{1}{\sqrt{1}} + \frac{1}{\sqrt{2}} + \frac{1}{\sqrt{3}} + \ldots + \frac{1}{\sqrt{n}} + \cdots.$

7. $\frac{1}{1-\frac{1}{2}} + \frac{1}{2-\frac{1}{3}} + \frac{1}{3-\frac{1}{4}} + \ldots + \frac{1}{n-\frac{1}{n+1}} + \cdots.$

8. $2 + \frac{3}{2}\cdot\frac{5}{4} + \frac{4}{3}\cdot\left(\frac{5}{4}\right)^2 + \ldots + \frac{n+1}{n}\left(\frac{5}{4}\right)^{n-1} + \cdots.$

Write the first three terms of the series whose nth term is given and test for convergence or divergence:

9. $\frac{1}{(2n-1)^2}$.
10. $\frac{1}{n!}$.
11. $\frac{1}{(2n)^3}$.
12. $\frac{1}{(2n-1)!}$.
13. $\frac{n}{n+1}$.
14. $\frac{2}{\sqrt[3]{n}}$.
15. $\frac{1}{2n\sqrt{n}}$.
16. $\frac{1}{\sqrt{n}-\frac{1}{n+1}}$.
17. $\frac{1}{2^n+n}$.
18. $\frac{1}{(2n)!}$.
19. $\frac{n^2+n}{2n^2+n+1}$.
20. $\frac{\sqrt{n}}{n^2}$.
21. $\frac{n(n+1)}{(n+2)(n+3)}$.
22. $\frac{1}{n^2+1}$.
23. $\frac{n+1}{n^2+1}$.
24. $\frac{1}{3^n+1}$.

★179. The Decimal Expression for a Real Number

We shall show that every real number can be expressed as a decimal (see section 14, chapter I).

THEOREM. *If r is any real number such that $0 < r < 1$, then r is the sum of an infinite series*

$$\frac{k_1}{10} + \frac{k_2}{10^2} + \frac{k_3}{10^3} + \ldots + \frac{k_n}{10^n} + \cdots \tag{1}$$

where k_n is an integer between 0 and 9 inclusive.

★ May be omitted without disturbing continuity.

Proof. Let l_1 be the smallest integer such that $l_1 \geqq 10r$. Let $k_1 = l_1 - 1$. Then

$$\frac{k_1}{10} < r \leqq \frac{k_1 + 1}{10}.$$

Let l_2 be the smallest integer such that $l_2 \geqq 10^2 r - 10k_1$. Let $k_2 = l_2 - 1$. Then

$$\frac{k_1}{10} + \frac{k_2}{10^2} < r \leqq \frac{k_1}{10} + \frac{k_2 + 1}{10^2}.$$

Similarly, determine the integer k_n such that

$$\frac{k_1}{10} + \frac{k_2}{10^2} + \dots + \frac{k_n}{10^n} < r \leqq \frac{k_1}{10} + \frac{k_2}{10^2} + \dots + \frac{k_n + 1}{10^n}.$$

It is not difficult to prove that $0 \leqq k_n \leqq 9$ for all n. The series (1) is convergent, since $\frac{k_n}{10^n} \leqq \frac{9}{10^n}$ and the series

$$\frac{9}{10} + \frac{9}{10^2} + \dots + \frac{9}{10^n} + \dots$$

is a convergent geometric series with ratio $\frac{1}{10}$. Since

$$r - \frac{1}{10^n} \leqq \frac{k_1}{10} + \frac{k_2}{10^2} + \dots + \frac{k_n}{10^n} < r,$$

we have $\left|\frac{k_1}{10} + \frac{k_2}{10^2} + \dots + \frac{k_n}{10^n} - r\right| \leqq \frac{1}{10^n}$. But $\lim_{n \to \infty} \frac{1}{10^n} = 0$.

Hence $$r = \lim_{n \to \infty} \left(\frac{k_1}{10} + \frac{k_2}{10^2} + \dots + \frac{k_n}{10^n}\right),$$

or $$r = \frac{k_1}{10} + \frac{k_2}{10^2} + \dots + \frac{k_n}{10^n} + \dots .$$

The series (1) is the decimal expression of r, and is usually written as

$$0.k_1k_2 \dots k_n \dots . \tag{2}$$

Thus, $\frac{1}{3} = 0.333 \dots = \frac{3}{10} + \frac{3}{10^2} + \dots + \frac{3}{10^n} + \dots .$

COROLLARY. *Every real number has a decimal expression.*

Proof. Every real number can be expressed as the sum of an integer plus a number r such that $0 \leqq r < 1$.

Example. $\sqrt{2} = 1.4142 \ldots = 1 + \frac{4}{10} + \frac{1}{10^2} + \frac{4}{10^3} + \frac{2}{10^4} + \ldots.$

Note. The decimal expression of a real number is not always unique. Thus

$$\frac{1}{4} = 0.25000 \ldots = \frac{2}{10} + \frac{5}{10^2} + \frac{0}{10^3} + \ldots + \frac{0}{10^n} + \ldots$$

and

$$\frac{1}{4} = 0.24999 \ldots = \frac{2}{10} + \frac{4}{10^2} + \frac{9}{10^3} + \ldots + \frac{9}{10^n} + \ldots.$$

It can be shown that it is unique unless there is a term after which the digits are all zeros or all nines.

180. The Ratio Tests

Let

(1) $$u_1 + u_2 + u_3 + \ldots + u_n + u_{n+1} + \ldots$$

be a series of positive terms. We shall consider only series for which the ratio u_{n+1}/u_n approaches a limit as $n \to \infty$. Let

(2) $$R = \lim_{n \to \infty} \frac{u_{n+1}}{u_n}.$$

THEOREM. *(a) If $R < 1$, the series (1) converges; (b) if $R > 1$, the series (1) diverges; (c) if $R = 1$, the test fails to tell us whether or not the series (1) converges.*

Proof of (a). Consider any number r such that $R < r < 1$. From (2) it follows that there exists a number N such that, for all $n \geqq N$, we have $u_{n+1}/u_n < r$. That is,

$$u_{N+1}/u_N < r, \quad \text{or } u_{N+1} < u_N r,$$
$$u_{N+2}/u_{N+1} < r, \text{ or } u_{N+2} < u_{N+1} r < u_N r^2,$$
$$u_{N+3}/u_{N+2} < r, \text{ or } u_{N+3} < u_N r^3,$$

and so on. Hence the terms of the series

(3) $$u_{N+1} + u_{N+2} + u_{N+3} + \ldots$$

are respectively less than the corresponding terms of the series

(4) $$u_N r + u_N r^2 + u_N r^3 + \ldots.$$

But (4) is a convergent geometric series with $r < 1$. Hence (3) converges by the comparison test, and (1) converges by the note in section 178.

Proof of (*b*). Consider any number r such that $1 < r < R$. By (2), there exists a number N such that for all $n \geqq N$ we have $u_{n+1}/u_n > r > 1$. Hence $u_{n+1} > u_n$ for all $n \geqq N$ and we cannot have $\lim_{n\to\infty} u_n = 0$. Hence by corollary 1, section 177, the series diverges.

Proof of (*c*). Consider the divergent harmonic series $1 + \frac{1}{2} + \ldots + \frac{1}{n} + \ldots$. Here $R = \lim_{n\to\infty} \frac{n}{n+1} = 1$. Consider the convergent p-series with $p = 2$, $1 + \frac{1}{2^2} + \frac{1}{3^2} + \ldots + \frac{1}{n^2} + \ldots$. Here also, $R = \lim_{n\to\infty} \frac{n^2}{(n+1)^2} = \lim_{n\to\infty} \frac{n^2}{n^2 + 2n + 1} = 1$. This proves (*c*).

Example 1. The series $1 + \frac{1}{2!} + \frac{1}{3!} + \ldots + \frac{1}{n!} + \ldots$ converges since $R = \lim_{n\to\infty} \frac{n!}{(n+1)!} = \lim_{n\to\infty} \frac{1}{n+1} = 0 < 1$.

EXERCISES

By means of the ratio tests, decide the convergence or divergence of each of the following series, applying other tests only if the ratio tests fail:

1. $\frac{1}{1 \cdot 2} + \frac{1}{2 \cdot 3} + \frac{1}{3 \cdot 4} + \ldots + \frac{1}{n(n+1)} + \ldots$.

2. $\frac{1}{2} + \frac{1}{2^2} + \frac{1}{2^3} + \ldots + \frac{1}{2^n} + \ldots$.

3. $3 + \frac{9}{2} + \frac{27}{3} + \ldots + \frac{3^n}{n} + \ldots$.

4. $1 + \frac{1}{\sqrt{2}} + \frac{1}{\sqrt{3}} + \ldots + \frac{1}{\sqrt{n}} + \ldots$.

5. $\frac{1}{2} + \frac{2}{4} + \frac{3}{8} + \ldots + \frac{n}{2^n} + \ldots$.

6. $\frac{1 \cdot 2}{2} + \frac{2 \cdot 3}{4} + \frac{3 \cdot 4}{8} + \ldots + \frac{n(n+1)}{2^n} + \ldots$.

7. $1 + \frac{1}{3!} + \frac{1}{5!} + \ldots + \frac{1}{(2n-1)!} + \ldots$.

8. $\frac{1}{2} + \frac{2}{4} + \frac{6}{8} + \ldots + \frac{n!}{2^n} + \ldots$.

9. $\frac{1}{3} + \frac{1 \cdot 3}{3 \cdot 6} + \frac{1 \cdot 3 \cdot 5}{3 \cdot 6 \cdot 9} + \ldots + \frac{1 \cdot 3 \ldots (2n - 1)}{3 \cdot 6 \ldots (3n)} + \ldots .$

10. $3 + \frac{3^2}{2!} + \frac{3^3}{3!} + \ldots + \frac{3^n}{n!} + \ldots .$

11. $2 + \frac{2^2}{2^2} + \frac{2^3}{3^2} + \ldots + \frac{2^n}{n^2} + \ldots .$

12. $\frac{1}{2!} + \frac{1}{4!} + \frac{1}{6!} + \ldots + \frac{1}{(2n)!} + \ldots .$

13. $\frac{1}{2 \cdot 1} + \frac{1}{2 \cdot 4} + \frac{1}{2 \cdot 9} + \ldots + \frac{1}{2n^2} + \ldots .$

14. $\frac{1}{1 \cdot 2} + \frac{1}{3 \cdot 4} + \frac{1}{5 \cdot 6} + \ldots + \frac{1}{(2n - 1) \cdot 2n} + \ldots .$

181. Series Whose Terms Are All Negative

Such a series obviously converges to $-S$ if and only if the series of positive terms formed by reversing all the signs converges to S.

182. Alternating Series

By an **alternating series** is meant a series whose successive terms alternate in sign.

THEOREM. *The alternating series**

$$u_1 - u_2 + u_3 - \ldots + (-1)^{n+1}u_n + \ldots \tag{1}$$

converges if (a) $\lim_{n \to \infty} u_n = 0$, *and (b)* $u_{n+1} < u_n$ *for all* $n = 1, 2, 3, \ldots$.

Proof. If $2m$ is any even positive integer, then the sum S_{2m} of the first $2m$ terms of (1) may be written in two ways:

$$S_{2m} = (u_1 - u_2) + (u_3 - u_4) + \ldots + (u_{2m-1} - u_{2m}); \tag{2}$$

$$S_{2m} = u_1 - (u_2 - u_3) - (u_4 - u_5) - \ldots - u_{2m}. \tag{3}$$

By hypothesis (*b*), each quantity in parentheses in (2) and (3) is positive. Therefore, (2) implies that S_{2m} increases as m increases; and (3) implies that $S_{2m} < u_1$ for all m. By theorem 2, section 176, S_{2m} approaches a limit S as m increases indefinitely.

Let S_{2m+1} be the sum of an odd number of terms of (1). Then

* The terms $u_1, u_2, u_3, \ldots$ are all positive.

$S_{2m+1} = S_{2m} + u_{2m+1}$. Therefore

$$\lim_{m\to\infty} S_{2m+1} = \lim_{m\to\infty} S_{2m} + \lim_{m\to\infty} u_{2m+1} = S + 0 = S$$

by the first paragraph of the proof, and hypothesis (*a*).

Since the partial sums S_n approach the limit S whether n is even or odd, we have $\lim_{n\to\infty} S_n = S$. This completes the proof.

COROLLARY. *Under the hypotheses of the theorem, the difference between a partial sum S_n and the sum S of the series is not greater than the next term u_{n+1} in absolute value. That is,* $|S - S_n| \leqq u_{n+1}$.

Proof. By arguments like those above, we have

$$|S - S_n| = u_{n+1} - (u_{n+2} - u_{n+3}) - (u_{n+4} - u_{n+5}) - \ldots \leqq u_{n+1}.$$

Example. The series $1 - \frac{1}{2} + \frac{1}{3} - \frac{1}{4} + \ldots + (-1)^{n+1}\frac{1}{n} + \ldots$ converges since (*a*) $\lim_{n\to\infty} \frac{1}{n} = 0$ and (*b*) $\frac{1}{n+1} < \frac{1}{n}$ for all $n = 1, 2, 3, \ldots$.

183. Absolute Convergence

A series

$$(1) \qquad u_1 + u_2 + u_3 + \ldots + u_n + \ldots$$

with some positive and some negative terms is said to **converge absolutely** if the series

$$(2) \qquad |u_1| + |u_2| + |u_3| + \ldots + |u_n| + \ldots$$

of the absolute values of its terms converges.

THEOREM. *If a series converges absolutely, it converges. That is, if (2) converges, so does (1).*

Proof. Let $S_n = u_1 + u_2 + \ldots + u_n$. Let P_n be the sum of the positive terms in the right member and let N_n be the sum of the absolute values of the negative terms in the right member. Then

$$(3) \qquad S_n = P_n - N_n.$$

Let $s_n = |u_1| + |u_2| + \ldots + |u_n|$. Then

$$s_n = P_n + N_n. \tag{4}$$

Since (2) converges to a sum s, we have $s_n = P_n + N_n \leqq s$. Therefore, $P_n \leqq s$ and $N_n \leqq s$. By theorem 2, section 176, P_n approaches a limit P and N_n approaches a limit N as n increases indefinitely. Therefore

$$\lim_{n\to\infty} S_n = \lim_{n\to\infty} (P_n - N_n) = P - N.$$

This completes the proof.

A series which converges but does not converge absolutely is called **conditionally convergent.**

Example. The series of the example of section 182 is conditionally convergent since the absolute values of its terms form the divergent harmonic series.

184. The General Ratio Tests

Let $u_1 + u_2 + u_3 + \ldots + u_n + \ldots$ be any series of real terms, and let

$$R = \lim_{n\to\infty} \left|\frac{u_{n+1}}{u_n}\right|.$$

We consider only series for which this limit exists.

THEOREM. *(a) If $R < 1$, the series converges absolutely. (b) If $R > 1$, the series diverges. (c) If $R = 1$, the test fails to decide whether or not the series converges.*

Proof of (a). From $R < 1$ it follows that the series $|u_1| + |u_2| + \ldots + |u_n| + \ldots$ converges. Hence the given series converges absolutely.

The proofs of (b) and (c) are like those of the theorem of section 180, and are left to the reader.

EXERCISES

Test for convergence and, if convergent, state whether absolutely or conditionally convergent:

1. $\frac{1}{2} - \frac{1}{4} + \frac{1}{8} - \ldots + \frac{(-1)^{n+1}}{2^n} + \ldots.$

2. $1 - \frac{1}{2!} + \frac{1}{3!} - \ldots + \frac{(-1)^{n+1}}{n!} + \ldots.$

3. $1 - \frac{1}{3} + \frac{1}{5} - \ldots + \frac{(-1)^{n+1}}{2n-1} + \ldots.$

4. $\frac{1}{1 \cdot 2} - \frac{1}{2 \cdot 3} + \frac{1}{3 \cdot 4} - \ldots + \frac{(-1)^{n+1}}{n(n+1)} + \ldots.$

5. $1 - \frac{1}{\sqrt{2}} + \frac{1}{\sqrt{3}} - \ldots + \frac{(-1)^{n+1}}{\sqrt{n}} + \ldots.$

6. $1 - \frac{1}{3!} + \frac{1}{5!} - \ldots + \frac{(-1)^{n+1}}{(2n-1)!} + \ldots.$

7. $\frac{1}{2} - \frac{2}{3} + \frac{3}{4} - \ldots + \frac{(-1)^{n+1}n}{n+1} + \ldots.$

8. $1 - \frac{1}{2!} + \frac{1}{4!} - \ldots + \frac{(-1)^{n+1}}{(2n-2)!} + \ldots.$

9. $\frac{1}{2} - \frac{1}{3} + \frac{1}{4} - \ldots + \frac{(-1)^{n+1}}{1+n} + \ldots.$

10. $1 - \frac{3}{2!} + \frac{3^2}{3!} - \ldots + \frac{(-1)^{n+1}3^{n-1}}{n!} + \ldots.$

185. Power Series

If $a_0, a_1, a_2, \ldots, a_n, \ldots$ are constants, then

$$a_0 + a_1x + a_2x^2 + \ldots + a_{n-1}x^{n-1} + \ldots, \tag{1}$$

and

$$a_0 + a_1(x-h) + a_2(x-h)^2 + \ldots + a_{n-1}(x-h)^{n-1} + \ldots, \tag{2}$$

where h is a constant, are called **power series** in x or in $x - h$ respectively. The general ratio test may be used to determine values of x for which a power series converges.

Example. Find all (real) value of x for which the series $x - \frac{x^2}{2} + \frac{x^3}{3} - \ldots + \frac{(-1)^{n+1}x^n}{n} + \ldots$ converges.

Solution. Since

$$R = \lim_{n\to\infty} \left|\frac{x^{n+1}}{n+1} \cdot \frac{n}{x^n}\right| = \lim_{n\to\infty} |x| \frac{n}{n+1} = |x| \lim_{n\to\infty} \frac{1}{1+\frac{1}{n}} = |x|,$$

we see that the series converges absolutely when $|x| < 1$, or $-1 < x < 1$, and diverges when $|x| > 1$, or $x > 1$ and $x < -1$. The ratio test fails when $|x| = 1$. Hence we must investigate the convergence of the series when $x = \pm 1$ separately. For $x = 1$, we find the series $1 - \frac{1}{2} + \frac{1}{3} - \ldots + \frac{(-1)^{n+1}}{n} + \ldots$ which we know converges conditionally (see section 182). For $x = -1$, we have the series $-1 - \frac{1}{2} - \frac{1}{3} - \ldots - \frac{1}{n} - \ldots$, which is the negative of the divergent harmonic series and hence diverges (see section 181). Hence the series converges for $-1 < x \leqq 1$, and diverges for all other values.

EXERCISES

Find all (real) values of x for which the series converges:

1. $1 + x + x^2 + \ldots + x^{n-1} + \ldots.$

2. $1 - \frac{x^2}{2!} + \frac{x^4}{4!} - \ldots + \frac{(-1)^{n+1}x^{2n-2}}{(2n-2)!} + \ldots.$

3. $x - \frac{x^3}{3!} + \frac{x^5}{5!} - \ldots + \frac{(-1)^{n+1}x^{2n-1}}{(2n-1)!} + \ldots.$

4. $1 + x + \frac{x^2}{2!} + \ldots + \frac{x^{n-1}}{(n-1)!} + \ldots.$

5. $x - \frac{x^3}{3} + \frac{x^5}{5} - \ldots + \frac{(-1)^{n+1}x^{2n-1}}{2n-1} + \ldots.$

6. $1 - \frac{x}{2} + \frac{x^2}{2^2} - \ldots + \frac{(-1)^{n+1}x^{n-1}}{2^{n-1}} + \ldots.$

7. $2x + 2^2x^2 + 2^3x^3 + \ldots + 2^nx^n + \ldots.$

8. $\frac{x}{1 \cdot 2} - \frac{x^2}{2 \cdot 3} + \frac{x^3}{3 \cdot 4} - \ldots + \frac{(-1)^{n+1}x^n}{n(n+1)} + \ldots.$

9. $x + \frac{x^2}{\sqrt{2}} + \frac{x^3}{\sqrt{3}} + \ldots + \frac{x^n}{\sqrt{n}} + \ldots.$

10. $x + 2x^2 + 3x^3 + \ldots + nx^n + \ldots.$

11. $(x-1) - \frac{(x-1)^2}{2} + \frac{(x-1)^3}{3} - \ldots + \frac{(-1)^{n+1}(x-1)^n}{n} + \ldots.$

12. $\frac{x-2}{1} + \frac{(x-2)^2}{2\sqrt{2}} + \frac{(x-2)^3}{3\sqrt{3}} + \ldots + \frac{(x-2)^n}{n\sqrt{n}} + \ldots.$

13. Show that the binomial series

$$1 + kx + \frac{k(k-1)}{2!}x^2 + \ldots + \frac{k(k-1)\ldots(k-n+1)}{n!}x^n + \ldots,$$

where k is any real number, converges for $-1 < x < 1$ and diverges* for $|x| > 1$.

* The behavior of this series at $x = \pm 1$ depends on the value of k, and requires much more advanced methods of study. A complete investigation of the binomial series when k is any complex number was made by N. H. Abel (Norwegian, 1802–1829), about 1826. See section 104.

186. The Use of Series in Computation

Many useful functions can be expressed as power series. For example, if x is the number of radians* in an angle, it can be shown, by more advanced methods, that

$$(1) \qquad \sin x = x - \frac{x^3}{3!} + \frac{x^5}{5!} - \dots + \frac{(-1)^{n+1}x^{2n-1}}{(2n-1)!} + \dots .$$

Since the series on the right converges for all x, and in particular for $x = 1$, we may compute the value of the sine of 1 radian by taking the first three terms of this series:

$$\begin{aligned} \sin 1 &= 1 - \frac{1}{3!} + \frac{1}{5!} - \frac{1}{7!} + \dots \\ &= 1 - \frac{1}{6} + \frac{1}{120} - \frac{1}{5040} + \dots \\ &= 0.8417 \text{ approximately.} \end{aligned}$$

By the corollary of section 182, the error committed is less than 1/5040, since we used only the first three terms. Since 1/5040 is less than 0.0005, the answer is correct to the nearest thousandth. See also section 165.

Infinite series are much used in the construction of tables and have many other uses in advanced mathematics, both pure and applied.

EXERCISES

1. It can be shown that if x is the number of radians in an angle, then $\cos x = 1 - \dfrac{x^2}{2!} + \dfrac{x^4}{4!} - \dots + \dfrac{(-1)^{n+1}x^{2n-2}}{(2n-2)!} + \dots$. Find the value of the cosine of 1 radian to the nearest thousandth.

2. Using the series (1) above, find to the nearest thousandth the value of the sine of 1/2 radian.

3. Using the series of exercise 1, find to the nearest thousandth the value of the cosine of 0.3 radian.

4. Using the first five terms of the binomial series (exercise 13, section 185), find the value of $(1.02)^{-10}$.

5. The value of e, the base of the system of natural logarithms, can be obtained from the series $1 + \dfrac{1}{1!} + \dfrac{1}{2!} + \dots + \dfrac{1}{(n-1)!} + \dots$. Calculate it using the first six terms, and round off the answer to two decimal places.

* One radian is equal to $180/\pi$ degrees.

6. The value of π can be obtained from the series

$$\pi = 16\left[\frac{1}{5} - \frac{1}{3 \cdot 5^3} + \frac{1}{5 \cdot 5^5} - \frac{1}{7 \cdot 5^7} + - \cdots\right] - 4\left[\frac{1}{239} - \frac{1}{3 \cdot 239^3} + \frac{1}{5 \cdot 239^5} - + \cdots\right].$$

Calculate π, using the first three terms in each bracket, and round off the answer to three decimal places.

Chapter XXVIII

Interpolation and curve fitting

187. Linear Interpolation

Suppose y is some unknown function of x; that is, $y = f(x)$. Suppose further that we know the values of y for certain values of x:

x	x_0	x_1	x_2	...	x_n
$y = f(x)$	y_0	y_1	y_2	...	y_n

The graph of $y = f(x)$ is then drawn through the points (x_0,y_0), (x_1,y_1), ..., (x_n,y_n).

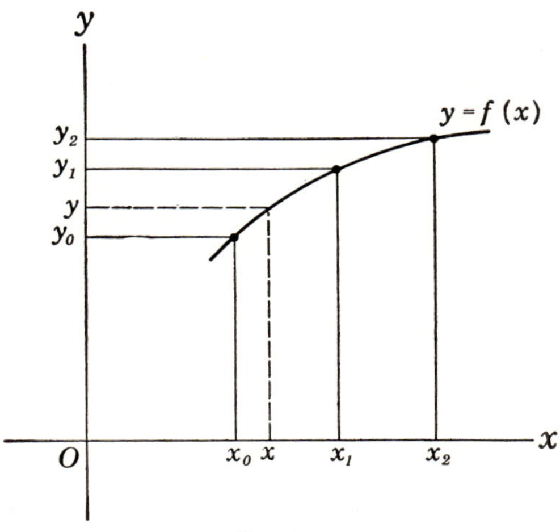

Fig. 67.

The act of estimating the values of y for values of x between those actually known or calculated or plotted is called **interpolation.**

A rough method of interpolation is to read from the graph sketched through the known points (Fig. 67).

The method commonly used in interpolating in tables is called the method of **proportional parts** or **linear interpolation.** It is illustrated in the following example. (See section 161 for further illustrations.)

Example. Suppose we have the following table:

x	1	2	3	4
y	4	6	2	-14

To find the value of y corresponding to $x = 2.75$, we say that 2.75 is 3/4 of the way from 2 to 3, so that the desired value of y will be approximately 3/4 of the way between 6 and 2, or $y = 3$. The work may be arranged as follows:

		x	y		
0.75		2	6		d
	1	2.75		-4	
		3	2		

$$\frac{d}{-4} = \frac{0.75}{1}, \; d = -3, \; y = 6 - 3 = 3.$$

This method amounts to replacing the actual curve, whatever it is, by the straight line joining the points (x_1,y_1) and (x_2,y_2)

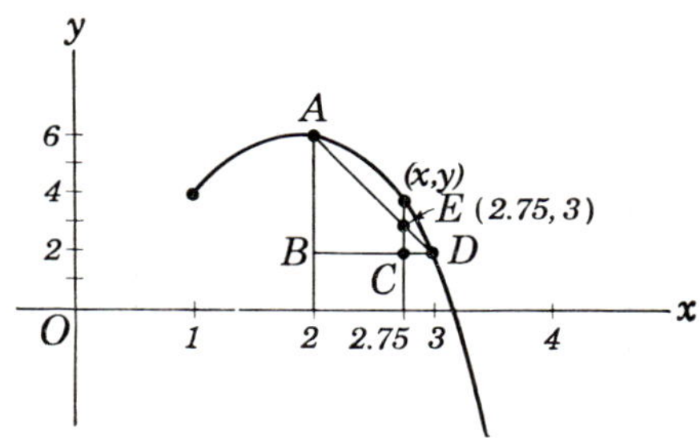

Fig. 68.

between which we are interpolating. In the example above, the straight line joining the points (2,6) and (3,2) is $y = -4x + 14$. When $x = 2.75$, the value of y determined from this line is $y = 3$.

The table of the example above was actually calculated from the function $y = 2 + 3x^2 - x^3$. The actual value of y corresponding to $x = 2.75$ therefore is $y = 3.890625$. The connection between the idea of proportional parts and the idea of approximating the curve by the straight line should be clear from the similar triangles ABD and ECD (Fig. 68), since $\frac{CE}{BA} = \frac{CD}{BD} = \frac{1}{4}$.

The accuracy of this method clearly depends on how closely the chord joining the points (x_1,y_1) and (x_2,y_2) approximates the actual curve. In practice this often depends on how close together are the values of x between which we are interpolating;* interpolating between values of x which are closer together is apt to be more reliable. If the values y_1 and y_2 are taken from a table, they are themselves not always exact since they may be subject to rounding-off errors.

EXERCISES

Using the table of the example above, calculate, by means of interpolation by proportional parts, the value of y corresponding to:

1. $x = 1.5$. **2.** $x = 2.75$. **3.** $x = 3\frac{1}{8}$. **4.** $x = 1.25$.
5. $x = 2.2$. **6.** $x = 2.5$.

7. Given that $\sqrt[3]{2.50} = 1.357$ and $\sqrt[3]{2.60} = 1.375$, find by linear interpolation the value of $\sqrt[3]{2.54}$ to three decimal places.

8. Given that $\sqrt[3]{4.60} = 1.663$ and $\sqrt[3]{4.70} = 1.675$, find by linear interpolation the value of $\sqrt[3]{4.68}$ to three decimal places.

9. From table I, find by linear interpolation the value of $\sqrt[3]{3.925}$ to four decimal places.

10. From table I, find by linear interpolation the value of $\sqrt[3]{66.75}$ to four decimal places.

188. Lagrange's Interpolation Formula

It is obviously plausible that replacing the actual (unknown) curve by a polynomial of degree two is likely to give a better approximation than the linear polynomial used in the preceding section, and that in general a polynomial of higher degree is likely to be better than one of lower degree.

THEOREM. *If $x_0, x_1, x_2, \ldots, x_r$ are $r + 1$ different values of x, and $y_0, y_1, y_2, \ldots, y_r$ are any $r + 1$ values of y, then there exists one and only one polynomial $f(x)$ of degree less than or equal to r*

* That this is not always the case may be seen from the example $y = 1/10x$, by interpolating by proportional parts between the points $(-0.1,-1)$ and $(0.5,0.2)$, obtaining the point $(0.2,-0.4)$, which is far from the true point $(0.2,0.5)$. See M. Richardson, *Fundamentals of Mathematics*, Macmillan, 1941.

such that the curve $y = f(x)$ passes through the $r + 1$ points (x_0,y_0), (x_1,y_1), . . ., (x_r,y_r).

Proof. By direct substitution, it is easily verified that a polynomial of degree r or less passing through the $r + 1$ points (x_0,y_0), (x_1,y_1), . . ., (x_r,y_r), where x_0, x_1, . . ., x_r are all distinct, is given by

$$(1)\quad \begin{cases} y = y_0 \dfrac{(x - x_1)(x - x_2) \dots (x - x_r)}{(x_0 - x_1)(x_0 - x_2) \dots (x_0 - x_r)} \\ \qquad + y_1 \dfrac{(x - x_0)(x - x_2) \dots (x - x_r)}{(x_1 - x_0)(x_1 - x_2) \dots (x_1 - x_r)} \\ \qquad + \dots + y_r \dfrac{(x - x_0)(x - x_1) \dots (x - x_{r-1})}{(x_r - x_0)(x_r - x_1) \dots (x_r - x_{r-1})}. \end{cases}$$

That there cannot be more than one such polynomial follows immediately from corollary 2, section 119, chapter XIX.

For example, the unique polynomial of degree 2 or less passing through the points (x_0,y_0), (x_1,y_1), (x_2,y_2), where x_0, x_1, x_2 are all distinct, is

$$y = y_0 \frac{(x - x_1)(x - x_2)}{(x_0 - x_1)(x_0 - x_2)} + y_1 \frac{(x - x_0)(x - x_2)}{(x_1 - x_0)(x_1 - x_2)} + y_2 \frac{(x - x_0)(x - x_1)}{(x_2 - x_0)(x_2 - x_1)}.$$

Example. The unique polynomial of degree 3 or less passing through the points (1,4), (2,6), (3,2), (4,−14) is

$$y = 4\frac{(x - 2)(x - 3)(x - 4)}{(1 - 2)(1 - 3)(1 - 4)} + 6\frac{(x - 1)(x - 3)(x - 4)}{(2 - 1)(2 - 3)(2 - 4)} + 2\frac{(x - 1)(x - 2)(x - 4)}{(3 - 1)(3 - 2)(3 - 4)} - 14\frac{(x - 1)(x - 2)(x - 3)}{(4 - 1)(4 - 2)(4 - 3)}$$

which reduces to $y = -x^3 + 3x^2 + 2$.

Note. Formula (1) is credited to J. L. Lagrange (French, 1736–1813), who made many remarkable contributions to the progress of mathematics, both pure and applied.

Other methods of determining this polynomial are given in the next two sections.

EXERCISES

1. (*a*) Determine the unique polynomial of degree $\leqq 3$ whose graph passes through the points given in the following table:

x	1	3	4	5
y	−2	14	40	86

(*b*) Use the polynomial obtained in (*a*) to interpolate for the value of y corresponding to $x = 2$.

2. Same questions as exercise 1 for the table

x	1	3	4	5
y	-2	4	13	26

3. Same questions as exercise 1 for the table

x	1	3	4	5
y	-1	3	5	7

4. (*a*) Find the unique polynomial of degree $\leqq 2$ whose graph passes through the points (1,4), (2,4), and (4,76).

(*b*) Use the polynomial obtained in (*a*) to find the values of y corresponding to $x = 3$ and to $x = 1.5$ respectively.

5. (*a*) Find the unique polynomial $f(x)$ of degree $\leqq 3$ such that $f(1) = 0$, $f(2) = -4$, $f(3) = -6$, and $f(5) = -4$.

(*b*) Find the values of $f(4)$ and $f(2.25)$.

6. (*a*) Find the unique polynomial $f(x)$ of degree $\leqq 3$ whose graph passes through the points given in the following table:

x	1	3	4	5
y	7	31	67	127

(*b*) Use the polynomial obtained in (*a*) to find the values of $f(2)$ and $f(3.3)$.

7. (*a*) Find the unique polynomial of the smallest possible degree whose graph passes through the points given in the following table:

x	1	2	3	4	5
y	9	18	73	234	585

(*b*) Use the polynomial obtained in (*a*) to find the value of y corresponding to $x = 2.2$.

8. (*a*) Find the unique polynomial of the smallest possible degree whose graph passes through the points given in the following table:

x	2	4	8
y	-2	22	118

(*b*) Using the polynomial obtained in (*a*), find the values of y corresponding to $x = 3$, $x = 5$, $x = 6$, and $x = 9$, respectively.

189. Method of Undetermined Coefficients

The unique polynomial discussed in the theorem of the preceding section may be determined by the **method of undetermined coefficients,** as in the following example.

Example 1. Find the unique polynomial of degree 3 or less whose graph passes through the four points (1,4), (2,6), (3,2), (4,−14).

Solution. Let $y = ax^3 + bx^2 + cx + d$. Since $x = 1$, $y = 4$ must satisfy this equation, we have

$$4 = a + b + c + d, \tag{1}$$

a linear equation in the coefficients a, b, c, and d. Each of the other three given points similarly yields a linear equation in a, b, c, and d:

$$6 = 8a + 4b + 2c + d, \tag{2}$$

$$2 = 27a + 9b + 3c + d, \tag{3}$$

$$-14 = 64a + 16b + 4c + d. \tag{4}$$

Solving this system of equations (1), (2), (3), and (4), we find $a = -1$, $b = 3$, $c = 0$, $d = 2$. Hence the desired polynomial is $y = -x^3 + 3x^2 + 2$.

The term *method of undetermined coefficients* is applied in many other situations in mathematics when the form of the answer is known but the values of the coefficients remain to be calculated. Compare, for example, chapter XXVI on partial fractions.

EXERCISES

1–8. Work exercises 1–8 of section 188 by means of the method of undetermined coefficients.

9. Determine the coefficients in $y = ax^4 + bx + c$ so that the graph passes through the points (1,−6), (2,7), and (−1,−2).

10. Determine the coefficients in $y = ax^2 + \frac{b}{x}$ so that the graph passes through the points (1,−9) and (3,23).

11. Determine the coefficients in $y = a + \frac{b}{x} + \frac{c}{x^2}$ so that the graph passes through the points (1,−1), (−1,23), and (3,5/3).

12. Determine the coefficients in $y = \frac{a}{x} + b + cx$ so that the graph passes through the points (1,5), (2,−1), and (3,−5).

190. The Gregory-Newton Interpolation Formula

Another method for determining the unique polynomial $y = f(x)$ of degree r or less passing through the $r + 1$ points (x_0,y_0), (x_1,y_1), . . ., (x_r,y_r), where x_0, x_1, . . ., x_r are all distinct, is often convenient.

The student should review sections 151 and 152, chapter XXI, at this point. From formula (6) of section 151, we have, for $k = 0, 1, 2, \ldots, r$,

$$(1) \quad y_k = y_0 + k\Delta y_0 + \frac{k(k-1)}{2!}\Delta^2 y_0 + \frac{k(k-1)(k-2)}{3!}\Delta^3 y_0 + \ldots + \frac{k(k-1)(k-2)\ldots(k-r+1)}{r!}\Delta^r y_0.$$

By theorem 2, section 152, all differences of order higher than the rth are equal to zero since $f(x)$ is a polynomial of degree r or less.

For simplicity, we now restrict ourselves to the case in which the values $x_0, x_1, \ldots, x_r$ are evenly spaced. Let h be the constant difference between the successive values $x_0, x_1, \ldots, x_r$. Thus $x_1 = x_0 + h$, $x_2 = x_0 + 2h$, $\ldots$, $x_r = x_0 + rh$. That is,

$$(2) \quad x_k = x_0 + kh \quad (k = 0, 1, \ldots, r)$$

or

$$(3) \quad k = \frac{x_k - x_0}{h}.$$

Hence

$$(4) \quad \begin{cases} k - 1 = \dfrac{x_k - x_0 - h}{h} \\ k - 2 = \dfrac{x_k - x_0 - 2h}{h} \\ \cdot \\ \cdot \\ \cdot \\ k - r + 1 = \dfrac{x_k - x_0 - [r-1]h}{h}. \end{cases}$$

Substituting (3) and (4) in (1), we get

$$(5) \quad y_k = y_0 + \frac{x_k - x_0}{h}\Delta y_0 + \frac{\left(\frac{x_k - x_0}{h}\right)\left(\frac{x_k - x_0 - h}{h}\right)}{2!}\Delta^2 y_0 + \frac{\left(\frac{x_k - x_0}{h}\right)\left(\frac{x_k - x_0 - h}{h}\right)\left(\frac{x_k - x_0 - 2h}{h}\right)}{3!}\Delta^3 y_0 + \ldots + \frac{\left(\frac{x_k-x_0}{h}\right)\left(\frac{x_k-x_0-h}{h}\right)\left(\frac{x_k-x_0-2h}{h}\right)\ldots\left(\frac{x_k-x_0-[r-1]h}{h}\right)}{r!}\Delta^r y_0.$$

Replacing x_k by x and simplifying, we find that the polynomial

$$(6)\quad y = f(x) = y_0 + \frac{x - x_0}{h}\Delta y_0 + \frac{(x - x_0)(x - x_0 - h)}{2!h^2}\Delta^2 y_0 + \frac{(x - x_0)(x - x_0 - h)(x - x_0 - 2h)}{3!h^3}\Delta^3 y_0 + \dots + \frac{(x - x_0)(x - x_0 - h)(x - x_0 - 2h)\dots(x - x_0 - [r-1]h)}{r!h^r}\Delta^r y_0$$

satisfies the requirements of the problem. It is of degree r or less, and for $k = 0, 1, \dots, r$ we have $y = f(x_k) = y_k$.

Example 1. Find the unique polynomial of degree 3 or less passing through the points (1,4), (2,6), (3,2), (4,−14).

Solution. We construct the successive differences of y_0, y_1, y_2, y_3:

y_k:	4,	6,	2,	−14
Δy_k:	2,	−4,	−16	
$\Delta^2 y_k$:	−6,	−12		
$\Delta^3 y_k$:	−6.			

Here $y_0 = 4$, $\Delta y_0 = 2$, $\Delta^2 y_0 = -6$, $\Delta^3 y_0 = -6$, $x_0 = 1$, and $h = 1$. Substituting in (6) we get

$$y = 4 + \frac{x-1}{1}\cdot 2 + \frac{(x-1)(x-2)}{2!1^2}(-6) + \frac{(x-1)(x-2)(x-3)}{3!1^3}(-6),$$

which reduces to $y = -x^3 + 3x^2 + 2$.

The formula (6) is often called Newton's formula, after Isaac Newton (1642–1727) although it was discovered by James Gregory (1638–1675).

Example 2. By the Gregory-Newton interpolation formula find an approximate value of $\sqrt[3]{28.8}$ given that $\sqrt[3]{27.0} = 3.000$, $\sqrt[3]{28.0} = 3.037$, $\sqrt[3]{29.0} = 3.072$, $\sqrt[3]{30.0} = 3.107$, approximately.

Solution. We form the table of differences:

y_k:	3.000	3.037	3.072	3.107
Δy_k:	0.037	0.035	0.035	
$\Delta^2 y_k$:	−0.002	0.000		
$\Delta^3 y_k$:	0.002.			

Here $h = 1$, $x_0 = 27$, $y_0 = 3$, $\Delta y_0 = 0.037$, $\Delta^2 y_0 = -0.002$, $\Delta^3 y_0 = 0.002$, and $x = 28.8$. From formula (6) we have

$$(1)\quad y = 3 + \frac{1.8}{1}(0.037) + \frac{(1.8)(0.8)}{2!1^2}(-0.002) + \frac{(1.8)(0.8)(-0.2)}{3!1^3}(0.002)$$

$$= 3 + 0.0666 - 0.00144 - 0.000096 = 3.065064$$

approximately. This answer is correct to the nearest thousandth. We have essentially approximated part of the curve $y = \sqrt[3]{x}$ by a cubic polynomial. If we had used linear interpolation between the values 3.037 and 3.072 we would have obtained 3.065. Thus in this example we have not gained in accuracy by increasing the degree of the approximating curve. This might have been foreseen from the fact that the first differences were nearly equal, suggesting that a first-degree polynomial would be almost a perfect fit. In general, however, one may expect better accuracy from higher degree curves. The problem of estimating the accuracy to be expected lies beyond the scope of this book.

EXERCISES

Use the Gregory-Newton interpolation formula in each of the following exercises:

1. (*a*) Find the unique polynomial of lowest degree whose graph passes through the points given in the following table:

x	2	3	4	5	6
y	−3	4	15	30	49

(*b*) Use the polynomial obtained in (*a*) to find the values of y corresponding to $x = 4.2$, and to $x = 7$.

2. (*a*) Find the unique polynomial of lowest degree whose graph passes through the points given in the following table:

x	1.5	2.1	2.7	3.3
y	1.0	3.4	5.8	8.2

(*b*) Use the polynomial obtained in (*a*) to find the values of y corresponding to $x = 2.4$ and $x = 4.0$.

3. (*a*) Find the unique polynomial of lowest degree whose graph passes through the points given in the table:

x	1	2	3	4	5	6
y	2	7	56	209	550	1187

(*b*) Use the polynomial obtained in (*a*) to find the value of y corresponding to $x = 2.2$.

4. (*a*) Find the unique polynomial of lowest degree whose graph passes through the points given in the table:

x	1	3	5	7
y	12	10	-40	-186

(*b*) Use the polynomial obtained in (*a*) to find the values of y corresponding to $x = 4$ and $x = 0.5$.

5. (*a*) Find the unique polynomial of lowest degree whose graph passes through the points given in the table:

x	1	3	5	7	9
y	6	-4	18	120	350

(*b*) Use the polynomial obtained in (*a*) to find the values of y corresponding to $x = 4$, and $x = 10$.

6. Find the unique polynomial $f(x)$ of lowest degree such that $f(1) = 12, f(2.5) = 16.5, f(4) = 12, f(5.5) = -1.5$.

7. Find the unique polynomial $f(x)$ of lowest degree such that $f(0.8) = 0.788, f(1.2) = 0.508, f(1.6) = 0.612, f(2) = 1.1$, and use it to find $f(1)$ and $f(1.4)$.

8. Work exercise 7 of section 188 by means of the Gregory-Newton formula.

9. Given that $22^3 = 10648.0$, $23^3 = 12167.0$, $24^3 = 13824.0$, approximate 23.3^3 by the method of example 2, above.

10. Given that $\sqrt[3]{670} = 8.750$, $\sqrt[3]{680} = 8.794$, $\sqrt[3]{690} = 8.837$, approximate $\sqrt[3]{684}$ by the method of example 2, above.

191. Curve Fitting and Empirical Equations

Suppose numerous observations are tabulated which relate two variables x and y:

x	x_0	x_1	x_2	...	x_m
y	y_0	y_1	y_2	...	y_m

Plotting the $m + 1$ points (x_0,y_0), (x_1,y_1), . . ., (x_m,y_m), we obtain a so-called scatter diagram (Fig. 69). To be sure, if the values x_0, x_1, . . ., x_m are all different, we can pass a curve $y = f(x)$, where $f(x)$ is a polynomial of degree not greater than m, through the $m + 1$ points *exactly*. But if our purpose is practical, we desire the simplest possible equation $y = f(x)$. Furthermore, in practice, the coordinates of the $m + 1$ given points are obtained by some sort of measurement and may therefore be assumed to be somewhat in error.* Hence we would prefer a simple equation which *nearly* goes through the given points to a complicated equation which goes through them *exactly*.

The pattern of points on the scatter diagram may suggest the form of a simple equation, such as (*a*) a linear function $y = ax + b$ or straight line, (*b*) a quadratic function $y = ax^2 + bx + c$ or

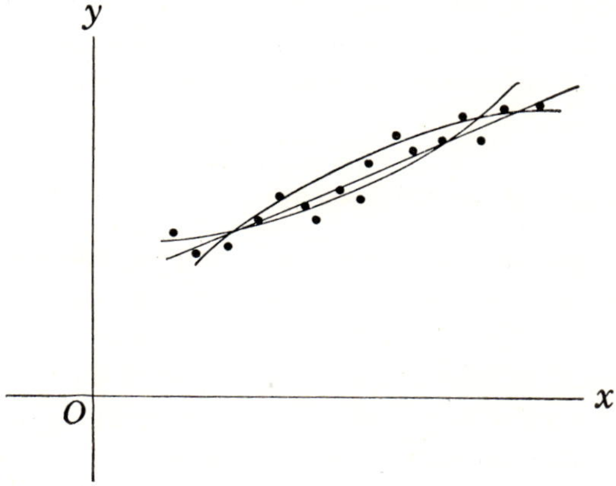

Fig. 69.

parabola, (*c*) direct variation as some power, as $y = cx^n$, (*d*) a polynomial of some definite degree, or (*e*) an exponential curve $y = cb^{ax}$. Of course, there may be doubt as to the appropriate form (see Fig. 69), and, of course, other forms besides those treated here are possible.†

If the form of the equation is decided upon, either for theoretical reasons or by intelligent guesswork, the next step is to determine the values of the coefficients in the equation so as to fit the given points well. Methods for doing this will be taken up later. If a straight line seems appropriate, a crude but quick

* It is actually assumed that any errors in the x's are negligible, while the y's are subject to error. This is a situation often attained in experimental practice, and the subtleties involved may well be passed over here.

† For further study along these lines, see Whittaker and Robinson, *Calculus of Observations*, Blackie and Sons, 1932; D. R. Hartree, *Numerical Analysis*, Oxford, 1952; F. B. Hildebrand, *Introduction to Numerical Analysis*, McGraw-Hill, 1956; A. S. Householder, *Principles of Numerical Analysis*, McGraw-Hill, 1953; Z. Kopal, *Numerical Analysis*, Wiley, 1955; W. E. Milne, *Numerical Calculus*, Princeton University Press, 1949; K. L. Nielsen, *Methods in Numerical Analysis*, Macmillan, 1956; Scarborough, *Numerical Mathematical Analysis*, Johns Hopkins Press, 1930; and textbooks on mathematical statistics.

method is to draw a line with a ruler so that it passes close to most of the points, and find approximate values of a and b from the graph itself. The process of forming an equation to fit empirical data approximately is called **curve fitting** and the resulting equation is often called an **empirical equation.** In practice the resulting equation may be used for interpolating, or for suggesting laws connecting the variables, or for prediction of further corresponding values of x and y, or for the determination of the values of unknown constants in laws whose form is theoretically known.

192. The Method of Differences

A convenient method of curve fitting uses differences and the Gregory-Newton formula. Suppose the given points (x_0,y_0), (x_1,y_1), $\ldots$, (x_m,y_m) are such that the values of x_0, x_1, $\ldots$, x_m are equally spaced.* If these points lay exactly on the graph of an equation $y = f(x)$ where $f(x)$ is a polynomial of degree r, then the differences of rth order, $\Delta^r y_0$, $\Delta^r y_1$, $\Delta^r y_2$, $\ldots$, would be exactly equal, and conversely (see theorems 1 and 2, section 152, chapter XXI). Therefore it is plausible that *if the differences of rth order, $\Delta^r y_0$, $\Delta^r y_1$, $\Delta^r y_2$, $\ldots$, are nearly equal, then the given points will nearly lie on the graph of an equation $y = f(x)$ where $f(x)$ is a polynomial of degree r. Hence construction of a table of differences will reveal whether the given points may be approximately fitted by a curve $y = f(x)$ where $f(x)$ is a polynomial of degree r.* In fact, the Gregory-Newton formula may be used to determine the desired polynomial, as in the following examples.

Example 1. Fit a curve to the points given by the table:

x	1	2	3	4	5	6	7
y	2.9	3.5	4.1	4.4	5.0	5.6	5.9

Solution. We obtain the table of differences:

y_k:	2.9	3.5	4.1	4.4	5.0	5.6	5.9
Δy_k:	0.6	0.6	0.3	0.6	0.6	0.3.	

The first differences are nearly equal. This suggests approximating the given points by a straight line or linear function. Plotting the

* Modifications of the Gregory-Newton formula which do not require this restriction can be found, for example, in Whittaker and Robinson, *Calculus of Observations,* Blackie and Sons, 1932.

seven given points on a graph bears out this conclusion. Hence we use the first two terms of formula (6), section 190,

$$y = y_0 + \frac{x - x_0}{h} \Delta y_0$$

with $y_0 = 2.9$, $x_0 = 1$, $h = 1$, $\Delta y_0 = 0.6$. We obtain

$$y = 2.9 + (x - 1)(0.6)$$

or

$$y = 0.6x + 2.3. \tag{1}$$

The student should plot this line and the given points on a graph.

Example 2. A projectile is thrown vertically upward from an initial height of 5 feet. Its height y, in feet, at the end of x seconds is observed for $x = 0, 1, \ldots, 5$, as in the following table:

x	0	1	2	3	4	5
y	5	115	193	240	255	238

Find an approximate formula expressing y in terms of x.

Solution. We obtain the difference table:

y:	5	115	193	240	255	238
Δy:	110	78	47	15	−17	
$\Delta^2 y$:	−32	−31	−32	−32.		

Since the differences of second order are nearly equal, we expect a quadratic polynomial to be a good fit. Hence we use the first three terms of formula (6), section 190,

$$y = y_0 + \frac{x - x_0}{h} \Delta y_0 + \frac{(x - x_0)(x - x_0 - h)}{2!h^2} \Delta^2 y_0$$

with $y_0 = 5$, $h = 1$, $x_0 = 0$, $\Delta y_0 = 110$, $\Delta^2 y_0 = -32$, obtaining

$$y = 5 + 110x + \frac{x(x - 1)}{2}(-32), \text{ or}$$

$$y = 5 + 126x - 16x^2. \tag{2}$$

EXERCISES

Use the method of differences to verify the appropriateness of the suggested degree of a polynomial fitting the tabulated data well, and to find such a polynomial; plot the graph of the calculated polynomial and the given points:

1. Fit a linear polynomial to the following data:

x	1	2	3	4	5	6
y	1.5	2.0	2.4	2.9	3.4	4.0

2. Fit a quadratic polynomial to the following data:

x	1	2	3	4	5	6
y	0.7	0.1	1.8	5.1	10.8	18.2

3. Fit a quadratic polynomial to the following data:

x	1	3	5	7	9	11
y	4.0	8.1	19.8	40.2	67.7	104.0

4. Fit a quadratic polynomial to the following data:

x	1	2	3	4	5	6
y	−19.7	−34.0	−54.6	−51.0	−53.2	−51.7

5. A thin wire is stretched by means of weights. Its length y, in inches, is observed for various weights x, in pounds, as in the following table:

x	1	$1\frac{1}{2}$	2	$2\frac{1}{2}$	3	$3\frac{1}{2}$	4
y	36.00	36.06	36.13	36.18	36.25	36.30	36.36

(*a*) Fit a straight line to the above data.

(*b*) What would the predicted length be for $x = 5$ pounds?

(*c*) Would it be safe to use the formula found in (*a*) to predict the length of the wire for $x = 1000$ pounds?

6. The cost y cents per item of manufacturing a certain article varies with the number x of hundreds produced as in the following table:

x	1	3	5	7
y	50.0	48.9	47.9	47.0

(*a*) Fit a linear polynomial $y = ax + b$ to the above data.

(*b*) Use the linear polynomial calculated in (*a*) to predict the value of y corresponding to $x = 10$.

(*c*) Would it be safe to use the polynomial obtained in (*a*) to predict the value of y corresponding to $x = 150$?

7. The height y, in feet, of a projectile fired vertically upward at the end of t seconds is given by the following table:

t	0	2	4	6	8	10
y	10	350	560	630	590	410

(*a*) Fit a quadratic polynomial $y = a + bt + ct^2$ to the above data.

(*b*) Use the polynomial calculated in (*a*) to predict the values of y corresponding to $t = 3$ and $t = 11$ respectively.

8. The relation between the number x of milligrams of maltose and its reducing effect on Fehling's solution as measured by the number y of milliliters of $0.1N$ sodium thiosulfate solution is given by the following table:*

x	10	20	30	40	50	60	70	80
y	1.89	3.63	5.33	6.95	8.90	10.48	12.25	13.80

(*a*) Fit a quadratic polynomial to the above data.

(*b*) Use the formula found in (*a*) to predict the value of y corresponding to $x = 100$.

9. The lift coefficient C_L of a certain airfoil section is related to the number α of degrees in the angle of attack (or angle of incidence) by the following table:

α	0	2	4	6
C_L	0.385	0.531	0.677	0.823

(*a*) Fit a linear polynomial $C_L = A\alpha + B$ to the above data.

(*b*) Use the polynomial calculated in (*a*) to predict the value of C_L corresponding to $\alpha = 8°$.

(*c*) Would it be safe to use the polynomial obtained in (*a*) to predict the value of C_L corresponding to $\alpha = 80°$?

10. Fit a cubic polynomial to the following data:

x	1	2	3	4	5	6
y	1.1	7.2	24.9	60.8	121.1	211.8

193. The Method of Averages

We shall explain this method of curve fitting by means of examples.

Example 1. As in example 1, section 192, we wish to fit a curve to the points

x	1	2	3	4	5	6	7
y	2.9	3.5	4.1	4.4	5.0	5.6	5.9

Solution. (*a*) By plotting the points (or else by examining a table of differences), we decide that a linear polynomial will suffice.

(*b*) Hence we must determine a and b so that

$$y = ax + b \tag{1}$$

* The author is indebted to Professor L. Sattler of the Department of Chemistry of Brooklyn College for supplying the data for this problem and for exercise 10, section 193.

will fit the given points well. Substituting the given values of x and y in (1), we get seven equations of first degree in a and b:

$$(2)\qquad \begin{cases} 2.9 = a + b \\ 3.5 = 2a + b \\ 4.1 = 3a + b \\ 4.4 = 4a + b \\ 5.0 = 5a + b \\ 5.6 = 6a + b \\ 5.9 = 7a + b. \end{cases}$$

In general, a system of more than two linear equations in two unknowns is inconsistent,* and we can hope to satisfy it at best only approximately, which is all we expect. Since there are two constants, a and b, to be determined, we divide the seven equations into two nearly equal groups, say the first four and the last three. Adding the equations in each group we get

$$(3)\qquad \begin{cases} 14.9 = 10a + 4b \\ 16.5 = 18a + 3b, \end{cases}$$

the first equation in (3) being the sum of the first four equations of (2), and the second equation of (3) being the sum of the last three equations of (2). Solving the system (3) for a and b and rounding off the solutions to the nearest tenth (since the data are given to the nearest tenth), we find $a = 0.5$ and $b = 2.5$, approximately. Hence

$$(4)\qquad y = 0.5x + 2.5$$

is an approximate fit. See Fig. 70. This answer differs only slightly from the solution obtained in section 192.

Example 2. A projectile is thrown vertically upward from an initial height of 5 feet. Its height y, in feet, at the end of x seconds is observed for $x = 0, 1, \ldots, 5$ as in the following table:

x	0	1	2	3	4	5
y	5	115	193	240	255	238

Find an approximate formula for y in terms of x.

Solution. (a) By plotting the given points (or else by examining a table of differences) we decide that a quadratic function, or parabola, will fit well.

* See section 144.

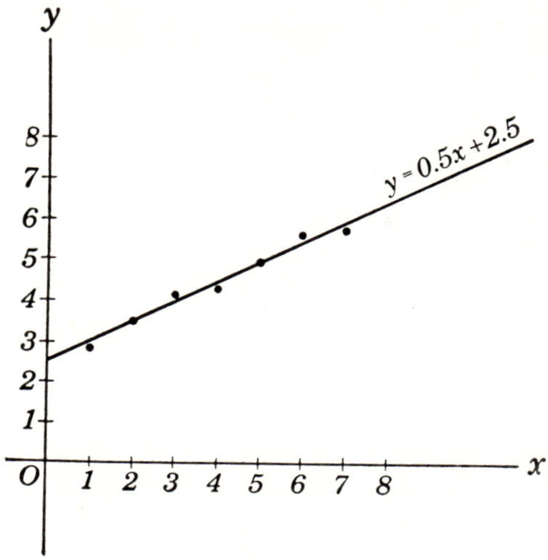

Fig. 70.

(*b*) Hence we must determine a, b, and c so that

$$y = a + bx + cx^2 \tag{5}$$

will fit the given six points approximately. Substituting the given values of x and y in (5), we get six equations of first degree in a, b, and c:

$$\begin{cases} 5 = a \\ 115 = a + b + c \\ 193 = a + 2b + 4c \\ 240 = a + 3b + 9c \\ 255 = a + 4b + 16c \\ 238 = a + 5b + 25c. \end{cases} \tag{6}$$

Since the system (6) has more equations than unknowns, it is likely to be inconsistent, but we may expect to satisfy it approximately. Since there are three constants, a, b, and c, to be determined, we divide the six equations into three groups, say the first two, the middle two, and the last two. Adding the equations in each group, we get

$$\begin{cases} 120 = 2a + b + c \\ 433 = 2a + 5b + 13c \\ 493 = 2a + 9b + 41c. \end{cases} \tag{7}$$

Solving the system (7), and rounding off the solutions to the nearest unit (since the data are given to the nearest unit), we have $a = 5$, $b = 126$, $c = -16$. Hence

$$y = 5 + 126x - 16x^2. \tag{8}$$

This answer is in accord with the result of example 2, section 192. Such exact agreement between the results of two different methods of curve fitting is not to be expected in every case.

Note 1. In this method it is not necessary that the values of x be evenly spaced as it was in section 192.

Note 2. In every case the problem reduces to solving approximately a system of linear equations with more equations than unknowns. Consider the system (2), to be specific. The difference between the right member and the left member of each equation may be called a **residual,** or **error,** or **deviation.** Denoting the residuals of the seven equations of (2) by $e_1, e_2, \ldots, e_7$, respectively, we have

$$\begin{aligned} e_1 &= a + b - 2.9 & e_5 &= 5a + b - 5.0 \\ e_2 &= 2a + b - 3.5 & e_6 &= 6a + b - 5.6 \\ e_3 &= 3a + b - 4.1 & e_7 &= 7a + b - 5.9. \\ e_4 &= 4a + b - 4.4 \end{aligned}$$

To demand that every residual be zero is unreasonable, since setting $e_1 = e_2 = \ldots = e_7 = 0$ yields a system of equations which is likely to be inconsistent. To demand that the sum (or average) of all the residuals be zero would not serve to determine a and b, since it would lead to only one equation connecting these two unknowns. Our method consists in demanding merely that

$$e_1 + e_2 + e_3 + e_4 = 0 \text{ and } e_5 + e_6 + e_7 = 0,$$

that is, that the averages

$$\frac{e_1 + e_2 + e_3 + e_4}{4} \text{ and } \frac{e_5 + e_6 + e_7}{3}$$

of the first four residuals and of the last three residuals respectively be zero. Similar remarks may be made for example 2, and hold in general. This accounts for the name "method of averages."

EXERCISES

By the method of averages, fit a polynomial of the indicated degree to the data; plot the graph of the calculated polynomial and the given points:

1–9. Exercises 1–9 of section 192.

10. A mixture of equal weights of dextrose and levulose when analyzed* yields the following equivalence:

x (mgs. sugar)	5	20	35	50	70	90
y (ml. 0.1N sodium thiosulfate)	1.59	8.00	14.52	18.89	22.69	25.22

Fit a quadratic polynomial $y = a + bx + cx^2$ to these data.

194. The Method of Least Squares

We shall explain this method by means of examples.

Example 1. As in example 1, sections 192 and 193, we wish to fit a curve to the points

x	1	2	3	4	5	6	7
y	2.9	3.5	4.1	4.4	5.0	5.6	5.9

Solution. (*a*) By plotting the points (or else by examining a table of differences), we decide that a linear function will suffice.

(*b*) Hence we must determine a and b so that

$$y = ax + b \tag{1}$$

will fit the given points well. Substituting the given values of x and y in (1), we get seven linear equations in a and b:

$$\left\{\begin{aligned} 2.9 &= a + b \\ 3.5 &= 2a + b \\ 4.1 &= 3a + b \\ 4.4 &= 4a + b \\ 5.0 &= 5a + b \\ 5.6 &= 6a + b \\ 5.9 &= 7a + b. \end{aligned}\right. \tag{2}$$

Multiplying each equation of (2) by the coefficient of a in it and adding the resulting equations, we get what is called the **normal equation corresponding to a;** multiplying each equation of (2)

* See footnote, page 469.

by the coefficient of b in it and adding the resulting equations, we get what is called the **normal equation corresponding to *b*.**

Thus,

$$\begin{array}{rcrl}
2.9 &=& a &+ \ \ b \\
7.0 &=& 4a &+ \ 2b \\
12.3 &=& 9a &+ \ 3b \\
17.6 &=& 16a &+ \ 4b \\
25.0 &=& 25a &+ \ 5b \\
33.6 &=& 36a &+ \ 6b \\
41.3 &=& 49a &+ \ 7b \\
\hline
\end{array}$$

(3) $139.7 = 140a + 28b$ (the normal equation corresponding to a).

Similarly, the normal equation corresponding to b is

(4) $$31.4 = 28a + 7b.$$

Solving the system consisting of the normal equations (3) and (4), and rounding off the answers to the nearest tenth (since the data are given to the nearest tenth), we get $a = 0.5$ and $b = 2.5$. Hence

(5) $$y = 0.5x + 2.5.$$

This is in accord with the result of example 1, section 193. (See Fig. 70, page 471.) Such exact agreement between the results of two different methods of curve fitting is not to be expected in every case.

Example 2. A projectile is thrown vertically upward from an initial height of 5 feet. Its height y, in feet, at the end of x seconds is observed for $x = 0, 1, \ldots, 5$ as in the following table:

x	0	1	2	3	4	5
y	5	115	193	240	255	238

Find an approximate formula for y in terms of x.

Solution. (*a*) By plotting the given points (or else by examining a table of differences), we decide that a quadratic function, or parabola, will fit well.

(*b*) Hence we must determine a, b, and c so that

(6) $$y = a + bx + cx^2$$

will fit the given six points approximately. Substituting the given values of x and y in (6), we get six linear equations in a, b, and c:

$$(7)\quad \begin{cases} 5 = a \\ 115 = a + b + c \\ 193 = a + 2b + 4c \\ 240 = a + 3b + 9c \\ 255 = a + 4b + 16c \\ 238 = a + 5b + 25c. \end{cases}$$

Since the system (7) has more equations than unknowns, it is likely to be inconsistent, but we may expect to satisfy it approximately.

Multiplying each equation of (7) by the coefficient of a in it, and adding the resulting equations, we get the **normal equation corresponding to a**; repeating this process with the coefficients of b and c, in turn, we get the normal equations corresponding to b and c, respectively. Thus the normal equations corresponding to a, b, and c are respectively

$$(8)\quad \begin{cases} 1046 = 6a + 15b + 55c \\ 3431 = 15a + 55b + 225c \\ 13077 = 55a + 225b + 979c. \end{cases}$$

Solving the system of normal equations (8), and rounding off the answers to the nearest unit (because the data are given to the nearest unit), we get $a = 5$, $b = 126$, $c = -16$. Hence

$$(9)\quad y = 5 + 126x - 16x^2.$$

This is in accord with the result of example 2, sections 192 and 193. Such exact agreement between the results of different methods of curve fitting is not to be expected in every case.

Note 1. In this method it is not necessary that the values of x be evenly spaced as it was in section 192.

Note 2. The problem of fitting a polynomial (or even certain other types of functions) reduces to solving approximately a system of linear equations with more equations than unknowns. To be specific, consider the system of three equations in two unknowns a and b:

$$(10)\quad \begin{aligned} m_1a + n_1b &= p_1 \\ m_2a + n_2b &= p_2 \\ m_3a + n_3b &= p_3. \end{aligned}$$

Let the **residuals,** or **errors,** or **deviations,** e_1, e_2, e_3, be defined by

$$(11) \qquad \begin{aligned} e_1 &= m_1a + n_1b - p_1, \; e_2 = m_2a + n_2b - p_2, \\ e_3 &= m_3a + n_3b - p_3. \end{aligned}$$

To demand that every residual be zero is unreasonable, since setting $e_1 = e_2 = e_3 = 0$ yields the system (10) which is likely to be inconsistent. We shall prove that *our method of solving the normal equations amounts to requiring that the sum of the squares of the residuals* $e_1^2 + e_2^2 + e_3^2$ *shall be a minimum.* This accounts for the name "least squares."

Proof. From (11) we find that

$$\begin{aligned} e_1^2 &= m_1^2a^2 + n_1^2b^2 + p_1^2 + 2m_1n_1ab - 2m_1p_1a - 2n_1p_1b \\ e_2^2 &= m_2^2a^2 + n_2^2b^2 + p_2^2 + 2m_2n_2ab - 2m_2p_2a - 2n_2p_2b \\ e_3^2 &= m_3^2a^2 + n_3^2b^2 + p_3^2 + 2m_3n_3ab - 2m_3p_3a - 2n_3p_3b. \end{aligned}$$

Adding these three equations, we find that the sum of the squares of the residuals is given by

$$(12) \qquad \begin{aligned} e_1^2 + e_2^2 + e_3^2 &= (m_1^2 + m_2^2 + m_3^2)a^2 \\ &+ (n_1^2 + n_2^2 + n_3^2)b^2 + (p_1^2 + p_2^2 + p_3^2) \\ &+ 2(m_1n_1 + m_2n_2 + m_3n_3)ab \\ &- 2(m_1p_1 + m_2p_2 + m_3p_3)a - 2(n_1p_1 + n_2p_2 + n_3p_3)b. \end{aligned}$$

The right member of (12) is a quadratic function of a and b which we may denote by $f(a,b)$.

We wish to determine a and b so that $f(a,b)$ will be minimum. Consider b to be a constant for a moment. Then $f(a,b)$ may be thought of as a quadratic function of a alone. By exercise 21, section 83, chapter XI, we get the minimum when a is equal to minus the coefficient of a divided by twice the coefficient of a^2. Therefore we must have, from (12),

$$a = -\frac{2(m_1n_1 + m_2n_2 + m_3n_3)b - 2(m_1p_1 + m_2p_2 + m_3p_3)}{2(m_1^2 + m_2^2 + m_3^2)}$$

or

$$(13) \qquad (m_1^2 + m_2^2 + m_3^2)a + (m_1n_1 + m_2n_2 + m_3n_3)b = m_1p_1 + m_2p_2 + m_3p_3.$$

But this is exactly the normal equation corresponding to a, derived from the equations (10).

Similarly, regarding a as constant and $f(a,b)$ as a quadratic function of b alone, we get

$$(14)\quad (m_1n_1 + m_2n_2 + m_3n_3)a + (n_1^2 + n_2^2 + n_3^2)b = n_1p_1 + n_2p_2 + n_3p_3$$

which is the normal equation corresponding to b, derived from the equations (10). Hence, in order for (12) to be minimum it is necessary that a and b satisfy the normal equations (13) and (14). To investigate whether this is also sufficient lies beyond the scope of this book.

Note 3. The term "best-fitting curve" is frequently used for the one obtained by the method of least squares. This usage arises from the fact that the criterion that the sum of the squares of the errors be minimum is closely related to other theoretical considerations occurring in the more advanced theory of probability and statistics.

EXERCISES

By the method of least squares, fit a polynomial of the indicated degree to the data; plot the graph of the calculated polynomial and the given points:

1–9. Exercises 1–9, section 192.

10. Exercise 10, section 193.

11. Fit a curve of the form $y = ax + \frac{b}{x}$ to the following data:

x	1	2	4	5	10
y	12.1	9.0	10.5	12.0	20.9

195. Exponential and Power Functions

Many laws of physics, chemistry, biology, etc., are of the **exponential form**

$$(1)\qquad y = ab^x$$

or of the **power form**

$$(2)\qquad y = ax^b.$$

(*a*) If the scatter diagram or any other information suggests that the exponential function (1) will fit, we proceed as follows.

From (1), we get

$$(3)\qquad \log y = \log a + x \log b.$$

Setting

$$(4) \qquad Y = \log y, \; A = \log a, \; B = \log b,$$

we get

$$(5) \qquad Y = A + Bx.$$

Hence we may fit a straight line to the points $(x_1, Y_1 = \log y_1)$, $(x_2, Y_2 = \log y_2)$, $\ldots$, $(x_m, Y_m = \log y_m)$ by any of the preceding methods. From the values of A and B thus found we calculate a and b from (4).

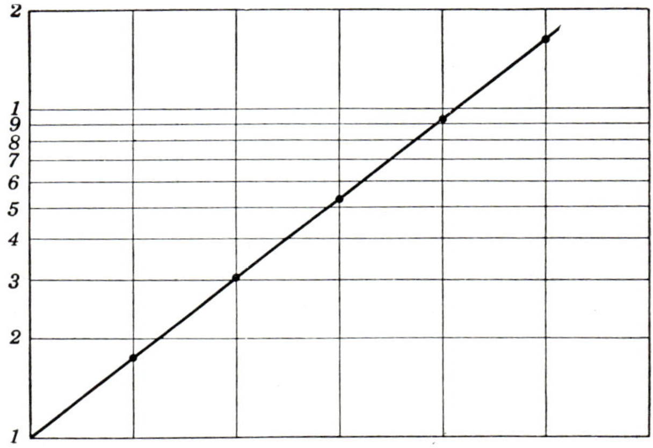

Fig. 71.

Note 1. Semi-logarithmic graph paper is helpful. Its use will remove the necessity of consulting a table of logarithms in order to plot equation (3) or (5). Furthermore, if the data can be plotted on it, and if the points (x_1,Y_1), (x_2,Y_2), $\ldots$, (x_m,Y_m) lie nearly in a straight line, the exponential form (1) is desirable.

Example 1. Fit a curve of the exponential form $y = ab^x$ to the following data:

x	0	1	2	3	4	5
y	100	175	305	535	940	1640

Solution. Let $Y = \log y$. Then, we have

x	0	1	2	3	4	5
Y	2	2.2430	2.4843	2.7284	2.9731	3.2148

By means of a graph (Fig. 71) of Y against x (or by calculating the first differences of the Y's), we see that a straight line will fit.

By the method of differences, we get $Y = 2 + 0.2430x$. From (4) we get $a = 100$, $b = 1.75$. Hence $y = 100(1.75)^x$.

(*b*) If the scatter diagram, or other information, suggests that the power function (2) will fit, we proceed as follows. From (2), we get

$$\log y = \log a + b \log x. \tag{6}$$

Let

$$Y = \log y, \; A = \log a, \; X = \log x. \tag{7}$$

Then

$$Y = A + bX. \tag{8}$$

A straight line may be fitted to the points $(X_1 = \log x_1, Y_1 = \log y_1)$, . . ., $(X_m = \log x_m, Y_m = \log y_m)$ by the methods of the preceding sections. The value of a can be determined from the resulting value of A by (7).

Note 2. Logarithmic (log-log) graph paper is helpful. Its use will remove the necessity of consulting a table of logarithms in order to plot equation (6) or (8). Furthermore, if the data are plotted on it and if the points (X_1,Y_1), . . ., (X_m,Y_m) lie nearly on a straight line, the power function (2) is desirable.

Example 2. Fit a curve of the power form $y = ax^b$ to the following data:

x	1	10	100	1000
y	100	3160	100,000	3,160,000

Solution. Setting $Y = \log y$ and $X = \log x$, we get

X	0	1	2	3
Y	2	3.4997	5.0000	6.4997

By means of a graph of Y against X (or by calculating the first differences of the Y's), we see that a straight line will fit. By the method of differences, we get $Y = 2 + 1.4997X$. From (7), we get $y = 100x^{1.4997}$.

EXERCISES

1. The number y of bacteria in a culture at the end of x hours is given by:

x	0	1	2	3
y	100	112	125	140

(*a*) Fit an exponential curve $y = ab^x$ to the above data.

(*b*) Use the formula calculated in (*a*) to predict the number of bacteria present at $x = 4$.

2. The number y of millions of population of a certain country at the end of x years is given by

x	0	5	10	15
y	4.50	5.47	6.66	8.10

(*a*) Fit an exponential curve $y = ab^x$ to the above data.

(*b*) Use the formula calculated in (*a*) to predict the population at $x = 20$.

3. Suppose that a body moves through a resisting medium so that its velocity y feet per minute at the end of x minutes is given by

x	0	1	2	3
y	100	76	57	43

(*a*) Fit an exponential curve of the form $y = ab^x$ to the above data.

(*b*) Use the formula calculated in (*a*) to predict the velocity at $x = 4$.

(*c*) Would it be safe to use the formula of (*a*) to predict the velocity at $x = 120$?

4. The period, y years, of revolution about the sun for each planet, and the radius, x millions of miles, of its orbit are given in the following table:

	Mercury	Venus	Earth	Mars	Jupiter	Saturn	Uranus	Neptune
x	36.0	67.2	92.9	142	483	886	1780	2790
y	0.24	0.62	1.00	1.88	11.9	29.5	84.0	165

Fit a power function $y = ax^b$ to the above data.

5. The pressure, p inches of mercury, of the atmosphere, under certain conditions, is related to the density ratio σ by the following table:

σ	1	0.862	0.738	0.630
p	29.9	24.3	19.5	15.7

Fit a power function $p = a\sigma^b$ to the above table.

6. Prove that if the sequence $x_1, x_2, x_3, \ldots, x_n, \ldots$ is a geometric progression, then the sequence $\log x_1, \log x_2, \log x_3, \ldots, \log x_n, \ldots$ is an arithmetic progression.

7. Prove that if the sequence $x_1, x_2, \ldots, x_n, \ldots$ is a geometric progression, then the sequence of differences of first order $\Delta x_1, \Delta x_2, \Delta x_3, \ldots, \Delta x_n, \ldots$ is a geometric progression.

8. Prove that if the values $x_1, x_2, \ldots, x_n, \ldots$ are in arithmetic progression with common difference h, and the corresponding values $y_1, y_2, \ldots, y_n$ are in geometric progression with common ratio r, then $y_n = ab^{x_n}$ where $a = y_1 r^{-x_1/h}$ and $b = r^{1/h}$.

9. Prove that if the values $x_1, x_2, \ldots, x_n, \ldots$ are in geometric progression with common ratio r, and the corresponding values $y_1, y_2, \ldots, y_n, \ldots$ are in geometric progression with common ratio s, then $y_n = ax_n^b$ where $b = \log s/\log r$ and $a = y_1/x_1^b$.

196. Extrapolation

The equations or curves arrived at by the various methods explained in this chapter are, in practice, used not merely for interpolation, or reading between the known values, but also for **extrapolation,** or reading beyond the range of known values. Needless to say, such predictions must be accompanied by an appropriate amount of scientific caution and skepticism. Thus, if one fitted a curve to the graph of some stock market index over a period of weeks and then used the curve to predict what would happen next week, he would be most unwise. If he had done that just before the crash of 1929, he might well have been ruined. On the other hand, if an astronomer thus predicts the orbit of a comet for the next few centuries from a very few observations, he is likely to be astonishingly accurate. But this is due to the overwhelming amount of corroborative evidence, both theoretical and empirical, which makes him sure of the form of the functions involved. Even in the physical sciences, a curve which gives good predictions for some limited range of values may be useless too far beyond that range. However, when properly used, methods of extrapolation are extremely helpful in the investigation of natural laws.

TABLE I

Powers — Roots — Reciprocals

n	n^2	$\sqrt{n}$	$\sqrt{10n}$	n^3	$\sqrt[3]{n}$	$\sqrt[3]{10n}$	$\sqrt[3]{100n}$	$1/n$
1.0	1.0000	1.0000	3.1623	1.0000	1.0000	2.1544	4.6416	1.0000
1.1	1.2100	1.0488	3.3166	1.3310	1.0323	2.2240	4.7914	.9091
1.2	1.4400	1.0954	3.4641	1.7280	1.0627	2.2894	4.9324	.8333
1.3	1.6900	1.1402	3.6056	2.1970	1.0914	2.3513	5.0658	.7692
1.4	1.9600	1.1832	3.7417	2.7440	1.1187	2.4101	5.1925	.7143
1.5	2.2500	1.2247	3.8730	3.3750	1.1447	2.4662	5.3133	.6667
1.6	2.5600	1.2649	4.0000	4.0960	1.1696	2.5198	5.4288	.6250
1.7	2.8900	1.3038	4.1231	4.9130	1.1935	2.5713	5.5397	.5882
1.8	3.2400	1.3416	4.2426	5.8320	1.2164	2.6207	5.6462	.5556
1.9	3.6100	1.3784	4.3589	6.8590	1.2386	2.6684	5.7489	.5263
2.0	4.0000	1.4142	4.4721	8.0000	1.2599	2.7144	5.8480	.5000
2.1	4.4100	1.4491	4.5826	9.2610	1.2806	2.7589	5.9439	.4762
2.2	4.8400	1.4832	4.6904	10.6480	1.3006	2.8020	6.0368	.4545
2.3	5.2900	1.5166	4.7958	12.1670	1.3200	2.8439	6.1269	.4348
2.4	5.7600	1.5492	4.8990	13.8240	1.3389	2.8845	6.2145	.4167
2.5	6.2500	1.5811	5.0000	15.6250	1.3572	2.9240	6.2996	.4000
2.6	6.7600	1.6125	5.0990	17.5760	1.3751	2.9625	6.3825	.3846
2.7	7.2900	1.6432	5.1962	19.6830	1.3925	3.0000	6.4633	.3704
2.8	7.8400	1.6733	5.2915	21.9520	1.4095	3.0366	6.5421	.3571
2.9	8.4100	1.7029	5.3852	24.3890	1.4260	3.0723	6.6191	.3448
3.0	9.0000	1.7321	5.4772	27.0000	1.4422	3.1072	6.6943	.3333
3.1	9.6100	1.7607	5.5678	29.7910	1.4581	3.1414	6.7679	.3226
3.2	10.2400	1.7889	5.6569	32.7680	1.4736	3.1748	6.8399	.3125
3.3	10.8900	1.8166	5.7446	35.9370	1.4888	3.2075	6.9104	.3030
3.4	11.5600	1.8439	5.8310	39.3040	1.5037	3.2396	6.9795	.2941
3.5	12.2500	1.8708	5.9161	42.8750	1.5183	3.2711	7.0473	.2857
3.6	12.9600	1.8974	6.0000	46.6560	1.5326	3.3019	7.1138	.2778
3.7	13.6900	1.9235	6.0828	50.6530	1.5467	3.3322	7.1791	.2703
3.8	14.4400	1.9494	6.1644	54.8720	1.5605	3.3620	7.2432	.2632
3.9	15.2100	1.9748	6.2450	59.3190	1.5741	3.3912	7.3061	.2564
4.0	16.0000	2.0000	6.3246	64.0000	1.5874	3.4200	7.3681	.2500
4.1	16.8100	2.0248	6.4031	68.9210	1.6005	3.4482	7.4290	.2439
4.2	17.6400	2.0494	6.4807	74.0880	1.6134	3.4760	7.4889	.2381
4.3	18.4900	2.0736	6.5574	79.5070	1.6261	3.5034	7.5478	.2326
4.4	19.3600	2.0976	6.6333	85.1840	1.6386	3.5303	7.6059	.2273
4.5	20.2500	2.1213	6.7082	91.1250	1.6510	3.5569	7.6631	.2222
4.6	21.1600	2.1448	6.7823	97.3360	1.6631	3.5830	7.7194	.2174
4.7	22.0900	2.1679	6.8557	103.823	1.6751	3.6088	7.7750	.2128
4.8	23.0400	2.1909	6.9282	110.592	1.6869	3.6342	7.8297	.2083
4.9	24.0100	2.2136	7.0000	117.649	1.6985	3.6593	7.8837	.2041
5.0	25.0000	2.2361	7.0711	125.000	1.7100	3.6840	7.9370	.2000
5.1	26.0100	2.2583	7.1414	132.651	1.7213	3.7084	7.9896	.1961
5.2	27.0400	2.2804	7.2111	140.608	1.7325	3.7325	8.0415	.1923
5.3	28.0900	2.3022	7.2801	148.877	1.7435	3.7563	8.0927	.1887
5.4	29.1600	2.3238	7.3485	157.464	1.7544	3.7798	8.1433	.1852

Powers — Roots — Reciprocals (Cont.)

n	n^2	$\sqrt{n}$	$\sqrt{10n}$	n^3	$\sqrt[3]{n}$	$\sqrt[3]{10n}$	$\sqrt[3]{100n}$	$1/n$
5.5	30.2500	2.3452	7.4162	166.375	1.7652	3.8030	8.1932	.1818
5.6	31.3600	2.3664	7.4833	175.616	1.7758	3.8259	8.2426	.1786
5.7	32.4900	2.3875	7.5498	185.193	1.7863	3.8485	8.2913	.1754
5.8	33.6400	2.4083	7.6158	195.112	1.7967	3.8709	8.3396	.1724
5.9	34.8100	2.4290	7.6811	205.379	1.8070	3.8930	8.3872	.1695
6.0	36.0000	2.4495	7.7460	216.000	1.8171	3.9149	8.4343	.1667
6.1	37.2100	2.4698	7.8102	226.981	1.8272	3.9365	8.4809	.1639
6.2	38.4400	2.4900	7.8740	238.328	1.8371	3.9579	8.5270	.1613
6.3	39.6900	2.5100	7.9372	250.047	1.8469	3.9791	8.5726	.1587
6.4	40.9600	2.5298	8.0000	262.144	1.8566	4.0000	8.6177	.1563
6.5	42.2500	2.5495	8.0623	274.625	1.8663	4.0207	8.6624	.1538
6.6	43.5600	2.5690	8.1240	287.496	1.8758	4.0412	8.7066	.1515
6.7	44.8900	2.5884	8.1854	300.763	1.8852	4.0615	8.7503	.1493
6.8	46.2400	2.6077	8.2462	314.432	1.8945	4.0817	8.7937	.1471
6.9	47.6100	2.6268	8.3066	328.509	1.9038	4.1016	8.8366	.1449
7.0	49.0000	2.6458	8.3666	343.000	1.9129	4.1213	8.8790	.1429
7.1	50.4100	2.6646	8.4261	357.911	1.9220	4.1408	8.9211	.1408
7.2	51.8400	2.6833	8.4853	373.248	1.9310	4.1602	8.9628	.1389
7.3	53.2900	2.7019	8.5440	389.017	1.9399	4.1793	9.0041	.1370
7.4	54.7600	2.7203	8.6023	405.224	1.9487	4.1983	9.0450	.1351
7.5	56.2500	2.7386	8.6603	421.875	1.9574	4.2172	9.0856	.1333
7.6	57.7600	2.7568	8.7178	438.976	1.9661	4.2358	9.1258	.1316
7.7	59.2900	2.7749	8.7750	456.533	1.9747	4.2543	9.1657	.1299
7.8	60.8400	2.7928	8.8318	474.552	1.9832	4.2727	9.2052	.1282
7.9	62.4100	2.8107	8.8882	493.039	1.9916	4.2908	9.2443	.1266
8.0	64.0000	2.8284	8.9443	512.000	2.0000	4.3089	9.2832	.1250
8.1	65.6100	2.8460	9.0000	531.441	2.0083	4.3267	9.3217	.1235
8.2	67.2400	2.8636	9.0554	551.368	2.0165	4.3445	9.3599	.1220
8.3	68.8900	2.8810	9.1104	571.787	2.0247	4.3621	9.3978	.1205
8.4	70.5600	2.8983	9.1652	592.704	2.0328	4.3795	9.4354	.1190
8.5	72.2500	2.9155	9.2195	614.125	2.0408	4.3968	9.4727	.1176
8.6	73.9600	2.9326	9.2736	636.056	2.0488	4.4140	9.5097	.1163
8.7	75.6900	2.9496	9.3274	658.503	2.0567	4.4310	9.5464	.1149
8.8	77.4400	2.9665	9.3808	681.472	2.0646	4.4480	9.5828	.1136
8.9	79.2100	2.9833	9.4340	704.969	2.0723	4.4647	9.6190	.1124
9.0	81.0000	3.0000	9.4868	729.000	2.0801	4.4814	9.6549	.1111
9.1	82.8100	3.0166	9.5394	753.571	2.0878	4.4979	9.6905	.1099
9.2	84.6400	3.0332	9.5917	778.688	2.0954	4.5144	9.7259	.1087
9.3	86.4900	3.0496	9.6436	804.357	2.1029	4.5307	9.7610	.1075
9.4	88.3600	3.0659	9.6954	830.584	2.1105	4.5468	9.7959	.1064
9.5	90.2500	3.0822	9.7468	857.375	2.1179	4.5629	9.8305	.1053
9.6	92.1600	3.0984	9.7980	884.736	2.1253	4.5789	9.8648	.1042
9.7	94.0900	3.1145	9.8489	912.673	2.1327	4.5947	9.8990	.1031
9.8	96.0400	3.1305	9.8995	941.192	2.1400	4.6104	9.9329	.1020
9.9	98.0100	3.1464	9.9499	970.299	2.1472	4.6261	9.9666	.1010
10.0	100.000	3.1623	10.000	1000.00	2.1544	4.6416	10.0000	.1000

TABLE II
Trigonometric Functions

Deg.	Sin	Cos	Tan	Deg.	Sin	Cos	Tan
0	0.0000	1.0000	0.0000	45	0.7071	0.7071	1.0000
1	0.0175	0.9998	0.0175	46	0.7193	0.6947	1.0355
2	0.0349	0.9994	0.0349	47	0.7314	0.6820	1.0724
3	0.0523	0.9986	0.0524	48	0.7431	0.6691	1.1106
4	0.0698	0.9976	0.0699	49	0.7547	0.6561	1.1504
5	0.0872	0.9962	0.0875	50	0.7660	0.6428	1.1918
6	0.1045	0.9945	0.1051	51	0.7771	0.6293	1.2349
7	0.1219	0.9925	0.1228	52	0.7880	0.6157	1.2799
8	0.1392	0.9903	0.1405	53	0.7986	0.6018	1.3270
9	0.1564	0.9877	0.1584	54	0.8090	0.5878	1.3764
10	0.1736	0.9848	0.1763	55	0.8192	0.5736	1.4281
11	0.1908	0.9816	0.1944	56	0.8290	0.5592	1.4826
12	0.2079	0.9781	0.2126	57	0.8387	0.5446	1.5399
13	0.2250	0.9744	0.2309	58	0.8480	0.5299	1.6003
14	0.2419	0.9703	0.2493	59	0.8572	0.5150	1.6643
15	0.2588	0.9659	0.2679	60	0.8660	0.5000	1.7321
16	0.2756	0.9613	0.2867	61	0.8746	0.4848	1.8040
17	0.2924	0.9563	0.3057	62	0.8829	0.4695	1.8807
18	0.3090	0.9511	0.3249	63	0.8910	0.4540	1.9626
19	0.3256	0.9455	0.3443	64	0.8988	0.4384	2.0503
20	0.3420	0.9397	0.3640	65	0.9063	0.4226	2.1445
21	0.3584	0.9336	0.3839	66	0.9135	0.4067	2.2460
22	0.3746	0.9272	0.4040	67	0.9205	0.3907	2.3559
23	0.3907	0.9205	0.4245	68	0.9272	0.3746	2.4751
24	0.4067	0.9135	0.4452	69	0.9336	0.3584	2.6051
25	0.4226	0.9063	0.4663	70	0.9397	0.3420	2.7475
26	0.4384	0.8988	0.4877	71	0.9455	0.3256	2.9042
27	0.4540	0.8910	0.5095	72	0.9511	0.3090	3.0777
28	0.4695	0.8829	0.5317	73	0.9563	0.2924	3.2709
29	0.4848	0.8746	0.5543	74	0.9613	0.2756	3.4874
30	0.5000	0.8660	0.5774	75	0.9659	0.2588	3.7321
31	0.5150	0.8572	0.6009	76	0.9703	0.2419	4.0108
32	0.5299	0.8480	0.6249	77	0.9744	0.2250	4.3315
33	0.5446	0.8387	0.6494	78	0.9781	0.2079	4.7046
34	0.5592	0.8290	0.6745	79	0.9816	0.1908	5.1446
35	0.5736	0.8192	0.7002	80	0.9848	0.1736	5.6713
36	0.5878	0.8090	0.7265	81	0.9877	0.1564	6.3138
37	0.6018	0.7986	0.7536	82	0.9903	0.1392	7.1154
38	0.6157	0.7880	0.7813	83	0.9925	0.1219	8.1443
39	0.6293	0.7771	0.8098	84	0.9945	0.1045	9.5144
40	0.6428	0.7660	0.8391	85	0.9962	0.0872	11.430
41	0.6561	0.7547	0.8693	86	0.9976	0.0698	14.301
42	0.6691	0.7431	0.9004	87	0.9986	0.0523	19.081
43	0.6820	0.7314	0.9325	88	0.9994	0.0349	28.636
44	0.6947	0.7193	0.9657	89	0.9998	0.0175	57.290
45	0.7071	0.7071	1.0000	90	1.0000	0.0000	——

TABLE III
Four-place Logarithm Tables

n	0	1	2	3	4	5	6	7	8	9
10	0000	0043	0086	0128	0170	0212	0253	0294	0334	0374
11	0414	0453	0492	0531	0569	0607	0645	0682	0719	0755
12	0792	0828	0864	0899	0934	0969	1004	1038	1072	1106
13	1139	1173	1206	1239	1271	1303	1335	1367	1399	1430
14	1461	1492	1523	1553	1584	1614	1644	1673	1703	1732
15	1761	1790	1818	1847	1875	1903	1931	1959	1987	2014
16	2041	2068	2095	2122	2148	2175	2201	2227	2253	2279
17	2304	2330	2355	2380	2405	2430	2455	2480	2504	2529
18	2553	2577	2601	2625	2648	2672	2695	2718	2742	2765
19	2788	2810	2833	2856	2878	2900	2923	2945	2967	2989
20	3010	3032	3054	3075	3096	3118	3139	3160	3181	3201
21	3222	3243	3263	3284	3304	3324	3345	3365	3385	3404
22	3424	3444	3464	3483	3502	3522	3541	3560	3579	3598
23	3617	3636	3655	3674	3692	3711	3729	3747	3766	3784
24	3802	3820	3838	3856	3874	3892	3909	3927	3945	3962
25	3979	3997	4014	4031	4048	4065	4082	4099	4116	4133
26	4150	4166	4183	4200	4216	4232	4249	4265	4281	4298
27	4314	4330	4346	4362	4378	4393	4409	4425	4440	4456
28	4472	4487	4502	4518	4533	4548	4564	4579	4594	4609
29	4624	4639	4654	4669	4683	4698	4713	4728	4742	4757
30	4771	4786	4800	4814	4829	4843	4857	4871	4886	4900
31	4914	4928	4942	4955	4969	4983	4997	5011	5024	5038
32	5051	5065	5079	5092	5105	5119	5132	5145	5159	5172
33	5185	5198	5211	5224	5237	5250	5263	5276	5289	5302
34	5315	5328	5340	5353	5366	5378	5391	5403	5416	5428
35	5441	5453	5465	5478	5490	5502	5514	5527	5539	5551
36	5563	5575	5587	5599	5611	5623	5635	5647	5658	5670
37	5682	5694	5705	5717	5729	5740	5752	5763	5775	5786
38	5798	5809	5821	5832	5843	5855	5866	5877	5888	5899
39	5911	5922	5933	5944	5955	5966	5977	5988	5999	6010
40	6021	6031	6042	6053	6064	6075	6085	6096	6107	6117
41	6128	6138	6149	6160	6170	6180	6191	6201	6212	6222
42	6232	6243	6253	6263	6274	6284	6294	6304	6314	6325
43	6335	6345	6355	6365	6375	6385	6395	6405	6415	6425
44	6435	6444	6454	6464	6474	6484	6493	6503	6513	6522
45	6532	6542	6551	6561	6571	6580	6590	6599	6609	6618
46	6628	6637	6646	6656	6665	6675	6684	6693	6702	6712
47	6721	6730	6739	6749	6758	6767	6776	6785	6794	6803
48	6812	6821	6830	6839	6848	6857	6866	6875	6884	6893
49	6902	6911	6920	6928	6937	6946	6955	6964	6972	6981
50	6990	6998	7007	7016	7024	7033	7042	7050	7059	7067
51	7076	7084	7093	7101	7110	7118	7126	7135	7143	7152
52	7160	7168	7177	7185	7193	7202	7210	7218	7226	7235
53	7243	7251	7259	7267	7275	7284	7292	7300	7308	7316
54	7324	7332	7340	7348	7356	7364	7372	7380	7388	7396

Four-place Logarithm Tables (Cont.)

n	0	1	2	3	4	5	6	7	8	9
55	7404	7412	7419	7427	7435	7443	7451	7459	7466	7474
56	7482	7490	7497	7505	7513	7520	7528	7536	7543	7551
57	7559	7566	7574	7582	7589	7597	7604	7612	7619	7627
58	7634	7642	7649	7657	7664	7672	7679	7686	7694	7701
59	7709	7716	7723	7731	7738	7745	7752	7760	7767	7774
60	7782	7789	7796	7803	7810	7818	7825	7832	7839	7846
61	7853	7860	7868	7875	7882	7889	7896	7903	7910	7917
62	7924	7931	7938	7945	7952	7959	7966	7973	7980	7987
63	7993	8000	8007	8014	8021	8028	8035	8041	8048	8055
64	8062	8069	8075	8082	8089	8096	8102	8109	8116	8122
65	8129	8136	8142	8149	8156	8162	8169	8176	8182	8189
66	8195	8202	8209	8215	8222	8228	8235	8241	8248	8254
67	8261	8267	8274	8280	8287	8293	8299	8306	8312	8319
68	8325	8331	8338	8344	8351	8357	8363	8370	8376	8382
69	8388	8395	8401	8407	8414	8420	8426	8432	8439	8445
70	8451	8457	8463	8470	8476	8482	8488	8494	8500	8506
71	8513	8519	8525	8531	8537	8543	8549	8555	8561	8567
72	8573	8579	8585	8591	8597	8603	8609	8615	8621	8627
73	8633	8639	8645	8651	8657	8663	8669	8675	8681	8686
74	8692	8698	8704	8710	8716	8722	8727	8733	8739	8745
75	8751	8756	8762	8768	8774	8779	8785	8791	8797	8802
76	8808	8814	8820	8825	8831	8837	8842	8848	8854	8859
77	8865	8871	8876	8882	8887	8893	8899	8904	8910	8915
78	8921	8927	8932	8938	8943	8949	8954	8960	8965	8971
79	8976	8982	8987	8993	8998	9004	9009	9015	9020	9025
80	9031	9036	9042	9047	9053	9058	9063	9069	9074	9079
81	9085	9090	9096	9101	9106	9112	9117	9122	9128	9133
82	9138	9143	9149	9154	9159	9165	9170	9175	9180	9186
83	9191	9196	9201	9206	9212	9217	9222	9227	9232	9238
84	9243	9248	9253	9258	9263	9269	9274	9279	9284	9289
85	9294	9299	9304	9309	9315	9320	9325	9330	9335	9340
86	9345	9350	9355	9360	9365	9370	9375	9380	9385	9390
87	9395	9400	9405	9410	9415	9420	9425	9430	9435	9440
88	9445	9450	9455	9460	9465	9469	9474	9479	9484	9489
89	9494	9499	9504	9509	9513	9518	9523	9528	9533	9538
90	9542	9547	9552	9557	9562	9566	9571	9576	9581	9586
91	9590	9595	9600	9605	9609	9614	9619	9624	9628	9633
92	9638	9643	9647	9652	9657	9661	9666	9671	9675	9680
93	9685	9689	9694	9699	9703	9708	9713	9717	9722	9727
94	9731	9736	9741	9745	9750	9754	9759	9763	9768	9773
95	9777	9782	9786	9791	9795	9800	9805	9809	9814	9818
96	9823	9827	9832	9836	9841	9845	9850	9854	9859	9863
97	9868	9872	9877	9881	9886	9890	9894	9899	9903	9908
98	9912	9917	9921	9926	9930	9934	9939	9943	9948	9952
99	9956	9961	9965	9969	9974	9978	9983	9987	9991	9996

TABLE IV

Compound Interest

Amount of One Dollar Principal at Compound Interest After n Periods

n	2%	2½%	3%	3½%	4%	4½%	5%	6%	7%
1	1.0200	1.0250	1.0300	1.0350	1.0400	1.0450	1.0500	1.0600	1.0700
2	1.0404	1.0506	1.0609	1.0712	1.0816	1.0920	1.1025	1.1236	1.1449
3	1.0612	1.0769	1.0927	1.1087	1.1249	1.1412	1.1576	1.1910	1.2250
4	1.0824	1.1038	1.1255	1.1475	1.1699	1.1925	1.2155	1.2625	1.3108
5	1.1041	1.1314	1.1593	1.1877	1.2167	1.2462	1.2763	1.3382	1.4026
6	1.1262	1.1597	1.1941	1.2293	1.2653	1.3023	1.3401	1.4185	1.5007
7	1.1487	1.1887	1.2299	1.2723	1.3159	1.3609	1.4071	1.5036	1.6058
8	1.1717	1.2184	1.2668	1.3168	1.3686	1.4221	1.4775	1.5938	1.7182
9	1.1951	1.2489	1.3048	1.3629	1.4233	1.4861	1.5513	1.6895	1.8385
10	1.2190	1.2801	1.3439	1.4106	1.4802	1.5530	1.6289	1.7908	1.9672
11	1.2434	1.3121	1.3842	1.4600	1.5395	1.6229	1.7103	1.8983	2.1049
12	1.2682	1.3449	1.4258	1.5111	1.6010	1.6959	1.7959	2.0122	2.2522
13	1.2936	1.3785	1.4685	1.5640	1.6651	1.7722	1.8856	2.1329	2.4098
14	1.3195	1.4130	1.5126	1.6187	1.7317	1.8519	1.9799	2.2609	2.5785
15	1.3459	1.4483	1.5580	1.6753	1.8009	1.9353	2.0789	2.3966	2.7590
16	1.3728	1.4845	1.6047	1.7340	1.8730	2.0224	2.1829	2.5404	2.9522
17	1.4002	1.5216	1.6528	1.7947	1.9479	2.1134	2.2920	2.6928	3.1588
18	1.4282	1.5597	1.7024	1.8575	2.0258	2.2085	2.4066	2.8543	3.3799
19	1.4568	1.5987	1.7535	1.9225	2.1068	2.3079	2.5270	3.0256	3.6165
20	1.4859	1.6386	1.8061	1.9898	2.1911	2.4117	2.6533	3.2071	3.8697
21	1.5157	1.6796	1.8603	2.0594	2.2788	2.5202	2.7860	3.3996	4.1406
22	1.5460	1.7216	1.9161	2.1315	2.3699	2.6337	2.9253	3.6035	4.4304
23	1.5769	1.7646	1.9736	2.2061	2.4647	2.7522	3.0715	3.8197	4.7405
24	1.6084	1.8087	2.0328	2.2833	2.5633	2.8760	3.2251	4.0489	5.0724
25	1.6406	1.8539	2.0938	2.3632	2.6658	3.0054	3.3864	4.2919	5.4274
26	1.6734	1.9003	2.1566	2.4460	2.7725	3.1407	3.5557	4.5494	5.8074
27	1.7069	1.9478	2.2213	2.5316	2.8834	3.2820	3.7335	4.8223	6.2139
28	1.7410	1.9965	2.2879	2.6202	2.9987	3.4297	3.9201	5.1117	6.6488
29	1.7758	2.0464	2.3566	2.7119	3.1187	3.5840	4.1161	5.4184	7.1143
30	1.8114	2.0976	2.4273	2.8068	3.2434	3.7453	4.3219	5.7435	7.6123
31	1.8476	2.1500	2.5001	2.9050	3.3731	3.9139	4.5380	6.0881	8.1451
32	1.8845	2.2038	2.5751	3.0067	3.5081	4.0900	4.7649	6.4534	8.7153
33	1.9222	2.2589	2.6523	3.1119	3.6484	4.2740	5.0032	6.8406	9.3253
34	1.9607	2.3153	2.7319	3.2209	3.7943	4.4664	5.2533	7.2510	9.9781
35	1.9999	2.3732	2.8139	3.3336	3.9461	4.6673	5.5160	7.6861	10.6766
36	2.0399	2.4325	2.8983	3.4503	4.1039	4.8774	5.7918	8.1473	11.4239
37	2.0807	2.4933	2.9852	3.5710	4.2681	5.0969	6.0814	8.6361	12.2236
38	2.1223	2.5557	3.0748	3.6960	4.4388	5.3262	6.3855	9.1543	13.0793
39	2.1647	2.6196	3.1670	3.8254	4.6164	5.5659	6.7048	9.7035	13.9948
40	2.2080	2.6851	3.2620	3.9593	4.8010	5.8164	7.0400	10.2857	14.9745
41	2.2522	2.7522	3.3599	4.0978	4.9931	6.0781	7.3920	10.9029	16.0227
42	2.2972	2.8210	3.4607	4.2413	5.1928	6.3516	7.7616	11.5570	17.1443
43	2.3432	2.8915	3.5645	4.3897	5.4005	6.6374	8.1497	12.2505	18.3444
44	2.3901	2.9638	3.6715	4.5433	5.6165	6.9361	8.5572	12.9855	19.6285
45	2.4379	3.0379	3.7816	4.7024	5.8412	7.2482	8.9850	13.7646	21.0025
46	2.4866	3.1139	3.8950	4.8669	6.0748	7.5744	9.4343	14.5905	22.4726
47	2.5363	3.1917	4.0119	5.0373	6.3178	7.9153	9.9060	15.4659	24.0457
48	2.5871	3.2715	4.1323	5.2136	6.5705	8.2715	10.4013	16.3939	25.7289
49	2.6388	3.3533	4.2562	5.3961	6.8333	8.6437	10.9213	17.3775	27.5299
50	2.6916	3.4371	4.3839	5.5849	7.1067	9.0326	11.4674	18.4202	29.4570

TABLE V

Compound Discount

Present Value of One Dollar Due at the End of n Periods

n	2%	2½%	3%	3½%	4%	4½%	5%	6%	7%
1	.98039	.97561	.97087	.96618	.96154	.95694	.95238	.94340	.93458
2	.96117	.95181	.94260	.93351	.92456	.91573	.90703	.89000	.87344
3	.94232	.92860	.91514	.90194	.88900	.87630	.86384	.83962	.81630
4	.92385	.90595	.88849	.87144	.85480	.83856	.82270	.79209	.76290
5	.90573	.88385	.86261	.84197	.82193	.80245	.78353	.74726	.71299
6	.88797	.86230	.83748	.81350	.79031	.76790	.74622	.70496	.66634
7	.87056	.84127	.81309	.78599	.75992	.73483	.71068	.66506	.62275
8	.85349	.82075	.78941	.75941	.73069	.70319	.67684	.62741	.58201
9	.83676	.80073	.76642	.73373	.70259	.67290	.64461	.59190	.54393
10	.82035	.78120	.74409	.70892	.67556	.64393	.61391	.55839	.50835
11	.80426	.76214	.72242	.68495	.64958	.61620	.58468	.52679	.47509
12	.78849	.74356	.70138	.66178	.62460	.58966	.55684	.49697	.44401
13	.77303	.72542	.68095	.63940	.60057	.56427	.53032	.46884	.41496
14	.75788	.70773	.66112	.61778	.57748	.53997	.50507	.44230	.38782
15	.74301	.69047	.64186	.59689	.55526	.51672	.48102	.41727	.36245
16	.72845	.67362	.62317	.57671	.53391	.49447	.45811	.39365	.33873
17	.71416	.65720	.60502	.55720	.51337	.47318	.43630	.37136	.31657
18	.70016	.64117	.58739	.53836	.49363	.45280	.41552	.35034	.29586
19	.68643	.62553	.57029	.52016	.47464	.43330	.39573	.33051	.27651
20	.67297	.61027	.55368	.50257	.45639	.41464	.37689	.31180	.25842
21	.65978	.59539	.53755	.48557	.43883	.39679	.35894	.29416	.24151
22	.64684	.58086	.52189	.46915	.42196	.37970	.34185	.27751	.22571
23	.63416	.56670	.50669	.45329	.40573	.36335	.32557	.26180	.21095
24	.62172	.55288	.49193	.43796	.39012	.34770	.31007	.24698	.19715
25	.60953	.53939	.47761	.42315	.37512	.33273	.29530	.23300	.18425
26	.59758	.52623	.46369	.40884	.36069	.31840	.28124	.21981	.17220
27	.58586	.51340	.45019	.39501	.34682	.30469	.26785	.20737	.16093
28	.57437	.50088	.43708	.38165	.33348	.29157	.25509	.19563	.15040
29	.56311	.48866	.42435	.36875	.32065	.27902	.24295	.18456	.14056
30	.55207	.47674	.41199	.35628	.30832	.26700	.23138	.17411	.13137
31	.54125	.46511	.39999	.34423	.29646	.25550	.22036	.16425	.12277
32	.53063	.45377	.38834	.33259	.28506	.24450	.20987	.15496	.11474
33	.52023	.44270	.37703	.32134	.27409	.23397	.19987	.14619	.10723
34	.51003	.43191	.36604	.31048	.26355	.22390	.19035	.13791	.10022
35	.50003	.42137	.35538	.29998	.25342	.21425	.18129	.13011	.09366
36	.49022	.41109	.34503	.28983	.24367	.20503	.17266	.12274	.08754
37	.48061	.40107	.33498	.28003	.23430	.19620	.16444	.11580	.08181
38	.47119	.39128	.32523	.27056	.22529	.18775	.15661	.10924	.07646
39	.46195	.38174	.31575	.26141	.21662	.17967	.14915	.10306	.07146
40	.45289	.37243	.30656	.25257	.20829	.17193	.14205	.09722	.06678
41	.44401	.36335	.29763	.24403	.20028	.16453	.13528	.09172	.06241
42	.43530	.35448	.28896	.23578	.19257	.15744	.12884	.08653	.05833
43	.42677	.34584	.28054	.22781	.18517	.15066	.12270	.08163	.05451
44	.41840	.33740	.27237	.22010	.17805	.14417	.11686	.07701	.05095
45	.41020	.32917	.26444	.21266	.17120	.13796	.11130	.07265	.04761
46	.40215	.32115	.25674	.20547	.16461	.13202	.10600	.06854	.04450
47	.39427	.31331	.24926	.19852	.15828	.12634	.10095	.06466	.04159
48	.38654	.30567	.24200	.19181	.15219	.12090	.09614	.06100	.03887
49	.37896	.29822	.23495	.18532	.14634	.11569	.09156	.05755	.03632
50	.37153	.29094	.22811	.17905	.14071	.11071	.08720	.05429	.03395

TABLE VI

Amount of an Annuity

Amount of an Annuity of One Dollar per Period After n Periods

n	2%	2½%	3%	3½%	4%	4½%	5%	6%	7%
1	1.0000	1.0000	1.0000	1.0000	1.0000	1.0000	1.0000	1.0000	1.0000
2	2.0200	2.0250	2.0300	2.0350	2.0400	2.0450	2.0500	2.0600	2.0700
3	3.0604	3.0756	3.0909	3.1062	3.1216	3.1370	3.1525	3.1836	3.2149
4	4.1216	4.1525	4.1836	4.2149	4.2465	4.2782	4.3101	4.3746	4.4399
5	5.2040	5.2563	5.3091	5.3625	5.4163	5.4707	5.5256	5.6371	5.7507
6	6.3081	6.3877	6.4684	6.5502	6.6330	6.7169	6.8019	6.9753	7.1533
7	7.4343	7.5474	7.6625	7.7794	7.8983	8.0192	8.1420	8.3938	8.6540
8	8.5830	8.7361	8.8923	9.0517	9.2142	9.3800	9.5491	9.8975	10.2598
9	9.7546	9.9545	10.1591	10.3685	10.5828	10.8021	11.0266	11.4913	11.9780
10	10.9497	11.2034	11.4639	11.7314	12.0061	12.2882	12.5779	13.1808	13.8164
11	12.1687	12.4835	12.8078	13.1420	13.4864	13.8412	14.2068	14.9716	15.7836
12	13.4121	13.7956	14.1920	14.6020	15.0258	15.4640	15.9171	16.8699	17.8885
13	14.6803	15.1404	15.6178	16.1130	16.6268	17.1599	17.7130	18.8821	20.1406
14	15.9739	16.5190	17.0863	17.6770	18.2919	18.9321	19.5986	21.0151	22.5505
15	17.2934	17.9319	18.5989	19.2957	20.0236	20.7841	21.5786	23.2760	25.1290
16	18.6393	19.3802	20.1569	20.9710	21.8245	22.7193	23.6575	25.6725	27.8881
17	20.0121	20.8647	21.7616	22.7050	23.6975	24.7417	25.8404	28.2129	30.8402
18	21.4123	22.3863	23.4144	24.4997	25.6454	26.8551	28.1324	30.9057	33.9990
19	22.8406	23.9460	25.1169	26.3572	27.6712	29.0636	30.5390	33.7600	37.3790
20	24.2974	25.5447	26.8704	28.2797	29.7781	31.3714	33.0660	36.7856	40.9955
21	25.7833	27.1833	28.6765	30.2695	31.9692	33.7831	35.7193	39.9927	44.8652
22	27.2990	28.8629	30.5368	32.3289	34.2480	36.3034	38.5052	43.3923	49.0057
23	28.8450	30.5844	32.4529	34.4604	36.6179	38.9370	41.4305	46.9958	53.4361
24	30.4219	32.3490	34.4265	36.6665	39.0826	41.6892	44.5020	50.8156	58.1767
25	32.0303	34.1578	36.4593	38.9499	41.6459	44.5652	47.7271	54.8645	63.2490
26	33.6709	36.0117	38.5530	41.3131	44.3117	47.5706	51.1135	59.1564	68.6765
27	35.3443	37.9120	40.7096	43.7591	47.0842	50.7113	54.6691	63.7058	74.4838
28	37.0512	39.8598	42.9309	46.2906	49.9676	53.9933	58.4026	68.5281	80.6977
29	38.7922	41.8563	45.2189	48.9108	52.9663	57.4230	62.3227	73.6398	87.3465
30	40.5681	43.9027	47.5754	51.6227	56.0849	61.0071	66.4388	79.0582	94.4608
31	42.3794	46.0003	50.0027	54.4295	59.3283	64.7524	70.7608	84.8017	102.0730
32	44.2270	48.1503	52.5028	57.3345	62.7015	68.6662	75.2988	90.8898	110.2182
33	46.1116	50.3540	55.0778	60.3412	66.2095	72.7562	80.0638	97.3432	118.9334
34	48.0338	52.6129	57.7302	63.4532	69.8579	77.0303	85.0670	104.1838	128.2588
35	49.9945	54.9282	60.4621	66.6740	73.6522	81.4966	90.3203	111.4348	138.2369
36	51.9944	57.3014	63.2759	70.0076	77.5983	86.1640	95.8363	119.1209	148.9135
37	54.0343	59.7339	66.1742	73.4579	81.7022	91.0413	101.6281	127.2681	160.3374
38	56.1149	62.2273	69.1594	77.0289	85.9703	96.1382	107.7095	135.9042	172.5610
39	58.2372	64.7830	72.2342	80.7249	90.4091	101.4644	114.0950	145.0585	185.6403
40	60.4020	67.4026	75.4013	84.5503	95.0255	107.0303	120.7998	154.7620	199.6351
41	62.6100	70.0876	78.6633	88.5095	99.8265	112.8467	127.8398	165.0477	214.6096
42	64.8622	72.8398	82.0232	92.6074	104.8196	118.9248	135.2318	175.9505	230.6322
43	67.1595	75.6608	85.4839	96.8486	110.0124	125.2764	142.9933	187.5076	247.7765
44	69.5027	78.5523	89.0484	101.2383	115.4129	131.9138	151.1430	199.7580	266.1209
45	71.8927	81.5161	92.7199	105.7817	121.0294	138.8500	159.7002	212.7435	285.7493
46	74.3306	84.5540	96.5015	110.4840	126.8706	146.0982	168.6852	226.5081	306.7518
47	76.8172	87.6679	100.3965	115.3510	132.9454	153.6726	178.1194	241.0986	329.2244
48	79.3535	90.8596	104.4084	120.3883	139.2632	161.5879	188.0254	256.5645	353.2701
49	81.9406	94.1311	108.5406	125.6018	145.8337	169.8594	198.4267	272.9584	378.9990
50	84.5794	97.4843	112.7969	130.9979	152.6671	178.5030	209.3480	290.3359	406.5289

TABLE VII

Present Value of an Annuity

Present Value of One Dollar per Period for n Periods

n	2%	2½%	3%	3½%	4%	4½%	5%	6%	7%
1	.9804	.9756	.9709	.9662	.9615	.9569	.9524	.9434	.9346
2	1.9416	1.9274	1.9135	1.8997	1.8861	1.8727	1.8594	1.8334	1.8080
3	2.8839	2.8560	2.8286	2.8016	2.7751	2.7490	2.7232	2.6730	2.6243
4	3.8077	3.7620	3.7171	3.6731	3.6299	3.5875	3.5460	3.4651	3.3872
5	4.7135	4.6458	4.5797	4.5151	4.4518	4.3900	4.3295	4.2124	4.1002
6	5.6014	5.5081	5.4172	5.3286	5.2421	5.1579	5.0757	4.9173	4.7665
7	6.4720	6.3494	6.2303	6.1145	6.0021	5.8927	5.7864	5.5824	5.3893
8	7.3255	7.1701	7.0197	6.8740	6.7327	6.5959	6.4632	6.2098	5.9713
9	8.1622	7.9709	7.7861	7.6077	7.4353	7.2688	7.1078	6.8017	6.5152
10	8.9826	8.7521	8.5302	8.3166	8.1109	7.9127	7.7217	7.3601	7.0236
11	9.7868	9.5142	9.2526	9.0016	8.7605	8.5289	8.3064	7.8869	7.4987
12	10.5753	10.2578	9.9540	9.6633	9.3851	9.1186	8.8633	8.3838	7.9427
13	11.3484	10.9832	10.6350	10.3027	9.9856	9.6829	9.3936	8.8527	8.3577
14	12.1062	11.6909	11.2961	10.9205	10.5631	10.2228	9.8986	9.2950	8.7455
15	12.8493	12.3814	11.9379	11.5174	11.1184	10.7395	10 3797	9.7122	9.1079
16	13.5777	13.0550	12.5611	12.0941	11.6523	11.2340	10.8378	10.1059	9.4466
17	14.2919	13.7122	13.1661	12.6513	12.1657	11.7072	11.2741	10.4773	9.7632
18	14.9920	14.3534	13.7535	13.1897	12.6593	12.1600	11.6896	10.8276	10.0591
19	15.6785	14.9789	14.3238	13.7098	13.1339	12.5933	12.0853	11.1581	10.3356
20	16.3514	15.5892	14 8775	14.2124	13.5903	13.0079	12.4622	11.4699	10.5940
21	17.0112	16.1845	15.4150	14.6980	14.0292	13.4047	12.8212	11.7641	10.8355
22	17.6580	16.7654	15.9369	15.1671	14.4511	13.7844	13.1630	12.0416	11.0612
23	18.2922	17.3321	16.4436	15.6204	14.8568	14.1478	13.4886	12.3034	11.2722
24	18.9139	17.8850	16.9355	16.0584	15.2470	14.4955	13.7986	12.5504	11.4693
25	19.5235	18.4244	17.4131	16.4815	15.6221	14.8282	14.0939	12.7834	11.6536
26	20.1210	18.9506	17.8768	16.8904	15.9828	15.1466	14.3752	13.0032	11.8258
27	20.7069	19.4640	18.3270	17.2854	16.3296	15.4513	14.6430	13.2105	11.9867
28	21.2813	19.9649	18.7641	17.6670	16.6631	15.7429	14.8981	13.4062	12.1371
29	21.8444	20.4535	19.1885	18.0358	16.9837	16.0219	15.1411	13.5907	12.2777
30	22.3965	20.9303	19.6004	18.3920	17.2920	16.2889	15.3725	13.7648	12.4090
31	22.9377	21.3954	20.0004	18.7363	17.5885	16.5444	15.5928	13.9291	12.5318
32	23.4683	21.8492	20.3888	19.0689	17.8736	16.7889	15.8027	14.0840	12.6466
33	23.9886	22.2919	20.7658	19.3902	18.1476	17.0229	16.0025	14.2302	12.7538
34	24.4986	22.7238	21.1318	19.7007	18.4112	17.2468	16.1929	14.3681	12.8540
35	24.9986	23.1452	21.4872	20.0007	18.6646	17.4610	16.3742	14.4982	12.9477
36	25.4888	23.5563	21.8323	20.2905	18.9083	17.6660	16.5469	14.6210	13.0352
37	25.9695	23.9573	22.1672	20.5705	19.1426	17.8622	16.7113	14.7368	13.1170
38	26.4406	24.3486	22.4925	20.8411	19.3679	18.0500	16.8679	14.8460	13.1935
39	26.9026	24.7303	22.8082	21.1025	19.5845	18.2297	17.0170	14.9491	13.2649
40	27.3555	25.1028	23.1148	21.3551	19.7928	18.4016	17.1591	15.0463	13.3317
41	27.7995	25.4661	23.4124	21.5991	19.9931	18.5661	17.2944	15.1380	13.3941
42	28.2348	25.8206	23.7014	21.8349	20.1856	18.7236	17.4232	15.2245	13.4524
43	28.6616	26.1664	23.9819	22.0627	20.3708	18.8742	17.5459	15.3062	13.5070
44	29.0800	26.5038	24.2543	22.2828	20.5488	19.0184	17.6628	15.3832	13.5579
45	29.4902	26.8330	24.5187	22.4955	20.7200	19.1563	17.7741	15.4558	13.6055
46	29.8923	27.1542	24.7754	22.7009	20.8847	19.2884	17.8801	15.5244	13.6500
47	30.2866	27.4675	25.0247	22.8994	21.0429	19.4147	17.9810	15.5890	13.6910
48	30.6731	27.7732	25.2667	23.0912	21.1951	19.5356	18.0772	15.6500	13.7305
49	31.0521	28.0714	25.5017	23.2766	21.3415	19.6513	18.1687	15.7076	13.7668
50	31.4236	28.3623	25.7298	23.4556	21.4822	19.7620	18.2559	15.7619	13.8007

TABLE VIII

Commissioners Standard Ordinary Mortality Table (1941)

Based on 1,000,000 Living at Age One

Age	Number Living	Deaths	Age	Number Living	Deaths
0	1023102	23102	**50**	810900	9990
1	1000000	5770	51	800910	10628
2	994230	4116	52	790282	11301
3	990114	3347	53	778981	12020
4	986767	2950	54	766961	12770
5	983817	2715	55	754191	13560
6	981102	2561	56	740631	14390
7	978541	2417	57	726241	15251
8	976124	2255	58	710990	16147
9	973869	2065	59	694843	17072
10	971804	1914	**60**	677771	18022
11	969890	1852	61	659749	18988
12	968038	1859	62	640761	19979
13	966179	1913	63	620782	20958
14	964266	1996	64	599824	21942
15	962270	2069	65	577882	22907
16	960201	2103	66	554975	23842
17	958098	2156	67	531133	24730
18	955942	2199	68	506403	25553
19	953743	2260	69	480850	26302
20	951483	2312	**70**	454548	26955
21	949171	2382	71	427593	27481
22	946789	2452	72	400112	27872
23	944337	2531	73	372240	28104
24	941806	2609	74	344136	28154
25	939197	2705	75	315982	28009
26	936492	2800	76	287973	27651
27	933692	2904	77	260322	27071
28	930788	3025	78	233251	26262
29	927763	3154	79	206989	25224
30	924609	3292	**80**	181765	23966
31	921317	3437	81	157799	22502
32	917880	3598	82	135297	20857
33	914282	3767	83	114440	19062
34	910515	3961	84	95378	17157
35	906554	4161	85	78221	15185
36	902393	4386	86	63036	13198
37	898007	4625	87	49838	11245
38	893382	4878	88	38593	9378
39	888504	5162	89	29215	7638
40	883342	5459	**90**	21577	6063
41	877883	5785	91	15514	4681
42	872098	6131	92	10833	3506
43	865967	6503	93	7327	2540
44	859464	6910	94	4787	1776
45	852554	7340	95	3011	1193
46	845214	7801	96	1818	813
47	837413	8299	97	1005	551
48	829114	8822	98	454	329
49	820292	9392	99	125	125

Answers to odd-numbered exercises

Section 1

1. (*a*) Must be true. (*b*) May be true or false. (*c*) May be true or false. (*d*) Must be false.
3. (*d*), (*e*). **5.** Sufficient. **7.** Yes. **9.** Both invalid.
11. All invalid. **13.** Invalid. **15.** Invalid. **17.** Valid.

Section 4

1. Closure for addition. **3.** Commutative for addition.
5. Associative for addition. **7.** Distributive. **9.** Distributive.
11. 14. **13.** 19. **15.** 45. **17.** 94. **19.** 82.

Section 5

1. $2 \cdot 4 = 8$. **3.** (*a*), (*c*), (*d*), (*e*). **5.** (*c*) $a > b$; (*d*) a divisible by b.

Section 6

3. $2 \cdot 2 \cdot 2 \cdot 3$. **5.** $3 \cdot 3 \cdot 17$. **7.** Prime. **9.** $3 \cdot 5 \cdot 11 \cdot 19$.
11. (*a*) 20. (*b*) 240. **13.** (*a*) 12. (*b*) 72. **15.** (*a*) 1. (*b*) 6450.
17. (*a*) 12. (*b*) 360. **19.** (*a*) 20. (*b*) 240. **21.** 3.
25. (*a*) Yes. (*b*) No. (*c*) Yes.

Section 7

1. $\frac{4}{3}$. **3.** $\frac{1}{3}$. **5.** $\frac{5a}{7d}$. **7.** $\frac{m + 3n}{mn}$. **9.** $\frac{20}{35}$. **11.** $\frac{ad}{bd}$. **13.** $\frac{1}{6}$.
15. $\frac{1}{28}$. **17.** $\frac{x}{w}$. **19.** 3.

Section 8

1. $\frac{5}{12}$. **3.** $\frac{7}{8}$. **5.** $\frac{1}{12}$. **7.** $\frac{1}{8}$. **9.** $\frac{17}{12}$. **11.** $\frac{1}{2}$. **13.** $\frac{3 + x}{3x}$
15. $\frac{a + bc}{3b}$. **17.** Less than. **19.** Greater than. **21.** (*b*) $x > y$.

Section 9

1. 6. **3.** -9. **5.** -10. **7.** -9. **9.** -10. **11.** 0. **13.** 30. **15.** -6. **17.** -6. **19.** -3.

Section 10

1. -9. **3.** 5. **5.** -5. **7.** -3. **9.** -3. **11.** 1. **13.** 5/2. **15.** 40. **17.** $-\frac{4}{15}$. **19.** $\frac{33}{5}$. **21.** $\frac{27}{2}$.

Section 11

1. Yes. **3.** Yes. **5.** Yes. **7.** Yes.

Section 12

1. -3. **3.** 5. **5.** 2. **7.** -2. **9.** 3.
11. (*a*) No, except when $a = 0$ or $b = 0$. (*b*) $a^2 + 2ab + b^2$.
13. No, except when $x = 0$ or $y = 0$. **15.** 5. **17.** a. **19.** x.
21. 1/2. **23.** 36. **25.** 28. **27.** 121. **29.** 108.

Section 14

1. (*a*) 2.236. (*b*) 2.24. **3.** (*a*) 1.442. (*b*) 1.44.
5. (*a*) 1.710. (*b*) 1.71. **13.** (*b*) No. **15.** True.

Section 15

1. 24. **3.** 29. **5.** 47. **7.** 248. **9.** 344. **11.** 5.24.
13. 15.166. **15.** 7.517. **17.** 1.803. **19.** 22.091.

Section 16

1. $4i$. **3.** $i\sqrt{7}$. **5.** xi. **7.** $20i$. **9.** $2 + 3i$. **11.** $4 + 8i$.
13. $8 + 3i$. **15.** $-1 - 3i$. **17.** -2. **19.** $1 - i$. **21.** $6 + 12i$.
23. $2 + 16i$. **25.** $11 - 16i$. **27.** $10 - 11i$. **29.** $18 + 14i$.
31. Complex, imaginary. **33.** Complex, imaginary, pure imaginary.
35. Complex, real, irrational, positive.
37. Complex, real, rational, negative.
39. Complex, real, rational, positive.
41. Complex, real, rational, integral, zero.

Section 23

31. a^6. **33.** $1/a^4$. **35.** $12a^5b^9$. **37.** $y^2/4x^5$. **39.** $-2x^8y^{12}$.
41. $b^{10}/2$. **43.** -72. **45.** 2. **47.** a^2b^3. **49.** x^2y^3.
51. False; 3^5. **53.** False; 3^5. **55.** True. **57.** False; $2 \cdot 5^3$.
59. False; 3^6. **61.** False; 4^6.

Section 24

1. (*a*) *B*, *C*; (*b*) 28. **3.** (*a*) *A*; (*b*) 4. **5.** (*a*) *B*; (*b*) 3/10.
7. (*a*) *B*, *C*; (*b*) $12 - \sqrt{2}$. **9.** (*a*) *B*, *C*; (*b*) 3. **11.** (*a*) -3; (*b*) 2.
13. (*a*) -1; (*b*) 6. **15.** (*a*) $-\frac{1}{2}$; (*b*) 6. **17.** (*a*) 10; (*b*) 0.

Section 25

1. $-x + 7y$. **3.** $-2x^2 + 2xy + y^2$. **5.** $x + 2y$.
7. $-x^2 - 2x + 5$. **9.** $2x^2 - 4x$. **11.** y. **13.** $x^3 - 2x^2 - x$.
15. $12x^2 - 5x + 3$. **17.** $5x - 10$.
19. (*a*) $(-x + y) + (2a - 3b)$; (*b*) $-(x - y) - (-2a + 3b)$.

Section 26

1. $4x^2 - 2x + 4$; 2. **3.** $3x + 4$; 1. **5.** $5x + 5$; 1.
7. $x^4y - x^3y + x^2y^2 + 2x - 3$; 5. **9.** Integer. **11.** No.

Section 28

7. Conditional. **9.** Conditional.

Section 29

1. 7. **3.** No root. **5.** No root.

Section 30

1. $x^2 - x + 4 = 0$; 2. **3.** $2x^2 + x + 8 = 0$; 2. **5.** $2x^3 + 7 = 0$; 3.

Section 31

1. 27. **3.** 3. **5.** -18. **7.** 19. **9.** 2. **11.** 2, -5.

Section 32

1. 8. **3.** -1. **5.** -7. **7.** $3n + 2$. **9.** $-3r + 2$. **11.** 0.
13. 5. **15.** $r^2 - 4$. **17.** 16. **19.** 1. **21.** 1.28. **23.** 9.
25. $\frac{4}{9}$. **27.** Does not exist. **29.** $\frac{4}{3}$.

Section 33

1. (*a*) 64 ft. (*b*) 400 ft. **3.** 14 sq. ft. **5.** $26.88. **7.** 95°F.
9. 113.0 sq. ft.

Section 34

3. 0.

Section 35

1. 4. **3.** -3. **5.** ± 3. **7.** None. **9.** 2, -3. **11.** 1, -2, 4.

Section 37

1. 2. **3.** -6. **5.** ± 1. **7.** $-1, 5$. **9.** 1, 3. **11.** 2, 3, -1.
13. $4/3, -1/2$. **15.** ± 1.4 approx.

Section 38

1. 0. **3.** 2. **5.** $-3/2$. **7.** ± 1. **9.** $1, -2$. **11.** 2.
13. None.

Section 40

1. $7x^2 - 4x - 1$. **3.** $3x^2 - 2xy + 2y^2$. **5.** $6x^3 - 3x^2y + 5xy^2 + 4y^2$.
7. $3x^2 + xy - 7y^2$. **9.** $-6x^2 + 4x + 4$. **11.** $-3x^3 + 3x^2y - 3y^3$.

Section 41

1. $-10xy + 6xz$. **3.** $2x^3 - 8x^2 + 6x$. **5.** $4x^3 + 8x^2y - 12xy^2$.
7. $x^2 - 6x + 8$. **9.** $6x^2 + 27x - 15$. **11.** $6x^2 - 11x + 4$.
13. $x^2 + 2xy + y^2$. **15.** $x^2 - y^2$. **17.** $x^3 - y^3$.
19. $4x^3 + 4x^2 - 21x + 9$. **21.** $a^2x^{2c} - b^2$. **23.** $a^2x^{2c} - 2abx^c + b^2$.
25. $6x^4 - 4x^3 + 3x^2 + 7x - 6$. **27.** $x^4 - y^4$. **29.** $x^4 - y^4$.

Section 42

1. $4x^6$. **3.** $3x^4y^8$. **5.** $6x^2 + 7x^4y^3$. **7.** $-4x^6 + 2x^4 - 1$.
9. $-5x^2y^2 + \frac{10}{3}x + 3$.

Section 43

1. $x - 6$. **3.** $2x + 3$. **5.** $x^2 - 5x + 6$. **7.** $x^2 + x - 6$.
9. $5x - 1$. **11.** $x^2 - 2xy + 4y^2$. **13.** $x^2 - 2xy + 3y^2$.
15. $x^2 - xy - 6y^2$. **17.** Quotient $x + 1$, remainder 5.
19. Quotient $6x^2 + 5x - 4$, remainder 5.
21. Quotient $2x + 1$, remainder $x - 5$.
23. Quotient $x + 1$, remainder $2x + 3$.
25. Quotient $x^3 + 3$, remainder 2.
27. Quotient $x - 3$, remainder $5x - 1$. **29.** $x^2 - xy + y^2$.
31. $x^2 + xy + y^2$. **33.** $x^3 - x^2y + xy^2 - y^3$. **35.** $x^2 - 4$.

Section 44

1. 39. **3.** 18. **5.** 82. **7.** 111. **9.** 220. **11.** 2001. **13.** 5.
15. 58. **17.** 14. **19.** 1100. **21.** 1100010. **23.** 111001.
25. 122. **27.** 113. **29.** 2122. **31.** 130. **33.** 1000. **35.** 110.
37. 1111. **39.** 11. **41.** (*a*) 112; 34. (*b*) 201. (*c*) 4413.
43. (*a*) 29. (*b*) 41. (*c*) 1002. (*d*) 25.
45. (*a*) 58. (*b*) 112. (*c*) 2011. (*d*) 4*t*.
47. (*a*) 131. (*b*) 245. (*c*) 11212. (*d*) *te*.
49. 143. (*b*) 263. (*c*) 12022. (*d*) *ee*.

Section 46

1. $6ax + 8ay$. **3.** $-9a^4b^2 + 3a^2b^3$. **5.** $x^2 - 25y^2$. **7.** $1 - 16x^6$.
9. $4x^2 + 12x + 9$. **11.** $4x^2 - 12xy + 9y^2$. **13.** $x^6 - 6x^3y^3 + 9y^6$.
15. $16x^4 - 40x^2y^3 + 25y^6$. **17.** $x^2 + 8x + 15$. **19.** $x^2 - 2x - 8$.
21. $8u^2 - 14u + 3$. **23.** $12u^2 + u - 6$. **25.** $2x^2 - xy - 15y^2$.
27. $8s^4 - 2s^2t^2 - 15t^4$. **29.** $x^2 + 4xy + 4y^2 - z^2$
31. $4x^2 + 12x + 9 - 16xy - 24y + 16y^2$. **33.** $x^2 + 4xy + 4y^2 - 9z^2$.
35. $9x^2 - 12xy + 4y^2 - 24x + 16y + 16$.
37. $x^2 + 2xy + y^2 - u^2 + 2uv - v^2$. **39.** $x^{2a} + 2x^ay^b + y^{2b}$.
41. $x^{2a} - y^{2b}$.

Section 47

1. $2a(x - 4a^2)$. **3.** $2xy(2x^2 - 3y^2 + 4xy)$. **5.** $(3x + 4y)(3x - 4y)$.
7. $(2 + 7xy)(2 - 7xy)$. **9.** $(9u + 8v)(9u - 8v)$. **11.** $(3x + 2y)^2$.
13. $(5x - 2y)^2$. **15.** $(xy - 8)^2$. **17.** $(2ax + b)^2$. **19.** $\left(x + \frac{b}{2a}\right)^2$.
21. $x(5x + 2y)(5x - 2y)$. **23.** $2x^2(3x + 2y)(3x - 2y)$.
25. $(x^2 + 9)(x + 3)(x - 3)$. **27.** $(3x + y)(9x^2 - 3xy + y^2)$.
29. $(x - 4y)(x^2 + 4xy + 16y^2)$.
31. $(2x + y)(4x^2 - 2xy + y^2)(2x - y)(4x^2 + 2xy + y^2)$.
33. $(x + 2y)(2a - b)$. **35.** $(x + 2y + 4)(x + 2y - 4)$.
37. $(3x + y - z)(3x - y + z)$. **39.** $(x - 3y - 2)^2$.
41. $(3x + 2y)(x - 4y)$. **43.** $2x^2y(y - 3x)(y^2 + 3xy + 9x^2)$.
45. $(x + y + 1)(x - y + 3)$. **47.** $(2x - y)(4x^2 + 8xy + 7y^2)$.
49. $(x + 2)(3x - 4y)(-x - 2y)$. **51.** $5xy(x^2 + 4y^2)(x + 2y)(x - 2y)$.
53. $3x^2y(x + 2y)(x^2 - 2xy + 4y^2)$.

Section 48

1. $(a + 5)(x - y)$. **3.** $(x - 4)(x - y)$. **5.** $(2x + 5)(x^2 - 2)$.
7. $(x - 3y)(x + 3y + 1)$. **9.** $(x + 4 + y)(x + 4 - y)$.
11. $(4x + y)(1 - 3x)$. **13.** $(x + 3y + z + 1)(x + 3y - z - 1)$.
15. $(x^2 - y^2 + xy)(x^2 - y^2 - xy)$. **17.** $(x^2 + 1 + x)(x^2 + 1 - x)$.
19. $(x^2 + 2 + 2x)(x^2 + 2 - 2x)$. **21.** $(x^2 + 3 + 2x)(x^2 + 3 - 2x)$.
23. $(x^2 - 3y^2 + xy)(x^2 - 3y^2 - xy)$.

Section 49

1. $(x - 2)(x - 6)$. **3.** $(x - 5)(x + 3)$. **5.** $(x + 6)(x - 3)$.
7. Prime. **9.** $(k - 18)(k + 2)$. **11.** $(x - 12y)(x + 2y)$.
13. $(m - 12)(m - 4)$. **15.** $(x - 3)(2x - 3)$. **17.** $(2m + 5n)(2m - n)$.
19. Prime. **21.** $(2x + 1)(4x - 3)$. **23.** $(4x - 3y)(x + 2y)$.
25. $x(x - 1)$. **27.** $(2x - y + 12)(2x - y - 3)$.
29. $(9x - 19)(6x - 9)$. **31.** $(x - 8)(x - 7)$.
33. $2xy^2(2x - y)(3x + 2y)$. **35.** $2x^2y(2x - 5y)(x + 2y)$.
37. $(x^2 + 6)(x^2 + 2)$. **39.** $(x - 2)(x^2 + 2x + 4)(x + 1)(x^2 - x + 1)$.
41. $(2x^2 + y^2)(x + 2y)(x - 2y)$.

Section 50

1. $(x + y)(x^4 - x^3y + x^2y^2 - xy^3 + y^4)$.
3. $(x + y)(x^2 - xy + y^2)(x - y)(x^2 + xy + y^2)$.
5. $(x^2 + 9y^2)(x + 3y)(x - 3y)$.
7. $(xy + 3)(x^4y^4 - 3x^3y^3 + 9x^2y^2 - 27xy + 81)$.
9. $(4x^2 + 1)(16x^4 - 4x^2 + 1)$.
11. $(2a + b)(4a^2 - 2ab + b^2)(2a - b)(4a^2 + 2ab + b^2)$.
13. $(16m^4 + n^4)(4m^2 + n^2)(2m + n)(2m - n)$.

Section 51

1. $2xy;\ 24x^3y^2$. **3.** $x^2y;\ 18x^4y^3$. **5.** $x - 3;\ 30(x - 3)$.
7. $(x - 2);\ 6(x + 2)(x - 2)$. **9.** $(x - y);\ (x - y)^2(x + y)$.
11. $(x - 3);\ (x - 3)(x + 3)(x - 2)$.
13. $(x + 3);\ (x + 3)(x - 3)(x^2 - 3x + 9)(x - 2)$.
15. $1;\ (x + 2)(x - 2)(x - 3)(x^2 - xy + y^2)$.
17. $(x - y);\ (x - y)^2(x + y)(x^2 + xy + y^2)$.
19. $1;\ (x^2 + y^2)(x + y)(x - y)(x + 2y)$.
21. $2x - 3;\ 4x(2x + 3)(2x - 3)(x + 2)$.
23. $(x - 2y);\ 6x(x - 2y)^2(x + 2y)$.

Section 52

1. $\frac{3}{5}$. **3.** $\frac{2y}{3x}$. **5.** $\frac{3xy^2}{2}$. **7.** $\frac{6}{5a}$. **9.** $\frac{5y^2}{3}$. **11.** $\frac{4x}{3}$.
13. $\frac{4(x - y)}{3}$. **15.** $\frac{x - 4}{x + 4}$. **17.** $\frac{4x}{x + 1}$. **19.** $\frac{x - 6}{x - 3}$.
21. $\frac{x^2 + xy + y^2}{x + y}$. **23.** $\frac{y - x}{3}$. **25.** $-\frac{9x^2 + 3x + 1}{3x + 1}$.
27. $-\frac{2(x + 2y)}{5(x^2 + 2xy + 4y^2)}$. **29.** $\frac{2x - 3y}{2(x - 3y)}$.

Section 53

1. $\frac{5}{4}$. **3.** 9. **5.** $\frac{8z^2}{5xy^2}$. **7.** $\frac{9ax}{bcz}$. **9.** $\frac{3y}{4a}$. **11.** $\frac{2(x + 3)}{3x}$.
13. $\frac{x - 3y}{x + y}$. **15.** $\frac{x^2}{x - 5}$. **17.** $\frac{x + 2}{x - 1}$. **19.** $\frac{1}{2}$.
21. $\frac{x + 2y}{x - 3y}$. **23.** $\frac{2x - y}{x + 5y}$. **25.** $\frac{(x + 2)^2(x - 2)}{3x - 2}$.
27. $(3x + y)^2$. **29.** 1.

Section 54

1. $\frac{1}{8}$. **3.** $\frac{11}{8}$. **5.** x. **7.** $\frac{2a^2 + 15}{6a}$. **9.** $\frac{bc + ac - ab}{abc}$.
11. $\frac{9x + 4}{6x^2}$. **13.** $\frac{6x^2 - 4x + 9}{6x}$. **15.** $\frac{x + 3}{x - 2}$. **17.** $\frac{3x^2 - 7x + 8}{(2x - 1)(x - 2)}$.
19. $\frac{4ab}{(a + b)(a - b)}$. **21.** $\frac{-2a}{(x - a)^2(x + a)}$. **23.** $\frac{1}{x - 3}$
25. $\frac{-2}{(x - y)(x - 2y)}$. **27.** $\frac{5}{(x + 9)(x - 6)}$. **29.** 0.
31. $\frac{-4a}{x + a}$. **33.** $\frac{x^2 - 2xy + 8y^2}{(x - 6y)(x + 4y)(x + 2y)}$.

Section 55

1. 7. **3.** $\frac{b - a}{b + a}$. **5.** $\frac{x^2 - y^2}{x^2 + y^2}$. **7.** $\frac{3}{2}$. **9.** $\frac{2ab}{b^2 - a^2}$. **11.** $\frac{xm - yp}{xp - ym}$
13. $\frac{x}{y}$. **15.** -1. **17.** $\frac{2a^2 - b^2}{b^2}$ **19.** $\frac{a - 2}{a + 2}$. **21.** $-\frac{x - 2}{x + 2}$.
23. c^2. **25.** $\frac{1}{m - n}$. **27.** $\frac{2a^2 - b^2}{b^2}$. **29.** $\frac{a^3 + 2a^2 + 7a + 8}{a^3 + 7a}$.
31. $2a^2 - 1$.

Section 56

1. 7. **3.** 6. **5.** 2. **7.** 4. **9.** 11/7. **11.** 5/4. **13.** 0.
15. $-1/9$. **17.** 51/19. **19.** -1. **21.** 5. **23.** 11/8. **25.** $-8/5$.
27. 15/4. **29.** 0.057. **31.** -1.9.

Section 57

1. 5. **3.** 14. **5.** 0. **7.** -13. **9.** No solution. **11.** 4/3.
13. 4. **15.** $-3/2$. **17.** No solution. **19.** No solution. **21.** 9/14.
23. No solution. **25.** -16.

Section 58

1. $5ab$. **3.** $a + 3b$. **5.** $\frac{4b}{a - 2}$ if $a \neq 2$. **7.** $\frac{2b}{2a - b}$ if $a \neq \frac{b}{2}$.
9. $2a + b$ if $2a \neq b$. **11.** $2a^2 + 3a - 6$, $a \neq 0$.
13. $a + b$ if $a \neq b$, $a \neq 0$, $b \neq 0$. **15.** $\frac{2ab}{a - b}$ if $a \neq b$, $a \neq 0$, $b \neq 0$.
17. $\frac{a + b}{2}$ if $a \neq b$. **19.** $\frac{3x - 2}{5}$. **21.** $\frac{9 - x}{8}$. **23.** $\frac{C - Ax}{B}$ if $B \neq 0$.
25. $v = \frac{c}{p}$. **27.** $b = \frac{2A}{h}$. **29.** $l = \frac{P - 2w}{2}$. **31.** $F = \frac{9}{5}C + 32$.

33. $h = \dfrac{3V}{\pi r^2}$. **35.** $m_2 = \dfrac{Fr^2}{Gm_1}$. **37.** $d = \dfrac{2(S - an)}{n(n - 1)}$, $n > 1$.

39. $r = \dfrac{nE - IR}{In}$. **41.** $w_2 = \dfrac{d_1 w_1}{d_2}$. **43.** $t = \dfrac{P - p}{pc}$.

45. $h = h_1 + \dfrac{1}{c}(p_1 - p)$. **47.** $C = \dfrac{C_1 C_2}{C_1 + C_2}$. **49.** $a = \dfrac{bc}{d}$.

Section 59

13. 4. **15.** 12. **17.** $\frac{5}{3}$. **19.** 8. **21.** -2. **23.** 2.
25. $y = 3x - 1$. **27.** $y = \frac{1}{4}x - \frac{5}{2}$. **29.** $y = -\frac{5}{2}x + \frac{1}{2}$.
31. $y = 4x + 3$. **33.** $y = -\frac{1}{2}x - \frac{3}{2}$. **35.** $y = 3x - 9$.
39. Parallel lines.

Section 60

The value for x is given first.

1. 4, -3. **3.** -3, 1. **5.** $\frac{1}{3}$, 1. **7.** 4, 0. **9.** -3, 2.
11. 1, -5. **13.** -3, 2. **15.** 1, $\frac{1}{2}$. **17.** -3, -1. **19.** $\frac{150}{13}$, $-\frac{30}{13}$.
21. $\frac{24}{17}$, $\frac{36}{17}$. **23.** $4a$, $3a$. **25.** $a - 2b$, $b - 3a$. **27.** $2a - 3$, $a + 1$.

29. $\dfrac{4b - 3a}{11}$, $\dfrac{5a - 3b}{11}$. **31.** $a + b$, $b - a$. **33.** $\dfrac{k_1 b_2 - k_2 b_1}{a_1 b_2 - a_2 b_1}$, $\dfrac{a_1 k_2 - a_2 k_1}{a_1 b_2 - a_2 b_1}$.

35. $\dfrac{2(a + b)}{3}$, $\dfrac{2c}{3}$. **37.** $2b$, $\dfrac{2b(a - b)}{c}$.

Section 61

1. Simultaneous; $x = 2$, $y = 0$. **3.** Inconsistent. **5.** Dependent.
7. Simultaneous; $x = 1$, $y = -2$. **9.** Inconsistent. **11.** Dependent.

Section 62

1. $x = 2$, $y = 3$, $z = 1$. **3.** $x = -1$, $y = -3$, $z = 2$.
5. $x = \frac{1}{2}$, $y = -\frac{3}{2}$, $z = 0$. **7.** $A = 14/75$, $B = -1/5$, $C = -13/75$.
9. $x = \dfrac{a}{2}$, $y = \dfrac{5a}{9}$, $z = \dfrac{-2a}{3}$. **11.** $x = 1$, $y = -1$, $z = 2$, $t = 3$.
13. $x = 1$, $y = 0$, $z = -1$, $t = 2$, $u = 3$.

Section 63

1. $x = -2$, $y = 1$. **3.** $x = -3$, $y = 4$. **5.** $x = -2$, $y = 1$.
7. $x = 3$, $y = -2$. **9.** $x = -\frac{1}{2}$, $y = \frac{1}{3}$. **11.** $x = \frac{2}{3}$, $y = \frac{3}{4}$.
13. $x = \frac{5}{3}$, $y = -\frac{5}{4}$. **15.** $x = \frac{1}{2}$, $y = \frac{1}{3}$, $z = -1$.
17. $x = \frac{1}{2}$, $y = -\frac{1}{2}$, $z = \frac{1}{3}$. **19.** $x = 1$, $y = -1$, $z = \frac{1}{2}$.

21. $x = 0$, $y = -\frac{4}{3}$. **23.** $x = \dfrac{a^2 + b^2}{a + b}$, $y = \dfrac{a^2 + b^2}{a - b}$.

Section 64

1. 11, 19. **3.** 35 yrs. **5.** 13 nickels, 11 quarters.
7. 14 nickels, 12 dimes, 6 quarters. **9.** 48 ft., 12 ft.
11. \$4500 at 4%, \$1500 at 3%. **13.** 89. **15.** 37, 38, 39.
17. 27, 29, 31. **19.** 2 2/3 days. **21.** 3 days. **23.** 5 5/23 hrs.
25. 2 hrs., 100 mi. **27.** 300 m.p.h., 30 m.p.h.
29. 17.5 m.p.h., 2.5 m.p.h. **31.** 4:21 9/11 o'clock. **33.** 48 lb.
35. 6 ft. **37.** 3 ft. from fulcrum on other side. **39.** $5\frac{5}{8}$ gal.
41. 32 lb. **43.** 14 lb at 70¢, 28 lb. at 40¢. **45.** 13 ft., 5 ft. **47.** 45.
49. 20, 40, 120 degrees.

Section 65

1. $\frac{3}{x^2}\cdot$ **3.** 3. **5.** $\frac{3y^3}{x^2}\cdot$ **7.** b/a. **9.** $\frac{b-a}{b+a}\cdot$ **11.** $\frac{2a}{b-2a}\cdot$
13. $\frac{xy}{y-x}\cdot$ **15.** $\frac{1}{xy}\cdot$ **17.** $\frac{x^2y^2}{y^2-x^2}\cdot$ **19.** $-\frac{1}{8}$. **21.** 8. **23.** 1.
25. 1/36. **27.** 1. **29.** 1/50. **31.** $8x/y$. **33.** $\frac{1}{8}x^6$.
35. $x^5 + 2x + 5$. **37.** $8y^2$. **39.** 10. **41.** 1. **43.** 720.

Section 66

1. 5×10^5. **3.** 3.47×10^6. **5.** 4.27×10^{-1}. **7.** 4.16×10^0.
9. 4.3×10^{-4}. **11.** 100,000. **13.** 0.00001. **15.** 37,800,000.
17. 0.0000576. **19.** 0.00000627. **21.** 120. **23.** 5×10^{14} per sec.
25. 220. **27.** 1.316385×10^{25} lb. **29.** 1.6×10^{-19}.

Section 67

1. 2. **3.** 1/4. **5.** 1/2. **7.** 6. **9.** 1/8. **11.** 1/1000.
13. 64. **15.** 100. **17.** 1000. **19.** 100. **21.** 1/5. **23.** $2a^2$.
25. x^{-7}. **27.** $a^{15/2}b$. **29.** $\frac{3}{2}x^2y^2$. **31.** $(\sqrt[12]{x})^{17}$. **33.** x.
35. $a + 2\sqrt{ab} + b$. **37.** $a + 3a^{2/3}b^{1/3} + 3a^{1/3}b^{2/3} + b$.

Section 69

1. 5. **3.** a. **5.** 6. **7.** 3. **9.** 2. **11.** 3. **13.** $3\sqrt{2}$.
15. $6\sqrt{2}$. **17.** $4\sqrt{5}$. **19.** $2\sqrt[3]{4}$. **21.** $\sqrt{2}/4$. **23.** $\sqrt{10}/4$.
25. $\sqrt[12]{5}$. **27.** $\sqrt{2}$. **29.** 2. **31.** $\frac{1}{2}$. **33.** $\frac{1}{2}$. **35.** -4.
37. -2. **39.** $4x^4y^5$. **41.** $2x^{18}y^4\sqrt{y}$. **43.** $-3a^2y^2\sqrt[3]{a^2}$.
45. $3x^3y^2\sqrt[3]{2xy^2}$. **47.** $\frac{a^3}{3b^4}\cdot$ **49.** $\frac{4}{5x^4y^3}\sqrt{y}$. **51.** $\frac{3x^2}{5y^4}\sqrt{2xy}$.
53. $\frac{2x^2}{y^2}\sqrt[3]{x^2}$. **55.** $x\sqrt{a^2+b^2}$. **57.** $4\sqrt{x^2+1}$. **59.** $\frac{1+3x}{x}\sqrt{x}$.
61. $\frac{3+2x}{2x}\sqrt{2x}$. **63.** 0.632. **65.** 1.5275. **67.** 1.472. **69.** 1.186.

Section 70

1. $6\sqrt{2}$. **3.** $7\sqrt[3]{5}$. **5.** $\sqrt{2}$. **7.** $3\sqrt{3}$. **9.** $7\sqrt{2}$.
11. $\frac{23\sqrt{3}}{3}$. **13.** $-5\sqrt{a}$. **15.** $(4a - 2b + 5)\sqrt{2x}$.
17. $(2a + 3b)\sqrt[3]{2x}$. **19.** $4\sqrt{3} + 4\sqrt{2}$. **21.** $\left(1 - \frac{a}{2}\right)\sqrt{ab}$.
23. $\frac{2y}{x^2 - y^2}\sqrt{x^2 - y^2}$. **25.** $-b/a$.

Section 71

1. $\sqrt{15}$. **3.** $4\sqrt{3}$. **5.** $12\sqrt{6}$. **7.** $24\sqrt{3}$. **9.** $2\sqrt[3]{6}$.
11. $6x\sqrt{y}$. **13.** $96x$. **15.** $48x^2y$. **17.** $24x$. **19.** $9(x^2 + y^2)$.
21. $(x + y)^3$. **23.** $\sqrt{6} + 2\sqrt{3}$. **25.** $3\sqrt{2} - 2\sqrt{3}$. **27.** 1.
29. $30 + 12\sqrt{6}$. **31.** $9 + 7\sqrt{5}$. **33.** $-5\sqrt{6}$. **35.** $6\sqrt{2}$.
37. $21 + 8\sqrt{2}$. **39.** 0. **41.** $43 + 12\sqrt{2}$. **43.** c/a.
45. $7 + x + 6\sqrt{x - 2}$. **47.** $2x - 2\sqrt{x^2 - 9}$.
49. $4x^2 + 18x + 9 - 12x\sqrt{2x - 1}$.

Section 72

1. $\frac{1}{2}\sqrt{10}$. **3.** $\sqrt{5}$. **5.** $x\sqrt{5x}$. **7.** $\frac{x\sqrt{10}}{6y}$. **9.** $\frac{\sqrt[3]{10}}{2}$.
11. $2 + \sqrt{3}$. **13.** $4(\sqrt{3} + \sqrt{2})$. **15.** $\frac{\sqrt{3} - 1}{2}$.
17. $\frac{\sqrt{15} + 3\sqrt{2}}{3}$. **19.** $2 - \sqrt{3}$. **21.** $\frac{34 - 7\sqrt{10}}{37}$.
23. $\frac{26 + 9\sqrt{15}}{7}$. **25.** $2 + \sqrt{3}$ **27.** $\frac{\sqrt{x - 1} + x - 1}{2 - x}$.
29. (*a*) 1.581, (*c*) 2.236.

Section 73

1. $\sqrt[10]{x^7}$. **3.** $x\sqrt[15]{x^4}$. **5.** $\sqrt[6]{250}$. **7.** $\sqrt[6]{32}$. **9.** $\sqrt[6]{7}$.
11. $\sqrt[12]{2}$. **13.** $y\sqrt[3]{xy^2}$.

Section 74

1. 2, -4. **3.** 3, -2. **5.** ± 7. **7.** $\pm 5/2$. **9.** 2, -9.
11. 18, -2. **13.** $1/2$, $-5/3$. **15.** $2/3$, $-5/2$. **17.** $3/8$, -1.
19. $3/4$, $-5/3$. **21.** $\pm\sqrt{5}$. **23.** $\pm\frac{1}{2}\sqrt{10}$. **25.** k, k.
27. $k/3$, $-4k/7$. **29.** ± 2, $\pm 2\sqrt{2}$. **31.** $\pm\sqrt{3}$, $\pm\sqrt{5}$. **33.** $\pm 3i$.
35. $\pm\frac{3}{2}i$. **37.** 16, -2. **39.** 4, 5. **41.** 7, 0. **43.** $-5/3$, 0.
45. $-5/2$, $-5/2$. **47.** $t = \pm\frac{1}{g}\sqrt{2sg}$. **49.** $r = \pm\frac{1}{\pi}\sqrt{A\pi}$.

Section 75

1. 18, −2. **3.** −2, −12. **5.** 1/2, −5/3. **7.** $\frac{4}{3}$, $-\frac{5}{6}$.
9. −3, −1/2. **11.** (*a*) $2 \pm \sqrt{3}$; (*b*) 3.732, 0.268.
13. (*a*) $\dfrac{5 \pm \sqrt{3}}{2}$; (*b*) 3.366, 1.634.
15. (*a*) $-1 \pm \sqrt{5}$; (*b*) 1.236, −3.236. **17.** $1 \pm 2i$.
19. $-\frac{1}{2} \pm \frac{1}{2}i\sqrt{3}$. **21.** (*a*) $\dfrac{5 \pm \sqrt{21}}{2}$; (*b*) 4.791, 0.209. **23.** 7, −10/3.
25. (*a*) $-3 \pm \sqrt{17}$; (*b*) 1.123, −7.123. **27.** $\dfrac{-p \pm \sqrt{p^2 - 4q}}{2}$.

Section 76

1–26. See section 75. **27.** 5, −2. **29.** 3/2, 3/2.
31. $\dfrac{\sqrt{3} \pm i\sqrt{13}}{4}$. **33.** $\dfrac{-\sqrt{2} \pm i\sqrt{30}}{4}$. **35.** $\dfrac{\sqrt{3} \pm \sqrt{21}}{3}$.
37. $h \pm \sqrt{h^2 - k}$. **39.** $\dfrac{-d \pm \sqrt{d^2 - ac}}{a}$. **41.** $a \pm \sqrt{b}$.
43. $2i \pm \sqrt{3}$. **45.** $\dfrac{-3i}{4}, \dfrac{i}{3}$.

Section 77

1. 64; real, rational, unequal. **3.** 169; real, rational, unequal.
5. 8; real, irrational, unequal. **7.** 0; real, rational, equal.
9. −92; imaginary, unequal. **11.** 49; real, rational, unequal.
13. −144; imaginary, unequal. **15.** 25; real, irrational, unequal.
17. −25; imaginary, unequal. **19.** 0; imaginary, equal.
21. 16; imaginary, unequal. **23.** 0; real, irrational, equal. **25.** 16.
27. ±12. **29.** 2. **31.** 4, 0. **33.** 5, 0. **35.** ±24.
37. (*a*) $k < 4$, (*b*) $k > 4$. **39.** (*a*) $k > 6$, $k < -6$, (*b*) $-6 < k < 6$.

Section 78

11. 4, −1. **13.** 3, −1. **15.** −1/2, −1/2. **17.** ±3/2.
19. None. **21.** 4, 0. **29.** Tangent. **31.** Two points.
33. Not at all. **35.** Two points. **37.** (*a*) $k = 1$; (*b*) $k < 1$; (*c*) $k > 1$.
39. (*a*) $k = \pm 4$; (*b*) $k > 4$, $k < -4$; (*c*) $-4 < k < 4$.
41. (*a*) $k = \pm 5$; (*b*) $k > 5$, $k < -5$; (*c*) $-5 < k < 5$.

Section 79

1. 8, 15. **3.** −5/3, −4. **5.** −1/3, −5/3. **7.** 0, −7/2.
9. −11/2, 0. **11.** $r_2 = -2$, $k = -12$. **13.** $r_2 = 5/8$, $k = 11$.
15. $h = 2$; roots 5/2, 7/2. **17.** roots $-\frac{2}{3}, \frac{7}{3}$; $h = -14$.
19. $h = 2$; roots ±3/2. **21.** $x^2 - 4x - 12 = 0$.

23. $6x^2 + x - 2 = 0$. **25.** $3x^2 - x - 10 = 0$. **27.** $x^2 - 12 = 0$.
29. $x^2 + 5 = 0$. **31.** $x^2 - 4x + 1 = 0$. **33.** $x^2 - 6x + 13 = 0$.
35. $16x^2 - 16x + 1 = 0$. **37.** $x^2 - 2ax + a^2 - b = 0$.

Section 80

1. 10, −1. **3.** −2. **5.** 0. **7.** −1/2, −4. **9.** 3, 2/3.
11. $\frac{-3 \pm \sqrt{3}}{2}$. **13.** $\pm \frac{a\sqrt{3}}{3}$. **15.** $\pm\sqrt{3}, \pm 2\sqrt{3}$.
17. **17.** ±2, ±5. **19.** 7, 1, $4 \pm \sqrt{2}$. **21.** 4, −2, 1, 1.
23. −2, $\pm i$. **25.** −2, $1 \pm i\sqrt{3}$, 1, $\frac{-1 \pm i\sqrt{3}}{2}$.

Section 81

1. 16. **3.** 5. **5.** 4/3. **7.** 4, 20. **9.** No root. **11.** −3.
13. 3/8. **15.** No root. **17.** 65/16. **19.** 1/2. **21.** −6.

Section 82

1. 17 ft., 3 ft. **3.** 8 ft., 15 ft. **5.** 4 in. **7.** 2 yd. **9.** 22, 24.
11. 17. **13.** 17. **15.** 73. **17.** 20 m.p.h., 30 m.p.h.
19. 16 m.p.h.; 1.5 and 2.5 hr. **21.** 2.5 sec. **23.** 1024 ft.
25. 1.5 sec., 2.25 sec. **27.** 6 min., 12 min. **29.** 3%.
31. $\frac{1}{2L}\left(-R \pm \sqrt{R^2 - [4L/C]}\right)$.

Section 83

1. Min. (4, −13). **3.** Max. (2, 5). **5.** Min. $(\frac{5}{2}, -\frac{11}{2})$.
7. Min. $(-\frac{1}{2}, 1)$. **9.** Max. (−1/2, 16.5). **11.** 15, 15. **13.** 14, 14.
15. 3 sec., 144 ft. **17.** 65 cents. **19.** 40 yd., 80 yd., 3200 sq. yd.

Section 84

1. Circle. **3.** Parabola. **5.** Ellipse. **7.** Hyperbola.
9. Nonexistent. **11.** One point. **13.** Two intersecting straight lines.
15. Two coincident straight lines. **17.** Hyperbola.

Section 85

1. $x = 4, y = 3; x = -4, y = -3$. **3.** $x = 0, y = 5; x = 4, y = 3$.
5. $x = 1, y = 3; x = 2, y = 0$. **7.** $x = 3, y = 4; x = 4, y = 3$.
9. $x = -4, y = 4; x = 2, y = 1$. **11.** $x = 3, y = 2; x = \frac{21}{4}, y = -\frac{19}{4}$.
13. $x = 0, y = 0; x = 2, y = 2$. **15.** $x = \pm 3, y = 4$.
17. $x = \pm 1, y = 3; x = \pm 1, y = -3$.
19. $x = 3, y = \pm 2; x = -3, y = \pm 2$.
21. $x = \pm 3, y = 1$; $x = \pm 8.5$ approx., $y = 8$.
23. $x = 4, y = \pm 3; x = -4, y = \pm 3$.

Section 86

13. $x = 4, y = 2; x = 2, y = -2.$ **15.** $x = 2, y = 1; x = 5, y = 4.$
17. $x = -3, y = 2; x = -1, y = -2.$
19. $x = 2, y = 3; x = -17/11, y = 7/11.$
23. $k = \sqrt{2}; x = y = \sqrt{2}/2.$

Section 87

1. $x = -3, y = 2; x = -1, y = -2.$
3. $x = -4, y = -1; x = -2, y = 3.$ **5.** $x = 0, y = 0; x = 0, y = 4.$
7. $x = \pm 3, y = \pm 2.$ **9.** $x = \pm 3, y = \pm \sqrt{2}.$
11. $x = \pm 3/2, y = \pm \sqrt{2}/2.$ **13.** $x = \pm 1/3, y = \pm \sqrt{3}/3.$
15. $x = 4, y = 3; x = 3, y = 4; x = -4, y = -3; x = -3, y = -4.$
17. $x = -4, y = -3; x = 4, y = 3; x = -5, y = 0; x = 0, y = 5.$
19. $x = 5, y = 4; x = -5, y = -4; x = 3\sqrt{2}, y = 3; x = -3\sqrt{2}, y = -3.$
21. $x = 2, y = 1; x = -2, y = -1; x = 3, y = 2; x = -3, y = -2.$
23. $x = \pm 1, y = 0; x = 2, y = -3; x = -2, y = 3.$
25. $x = 1, y = 2; x = -1, y = -2; x = 1/2, y = 5/2; x = -1/2, y = -5/2.$
27. $x = 2, y = 1; x = 1, y = 2; x = -2, y = -1; x = -1, y = -2.$

Section 88

1. 13, 16. **3.** 35 ft., 12 ft. **5.** 19 ft., 16 ft. **7.** 30 m.p.h.; 60 miles.
9. 7.5 ft., 18 ft. **11.** 28 ft., 18 ft. **13.** 32 items at 15 cents each.
15. 32 in., 24 in.

Section 89

1. 1/4. **3.** 4/3. **5.** $2x^2/3$. **7.** $\frac{x-3}{2}$. **9.** 3/5. **11.** 3/4.
13. 1/2. **15.** 5. **17.** 0. **19.** $-1/10$. **21.** 95. **23.** 35.
25. b^2. **27.** 1. **29.** ± 4. **31.** $\pm 1/6$. **33.** $\pm \sqrt{ab}$.
35. $\pm(x - 2)$. **37.** 12. **39.** 1/32. **41.** $b^2/2$. **43.** 12 in., 18 in.
45. 22 in., 55 in., 66 in. **47.** $13\frac{1}{3}$ in., 16 in. **49.** 38 ft.

Section 90

1. (*a*) $y = 4x$; (*b*) 24; (*c*) tripled. **3.** (*a*) $y = \frac{1}{2}uv$; (*b*) 7.
5. (*a*) $r = \frac{1}{6}v^2$; (*b*) 41,667 ft.; (*c*) 600 ft./sec. **7.** 95.2 lb. approx.
9. 400 ft. **11.** (*a*) $V = \frac{1}{3}\pi r^2 h$; (*b*) 8π cu. in. **13.** 3 sec.
15. 300 cu. ft. **17.** 13 amp.

Section 91

1. 120. **3.** 20. **5.** (*a*) 120; (*b*) 216. **7.** 1320. **9.** 216.
11. 180. **13.** (*a*) 240; (*b*) 480. **15.** (*a*) 604,800; (*b*) 10,000,000.

17. (*a*) 840; (*b*) 120; (*c*) 240. **19.** 36. **21.** (*a*) 325; (*b*) 130; (*c*) 282. **23.** 54.

Section 92

1. 5040. **3.** 56. **5.** 72. **7.** 210. **9.** 20. **11.** 120. **13.** 14,400. **15.** (*a*) 576; (*b*) 2880. **17.** 325. **19.** 195. **21.** 162.

Section 93

1. 15,120. **3.** 12,600. **5.** 35. **7.** 210.

Section 94

1. 24. **3.** 60. **5.** 12. **7.** (*a*) 48; (*b*) 72.

Section 95

1. 15. **3.** 56. **5.** 45. **7.** 210. **9.** 84. **11.** 4. **13.** 20. **15.** 350. **17.** (*a*) 3,003; (*b*) 1,001; (*c*) 2,002. **19.** (*a*) 112; (*b*) 196. **21.** 63,000. **23.** 5,040. **25.** (*a*) 28; (*b*) 7. **27.** 2,520. **29.** 7,200. **31.** 10.

Section 96

1. 1/6. **3.** 1/36. **5.** (*a*) 1/4; (*b*) 1/13; (*c*) 1/52. **7.** (*a*) $\frac{1}{4}$; (*b*) $\frac{1}{2}$; (*c*) $\frac{1}{4}$. **9.** 2/5. **11.** 2/3. **13.** 1/4. **15.** 1/21. **17.** (*a*) 2/5; (*b*) 1/2. **19.** (*a*) 2/13; (*b*) 4/13. **21.** 1/20. **23.** (*a*) 1/35; (*b*) 1/35. **25.** (*a*) 1/24; (*b*) 23/24; (*c*) 1/4. **27.** (*a*) 7/18; (*b*) 11/18; (*c*) 4/9; (*d*) 5/9. **29.** 2/3.

Section 97

1. 93.3% **3.** 46.8% **5.** 50.1%

Section 98

1. 1/36. **3.** (*a*) 1/28,561; (*b*) 1/270,725. **5.** (*a*) 1/6; (*b*) 5/42. **7.** 25/216. **9.** (*a*) 15/128; (*b*) 121/128. **11.** (*a*) 125/3,888; (*b*) 23/648; (*c*) 3,875/3,888. **13.** (*a*) 1/7,776; (*b*) 1/60,466,176; (*c*) No. **15.** 11/12. **17.** 27 to 8. **19.** 7/10. **21.** 61/125. **23.** (*a*) 91/216; (*b*) 125 to 91. **25.** 0.057395628 or about 6%. **27.** (*a*) 53,760/9,765,625 or about 1/2 of 1%; (*b*) 62,201/9,765,625 or about 2/3 of 1%. **29.** (*a*) 2/5; (*b*) 7/15; (*c*) 2/15; (*d*) 13/15. **31.** 40/243.

Section 99

1. (*a*) 2, 4, 6, 8, 10; (*b*) 20; (*c*) $2k$; (*d*) $2(k + 1) = 2k + 2$.
3. (*a*) 1, 4, 9, 16, 25; (*b*) 100; (*c*) k^2; (*d*) $(k + 1)^2 = k^2 + 2k + 1$.

5. (*a*) 2, 6, 12, 20, 30; (*b*) 110; (*c*) $k(k + 1) = k^2 + k$; (*d*) $(k + 1)(k + 2) = k^2 + 3k + 2$.

7. (*a*) $\frac{1}{2}, \frac{-1}{6}, \frac{1}{12}, \frac{-1}{20}, \frac{1}{30}$; (*b*) $\frac{-1}{110}$; (*c*) $\frac{(-1)^{k+1}}{k(k + 1)}$; (*d*) $\frac{(-1)^{k+2}}{(k + 1)(k + 2)}$.

Section 102

1. $x^7 + 7x^6h + 21x^5h^2 + 35x^4h^3 + 35x^3h^4 + 21x^2h^5 + 7xh^6 + h^7$.
3. $p^5 + 5p^4q + 10p^3q^2 + 10p^2q^3 + 5pq^4 + q^5$.
5. $32a^{15} + 240a^{12}b^2 + 720a^9b^4 + 1{,}080a^6b^6 + 810a^3b^8 + 243b^{10}$.
7. $a^3 - 3a^2b + 3ab^2 - b^3$.
9. $a^5 - 5a^4b + 10a^3b^2 - 10a^2b^3 + 5ab^4 - b^5$.
11. $x^4/y^4 - 4x^2/y^3 + 6/y^2 - 4/x^2y + 1/x^4$.
13. $x^8/16 - x^5 + 6x^2 - 16/x + 16/x^4$. **15.** $680x^{14}h^3$. **17.** $3{,}003x^{10}h^5$.
19. $\frac{15}{2}a^{14}b^6$. **21.** $70a^4b^4$. **23.** $-63x^8$. **25.** $-198/x^4$.
27. $-\frac{960x^{11}}{y^7}$. **29.** $\frac{105x^6}{32}$. **31.** $-112x^7$. **33.** $90{,}720x^4$.
35. 1.094. **37.** 0.942. **39.** 1.243. **41.** 0.868.

Section 104

1. $1 + \frac{1}{2}x - \frac{1}{8}x^2 + \frac{1}{16}x^3 + \ldots$. **3.** $1 + \frac{1}{3}x - \frac{1}{9}x^2 + \frac{5}{81}x^3 + \ldots$.
5. $1 + x + x^2 + x^3 + \ldots$. **7.** 4.12. **9.** 4.02. **11.** 5.29.
13. 7.071. **15.** 2.924.

Section 105

13. $-i$. **15.** 1. **17.** -1. **19.** $3 - 4i$. **21.** $5 + 2i$.
23. $-3 - 2i$. **25.** $-1 - 2i\sqrt{2}$. **27.** $7 - 3i$. **29.** $-3 + 2i$.
43. $x = 3, y = -4$. **45.** $x = 4, y = 0$. **47.** $x = -4, y = 3$.
49. $x = 4, y = 3$. **51.** $x = 2, y = -3$.

Section 106

1. $5i$. **3.** $5i$. **5.** -20. **7.** -8. **9.** $-12i$. **11.** $108i$.
31. 11. **33.** $-1 + 7i\sqrt{2}$. **35.** $2 - 10i\sqrt{2}$. **37.** 1. **39.** i.
41. $-i$. **43.** $-i$. **45.** $\frac{1}{13} + \frac{8}{13}i$. **47.** $-\frac{18}{13} + \frac{14}{13}i$. **49.** $\frac{1}{2} - \frac{3}{4}i$.
51. $-i$. **53.** $\frac{6}{5} - \frac{3}{5}i$. **55.** i. **57.** $\frac{2}{7} - \frac{\sqrt{3}}{7}i$. **59.** $\frac{1}{18} - \frac{5\sqrt{5}}{18}i$.
61. $-\frac{5}{29} - \frac{2}{29}i$. **63.** $\frac{3}{11} + \frac{\sqrt{2}}{11}i$. **65.** $\frac{1}{2} - \frac{1}{2}i$. **67.** $-\frac{3}{13} - \frac{2}{13}i$.
69. $\frac{1}{4}i$. **71.** $\frac{2}{7} - \frac{\sqrt{3}}{7}i$. **73.** 0.

Section 109

1. $1/2, \sqrt{3}/2, \sqrt{3}/3$. **3.** $\sqrt{3}/2, 1/2, \sqrt{3}$. **5.** $\sqrt{2}/2, -\sqrt{2}/2, -1$.
7. $-1/2, -\sqrt{3}/2, \sqrt{3}/3$. **9.** $-\sqrt{3}/2, -1/2, \sqrt{3}$.

11. $-\sqrt{2}/2, \sqrt{2}/2, -1$. **13.** $1/2, \sqrt{3}/2, \sqrt{3}/3$.
15. $1/2, -\sqrt{3}/2, -\sqrt{3}/3$. **17.** 0, 1, 0. **19.** 0, −1, 0.
21. 0, 1, 0. **23.** 0.3420, 0.9397, 0.3640.
25. 0.9397, −0.3420, −2.747. **27.** −0.3420, −0.9397, 0.3640.
29. −0.9397, 0.3420, −2.747. **31.** 0.3420, 0.9397, 0.3640.
33. −0.7660, −0.6428, 1.192. **35.** 0.4067, −0.9135, −0.4452.
37. 0.8192, −0.5736, −1.428. **39.** −0.9063, −0.4226, 2.145.

Section 110

1. $\sqrt{3}+i$. **3.** $4\sqrt{2}+4i\sqrt{2}$. **5.** $-\frac{\sqrt{2}}{2}+\frac{\sqrt{2}}{2}i$. **7.** $3i$.
9. −4. **11.** 3. **13.** $-2\sqrt{2}-2i\sqrt{2}$. **15.** $-5i$.
17. $\sqrt{3}-i$. **19.** $2(\cos 60^\circ + i \sin 60^\circ)$.
21. $4\sqrt{2}(\cos 45^\circ + i \sin 45^\circ)$. **23.** $4(\cos 120^\circ + i \sin 120^\circ)$.
25. $2(\cos 135^\circ + i \sin 135^\circ)$. **27.** $3(\cos 90^\circ + i \sin 90^\circ)$.
29. $6(\cos 0^\circ + i \sin 0^\circ)$. **31.** $3(\cos 180^\circ + i \sin 180^\circ)$.
33. $5(\cos 270^\circ + i \sin 270^\circ)$. **35.** $4(\cos 330^\circ + i \sin 330^\circ)$.
37. $8[\cos (0^\circ + k \cdot 360^\circ) + i \sin (0^\circ + k \cdot 360^\circ)], k = 0, \pm 1, \pm 2, \ldots$.
39. $8[\cos (90^\circ + k \cdot 360^\circ) + i \sin (90^\circ + k \cdot 360^\circ)], k = 0, \pm 1, \pm 2, \ldots$.
41. $2[\cos (330^\circ + k \cdot 360^\circ) + i \sin (330^\circ + k \cdot 360^\circ)], k = 0, \pm 1, \pm 2, \ldots$.
43. $1.8794 + 0.6840i$. **45.** $-0.9848 - 0.1736i$.
47. (*a*) 0°; (*b*) 180°; (*c*) 90°; (*d*) 270°.

Section 111

1. (*a*) $6(\cos 180^\circ + i \sin 180^\circ)$; (*b*) −6.
3. (*a*) $12(\cos 135^\circ + i \sin 135^\circ)$; (*b*) $-6\sqrt{2}+6i\sqrt{2}$.
5. (*a*) $2(\cos 90^\circ + i \sin 90^\circ)$; (*b*) $2i$.
7. (*a*) $5(\cos 120^\circ + i \sin 120^\circ)$; (*b*) $-\frac{5}{2}+\frac{5\sqrt{3}}{2}i$.

Section 112

1. $27(\cos 180^\circ + i \sin 180^\circ) = -27$.
3. $8(\cos 150^\circ + i \sin 150^\circ) = -4\sqrt{3}+4i\sqrt{3}$.
5. $32(\cos 75^\circ + i \sin 75^\circ)$. **7.** $64(\cos 270^\circ + i \sin 270^\circ) = -64i$.
9. $\frac{1}{16}(\cos 120^\circ + i \sin 120^\circ) = -\frac{1}{32}+\frac{\sqrt{3}}{32}i$. **11.** 16.
13. $-2-2i$. **15.** 1. **17.** −16. **19.** −1.

Section 113

1. $\sqrt{2}+i\sqrt{2}, -\sqrt{2}-i\sqrt{2}$. **3.** $2\sqrt{3}+2i, -2\sqrt{3}+2i, -4i$.
5. $1+i\sqrt{3}, -2, 1-i\sqrt{3}$.
7. $1, \frac{1}{2}+\frac{\sqrt{3}}{2}i, -\frac{1}{2}+\frac{\sqrt{3}}{2}i, -1, -\frac{1}{2}-\frac{\sqrt{3}}{2}i, \frac{1}{2}-\frac{\sqrt{3}}{2}i$.

9. 2, 2(cos 72° + i sin 72°), 2(cos 144° + i sin 144°), 2(cos 216° + i sin 216°), 2(cos 288° + i sin 288°).
11. $\sqrt{2} + i\sqrt{2}$, 2(cos 165° + i sin 165°), 2(cos 285° + i sin 285°).
13. 1, $-\frac{1}{2} + i\frac{\sqrt{3}}{2}$, $-\frac{1}{2} - i\frac{\sqrt{3}}{2}$.
15. $\sqrt{3} + i$, $-1 + i\sqrt{3}$, $-\sqrt{3} - i$, $1 - i\sqrt{3}$. **17.** 2, $2i$, -2, $-2i$.
19. $2\sqrt{3} + 2i$, $-2\sqrt{3} - 2i$.
21. $\sqrt{2} + i\sqrt{2}$, 2(cos 105° + i sin 105°), 2(cos 165° + i sin 165°), $-\sqrt{2} - i\sqrt{2}$, 2(cos 285° + i sin 285°), 2(cos 345° + i sin 345°).

Section 114

1. 3, $-\frac{3}{2} + \frac{3i\sqrt{3}}{2}$, $-\frac{3}{2} - \frac{3i\sqrt{3}}{2}$. **3.** $\sqrt{3} - i$, $-\sqrt{3} - i$, $2i$.
5. $\sqrt{2} + i\sqrt{2}$, $-\sqrt{2} - i\sqrt{2}$. **7.** $\frac{3\sqrt{3}}{2} + \frac{3}{2}i$, $-\frac{3\sqrt{3}}{2} + \frac{3}{2}i$, $-3i$.
9. $\sqrt{3} + i$, $-1 + i\sqrt{3}$, $-\sqrt{3} - i$, $1 - i\sqrt{3}$.
11. $\sqrt{2} + i\sqrt{2}$, 2(cos 165° + i sin 165°), 2(cos 285° + i sin 285°).

Section 116

1. 2. **3.** -1. **5.** 8. **7.** -51. **9.** True. **11.** True.
13. False. **15.** True. **17.** False. **19.** True. **21.** True.
23. False.
25. (*a*) True. (*b*) True unless $y = 0$. (*c*) True. (*d*) True. (*e*) True unless $y = 0$. (*f*) True unless $y = 0$.
27. No. **29.** 3.

Section 117

1. Quotient = $4x^2 + 8x + 14$; remainder = 33.
3. Quotient = $x^2 + x - 3$; remainder = 2.
5. Quotient = $2x^2 + 4x + 6$; remainder = 8.
7. Quotient = $6x^2 - 2x - \frac{1}{3}$; remainder = $\frac{10}{9}$. **9.** 22, -18.
11. $\frac{5}{8}$, 0. **13.** -6.976, -0.916. **15.** False.
17. True; $2x^2 - 7x + 6$. **19.** False.
21. True; $x^4 - 2x^3 + 4x^2 - 8x + 16$. **23.** False. **25.** True.
27. False. **29.** True. **31.** True. **33.** True. **35.** $k = -7$.

Section 118

1. 3, 2, -1. **3.** -2, $-1 \pm i\sqrt{3}$. **5.** -3, $-1/2$, 2, 3.
7. 1, $-2/3$, $\frac{-5 \pm \sqrt{21}}{2}$. **9.** 3/2, 1/3, $\pm i\sqrt{2}$.

Section 119

1. 3, 1/2, −3. **3.** 1, 1, $-1 \pm i\sqrt{3}$. **5.** −3, −1/2, 2/3.
7. 0, 1, −2, $\pm\sqrt{3}$. **9.** $x^3 - 7x + 6 = 0$.
11. $2x^3 - 3x^2 - 11x + 6 = 0$. **13.** $x^3 - 4x^2 + 4x = 0$.
15. $4x^4 + 8x^3 - 11x^2 + 3x = 0$. **17.** $x^4 - 3x^2 = 0$.
19. $x^4 - 2x^3 + 2x^2 - 8x - 8 = 0$. **21.** $x^4 - 6x^3 + 14x^2 - 22x + 5 = 0$.
23. $6x^4 - 7x^3 + 26x^2 - 28x + 8 = 0$. **25.** $x^3 - 6x^2 + 12x - 8 = 0$.
27. $x^4 - 3x^2 + 2x = 0$. **29.** 1/2, 1/2, $-\frac{1}{2} \pm \frac{1}{2}i\sqrt{3}$.
31. −2/3, −2/3, $1 \pm i$.
33. $x^3 + (-2 - 2i)x^2 + (-5 + 2i)x + (6 + 12i) = 0$.

Section 121

1. $3 - 2i$. **3.** $-3 + 3i$. **5.** $-3i$. **7.** $x^3 - 7x^2 + 15x - 25 = 0$.
9. $4x^4 - 4x^3 + x^2 + 6x + 2 = 0$. **11.** $1 \pm i$, $\pm i$.
13. $\pm i$, $\dfrac{-3 \pm \sqrt{5}}{2}$. **15.** $\pm i$, $\pm i$, $\dfrac{-3 \pm \sqrt{5}}{2}$. **21.** False.
23. $x^3 - 5x^2 + 5x - 1 = 0$. **25.** $x^4 - 4x^3 + 2x^2 + 4x = 0$.
27. $2 \pm \sqrt{3}$, $\pm i$.

Section 122

1. 1, 3, −2. **3.** 2, 3, 3/2. **5.** 2, −3, 4/3. **7.** 1, −1/2, 3/4.
9. 2/3, −1/3, $\pm i$. **11.** −1/2, −1/2, $\dfrac{1 \pm \sqrt{5}}{2}$. **13.** None.
15. 0, 2, $\pm i$. **17.** 0, −1/2, $-1 \pm i$. **19.** 4, $-1 \pm i$.
21. 8, $\dfrac{1 \pm i\sqrt{7}}{2}$. **23.** 4, $1 \pm \sqrt{5}$. **25.** $(x + 2)(x - 3)(x + 4)$.
27. $(2x + 1)(3x - 4)(3x - 4)$. **29.** $(2x - 1)(x - 3 - 2i)(x - 3 + 2i)$.
31. $(2x + 1)(2x + 1)(x + 1 - i)(x + 1 + i)$. **33.** $(2x - 1)(x^2 + 4)$.
35. $(2x + 3)(x^2 - 2x + 5)$. **37.** $(2x - 1)(3x + 2)(x^2 + 2x + 4)$.
39. False.

Section 123

21. 1, 6, −2. **23.** ±1.4. **25.** No.

Section 124

1. $x^3 + 2x^2 - 5x - 6 = 0$. **3.** $x^3 + 4x^2 + x - 6 = 0$.
5. $9x^4 + 3x^3 + 7x^2 + 3x - 2 = 0$. **7.** $3x^4 + 2x^2 + 5 = 0$.
9. −4, −2/5. **11.** −1/2, −4/3.

Section 125

1. 2, −2. **3.** 2, −1. **5.** 2, 2. **7.** −3, +1. **9.** None.
11. 3, −1. **13.** None. **15.** ±1.

Section 126

1. (*a*) 1 pos., 2 imag. (*b*) Same.
3. (*a*) 1 pos., 1 neg., 2 imag.; or 4 imag. (*b*) 1 pos., 1 neg., 2 imag.
5. (*a*) 1 pos., 1 neg., 2 imag.; or 4 imag. (*b*) 1 pos., 1 neg., 2 imag.
7. (*a*) 2 pos., 2 imag.; or no pos., 4 imag. (*b*) Same.
9. (*a*) 4 imag.; (*b*) Same.
11. (*a*) 2 pos., 2 neg.; or 2 pos., 2 imag.; or 2 neg., 2 imag.; or 4 imag.; or 1 pos., 1 neg., 2 imag.; (*b*) Same except 1 pos., 1 neg., 2 imag. now ruled out.
13. (*a*) 3 pos.; or 1 pos., 2 imag. (*b*) Same.
15. (*a*) 1 neg.; 4 imag.; (*b*) Same.
17. (*a*) 1 pos., 1 neg., 4 imag.; or 6 imag.; (*b*) 1 pos., 1 neg., 4 imag.
19. (*a*) 1 zero root, 2 imag.; (*b*) Same. **27.** 2, -1. **39.** 5, 6, 7 in.
41. 2 in., or $5 - \sqrt{10}$ in.; two solutions. **43.** 2 in.

Section 127

1. 1.73. **3.** 1.15. **5.** 1.59. **7.** -1.24. **9.** 1.82.

Section 128

11. 1.236. **13.** 1.442. **15.** 2.351.

Section 129

1. $x^3 - 7x + 6 = 0$. **3.** $x^3 + 2x^2 - 3x = 0$.
5. $2x^3 - 17x^2 + 40x - 25 = 0$. **7.** $x^3 - 1.4x^2 - 1.68x + 1.728 = 0$.
9. $x^3 - 3.94x^2 + 0.8412x + 6.018408 = 0$.
11. $x^3 - 6x^2 - 16x + 96 = 0$. **13.** $x^3 - 40x^2 + 100x + 6{,}000 = 0$.
15. $x^4 + 78x^3 + 2{,}144x^2 + 21{,}922x - 7{,}249 = 0$.

Section 130

17. 3.732. **19.** 1.236, 1.414. **21.** 3.332. **23.** -3.503.
25. 0.268, 2.414, 3.732, -0.414. **27.** 0.92. **29.** -0.4, -40.6.

Section 131

1. 3/2, -2. **3.** Roots 2, 3, 3/2; $k = 27$.
5. Roots 4, $(-1 \pm i\sqrt{3})/2$; $k = -3$.
7. Roots 1, 1, -2, -2; $k = -4$. **9.** Roots 2, -1, 3/2; $h = -1$, $k = 6$.

Section 133

1. $-1, -\dfrac{1}{2} \pm i\dfrac{\sqrt{3}}{2}$. **3.** $1 + \omega$, $1 + \omega$, $2\omega^2$.
5. $1 + \sqrt[3]{4} + \sqrt[3]{2}$, $1 + \omega\sqrt[3]{4} + \omega^2\sqrt[3]{2}$, $1 + \omega^2\sqrt[3]{4} + \omega\sqrt[3]{2}$ where ω is an imaginary cube root of 1.

Section 134

1. $\frac{1}{2}(-\sqrt{2} \pm \sqrt{2\sqrt{2} - 2})$, $\frac{1}{2}(\sqrt{2} \pm i\sqrt{2\sqrt{2} + 2})$.
3. $-1 \pm \sqrt{2}$, $1 \pm 2i$.

Section 135

1. (*a*) 10. (*c*) -1. (*e*) -2. **39.** 7. **41.** 2, 3.

Section 136

1. -10. **3.** -28. **5.** 3. **7.** 3. **9.** 1, $-11/3$.
11. $\begin{vmatrix} b & c \\ e & f \end{vmatrix}$. **13.** $\begin{vmatrix} b & c \\ h & k \end{vmatrix}$. **15.** $\begin{vmatrix} a & b \\ d & e \end{vmatrix}$.
19. $a_1k_2c_3 - a_1k_3c_2 - a_2k_1c_3 + a_3k_1c_2 + a_2k_3c_1 - a_3k_2c_1$.

Section 137

11. $i_1 = 2$ amp., $i_2 = 2$ amp., $i_3 = 0$ amp.

Section 138

1. 14 square units. **3.** 13 square units. **7.** $y = x + 1$. **9.** $x = 2$.

Section 139

1. $a_1b_2 - a_2b_1$.

Section 140

3. $x = 6$. **5.** $\begin{vmatrix} 0 & -5 & -11 \\ 0 & 17 & 21 \\ 1 & 3 & 4 \end{vmatrix}$. **7.** $\begin{vmatrix} 2 & -1 & 0 & 3 \\ 0 & 0 & -1 & 0 \\ 5 & 4 & 2 & -6 \\ 5 & 5 & 1 & -1 \end{vmatrix}$.

Section 141

7. -17. **9.** -5. **11.** 0. **13.** 132. **15.** 17. **17.** 0.
19. (*a*) $(r - s)^{n-1}[r + (n - 1)s]$. (*b*) $(-1)^{n+1}n$.

Section 142

1. $x = 1, y = -1, z = 2, t = 1/3$. **3.** $x = 1/2, y = 3/2, z = 0, t = 1$.
5. $x = -2, y = 1, z = -1, t = 3$. **7.** $x = 3, y = 1/2, z = -1, t = 2$.

Section 143

1. $x = m, y = 0, z = -m$; $x = 1, y = 0, z = -1$; or $x = 2, y = 0, z = -2$.
3. $x = m, y = -2m, z = 2m$; $x = 1, y = -2, z = 2$; or $x = 2, y = -4, z = 4$.
5. No non-trivial solutions. **7.** $k = 1$.

Section 144

3. $x = 5, y = 2.$ **5.** Inconsistent. **7.** Inconsistent.
9. $x = a + 2, y = 4a - 2, z = a,$ where a is arbitrary.
11. Inconsistent.

Section 145

1. $b_0a_1^2 - b_1a_0a_1 + b_2a_0^2.$

Section 146

1. $(x = 0, y = 0), (x = -1, y = 1), (x = 9, y = 6), (x = 2, y = -1).$
3. $(x = 0, y = 0), (x = -2, y = 2), (x = 3, y = -1).$
5. $(x = 0, y = 0), (x = 1, y = 1).$

Section 147

1. $d = 3, l = 44, s_{15} = 345.$ **3.** $d = -3, l = -11, s_8 = -4.$
5. $d = 5/3, l = 61/3, s_{12} = 134.$ **7.** 12,960. **9.** (*a*) 27; (*b*) 1,026.
11. (*a*) 19; (*b*) 2,508. **13.** 5, 7.1, 9.2. **11.** 3, 13.4.
15. -2.5, 0, 2.5, 5, 7.5, 10, 12.5. **17.** $n(n + 1).$ **19.** $2n(n + 1).$
21. $a = 4, l = 40.$ **23.** $d = 4, s_n = 738.$ **25.** $n = 11, s_n = 165.$
27. $a = 3, s_n = 429.$ **29.** $d = 4, n = 18.$ **31.** $d = \dfrac{l^2 - a^2}{2s_n - a - l}.$
33. $d = \dfrac{2(nl - s_n)}{n^2 - n}.$ **35.** $a = \dfrac{2s_n - n(n - 1)d}{2n}.$ **37.** 73.
39. \$6,750. **41.** (*a*) \$3040; (*b*) \$23,200.
43. $a = p + q, d = p, l = pn + q, s_n = \dfrac{n}{2}[2q + (n + 1)p].$

Section 148

1. $r = 3, l = 162, s_5 = 242.$ **3.** $r = -2, l = -96, s_6 = -63.$
5. $r = -1/2, l = 3/16, s_5 = 33/16.$ **7.** 10, 20, 40, 80.
9. -54, 18, -6, 2. **11.** $n = 6, s_n = 1{,}008.$ **13.** $n = 5, s_n = 53\frac{7}{9}.$
15. 32. **17.** $\pm 8.$ **19.** 2. **21.** 256. **23.** 372 ft. **25.** \$2,342.56.
27. The geometric mean or 40%. **29.** 64/729.

Section 149

1. 1/23. **3.** 1/9, 1/14, 1/19. **5.** 1/10, 1/6, 1/2. **7.** 20/7.
9. 72/59.

Section 150

1. 3/2. **3.** 2/3. **5.** 9. **7.** 2/3. **9.** 2/11. **11.** 1/6.
13. 3/13. **15.** 23/11. **17.** 120 ft. **19.** 280 ft.

Section 152

1. $n(n + 1)(n + 2)/3$. **3.** $n(n + 1)(n + 2)/6$.
5. $n(n + 1)(2n + 1)(3n^2 + 3n - 1)/30$.
7. (*a*) $n(n + 1)(n + 2)/6$; (*b*) 1,140.
9. (*a*) $n(n + 1)(2n + 1)/6$; (*b*) 2,109.
11. (*a*) $n^2 - n + 1$;
(*b*) $n^2 - n + 1 + (n - 1)(n - 2)(n - 3)(n - 4)(n - 5)$; (*c*) No.
13. (*a*) $n^3 - n^2 + 1$;
(*b*) $n^3 - n^2 + 1 + (n - 1)(n - 2)(n - 3)(n - 4)(n - 5)$; (*c*) No.

Section 155

1. $x < 4$. **3.** $x < 3$. **5.** $x \geqq 13/14$. **7.** $-2 < x < 2$, or $|x| < 2$.
9. $x \geqq 4$, $x \leqq -4$, or $|x| \geqq 4$. **11.** $2 < x < 3$. **13.** $x > 4$, $x < 2$.
15. $-2 < x < 1$. **17.** $2 \leqq x \leqq 3$. **19.** $x < -1$, $3 < x < 4$.
21. $2 \leqq x \leqq 4$.

Section 156

21. $x \geqq 1$, $x \leqq -4$. **23.** $-3 \leqq x \leqq 1$.

Section 157

1. $x = 3$, $y = 2$. **3.** $x = 4$, $y = 1$. **5.** $x = 7/2$, $y = 3$.
7. $x = 16/7$, $y = 36/7$.

Section 158

1. $\log_5 25 = 2$. **3.** $\log_{10} 100 = 2$. **5.** $\log_8 2 = 1/3$.
7. $\log_5 \frac{1}{25} = -2$. **9.** $\log_5 1 = 0$. **11.** $2^4 = 16$. **13.** $10^{-2} = 0.01$.
15. $10^0 = 1$. **17.** $5^3 = 125$. **19.** $10^{-1} = 0.1$. **21.** 6. **23.** 2/3.
25. 1. **27.** 0. **29.** 3/2. **31.** 3. **33.** 5. **35.** 4/3.
37. $-1/3$. **39.** $-1/3$. **41.** 4. **43.** 1/64. **45.** 2. **47.** 2.
49. 4.

Section 159

1. 1.1461. **3.** 0.5441. **5.** 1.0791. **7.** -0.6990. **9.** 0.1505.
11. -0.0416. **13.** 0.5411. **15.** 0.6990. **17.** 1.1128. **19.** 0.9065.

Section 160

1. 2.0934. **3.** 8.0934 − 10. **5.** 1.5977. **7.** 8.6385 − 10.
9. 6.5391 − 10. **11.** 2.5563. **13.** 7.6990 − 10. **15.** 33.2.
17. 0.332. **19.** 49.8. **21.** 0.545. **23.** 9,370.
25. (*a*) $10^{2.9355} = 862$ approx.; (*b*) $\sqrt[10,000]{10^{29,355}} = 862$ approx.

Section 161

1. 2.3824. **3.** 0.4709. **5.** 8.9301 − 10. **7.** 9.6301 − 10.
9. 0.6037. **11.** 155.5. **13.** 3.447. **15.** 0.6303. **17.** 0.08008.
19. 0.02044.

Section 162

1. 3.21. **3.** 14.20. **5.** 5.000. **7.** One. **9.** Two.
11. $3.15 \leqq 3.2 < 3.25$, $3.195 \leqq 3.20 < 3.205$.

Section 163

1. 1.72. **3.** 2.59. **5.** 0.783. **7.** 40.7. **9.** 128×10.
11. 0.205. **13.** 10.82. **15.** 6.321. **17.** $-3{,}168 \times 10$. **19.** 2.663.
21. 1.93. **23.** 0.0545. **25.** $229.70. **27.** $675.60.
29. 18 years approx. **31.** 197,000,000 sq. mi.
33. $10,720,000. approx. **35.** 2 secs. **37.** (*a*) $328; (*b*) $107.
39. 120,000 sq. in. approx. **41.** 5.337.

Section 164

1. 9.6925 − 10. **3.** 0.0501. **5.** 7.2532 − 10.

Section 165

1. (*a*) 1.099; (*b*) 0.477. **3.** (*a*) 0.405; (*b*) 0.176. **5.** $\log_e x = 1.386$.
7. $\log_{10} x = 0.8407$.

Section 166

3. 1.43. **5.** −0.690. **7.** 175. **9.** $n = \dfrac{\log A - \log P}{\log (1 + r)}$.
11. $n = \dfrac{\log (si + 1)}{\log (1 + i)}$. **13.** $n = \dfrac{\log c - \log p}{\log v}$. **17.** 1.48 mg.
19. 0.0604 grams. **21.** 4. **23.** 1/2.

Section 167

1. (*a*) $60; (*b*) $560. **3.** (*a*) $23.44; (*b*) $773.44.
5. (*a*) $237.50; (*b*) $5\frac{26}{100}$% approx. **7.** (*a*) $932.50; (*b*) $4\frac{82}{100}$% approx.
9. $208.33. **11.** $1,239.
13. (*a*) $P = A/(1 + rn)$; (*b*) $(A - P)/Pn$; (*c*) $(A - P)/Pr$.

Section 168

1. (*a*) $657.95; (*b*) $659.75. **3.** (*a*) $339.96; (*b*) $346.07.
5. (*a*) $337.78; (*b*) $336.49. **7.** (*a*) $489.78; (*b*) $488.10. **9.** 27.
11. 17. **13.** 32. **15.** 4% per annum. **17.** 5% per annum.

Section 169

1. 4.04%. **3.** 6.09%. **5.** 8.16%. **7.** 4.06%. **9.** 6.14%.

Section 170

1. (*a*) \$3,519.42; (*b*) \$2,494.98. **3.** \$5,236.52. **5.** \$8,113.62.
7. \$241.67. **11.** \$679.34.

Section 171

1. 20. **3.** 185. **5.** 1. **7.** 6.

Section 172

1. $x - 3$. **3.** $x - 2$. **5.** $2x - 1$. **7.** 1.

Section 173

1. $\dfrac{5}{2x-1} - \dfrac{2}{x+2}$. **3.** $\dfrac{3}{2x+3} - \dfrac{2}{x-2}$.

5. $\dfrac{1}{x+1} + \dfrac{2}{x-2} - \dfrac{1}{x+3}$. **7.** $x^2 + 1 + \dfrac{3}{x+2} - \dfrac{1}{x+3}$.

9. $5 + \dfrac{1}{x+1} + \dfrac{2}{x+2} + \dfrac{3}{x+3}$. **11.** $\dfrac{3}{x+1} + \dfrac{2}{(x+1)^2}$.

13. $\dfrac{2}{x-2} - \dfrac{1}{(x-2)^2} + \dfrac{1}{(x-2)^3}$. **15.** $\dfrac{1}{2x+1} + \dfrac{2}{x+3} - \dfrac{1}{(x+3)^2}$.

17. $\dfrac{2}{x+1} - \dfrac{1}{(x+1)^2} + \dfrac{3}{x-2} + \dfrac{1}{(x-2)^2}$.

19. $3 + \dfrac{6}{2x+4} + \dfrac{1/2}{x-4} - \dfrac{1/2}{(x-4)^2}$. **21.** $\dfrac{2x-3}{x^2+x+1} - \dfrac{1}{x-1}$.

23. $\dfrac{3x-1}{x^2-x+1} + \dfrac{2}{x+1}$. **25.** $\dfrac{3}{x} + \dfrac{2x-1}{x^2+x+3}$.

27. $\dfrac{2}{x^2+1} - \dfrac{1}{2x^2-x+3}$. **29.** $x + 2 + \dfrac{2x-1}{x^2+3} - \dfrac{x-3}{x^2-x+2}$.

31. $4 + \dfrac{2}{x} - \dfrac{3}{x^2} + \dfrac{x-2}{x^2+x+2}$. **33.** $\dfrac{1}{x^2+1} + \dfrac{2x-3}{(x^2+1)^2}$.

35. $\dfrac{2}{x} + \dfrac{x}{x^2+3} + \dfrac{3}{(x^2+3)^2}$. **37.** $\dfrac{1}{(x^2+1)^2} + \dfrac{3x+1}{(x^2+2)^2}$.

39. $\dfrac{1}{x-1} + \dfrac{2x}{x^2+1} - \dfrac{1}{(x^2+1)^2}$.

Section 176

1. (*a*) 3.1, 3.01, 3.001, 3.0001; (*b*) 3. **3.** (*a*) $\frac{3}{2}, \frac{9}{4}, \frac{15}{8}, \frac{33}{16}$; (*b*) 2.
5. (*a*) $\frac{1}{2}, \frac{5}{12}, \frac{5}{13}, \frac{17}{46}$; (*b*) $\frac{1}{3}$. **7.** (*a*) $\frac{1}{2}, \frac{2}{3}, \frac{3}{4}, \frac{4}{5}$; (*b*) 1.
9. (*a*) $\frac{1}{3}, \frac{3}{5}, \frac{7}{9}, \frac{15}{17}$; (*b*) 1. **11.** (*a*) $\frac{4}{5}, \frac{9}{7}, \frac{16}{9}, \frac{25}{11}$; (*b*) diverges.

Section 178

9. $1 + \frac{1}{9} + \frac{1}{25} + \ldots$; convergent.
11. $\frac{1}{8} + \frac{1}{64} + \frac{1}{216} + \ldots$; convergent.
13. $\frac{1}{2} + \frac{2}{3} + \frac{3}{4} + \ldots$; divergent.
15. $\dfrac{1}{2 \cdot 1^{\frac{3}{2}}} + \dfrac{1}{2 \cdot 2^{\frac{3}{2}}} + \dfrac{1}{2 \cdot 3^{\frac{3}{2}}} + \ldots$; convergent.
17. $\frac{1}{3} + \frac{1}{6} + \frac{1}{11} + \ldots$; convergent. **19.** $\frac{1}{2} + \frac{5}{11} + \frac{6}{11} + \ldots$; divergent.
21. $\dfrac{1 \cdot 2}{3 \cdot 4} + \dfrac{2 \cdot 3}{4 \cdot 5} + \dfrac{3 \cdot 4}{5 \cdot 6} + \ldots$; divergent.
23. $1 + \frac{3}{5} + \frac{4}{10} + \ldots$; divergent.

Section 180

1. Convergent. **3.** Divergent. **5.** Convergent. **7.** Convergent.
9. Convergent. **11.** Divergent. **13.** Ratio test fails; convergent.

Section 184

1. Absolutely convergent. **3.** Conditionally convergent.
5. Conditionally convergent. **7.** Divergent.
9. Conditionally convergent.

Section 185

1. $-1 < x < 1$. **3.** All real values of x. **5.** $-1 \leqq x \leqq 1$.
7. $-\frac{1}{2} < x < \frac{1}{2}$. **9.** $-1 \leqq x < 1$. **11.** $0 < x \leqq 2$.

Section 186

1. 0.540. **3** 0.955. **5.** 2.72.

Section 187

1. $y = 5$. **3.** $y = 0$. **5.** 5.2. **7.** 1.364. **9.** 1.5774.

Section 188

1. (*a*) $y = x^3 - 2x^2 + 3x - 4$; (*b*) $y = 2$.
3. (*a*) $y = 2x - 3$; (*b*) $y = 1$.
5. (*a*) $y = x^2 - 7x + 6$; (*b*) $f(4) = -6, f(2.25) = -4.6875$.
7. (*a*) $y = x^4 - 2x^2 + 10$; (*b*) $y = 23.7456$.

Section 189

9. $y = x^4 - 2x - 5$. **11.** $y = 5 - \dfrac{12}{x} + \dfrac{6}{x^2}$.

Section 190

1. (*a*) $y = 2x^2 - 3x - 5$; (*b*) 17.68, 72.
3. (*a*) $y = x^4 - 3x^2 - x + 5$; (*b*) $y = 11.7056$.
5. (*a*) $y = x^3 - 5x^2 + 2x + 8$; (*b*) $y = 0, 528$.
7. $f(x) = 1.2x^2 - 3.1x + 2.5, f(1) = 0.6, f(1.4) = 0.512$. **9.** 12,649.61.

Section 192

1. $y = 0.5x + 1.0$. **3.** $y = 1.0x^2 - 1.8x + 4.8$.
5. (*a*) $y = 0.12x + 35.88$; (*b*) 36.48; (*c*) No.
7. (*a*) $y = 10 + 202.5t - 16.25t^2$; (*b*) 471; 271.
9. (*a*) $C_L = 0.073\alpha + 0.385$; (*b*) 0.969; (*c*) No.

Section 193

1. $y = 0.5x + 1.0$. **3.** $y = 1.0x^2 - 1.9x + 5.0$.
5. (*a*) $y = 0.12x + 35.88$; (*b*) 36.48; (*c*) No.
7. (*a*) $y = 13 + 199t - 16t^2$; (*b*) 466; 266.
9. (*a*) $C_L = 0.073\alpha + 0.385$; (*b*) 0.969; (*c*) No.

Section 194

1. $y = 0.5x + 1.0$. **3.** $y = 1.0x^2 - 1.9x + 5.0$.
5. (*a*) $y = 0.12x + 35.88$; (*b*) 36.48; (*c*) No.
7. (*a*) $y = 11 + 200t - 16t^2$; (*b*) 467; 275.
9. (*a*) $C_L = 0.073\alpha + 0.385$; (*b*) 0.969; (*c*) No. **11.** $y = 2.0x + \frac{10.1}{x}$.

Section 195

1. (*a*) $y = 100(1.12)^x$; (*b*) 157.
3. (*a*) $y = 100(0.76)^x$; (*b*) 33 ft./min.; (*c*) No. **5.** $p = 29.9\sigma^{1.4}$.

Index

P

Q

R